T0290911

Fluid Power Engineering

About the Author

Mahmoud Galal Rabie, Ph.D. is a professor of mechanical engineering. He has worked as:

- Professor of mechanical engineering at the Manufacturing Engineering and Production Technology Department of the Modern Academy for Engineering and Technology, Cairo, Egypt.
- Professor of mechanical engineering and head of the Aircraft Mechanical Engineering Department at the Military Technical College, Cairo, Egypt.
- Aircraft maintenance and repair engineer at the Egyptian Airforce.
- Instructor, teaching vocational training programs in the field of Fluid Power technology, covering more than 3747 trainees, engineers, and technicians.
- Consultant for major industrial firms in Egypt for Fluid Power Systems construction, operation, fault diagnostics, and repair.

He is the author and coauthor of 78 papers published in refereed international journals and presented at refereed conferences and the supervisor of 41 M.Sc. and Ph.D. theses.

Fluid Power Engineering

M. Galal Rabie, Ph.D.

Former Professor at the Military Technical College, Cairo, Egypt

Second Edition

New York Chicago San Francisco
Athens London Madrid
Mexico City Milan New Delhi
Singapore Sydney Toronto

McGraw Hill books are available at special quantity discounts to use as premiums and sales promotions, or for use in corporate training programs. To contact a representative, please visit the Contact Us pages at www.mhprofessional.com.

Fluid Power Engineering, Second Edition

1 2 3 4 5 6 7 8 9 LKV 29 28 27 26 25 24 23

Library of Congress Control Number: 2023942911

ISBN 978-1-265-51547-8
MHID 1-265-51547-6

Sponsoring Editor
Lara Zoble

Editorial Supervisor
Patty Mon

Project Manager
Revathi Viswanathan, KnowledgeWorks Global Ltd.

Acquisitions Coordinator
Olivia Higgins

Copy Editor
Girish Sharma

Proofreader
Manish Tiwari

Indexer
Edwin Durbin

Production Supervisor
Lynn M. Messina

Composition
KnowledgeWorks Global Ltd.

Illustration
KnowledgeWorks Global Ltd.

Art Director, Cover
Jeff Weeks

To my wife, Fatima Raafat
In recognition of the active support
she gave me throughout my professional career

For MATLAB® and Simulink® product information, or information on other related products, please contact:

The MathWorks, Inc.
3 Apple Hill Drive
Natick, MA 01760-2098 USA
Tel: (508) 647-7000
Fax: (508) 647-7001
E-mail: info@mathworks.com
Web: www.mathworks.com

Contents

Preface

This book deals with the construction, principles of operation, and calculation of fluid power systems. Special attention is paid to building a solid theoretical background that enables the reader to further study and analyze the steady state and dynamic performance of the diverse fluid power elements and systems. In addition to the mathematical treatment and theory, the book includes case studies of diverse elements of industrial, mobile, and aeronautical hydraulic power systems. Most of these elements are presented by detailed constructional drawings. The features of their design and performance are discussed. Moreover, this edition includes 1.18 GB of digital ancillaries containing 419 files and distributed in 101 folders. These ancillaries support learning, teaching, research, and vocational training. The ancillaries cover PowerPoint presentations with full-colored slides, MATLAB-SIMULINK programs, movies, animations, Automation Studio projects, and solutions to numerical problems. The ancillaries include conveniently selected topics from fluid mechanics and automatic control to enrich the theoretical background.

The book includes 12 chapters:

Chapter 1: Introduction to Hydraulic Power Systems

Chapter 2: Hydraulic Oils and Theoretical Background

Chapter 3: Hydraulic Transmission Lines

Chapter 4: Hydraulic Pumps

Chapter 5: Hydraulic Control Valves

Chapter 6: Accessories

Chapter 7: Hydraulic Actuators

Chapter 8: Hydraulic Servo Actuators

Chapter 9: Electrohydraulic Servovalve Technology

Chapter 10: Modeling and Simulation of Electrohydraulic Servosystems

Chapter 11: Electrohydraulic Proportional Valves Technology

Chapter 12: Introduction to Pneumatic Systems

Competences

The basic definition of competency is the ability to do something successfully and efficiently. This edition of fluid power engineering and its ancillaries aim at developing/reinforcing readers' competencies in the field of fluid power engineering. On successful completion of the study of the text and ancillaries thru one or more courses, the applicants must be able to:

1. Classify and compare mechanical, electrical, hydraulic, and pneumatic power systems.
2. Classify and compare the hydraulic fluids and evaluate the impact of fluid properties on the components and systems performance. Explain the cavitation phenomenon and its effect on the pump and other elements.
3. Explain and apply the mathematical background to develop mathematical models describing the dynamic behavior of highly nonlinear systems, mainly the control valves, accessories, and transmission lines.
4. Explain the construction, operation, and specifications of the basic components of hydraulic power and electrohydraulic proportional and servo systems; namely the pumps, valves, actuators, transmission lines, accessories, hydraulic servo actuators, electrohydraulic servo, and proportional valves.
5. Identify the standard symbols of hydraulic systems and use the available computer packages, such as the Automation Studeo™, to design the hydraulic circuits of targeted systems.
6. Use the MATLAB (SIMULINK) package efficiently to simulate the steady state and transient operation of the targeted systems and their basic elements, considering the background of the applicants; undergraduate, graduate, or researcher.
7. Solve limited operational problems related to the hydraulic power systems and their basic elements.
8. Explain the effects of air properties; air compressibility, air density, and air viscosity, on the pneumatic system performance.
9. Use the principles of control engineering to solve the precision and stability problems in electrohydraulic servo-systems and judge the effect of the implementation of PID controller on the performance of the electrohydraulic system.

Acknowledgments

I am indebted to my colleagues Prof. Dr. Ibrahim Saleh, Prof. Dr. Saad Kassem, and Dr. Yahya Elattar for the fruitful, and stimulating discussions we had, and for their objective comments on the book as a whole.

I would also like to express my gratitude to Bosch Rexroth AG, Norgren Ltd., Moog Inc., Famic Technologies Inc., Olaer Group, Geeplus Inc. Ltd., and Colmar Technik Spa for their kind support and permission to use their illustrations in this book.

Finally, I would like to extend my appreciation and gratitude to the staff of McGraw Hill Professional, especially Lara Zoble, senior editor, Lynn M. Messina, senior production supervisor; and Jeff Weeks, senior art director.

I would also like to thank Revathi Viswanathan, project manager, and her team at KnowledgeWorks Global Ltd.; Girish Sharma for copy editing; Edwin Durbin for creating the index; Manish Tiwari for proofreading; and RR Donnelley for printing and binding.

M. Galal Rabie, Ph.D.

CHAPTER 1

Introduction to Hydraulic Power Systems

1.1 Introduction

God created the first and most wonderful hydraulic system. It includes a double pump delivering a fluid flow rate of about 10 L/min at 0.16 bar maximum pressure. This pump feeds a piping network stretching more than 100,000 km. That's nearly two and a half times around the Earth. It operates continuously for a very long time, mostly maintenance free. It is the human blood circulatory system. By the age of 50 years, the hearts of 10 men should have pumped a volume of blood equaling that of the great Egyptian pyramid (2,600,000 m^3).

As for the hydraulic power systems developed by man, their history started practically 350 years ago. In 1647, Blaise Pascal published the fundamental law of hydrostatics: "Pressure in a fluid at rest is transmitted in all directions." In 1738, Bernoulli published his book *Hydrodynamica*, which included his kinetic-molecular theory of gases, the principle of jet propulsion, and the law of the conservation of energy. By the middle of the nineteenth century, fluid power started playing an important role in both the industrial and civil fields. In England, for example, many cities had central industrial hydraulic distribution networks, supplied by pumps driven by steam engines.

Before the universal adoption of electricity, hydraulic power was a sizable competitor to other energy sources in London. The London Hydraulic Power Company generated hydraulic power for everything from dock cranes and bridges to lifts in private households in Kensington and Mayfair. In the 1930s, during the glory days of hydraulic power, a 12 m^3/min average flow rate of water was pumped beneath the streets of London, raising and lowering almost anything that needed to be moved up and down. As a power source, hydraulic power was cheap, efficient, and easily transmitted through 300 km of underground cast-iron piping.

However, as electricity became cheaper and electronically powered equipment grew increasingly sophisticated, so industry and private citizens began to abandon hydraulic power.

High-pressure fluid power systems were put into practical application in 1925, when Harry Vickers developed the balanced vane pump. Today, fluid power systems dominate most of the engineering fields, partially or totally.

1.2 The Classification of Power Systems

Power systems are used to transmit and control power. This function is illustrated by Fig. 1.1. The following are the basic parts of a power system.

1. Source of energy, delivering mechanical power of rotary motion. Electric motors and internal combustion engines (ICE) are the most commonly used power sources. For special applications, steam turbines, gas turbines, or hydraulic turbines are used.
2. Energy transmission, transformation, and control elements.
3. Load requiring mechanical power of either rotary or linear motion.

In engineering applications, there exist different types of power systems: mechanical, electrical, and fluid. Figure 1.2 shows the classification of power systems.

1.2.1 Mechanical Power Systems

The mechanical power systems use mechanical elements to transmit and control the mechanical power. The drive train of a small car is a typical example of a mechanical

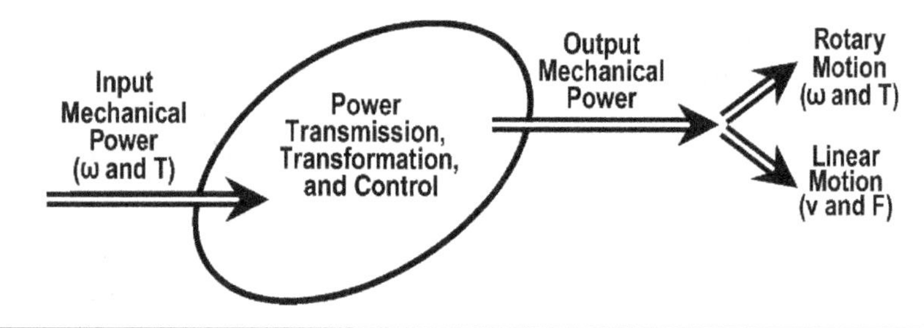

FIGURE 1.1 The function of a power system.

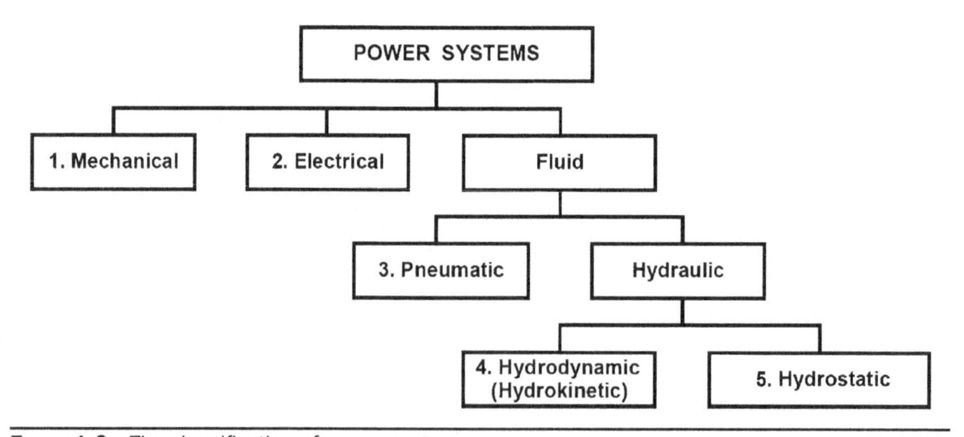

FIGURE 1.2 The classification of power systems.

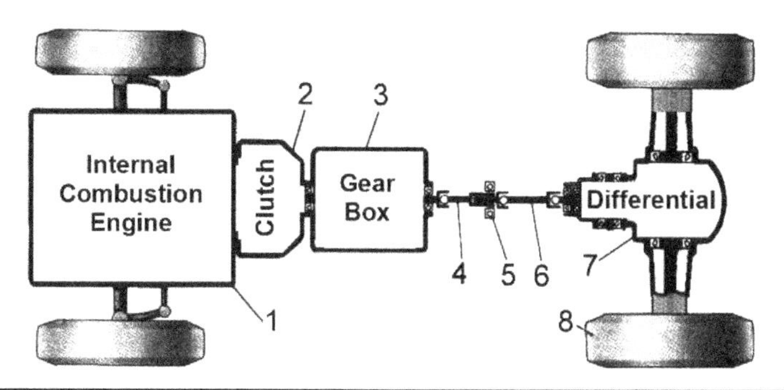

FIGURE 1.3 An automotive drive train.

power system (see Fig. 1.3). The gearbox (3) is connected to the engine (1) through the clutch (2). The input shaft of the gear box turns at the same speed as the engine. Its output shaft (4) turns at different speeds, depending on the selected gear transmission ratio. The power is then transmitted to the wheels (8) through the universal joints (5), drive shaft (6), and differential (7).

When compared with other power systems, mechanical power systems have advantages such as relatively simple construction, maintenance, and operation, as well as low cost. However, their power-to-weight ratio is minimal, the power transmission distance is too limited, and the flexibility and controllability are poor.

1.2.2 Electrical Power Systems

Electrical power systems solve the problems of power transmission distance and flexibility, and improve controllability. Figure 1.4 illustrates the principal of operation of electrical power systems. These systems offer advantages such as high flexibility and a very long power transmission distance, but they produce mainly rotary motion. Rectilinear motion, of high power, can be obtained by converting the rotary motion into rectilinear motion by using a suitable gear system or by using a drum and wire. However, holding the load position requires a special braking system.

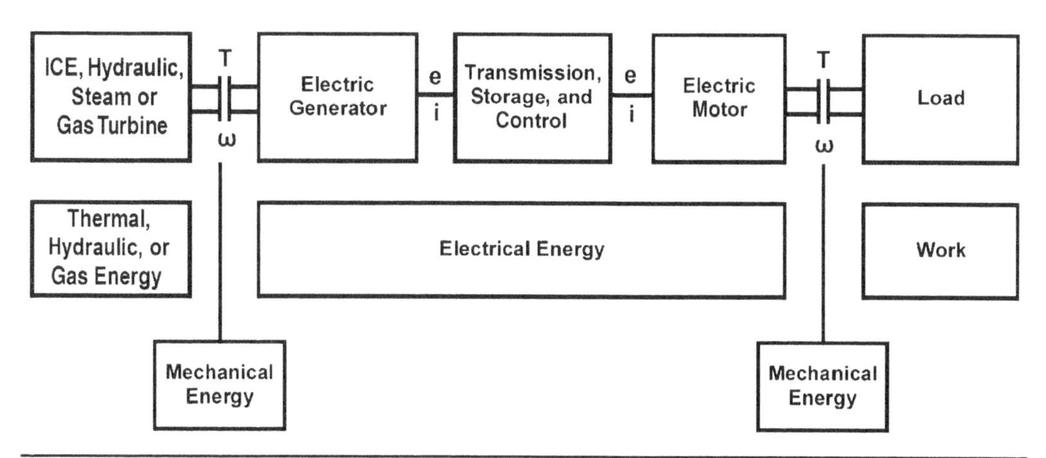

FIGURE 1.4 Power transmission in an electrical power system.

1.2.3 Pneumatic Power Systems

Pneumatic systems are power systems using compressed air as a working medium for the power transmission. Their principle of operation is similar to that of electric power systems. The air compressor converts the mechanical energy of the prime mover into mainly pressure energy of compressed air. This transformation facilitates the transmission and control of power. An air preparation process is needed to prepare the compressed air for use. The air preparation includes filtration, drying, and the adding of lubricating oil mist. The compressed air is stored in the compressed air reservoirs and transmitted through rigid and/or flexible lines. The pneumatic power is controlled by means of a set of pressure, flow, and directional control valves. Then, it is converted to the required mechanical power by means of pneumatic cylinders and motors (expanders). Figure 1.5 illustrates the process of power transmission in pneumatic systems.

1.2.4 Hydrodynamic Power Systems

The hydraulic power systems transmit mechanical power by increasing the energy of hydraulic liquids. Two types of hydraulic power systems are used: hydrodynamic and hydrostatic.

Hydrodynamic (also called hydrokinetic) power systems transmit power by increasing mainly the kinetic energy of liquid. Generally, these systems include a rotodynamic pump, a turbine, and additional control elements. The applications of hydrodynamic power systems are limited to rotary motion. These systems replace the classical mechanical transmission in the power stations and vehicles due to their high power-to-weight ratio and better controllability.

There are two main types of hydrodynamic power systems: hydraulic coupling and torque converter.

A hydraulic coupling (see Fig. 1.6) is essentially a fluid-based clutch. It consists of a pump (2), driven by the input shaft (1), and a turbine (3), coupled to the output shaft (4). When the pump impeller rotates, the oil flows to the turbine at high speed. The oil then impacts the turbine blades, where it loses most of the kinetic energy it gained from the pump. The oil re-circulates in a closed path inside the coupling and the power is transmitted from the input shaft to the output shaft. The input torque is practically equal to the output torque.

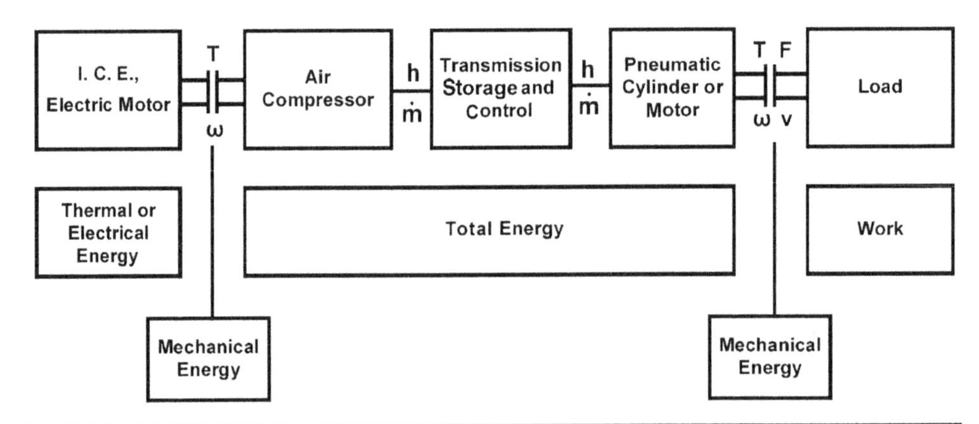

FIGURE 1.5 Power transmission in a pneumatic power system.

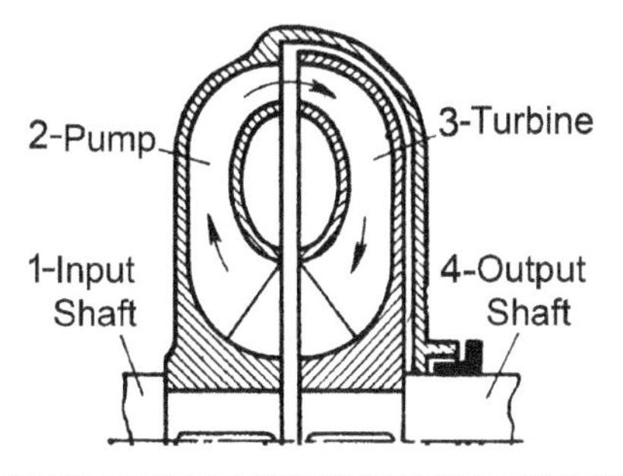

Figure 1.6 Hydraulic coupling.

The torque converter is a hydraulic coupling with one extra component: the stator, also called the reactor (5). (See Fig. 1.7.) The stator consists of a series of guide blades attached to the housing. The torque converters are used where it is necessary to control the output torque and develop a transmission ratio, other than unity, keeping acceptable transmission efficiency.

1.2.5 Hydrostatic Power Systems

In the hydrostatic power systems, the power is transmitted by increasing mainly the pressure energy of liquid. These systems are widely used in industry, mobile equipment, aircrafts, ship control, and others. This text deals with the hydrostatic power systems, which are commonly called *hydraulic power systems*. Figure 1.8 shows the operation principle of such systems.

The concepts of hydraulic energy, power, and power transformation are simply explained in the following: Consider a forklift that lifts a load vertically for a distance y during a time period Δt (see Fig. 1.9). To fulfill this function, the forklift acts on the load by a vertical force F. If the friction is negligible, then in the steady state, this force equals the total weight of the displaced parts ($F = mg$). The work done by the forklift is

$$W = Fy \tag{1.1}$$

By the end of the time period, Δt, the potential energy of the lifted body is increased by E, where

$$E = mgy = Fy \tag{1.2}$$

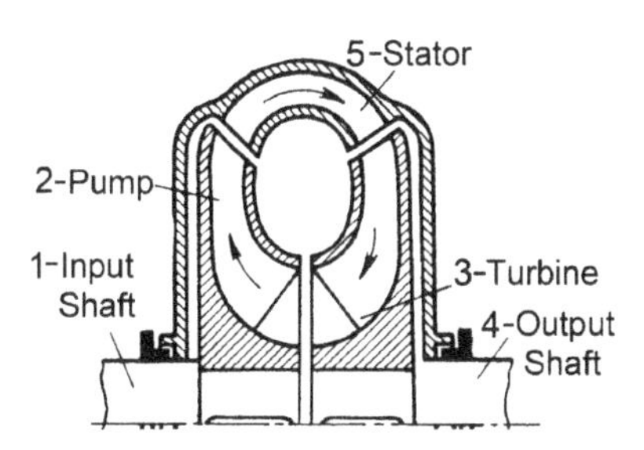

Figure 1.7 Torque converter.

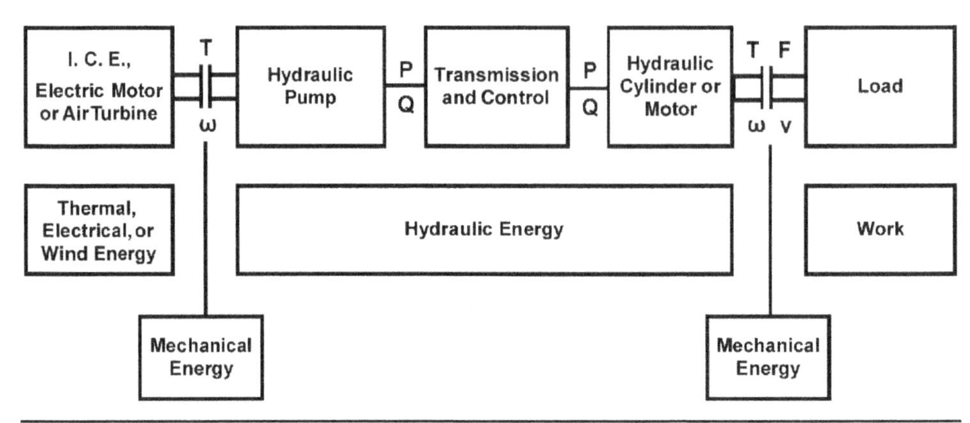

FIGURE 1.8 Power transmission in a hydraulic power system.

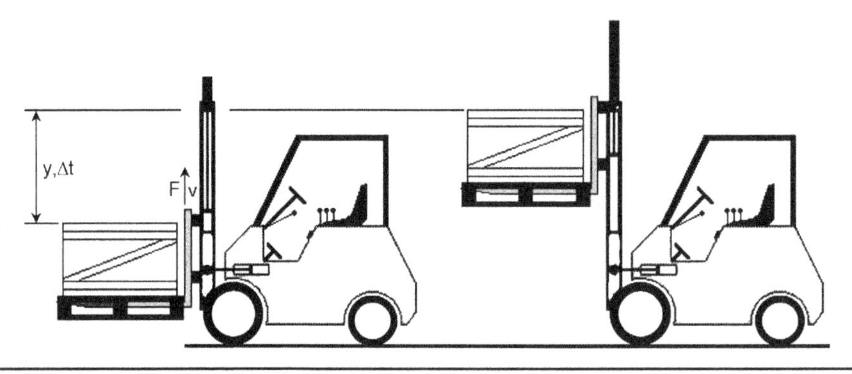

FIGURE 1.9 Load lifting by a forklift.

where E = Gained potential energy, J
F = Vertically applied force, N
g = Coefficient of gravitational force, m/s^2
m = Mass of lifted body, kg
W = Work, J
y = Vertical displacement, m

This amount of energy (E) is gained during time period Δt. The energy delivered to the lifted body per unit of time is the delivered power N, where

$$N = Fy/\Delta t = Fv \tag{1.3}$$

where N = Mechanical power delivered to the load, W
v = Lifting speed, m/s

The load is lifted by a hydraulic cylinder. This cylinder acts on the lifted body by a force F and drives it with a speed v. Figure 1.10 illustrates the action of the hydraulic cylinder. It is a single acting cylinder which extends by the pressure force and retracts by the body weight. The pressurized oil flows to the hydraulic cylinder at a flow rate Q (volumetric flow rate, m^3/s) and its pressure is p. Neglecting the friction in the cylinder, the pressure force which drives the piston in the extension direction is given by $F = pA_p$.

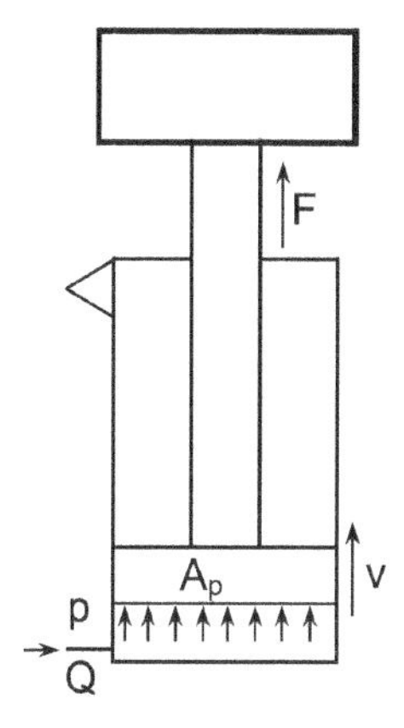

FIGURE 1.10 Lifting a body vertically by a hydraulic cylinder.

During the time period, Δt, the piston travels vertically a distance y. The volume of oil that entered the cylinder during this period is $V = A_p y$.

Then, the oil flow rate that entered the cylinder is

$$Q = \frac{V}{\Delta t} = \frac{A_p y}{\Delta t} = A_p v \tag{1.4}$$

Assuming an ideal cylinder, then the hydraulic power inlet to the cylinder is

$$N = Fv = pA_p Q / A_p = Qp \tag{1.5}$$

where A_p = Piston area, m^2
p = Pressure of inlet oil, Pa
Q = Flow rate, m^3/s
V = Piston swept volume, m^3

The mechanical power delivered to the load equals the hydraulic power delivered to the cylinder. This equality is due to the assumption of zero internal leakage and zero friction forces in the cylinder. The assumption of zero internal leakage is practical, for normal conditions. However, for aged seals, there may be non-negligible internal leakage. A part of the inlet flow leaks and the speed v becomes less than (Q/A_p). Also, a part of the pressure force overcomes the friction forces. Thus, the mechanical power output from the hydraulic cylinder is actually less than the input hydraulic power ($Fv < Qp$).

1.3 Basic Hydraulic Power Systems

Figure 1.11 shows the circuit of a simple hydraulic system, drawn in both functional-sectional schemes and standard hydraulic symbols. The function of this system is summarized in the following:

1. The prime mover supplies the system with the required mechanical power. The pump converts the input mechanical power to hydraulic power.

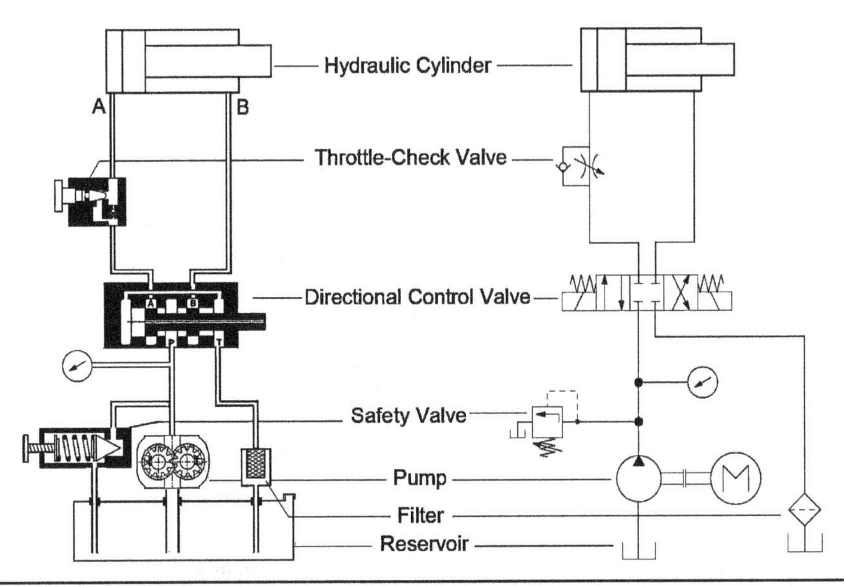

FIGURE 1.11 Hydraulic system circuit, schematic, and symbolic drawings.

2. The energy-carrying liquid is transmitted through the hydraulic transmission lines: pipes and hoses. The hydraulic power is controlled by means of valves of different types. This circuit includes three different types of valves: a pressure control valve, a directional control valve, and a flow control (throttle-check) valve.
3. The controlled hydraulic power is communicated to the hydraulic cylinder, which converts it to the required mechanical power. Generally, the hydraulic power systems provide both rotary and linear motions.

1.4 The Advantages and Disadvantages of Hydraulic Systems

The main advantages of the hydraulic power systems are the following:

1. High power-to-weight ratio.
2. Self-lubrication.
3. There is no saturation phenomenon in the hydraulic systems compared with saturation in electric machines. The maximum torque of an electric motor is proportional to the electric current, but it is limited by the magnetic saturation.
4. High force-to-mass and torque-to-inertia ratios, which result in high acceleration capability and a rapid response of the hydraulic motors.
5. High stiffness of the hydraulic cylinders, which allows stopping loads at any intermediate position.
6. Simple protection against overloading.
7. Possibility of energy storage in hydraulic accumulators.

8. Flexibility of transmission compared with mechanical systems.
9. Availability of both rotary and rectilinear motions.
10. Safe regarding explosion hazards.

Hydraulic power systems have the following disadvantages:

1. Hydraulic power is not readily available, unlike electrical. Hydraulic generators are therefore required.
2. High cost of production due to the requirements of small clearances and high precision production process.
3. High inertia of transmission lines, which increases their response time.
4. Limitation of the maximum and minimum operating temperature.
5. Fire hazard when using mineral oils.
6. Oil filtration problems.

1.5 Comparing Power Systems

Table 1.1 shows a brief comparison of the different power systems, while Table 1.2 gives the power variables in mechanical, electrical, and hydraulic systems.

System Property	Mechanical	Electrical	Pneumatic	Hydraulic
Input energy source	ICE and electric motor	ICE and hydraulic, air or steam turbines	ICE, electric motor, and pressure tank	ICE, electric motor, and air turbine
Energy transfer element	Mechanical parts, levers, shafts, gears	Electrical cables and magnetic field	Pipes and hoses	Pipes and hoses
Energy carrier	Rigid and elastic objects	Flow of electrons	Air	Hydraulic liquids
Power-to-weight ratio	Poor	Fair	Best	Best
Torque/inertia	Poor	Fair	Good	Best
Stiffness	Good	Poor	Fair	Best
Response speed	Fair	Best	Fair	Good
Dirt sensitivity	Best	Best	Fair	Fair
Relative cost	Best	Best	Good	Fair
Control	Fair	Best	Good	Good
Motion type	Mainly rotary	Mainly rotary	Linear or rotary	Linear or rotary

Table 1.1 Comparison of Power Systems

	Effort		Flow		Power	
	Variable	Units	Variable	Units	Variable	Units
Mech. Linear	Force, F	N	Velocity, v	m/s	$N = Fv$	W
Mech. Rotary	Torque, T	Nm	Angular speed, ω	rad/s	$N = \omega T$	W
Electrical (DC)	Electric potential, e	V	Electric current, i	A	$N = ei$	W
Hydraulic	Pressure, P	Pa	Flow rate, Q	m^3/s	$N = PQ$	W

TABLE 1.2 Effort, Flow, and Power Variables of Different Power Systems

1.6 Exercises

1. State the function of the power systems.

2. Discuss briefly the principle of operation of the different power systems giving the necessary schemes.

3. Draw the circuit of a simple hydraulic system, in standard symbols, and explain briefly the function of its basic elements.

4. State the advantages and disadvantages of hydraulic power systems.

5. Draw the circuit of a simple hydraulic system, including a pump, directional control valves, hydraulic cylinder, relief valve, and pressure gauge. State the function of the individual elements and discuss in detail the power transmission and transformation in the hydraulic power systems.

6. The given figure shows the extension mode of a hydraulic cylinder. Neglecting the losses in the transmission lines and control valves, calculate the loading force, F, returned flow rate, Q_t, piston speed, v, cylinder output mechanical power, N_m, and pump output hydraulic power, N_h. Comment on the calculation results, given

Delivery line pressure $P = 200$ bar

Pump flow rate $Q_p = 40$ L/min

Piston diameter $D = 100$ mm

Piston rod diameter $d = 70$ mm

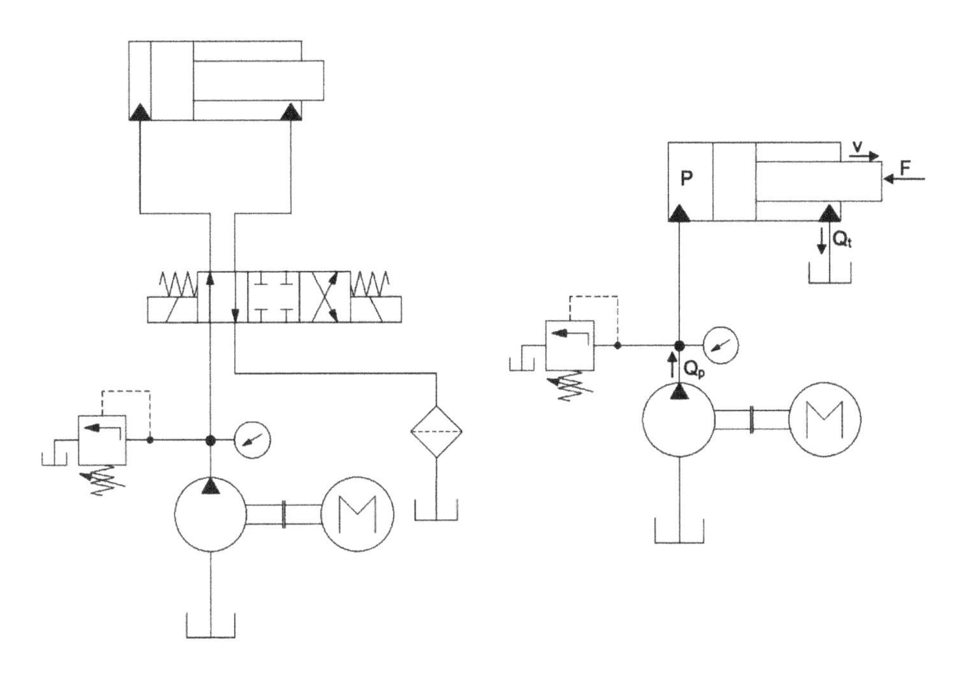

7. The given figure shows the extension mode of a hydraulic cylinder, in differential connection. The losses in the transmission lines and control valves were neglected. Calculate the loading force, F, inlet flow rate, Q_{in}, returned flow rate, Q_{out}, piston speed, v, cylinder output mechanical power, N_m, and pump output hydraulic power, N_h. Comment on the calculation results compared with the case of problem 6, given

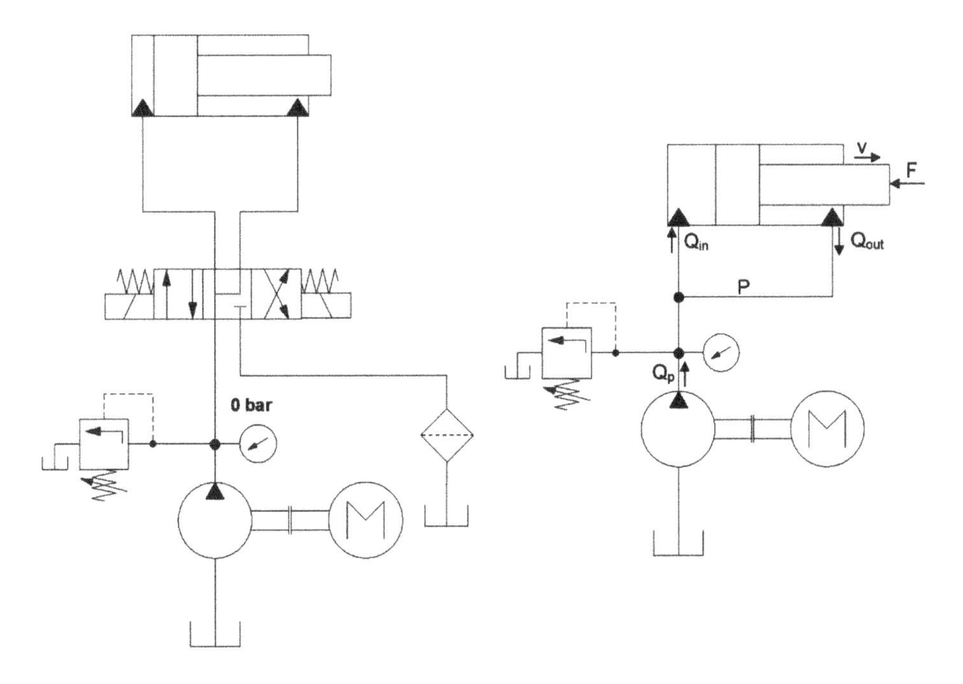

Delivery line pressure $P = 200$ bar

Pump flow rate $Q_p = 40$ L/min

Piston diameter $D = 100$ mm

Piston rod diameter $d = 70$ mm

8. Shown is the load-lowering mode of a hydraulic system. The lowering speed is controlled by means of a throttle-check valve. The hydraulic circuit is redrawn, neglecting the system elements' hydraulic losses, except the throttle valve. Discuss the construction and operation of this system. Calculate the pressure in the cylinder rod side P_C, the inlet flow rate Q_{in}, outlet flow rate Q_{out}, pump flow rate Q_P, pump output power N_h, and the area of the throttle valve A_t.

The flow rate through the throttling element is given by: $Q = C_d A_t \sqrt{2\Delta P/\rho}$, where

Q = Flow rate, m^3/s

C_d = Discharge coefficient

A_t = Throttle area, m^2

ΔP = Pressure difference, Pa

ρ = Oil density, kg/m^3

Given

Pump exit pressure = 30 bar

Piston speed = 0.07 m/s

Piston area $A_p = 78.5$ cm^2

Piston rod side area $A_r = 40$ cm^2

Oil density = 870 kg/m^3

Discharge coefficient = 0.611

Safety valve is pre-set at 350 bar

Weight of the body = 30 kN

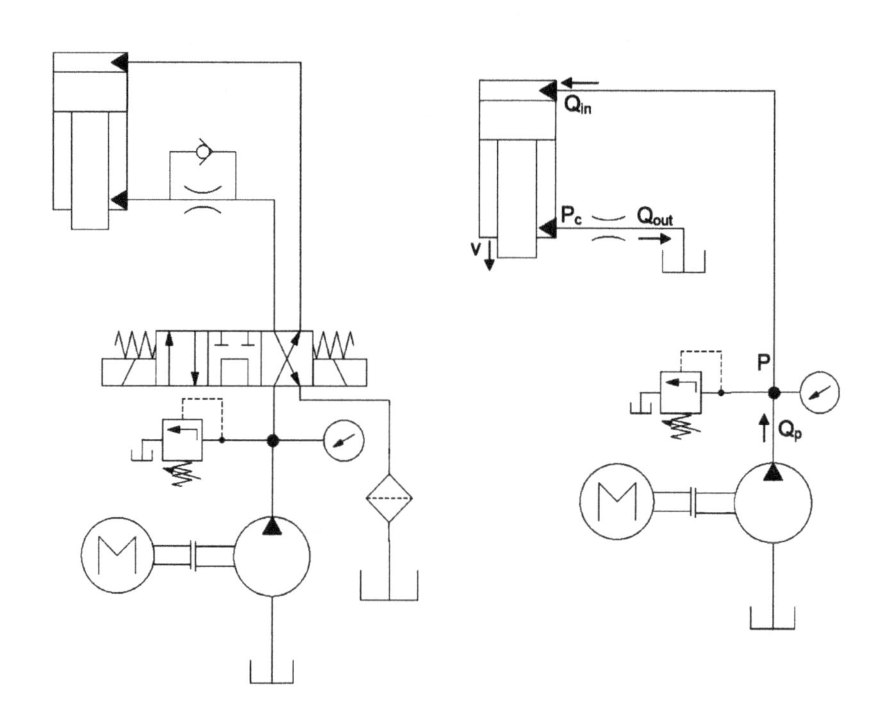

9. Shown is the circuit of problem 8 in the load-lifting mode. For the same pump flow rate, safety valve setting, and dimensions, calculate the maximum load that the system can lift. Calculate all of the system operating parameters in this mode. Neglect the hydraulic losses in the system elements, except for the throttle valve.

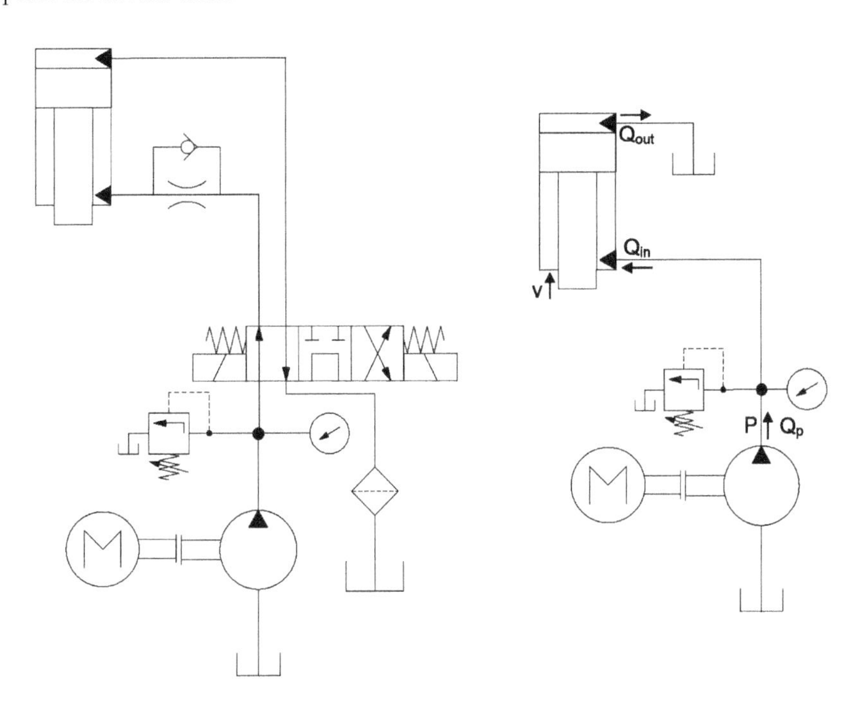

1.7 Nomenclature

A_p = Piston area, m^2
E = Gained potential energy, J
F = Vertically applied force, N
g = Coefficient of gravitational force, m/s^2
m = Mass of lifted body, kg
N = Mechanical power delivered to the load, W
p = Pressure, Pa
Q = Flow rate, m^3/s
v = Lifting speed, m/s
V = Piston swept volume, m^3
W = Work, J
y = Vertical displacement, m

CHAPTER 2
Hydraulic Oils and Theoretical Background

2.1 Introduction

Hydraulic fluids are used in hydrostatic power systems to transmit power. The power transmission is carried out by increasing, mainly, the pressure energy of the fluid. In addition to the *power transmission*, the hydraulic fluids serve to *lubricate* the contact surfaces, *cool* different elements, and *clean* the system. Water was the first fluid used for the transmission of fluid power. The main advantages of water as a hydraulic fluid are its availability, low cost, and fire resistance. On the other hand, water is of poor lubricity, has a narrow range of working temperature, and has a high rust-promoting tendency. These disadvantages limited its use to very special systems.

Although mineral oils were readily available at the beginning of the twentieth century, they were not practically used in hydraulic systems until the 1920s. In the 1940s, additives were first used to improve the physical and chemical properties of hydraulic mineral oils. The first additives were developed to counter rust and oxidation. However, mineral oils are highly flammable, and fire risk increases when operating at high temperatures. This has led to the development of fire-resistant fluids that are mainly water-based, with limitations on the operating conditions. The need for extremes of operating temperatures and pressures led to the development of synthetic fluids.

This chapter is dedicated to studying the properties of hydraulic fluids and their effect on a system's performance. It also explores the theoretical background needed for studying the topics of this text.

2.2 Basic Properties of Hydraulic Oils

2.2.1 Viscosity

Definitions and Formulas

Viscosity is the name given to the characteristic of a fluid, and describes the resistance to the laminar movement of two neighboring fluid layers against each other. Simply, *viscosity is the resistance to flow*. It results from the cohesion and interaction between molecules.

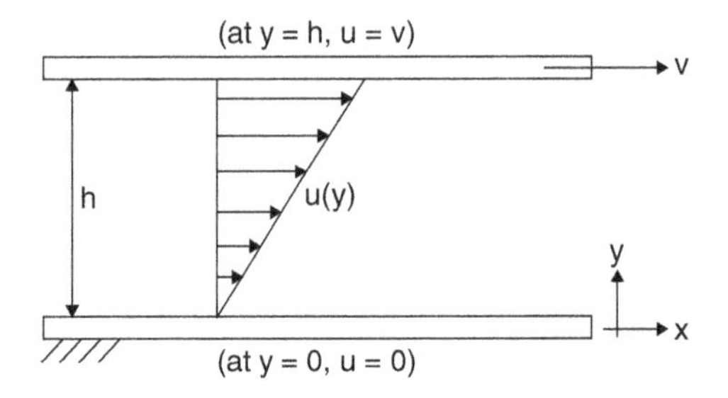

FIGURE 2.1 Velocity variation for a fluid between two near parallel plates.

Consider a fluid between two infinite plates (see Fig. 2.1). The lower plate is fixed, while the upper plate is moving at a steady speed v. The upper plate is subjected to a friction force to the left since it is doing work trying to drag the fluid along with it to the right. The fluid at the top of the channel will be subjected to an equal and opposite force. Similarly, the lower plate will be subjected to a friction force to the right since the fluid is trying to pull the plate along with it to the right. The fluid is subjected to shear stress, τ, given by Newton's law of viscosity.

$$\tau = \mu \frac{du}{dy} \tag{2.1}$$

The coefficient of *dynamic viscosity*, μ, is the shearing stress necessary to induce a unit flow velocity gradient in a fluid. In actual measurement, the viscosity coefficient of a fluid is obtained from the ratio of shearing stress to shearing rate.

$$\mu = \frac{\tau}{du/dy} \tag{2.2}$$

where τ = Shear stress, N/m^2
du/dy = Velocity gradient, s^{-1}
u = Fluid velocity, m/s
y = Displacement perpendicular to the velocity vector, m
μ = Coefficient of dynamic viscosity, Ns/m^2; μ is often expressed in poise (P), where $1\ P = 0.1\ Ns/m^2$

For Newtonian fluids, the coefficient of dynamic viscosity, μ, is independent of du/dy. However, it changes with temperature and pressure.

Kinematic viscosity, ν, is defined as the ratio of the dynamic viscosity to the density.

$$\nu = \frac{\mu}{\rho} \tag{2.3}$$

where ν = Kinematic viscosity, m^2/s
ρ = Oil density, kg/m^3

The kinematic viscosity, ν, is often expressed in stokes (St), where $1\ St = 10^{-4}\ m^2/s$, or in centistokes (cSt), where $1\ cSt = 10^{-6}\ m^2/s = 1\ mm^2/s$.

The viscosity units may be given in Redwood or Saybolt seconds, or in degrees Engler, according to the measuring method. These units are no longer used, but conversion tables are available.

The oil viscosity is affected by its temperature, as shown in Fig. 2.2. It decreases with the increase in temperature. Therefore, the viscosity is stated at a standard temperature (40°C for the ISO specification). A hydraulic fluid referred to as VG32 has a viscosity of 32 cSt at 40°C.

It is important to keep the oil viscosity within a certain range during the system's operation; otherwise, the operating conditions will change with temperature. The viscosity index (VI) of oil is a number used in industry to indicate the effect of temperature variation on the viscosity of the oil. A low VI signifies a relatively large change of viscosity with temperature variation. On the other hand, a high VI means relatively little change in viscosity over a wide temperature range. The best oil is the one that maintains constant viscosity throughout temperature changes. An example of the importance of the VI is the need for high VI hydraulic oil for military aircraft. These systems are exposed to a wide variation of atmospheric temperatures, ranging from below −18°C at high altitudes to over 45°C on the ground. For proper operation of these systems, the hydraulic fluid must have sufficiently high VI to perform its functions at the extremes of the temperature range.

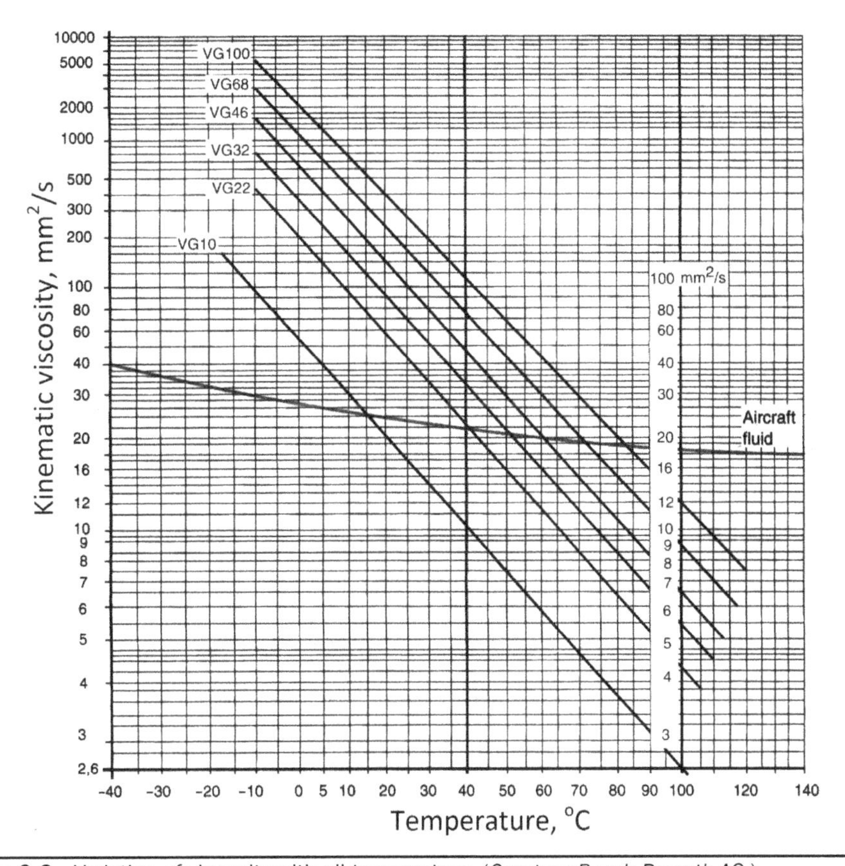

FIGURE 2.2 Variation of viscosity with oil temperature. (*Courtesy Bosch Rexroth AG.*)

The acceptable range of viscosity and operating temperature is determined by the manufacturers for each component. For gear pump, for example, the fluid temperature range is –15 to 80°C, and the recommended viscosity range is 10 to 300 cSt.

The effect of oil pressure on the viscosity is much less than that of temperature. The viscosity of fluids increases as its pressure increases (see Fig. 2.3). These characteristics must be taken into account when planning hydraulic systems which operate at wide pressure ranges. However, when operating at pressure levels within 300 bar, the effect of pressure on the oil viscosity is usually negligible.

Effect of Viscosity on Hydraulic System Operation

The oil viscosity influences the function of hydraulic power systems as it introduces resistance to fluid flow and to the motion of bodies moving in the fluid. Herein, the following effects are studied:

- Hydraulic losses in transmission lines
- Resistance to fluid flow in narrow conduits
- Viscous friction forces and damping effect

Hydraulic Losses in Transmission Lines, Hydraulic Resistance In hydraulic transmission lines, the flow may be laminar or turbulent depending on the ratio of the inertia forces to the viscous friction forces. This ratio is evaluated by the Reynolds number, Re. For laminar flow (see Fig. 2.4), the pressure losses in the line are calculated using the following relation (see App. 2B at the end of this chapter):

$$\Delta P = \lambda \frac{L}{D} \frac{\rho v^2}{2} \tag{2.4}$$

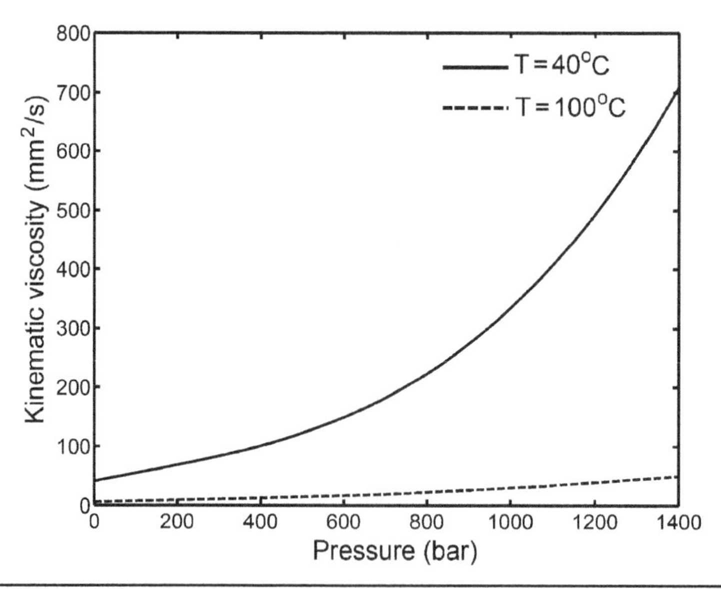

FIGURE 2.3 Variation of the kinematic viscosity of a typical mineral-based oil with pressure and temperature.

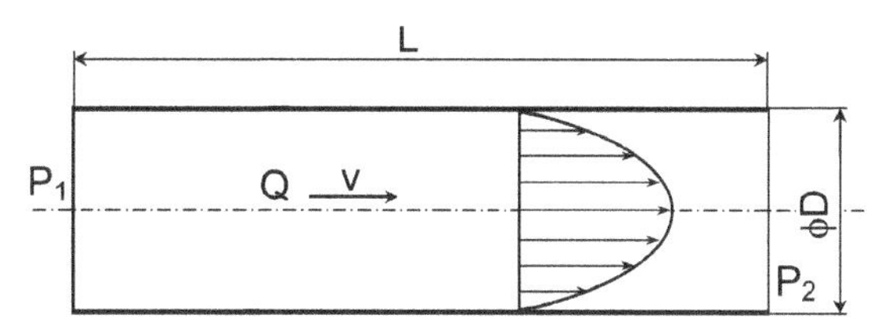

Figure 2.4 Laminar flow in pipeline.

$$v = 4Q/\pi D^2 \tag{2.5}$$

$$\lambda = 64/\mathrm{Re} \tag{2.6}$$

$$\mathrm{Re} = vD/\nu = \rho v D/\mu \tag{2.7}$$

where D = Inner pipe diameter, m
L = Pipe length, m
Re = Reynolds number
v = Mean fluid velocity, m/s
ΔP = Pressure losses in the pipe line, Pa
λ = Friction coefficient for laminar flow
ρ = Oil density, kg/m^3

The following expression for the pressure losses ΔP was obtained by substituting Eqs. (2.5) to (2.7) in Eq. (2.4):

$$\Delta P = \frac{128\mu L}{\pi D^4} Q = RQ \tag{2.8}$$

The term R expresses the resistance of the hydraulic transmission line. Its effect is equivalent to that of the electric resistance. Both of them dissipate energy and both are described by the same mathematical relation ($e = Ri$). The power loss ΔN in the pipeline is given by

$$\Delta N = QP_1 - QP_2 = Q\Delta P = RQ^2 = \frac{128\mu L}{\pi D^4} Q^2 \tag{2.9}$$

Resistance to Fluid Flow in Narrow Conduits

Internal Leakage in Hydraulic Elements Hydraulic power systems operate at pressure levels up to 700 bar. The internal leakage in hydraulic elements is one of the problems resulting from the operation at high-pressure levels and the increased clearances due to wear.

Figure 2.5 shows the internal leakage through a radial clearance between two concentric cylindrical bodies, a spool and sleeve, for example. Considering the shown fluid element in the radial clearance, and neglecting the minor losses at the inlet and outlet, and assuming a concentric stationary spool, an expression for the leakage flow rate can be deduced as follows. In the steady state, the fluid element speed is constant and the forces acting on it are in equilibrium. These forces are the pressure forces and the friction forces acting on the internal and external surfaces of the fluid element.

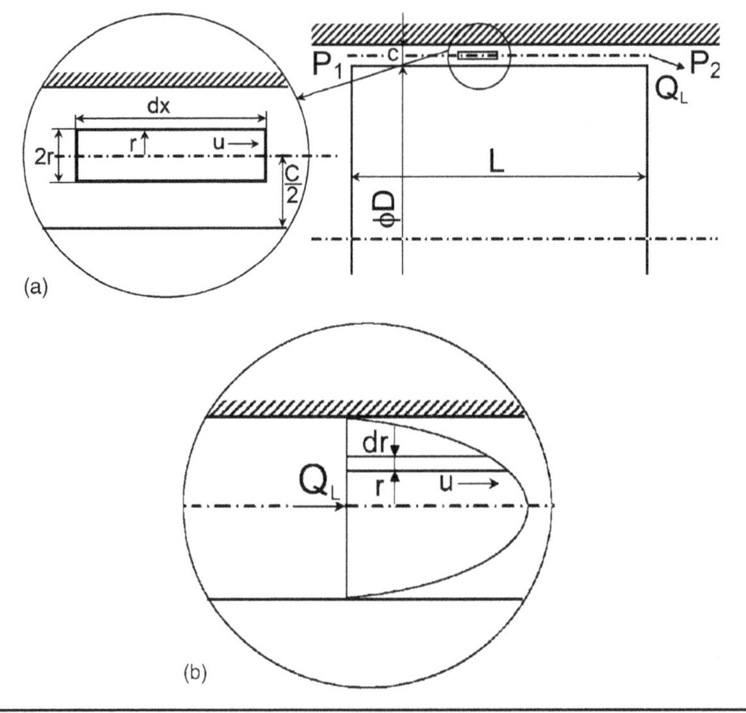

FIGURE 2.5 (a) Leakage fluid flow through a radial clearance and (b) laminar fluid velocity profile in a radial clearance.

The pressure force is $$F_P = 2r\pi D dP \tag{2.10}$$

The friction force is $$F_\tau = 2\pi D dx\,\tau \tag{2.11}$$

$$r = 0.5c - y \text{ then } \frac{du}{dy} = -\frac{du}{dr} \tag{2.12}$$

For Newtonian fluid, the shear stress is $\tau = \mu\dfrac{du}{dy} = -\mu\dfrac{du}{dr}$ (2.13)

Since $F_P = F_\tau$ then $\dfrac{du}{dr} = -\dfrac{r}{\mu}\dfrac{dP}{dx}$ or $du = -\dfrac{r}{\mu}\dfrac{dP}{dx}dr$ (2.14)

The pressure gradient dP/dx is constant.

$$\frac{dP}{dx} = \frac{\Delta P}{L} \text{ where } \Delta P = P_1 - P_2 \tag{2.15}$$

The velocity distribution in the radial clearance is found by integrating Eq. (2.14).

$$u = \int -\frac{r}{\mu}\frac{dP}{dx}dr + a = a - \frac{r^2}{2\mu}\frac{dP}{dx} \tag{2.16}$$

If the fluid velocity at the boundaries is zero, then

$$u = 0 \text{ for } r = \pm c/2 \tag{2.17}$$

By substitution from Eqs. (2.15) and (2.17) into Eq. (2.16), the following expression for the velocity distribution is obtained:

$$u = \frac{1}{2\mu}\frac{\Delta P}{L}\left(\frac{c^2}{4} - r^2\right) \tag{2.18}$$

The leakage flow rate, Q_L, is then found as follows:

$$Q_L = \int_{-c/2}^{c/2} u\pi D\,dr = \frac{\pi D c^3}{12\mu L}\Delta P \tag{2.19}$$

or

$$\Delta P = \frac{12\mu L}{\pi D c^3} Q_L = R_L Q_L \tag{2.20}$$

where
a = Constant, m/s
c = Radial clearance, m
D = Spool diameter, m
F_p = Pressure force acting on the fluid element, N
F_τ = Shear force acting on the fluid element, N
L = Length of leakage path, m
Q_L = Leakage flow rate, m^3/s
r = Radial distance from the midpoint of clearance, m
R_L = Resistance to leakage, Ns/m^5
y = Distance between the element side surface and solid boundary, m
u = Oil speed in the clearance, m/s
ΔP = Pressure difference across the radial clearance, Pa

It is important to note that the leakage is inversely proportional to the viscosity, μ, and directly proportional to the cube of radial clearance. If the radial clearance is doubled due to wear, the internal leakage increases eight times. The power loss due to leakage is given by

$$\Delta N = Q_L \Delta P = \frac{\pi D c^3}{12\mu L}\Delta P^2 = \frac{\Delta P^2}{R_L} \quad \text{or} \quad \Delta N = \left(\frac{12\mu L}{\pi D c^3}\right) Q_L^2 = R_L Q_L^2 \tag{2.21}$$

The internal leakage reduces the effective flow rates and increases the power losses. The dissipated power ΔN is converted to heat and leads to serious oil overheating problems. Therefore, it is important to keep the oil viscosity within the predetermined limits over the whole operating temperature range. This is fulfilled by using hydraulic oils of convenient viscosity index and implementation of oil coolers.

Fluid Flow in an Eccentric Mounting Radial Clearance In the case of eccentric mounting, the radial clearance thickness is not constant (see Fig. 2.6). The flow rate through a narrow radial clearance is given by

$$Q = \frac{\pi D c^3}{12\mu L}\left[1 + \frac{3}{2}\left(\frac{\varepsilon}{c}\right)^3\right]\Delta P \tag{2.22}$$

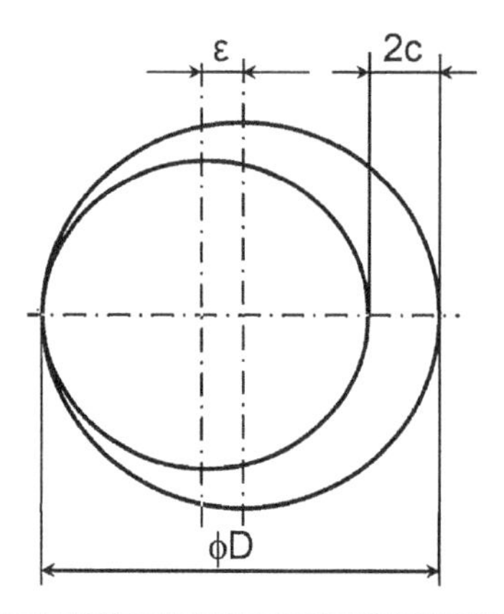

FIGURE 2.6 Radial clearance with eccentric mounting.

Fluid Flow in a Long-Thin-Slot Orifice In the case of a long-thin-slot orifice, the fluid flow is laminar. The following expression gives the fluid flow rate in this orifice (see Fig. 2.7).

$$Q = \frac{bh^3}{12\mu L}\Delta P \tag{2.23}$$

Fluid Flow in the Clearance Between a Circular Nozzle and a Plane Surface The flow rate in the clearance between the plane face of a nozzle and a plane surface (see Fig. 2.8), is given by the following relation:

$$Q = \frac{\pi(d_1 + d_2)h^3}{12(d_1 - d_2)\mu}\Delta P \tag{2.24}$$

Viscous Friction and Damping Effect The parts moving in oil are subjected to viscous friction forces due to the shear stress resulting from the oil viscosity. Figure 2.9 shows a spool moving axially. The cylindrical surface of the spool is subjected to a shear stress, τ. The

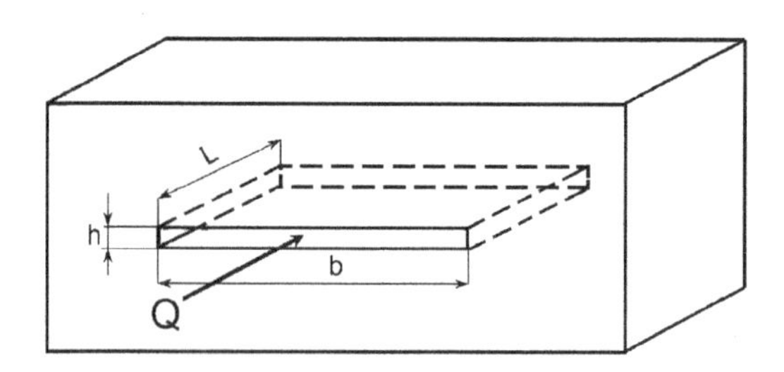

FIGURE 2.7 Long-thin-slot orifice.

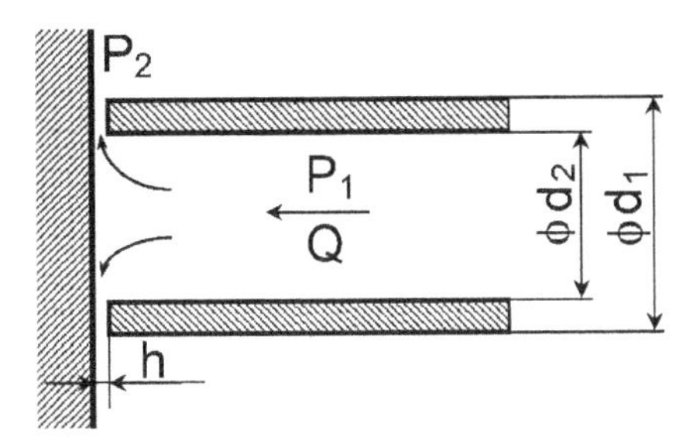

FIGURE 2.8 Nozzle controlled by a plane surface at distance h.

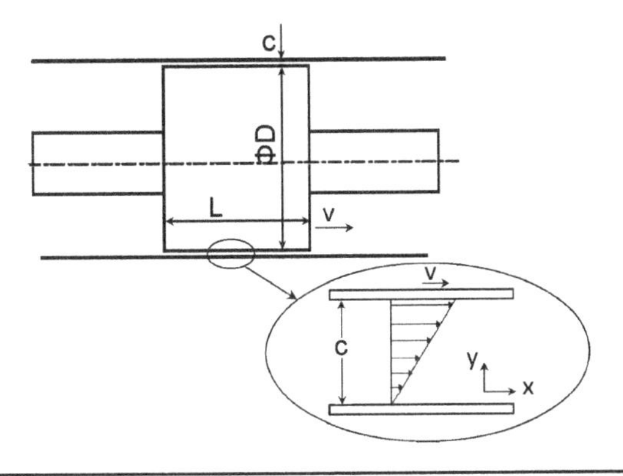

FIGURE 2.9 Velocity distribution in the spool radial clearance.

velocity distribution is assumed to be linear in the small radial clearance: c = 2 to 10 μm. An expression for the friction forces is deduced as follows:

$$\tau = \mu \frac{du}{dy} = \mu \frac{v}{c} \tag{2.25}$$

$$F = \pi D L \tau = \frac{\pi \mu D L}{c} v = f_v v \tag{2.26}$$

$$f_v = \frac{\pi \mu D L}{c} \tag{2.27}$$

where F = Friction force, N
f_v = Friction coefficient, Ns/m
L = Length of spool land, m

Assignment Derive an expression for the viscous friction coefficient if the spool is rotated by an angular speed ω under the action of a torque ($T = f_\omega \omega$).

The effect of viscous damping may be demonstrated by studying the motion of the spool of a directional control valve (see Fig. 2.10). The spool moves under the action of

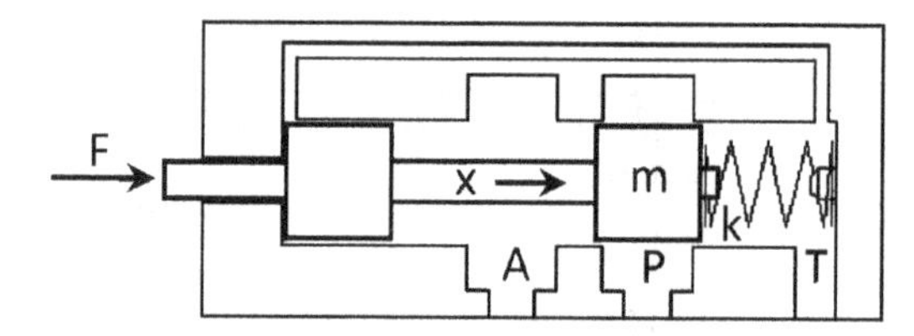

Figure 2.10 A spring-loaded spool valve.

the driving force (F), the spring force, and the friction force. The spool motion is described by the following equation:

$$F = m\frac{d^2x}{dt^2} + f_v\frac{dx}{dt} + kx \tag{2.28}$$

where F = Driving force, N
k = Spring stiffness, N/m
m = Mass of moving parts, kg
x = Spool displacement, m

Assuming zero initial conditions, and applying Laplace transformation to the equation of motion, the following transfer function can be deduced:

$$G(s) = \frac{X(s)}{F(s)} = \frac{1}{ms^2 + f_v s + k} = \frac{K}{\frac{s^2}{\omega_n^2} + \frac{2\zeta}{\omega_n}s + 1} \tag{2.29}$$

where ζ = Damping coefficient, proportional to the oil viscosity; $\zeta = \frac{\pi\mu DL}{2c\sqrt{km}}$
k = Spring stiffness, N/m
K = Gain, $K = 1/k$, m/N
ω_n = Natural frequency, $\omega_n = \sqrt{k/m}$, rad/s

2.2.2 Oil Density

Definition

The density is the mass per unit volume: $\rho = m/V$. The hydraulic oils are of low compressibility and volumetric thermal expansion. Therefore, under ordinary operating conditions, the oil density is practically constant. The density of mineral hydraulic oils ranges from 850 to 900 kg/m^3. The oil density affects both the transient and steady state operations of the hydraulic systems. The hydraulic losses in throttling elements and transmission lines are dominated mainly by the inertia and friction losses. The effect of oil inertia on these elements is discussed in this chapter.

Effect of Density on Hydraulic System Operation

Orifice Flow Orifices, short-tube or sharp-edged, are a basic means of control in fluid power systems. This section aims at deriving equations for the flow rate of fluid through orifices and evaluating the effect of fluid viscosity and inertia. In most cases, the orifice

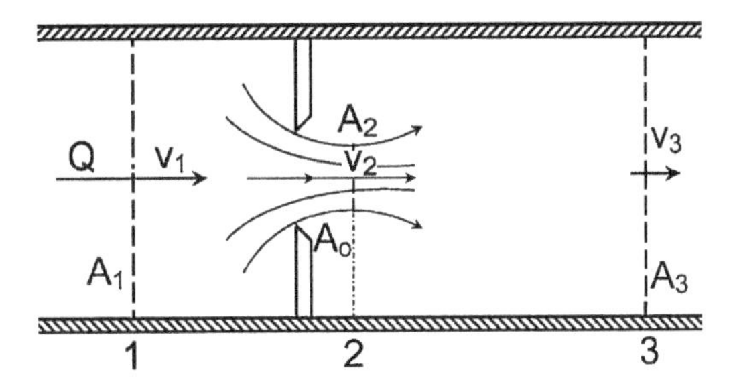

Figure 2.11 Flow through a sharp-edged orifice.

flow occurs at high Reynolds numbers. Such flow is referred to as turbulent flow, but this term does not have quite the same meaning as the pipe flow.

Referring to Fig. 2.11, the fluid particles are accelerated from velocity v_1 at section 1 to the jet velocity, v_2, at section 2 through a *sharp-edged* orifice. The fluid flow between sections 1 and 2 is nearly streamlined or potential flow, which justifies the application of Bernoulli's equation between these two sections.

$$v_2^2 - v_1^2 = \frac{2}{\rho}(P_1 - P_2) \tag{2.30}$$

The term $(P_1 - P_2)$ is the pressure difference required to accelerate the fluid from the lower upstream velocity (v_1) to the higher jet velocity (v_2). The kinetic energy of the jet is not recovered. It is converted into thermal energy, increasing the fluid temperature, and the pressures at sections 2 and 3 are practically equal.

The area of the jet (A_2) is smaller than the orifice area (A_0) due to the fluid inertia. The point along the jet where the area becomes a minimum is called vena contracta. The contraction coefficient is defined as

$$C_c = A_2/A_0 \tag{2.31}$$

Assuming an incompressible fluid, the application of the continuity equation yields

$$A_1 v_1 = A_2 v_2 \tag{2.32}$$

Considering Eqs. (2.30) to (2.32), the following expression is deduced for the jet velocity, v_2:

$$v_2 = \frac{1}{\sqrt{1-(A_2/A_1)^2}}\sqrt{\frac{2}{\rho}(P_1 - P_2)} \tag{2.33}$$

Actually, the jet velocity is slightly less than that calculated by Eq. (2.33), due to the losses caused by the viscous friction. This friction is taken into consideration by introducing the velocity coefficient C_v, ranging from 0.97 to 0.99, and defined as

$$C_v = \frac{\text{Actual velocity at vena contracta}}{v_2} \tag{2.34}$$

The flow rate through the orifice is thus given by the following expression:

$$Q = A_2 C_v v_2 = \frac{C_v A_2}{\sqrt{1-(A_2/A_1)^2}} \sqrt{\frac{2}{\rho}(P_1 - P_2)} \tag{2.35}$$

or

$$Q = C_d A_0 \sqrt{\frac{2}{\rho}(P_1 - P_2)} \tag{2.36}$$

where the discharge coefficient, C_d, is given by

$$C_d = \frac{C_v A_2/A_0}{\sqrt{1-(A_2/A_1)^2}} \quad \text{or} \quad C_d = \frac{C_v C_c}{\sqrt{1-C_c^2(A_0/A_1)^2}} \tag{2.37}$$

where C_c = Contraction coefficient depends on the geometry of the hole
C_d = Discharge coefficient, typically = 0.6 to 0.65
C_v = Velocity coefficient, typically = 0.97 to 0.99
v = Average fluid velocity, m/s

The discharge coefficient depends mainly on the contraction coefficient and the orifice geometry. For a round orifice, the contraction coefficient can be calculated using the following expression given by Merritt (1967):

$$C_c \left\{ 1 + \frac{2}{\pi}\left(\frac{D}{C_c d} - \frac{C_c d}{D}\right) \tan^{-1}\left(\frac{C_c d}{D}\right) \right\} = 1 \tag{2.38}$$

where D = Pipe diameter, m
d = Orifice diameter, m

The variation of the contraction coefficient with the diameter ratio (d/D) is shown in Fig. 2.12. For a sharp-edged orifice, the friction losses are negligible: $C_v = 1$. Therefore,

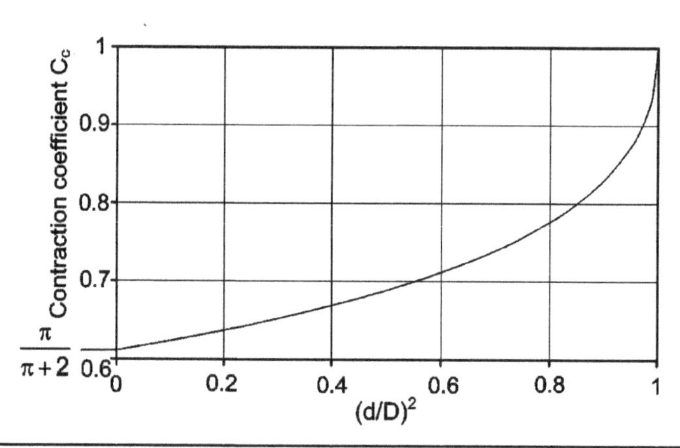

Figure 2.12 Effect of the diameter's ratio on the contraction coefficient of a sharp-edged orifice, calculated.

if the orifice diameter is much less than the pipe diameter ($d \ll D$), then $\tan^{-1}(C_c d/D) = C_c d/D$ and Eqs. (2.37) and (2.38) yield:

$$C_d = C_c = \pi/(\pi + 2) = 0.611 \tag{2.39}$$

Sharp-edged orifices are preferred because of their predictable characteristics. The pressure and power losses in these orifices are dominated by the fluid inertia. The fluid viscosity has no significant effect on their operation ($C_v \approx 1$). Therefore, the flow rate through sharp-edged orifices is viscosity-independent. On the other hand, for economy reasons, short tube orifices are widely used, especially as fixed restrictors in the flow passages. The fluid flow through a short tube orifice is subjected to the friction losses as well as the minor losses at inlet and outlet. Therefore, these orifices are viscosity-dependent. The discharge coefficient of a short tube orifice depends on the Reynolds number and the orifice geometry. The following expressions are used to calculate the discharge coefficient for the turbulent and laminar regions [Merritt (1967)]:

$$C_d = \left[1.5 + 13.74\left(\frac{L}{\text{Re}\,d}\right)^{\frac{1}{2}}\right]^{-\frac{1}{2}} \qquad \text{For } \frac{\text{Re}\,d}{L} > 50 \tag{2.40}$$

$$C_d = \left[2.28 + \frac{64L}{\text{Re}\,d}\right]^{-\frac{1}{2}} \qquad \text{For } \frac{\text{Re}\,d}{L} < 50 \tag{2.41}$$

The variation of the discharge coefficient with ($\text{Re}\,d/L$), calculated using these expressions, is plotted in Fig. 2.13.

Local Losses The local losses, also called minor losses, result from a rapid variation in the magnitude or direction of the velocity vector. The throttling elements, elbows, and T connections are typical local loss elements. Equation (2.42) gives the pressure losses in a local loss element. The local losses are directly proportional to the fluid

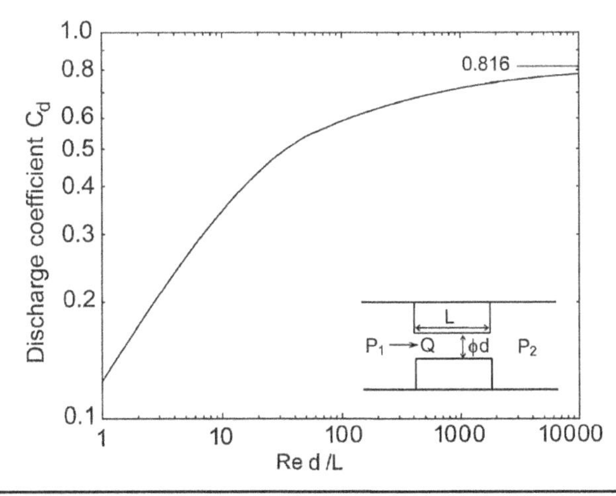

Figure 2.13 Variation of the discharge coefficient of a short tube orifice with Reynolds number and orifice dimensions, calculated.

density, and the local loss coefficient ζ is determined mainly by the geometry of the local loss feature.

$$\Delta P = \zeta \frac{\rho v^2}{2} \tag{2.42}$$

Hydraulic Inertia The dynamic behavior of hydraulic transmission lines is affected by the fluid inertia, compressibility, and resistance. The effect of fluid inertia can be formulated, assuming incompressible nonviscous flow, as follows:

Figure 2.14 shows a single oil lump, subjected to a pressure difference: $\Delta P = P_1 - P_2$. An expression for the line inertia is deduced, based on Newton's second law, as follows:

$$\mathrm{F = ma} \tag{2.43}$$

$$\Delta P A = \rho A L \frac{dv}{dt} \tag{2.44}$$

$$v = \frac{Q}{A} \tag{2.45}$$

Then,

$$\Delta P = \frac{\rho L}{A}\frac{dQ}{dt} = I\frac{dQ}{dt} \tag{2.46}$$

The hydraulic inertia, I, is given by

$$I = \frac{\rho L}{A} = \frac{4\rho L}{\pi D^2} \tag{2.47}$$

where A = Line cross-sectional area, m^2
D = Line inner diameter, m
I = Line inertia, kg/m^4
L = Line length, m

The hydraulic inertia affects the transient response of the hydraulic transmission lines, but it has no significant effect on its steady state behavior.

2.2.3 Oil Compressibility

Definition

Liquids are of very low compressibility, while gases are highly compressible. Therefore, liquids are usually assumed incompressible. But this assumption is applied when the liquid compressibility has no significant effect on the performance of the studied system.

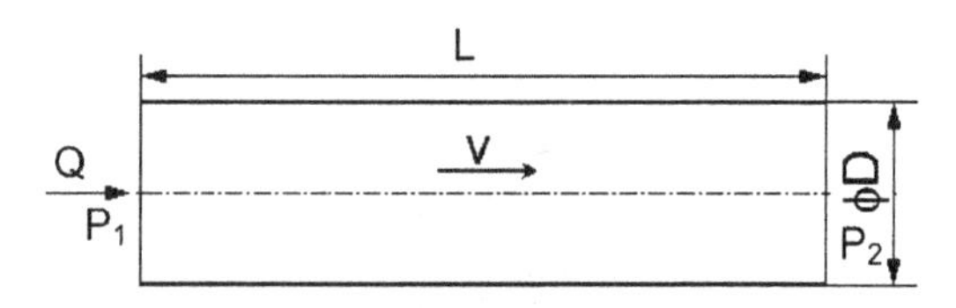

FIGURE 2.14 Single oil lump, subjected to pressure difference $\Delta P = P_1 - P_2$.

The liquid compressibility is defined as the ability of liquid to change its volume when its pressure varies. For pure liquid, the relation between the liquid volume and pressure variations is described by the following formula:

$$B = -\frac{\Delta P}{\Delta V/V} = -\frac{dP}{dV/V} \quad \text{or} \quad \frac{B}{V} = -\frac{dP}{dV} = -\frac{dP}{dt}\Big/\frac{dV}{dt} \tag{2.48}$$

where ΔP = Pressure variation, Pa
ΔV = Change in volume due to pressure variation, m^3
V = Initial liquid volume, m^3
B = Bulk modulus of liquid, typically $B = 1$ to 2 GPa for mineral oils

The hydraulic oil compressibility has a direct impact on the transient behavior of the hydraulic system. Generally, the reduction of oil volume by 1% requires an increase of its pressure by 10 to 20 MPa.

The bulk modulus of hydraulic oil is affected by the oil pressure and temperature. Figures 2.15 and 2.16 illustrate the effect of the oil pressure and temperature on the bulk modulus of typical mineral-based industrial hydraulic oil. The bulk modulus increases with the pressure increase and decreases with the temperature increase. The increase of oil pressure from zero to 300 bar increases the bulk modulus by about 21%, while the temperature increase by 40°C decreases the bulk modulus by 25%. However, the bulk modulus is usually assumed constant when working at the ordinary range of pressures and temperatures.

The bulk modulus of pure oil is nearly constant when operating at a certain temperature and pressure. However, when the oil includes bubbles of gases, air, or vapors, the bulk modulus of this mixture decreases due to the high compressibility of gases. If the total volume of mixture is V_T, the gas volume is αV_T, and the oil volume is $(1 - \alpha)V_T$, an equivalent bulk modulus B_e of the mixture is deduced as follows:

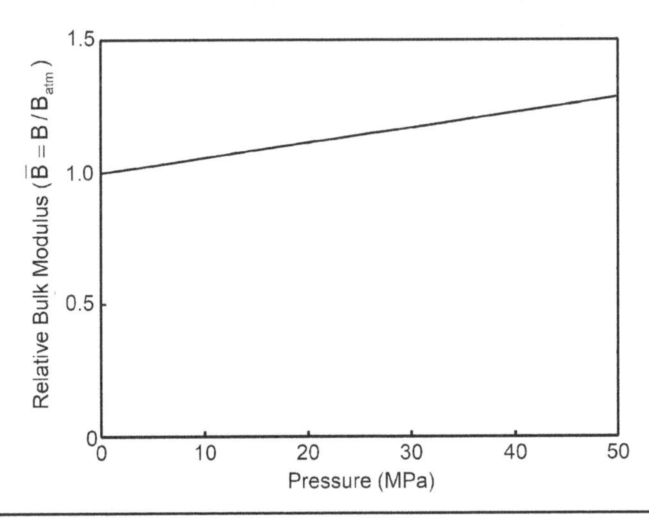

FIGURE 2.15 Effect of the pressure on the bulk modulus of a typical mineral-based hydraulic oil at constant temperature.

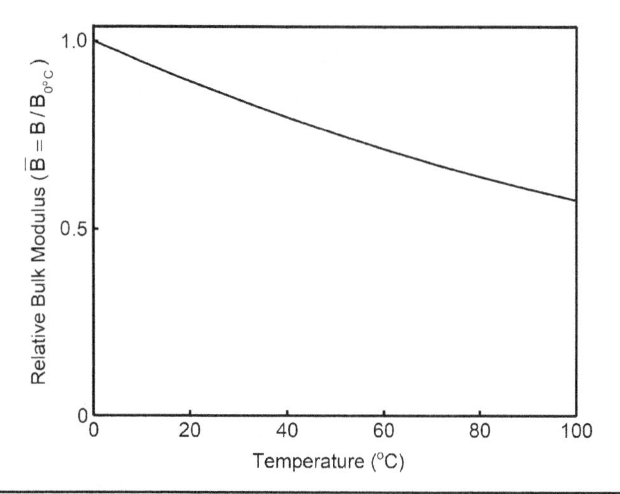

FIGURE 2.16 Effect of the temperature on the bulk modulus of a typical mineral-based hydraulic oil at constant pressure.

The compression of gases is governed by

$$PV^n = \text{const} \tag{2.49}$$

Then,

$$V^n dP + nV^{n-1} P dV = 0 \tag{2.50}$$

or

$$B_g = -\frac{dP}{dV/V} = nP \tag{2.51}$$

Assuming that the oil-gas mixture is subjected to a pressure variation ΔP, then the variations of volumes are

$$\Delta V_g = -\frac{\alpha V_T}{nP} \Delta P \tag{2.52}$$

$$\Delta V_o = -\frac{(1-\alpha)V_T}{B} \Delta P \tag{2.53}$$

$$B_e = -\frac{\Delta P}{(\Delta V_o + \Delta V_g)/V_T} \tag{2.54}$$

Thus,

$$B_e = \frac{nBP}{nP(1-\alpha) + B\alpha} \tag{2.55}$$

and

$$\bar{B} = \frac{B_e}{B} = \frac{nP}{nP(1-\alpha) + B\alpha} \tag{2.56}$$

or

$$\frac{1}{B_e} = \frac{\alpha}{nP} + \frac{1-\alpha}{B} \tag{2.57}$$

where α = Ratio of gases volume to the total volume, at atmospheric pressure
n = Polytropic exponent = 1 to 1.4
P = System absolute pressure, Pa
ΔV_g = Change in gas volume due to compressibility, m^3
ΔV_o = Change in oil volume due to compressibility, m^3
B_e = Equivalent bulk modulus of mixture, Pa

The nondimensional bulk modulus, $\bar{B} = B_e/B$, was calculated using Eq. (2.56). The calculation results are plotted in Fig. 2.17. This figure shows that the entrained air reduces considerably the bulk modulus of the mixture. Moreover, the entrained air results in noise, shuddering movements, and a large rise in temperature, in addition to the diesel effect. The diesel effect is the spontaneous combustion of an air-gas mixture. If the mineral oil, containing air bubbles, is rapidly compressed, the air bubbles become so hot that a spontaneous combustion may occur. Hence, at specific points, a large increase in temperature and pressure occurs, which may damage the seals on the hydraulic components. In addition, the age of the fluid will be reduced.

Effect of Oil Compressibility; Hydraulic Capacitance

Oil compressibility makes an important contribution to the dynamic behavior of the hydraulic control systems. The transient pressure variations, and consequently the transient variation of flow rates, forces, and accelerations, are highly affected by the oil compressibility. Herein, the effect of oil compressibility on the dynamics of hydraulic transmission lines is discussed. The effect of compressibility can be formulated by considering a single oil lump in a hydraulic transmission line. The system is subjected only to the effect of oil compressibility. The pressure in the line is P, the inlet flow rate is Q_1 and the outlet flow rate is Q_2 (see Fig. 2.18).

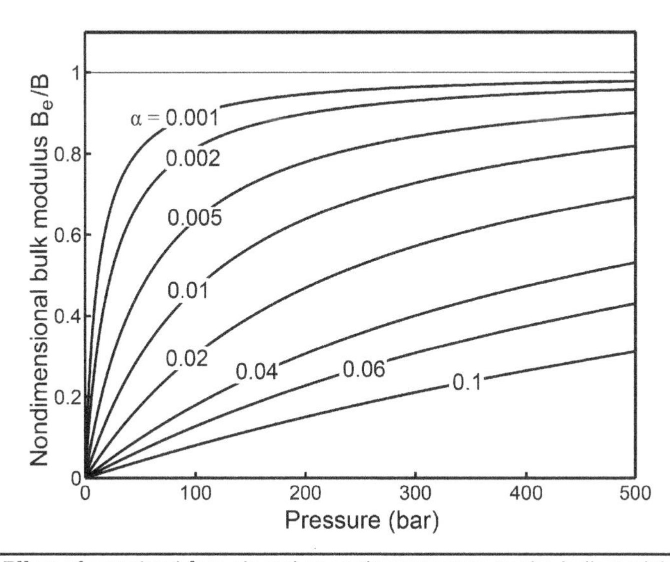

Figure 2.17 Effect of entrained free air and operating pressure on the bulk modulus of a typical mineral-based hydraulic oil, calculated.

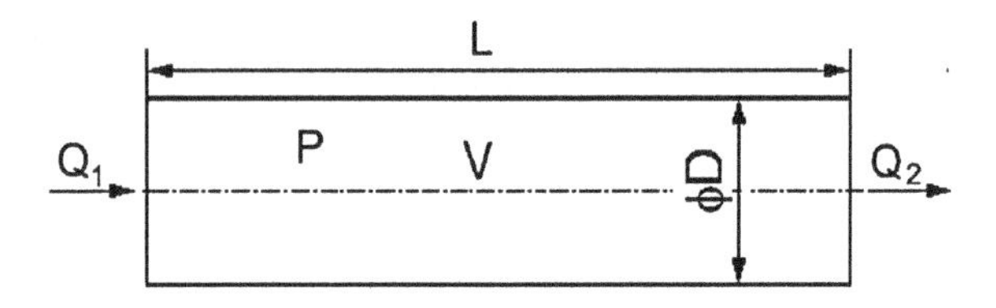

FIGURE 2.18 Single oil lump, subjected to oil compressibility.

Variation of volume of oil in line due to inlet and outlet flow rates, ΔV_Q, is

$$\Delta V_Q = \int \left(\sum Q_{\text{in}} - \sum Q_{\text{out}}\right) dt \tag{2.58}$$

Variation of volume of oil in the line due to the compressibility effect, ΔV_c, is

$$\Delta V_C = -\frac{V}{B}\Delta p \tag{2.59}$$

Then, referring to Eq. (2.48), $$\frac{dV_C}{dt} = -\frac{V}{B}\frac{dP}{dt} \tag{2.60}$$

$$\Delta V_Q + \Delta V_C = \Delta V_L \tag{2.61}$$

The variation of volume of line ΔV_L depends on the line material, wall thickness, diameter and system pressure: $\Delta V_L = f(P)$. Assuming rigid wall boundaries, then $\Delta V_L = 0$.

or $$\frac{dV_Q}{dt} + \frac{dV_C}{dt} = 0 \tag{2.62}$$

Then, $$\sum Q_{\text{in}} - \sum Q_{\text{out}} - \frac{V}{B}\frac{dP}{dt} = 0 \tag{2.63}$$

The application of Eq. (2.63) to the single lump flow in a pipe (see Fig. 2.18) yields

$$Q_1 - Q_2 - \frac{V}{B}\frac{dP}{dt} = 0 \tag{2.64}$$

or $$\Delta P = \int \frac{B}{V}(Q_1 - Q_2)\,dt \quad \text{or} \quad P = P_i + \int \frac{B}{V}(Q_1 - Q_2)\,dt \tag{2.65}$$

This form of continuity equation is widely applied in the fluid power system's analysis. The difference $(Q_1 - Q_2 = Q_C)$ is due to oil compressibility, where

$$Q_C = \frac{V}{B}\frac{dP}{dt} = C\frac{dP}{dt} \tag{2.66}$$

where $$C = \frac{\pi D^2 L}{4B} \tag{2.67}$$

The term C is called the hydraulic capacitance of the line. This capacitance is analogous to the electric capacitance since it has an energy storing effect and is described mathematically by the same expression: $i = C\,de/dt$.

Sometimes, the pipe wall deformation is not negligible. In such cases, it should be taken into consideration when calculating the hydraulic capacitance. The variation of volume of pipe line, ΔV_L, depends on the pipe length, the line material, wall thickness, diameter, and system pressure. The walls deform due to the combined effect of the radial and axial pressure forces. An expression for this volumetric variation due to a pressure increment ΔP is derived as follows.

Figure 2.19 illustrates the effect of pressure forces in the radial direction. The volume variation due to the radial wall deformation, ΔV_{LR}, can be calculated as follows:

$$\sigma = \frac{\Delta PDL}{2hL} = \frac{\Delta PD}{2h} = E\varepsilon_r \tag{2.68}$$

$$\varepsilon_r = \frac{\pi(D+\Delta D) - \pi D}{\pi D} = \frac{\Delta D}{D} \tag{2.69}$$

$$\Delta D = \frac{\Delta PD^2}{2Eh} \tag{2.70}$$

$$\Delta V_{LR} = \frac{\pi}{4} L\{(D+\Delta D)^2 - D^2\} = \frac{\pi L}{4} \Delta D(2D + \Delta D) \tag{2.71}$$

Considering the actual dimensions and parameters of transmission lines used in hydraulic power systems, the term ΔD is negligible compared to $2D$, then, Eq. (2.71) becomes

$$\Delta V_{LR} = \frac{\pi L}{4} \Delta D(2D) = \frac{\pi}{4} D^2 L\left(\frac{D}{Eh}\right)\Delta P \tag{2.72}$$

Figure 2.20 illustrates the effect of pressure forces in the axial direction. The volumetric variation due to axial wall deformation, V_{LA}, can be calculated as follows:

$$\sigma = \frac{\pi D^2 \Delta P}{4\pi Dh} = \frac{D\Delta P}{4h} = E\varepsilon_a \tag{2.73}$$

$$\varepsilon_a = \frac{\Delta L}{L} \tag{2.74}$$

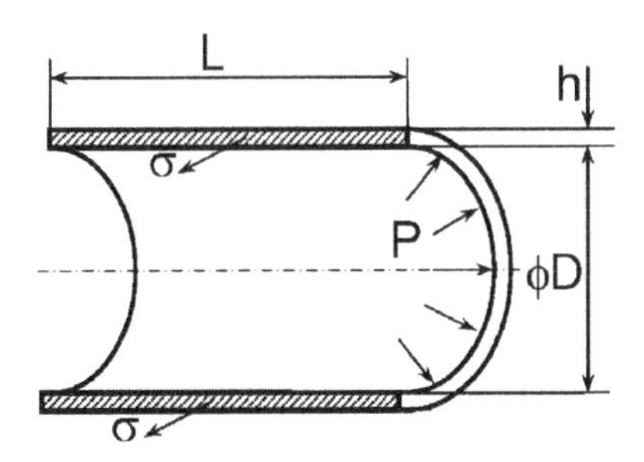

Figure 2.19 Radial pipe wall deformation.

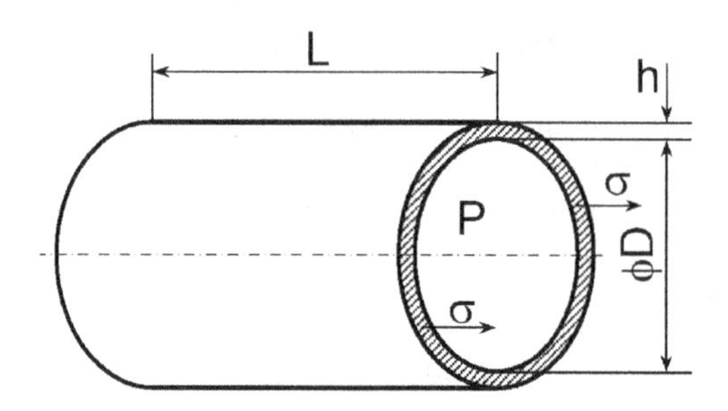

Figure 2.20 Axial pipe wall deformation.

$$\Delta L = \frac{\Delta PDL}{4Eh} \tag{2.75}$$

$$\Delta V_{LA} = \frac{\pi}{4} D^2 \Delta L = \frac{\pi}{4} D^2 L \left(\frac{D}{4Eh} \right) \Delta P \tag{2.76}$$

The total variation of line volume is

$$\Delta V_L = \Delta V_{LR} + \Delta V_{LA} = \frac{\pi}{4} D^2 L \left(\frac{5D}{4Eh} \right) \Delta P = V \Delta P \left(\frac{5D}{4Eh} \right) \tag{2.77}$$

By substituting Eqs. (2.58), (2.59), and (2.77) in Eq. (2.61), the following relation is obtained:

$$\int \left(\sum Q_{in} - \sum Q_{out} \right) dt - \frac{V}{B} \Delta P = V \Delta P \left(\frac{5D}{4Eh} \right) \tag{2.78}$$

$$\int \left(\sum Q_{in} - \sum Q_{out} \right) dt = V \left(\frac{1}{B} + \frac{5D}{4Eh} \right) \Delta P = C \Delta P \tag{2.79}$$

where

$$C = V \left(\frac{1}{B} + \frac{5D}{4Eh} \right) \tag{2.80}$$

Example 2.1 Apply the continuity equation to both chambers of the shown hydraulic cylinder.

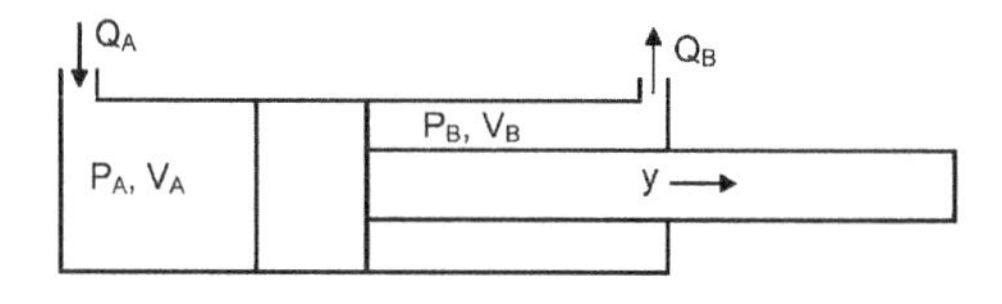

$$Q_A - A_P \frac{dy}{dt} = \frac{V_A}{B} \frac{dP_A}{dt}, \text{ where } V_A = V_{Ao} + A_p y$$

$$A_r \frac{dy}{dt} - Q_B = \frac{V_B}{B} \frac{dP_B}{dt} \text{ where } V_B = V_{Bo} - A_r y$$

where A_p = Piston area, m^2
A_r = Piston rod side area, m^2
V_A = Volume of oil in chamber A, m^3
V_{Ao} = Initial volume of piston side chamber, m^3
V_B = Volume of oil in chamber B, m^3
V_{Bo} = Initial volume of rod side chamber, m^3
y = Piston displacement, m

2.2.4 Thermal Expansion

The hydraulic liquids are subjected to volumetric thermal expansion. Generally, the volume of liquids changes with temperature as follows:

$$\Delta V_T = \alpha V \Delta T \tag{2.81}$$

where V = Initial volume of oil, m^3
α = Thermal expansion coefficient, for typical mineral oil $\alpha = 0.0007$ K^{-1}
ΔT = Temperature variation, °C
ΔV_T = Oil volume variation due to thermal expansion, m^3

In the case of oil trapped in a rigid vessel for a long period, the pressure may increase to enormous values due to the rise in oil temperature. Considering a volume of oil trapped in a hydraulic cylinder, the thermal expansion of the cylinder is negligible with respect to that of the oil. Therefore, the oil volume remains constant, even when its temperature increases, due to the combined effect of the oil compressibility and its thermal expansion. The change in pressure (ΔP) due to temperature variation (ΔT) is deduced as follows. Assuming that the total volume of oil is constant, then $\Delta V = 0$.

$$\Delta V = \Delta V_T + \Delta V_C = 0 \tag{2.82}$$

$$\Delta V_C = -\frac{V}{B} \Delta p \tag{2.83}$$

$$\alpha V \Delta T - \frac{V}{B} \Delta P = 0 \tag{2.84}$$

or

$$\Delta P = \alpha B \Delta T \tag{2.85}$$

where ΔV_C = Variation of volume due to oil compressibility, m^3

Example 2.2 A liquid of ($\alpha = 0.0007$) and ($B = 1$ GPa), trapped in a rigid vessel, is subjected to an increase in temperature of 50°C. Assuming that the vessel is of negligible thermal expansion, the oil pressure increases by 35 MPa.

2.2.5 Vapor Pressure

All liquids tend to evaporate by projecting molecules into the space above their surface. If this is a confined space, the partial pressure exerted by the molecules increases until the rate at which molecules reenter the liquid is equal to the rate at which they leave. In this equilibrium condition, the vapor pressure is the saturated vapor pressure (SVP). The molecular activity increases with temperature. Also, the reduction of pressure to values less than SVP induces a rapid rate of evaporation: boiling. Higher-saturated vapor pressure leads to a higher rate of evaporation. The *vapor pressure* is defined as the pressure at which a liquid will boil. It increases with the increase of oil temperature.

The reduction of the liquid pressure to the SVP leads to cavitation in a hydraulic system. The term *cavitation* refers to the formation and collapse of vapor-filled cavities in the liquid. The collapse of vapor cavities leads to very great local oil velocities. In the case of displacement pumps, the cavitation occurs when the suction pressure decreases below the vapor pressure of oil. The vapor cavities that form in the suction line of the pump collapse as they travel through the pump to its high-pressure zones. The collapse, under the action of high pressure, results in very high local velocities and great impact forces and pressures (up to 7000 bar), which cause the erosion of the pump elements. (See Sec. 4.4.)

2.2.6 Lubrication and Anti-Wear Characteristics

The fluid must be capable of covering the contact surfaces of all moving parts with a thin and continuous lubricating film. The lubricating film may be destroyed, as a result of high loading forces, insufficient oil delivery, and low viscosity. This would result in wear due to fretting. The lubricating power and film strength of a liquid are directly related to its chemical nature and can be improved by the addition of certain chemical agents. In addition to the wear due to fretting, there is also wear due to fatigue, abrasion, and corrosion.

2.2.7 Compatibility

The fluid must be fully compatible with other materials used in the hydraulic system, such as those used for bearings, seals, paints, and so on. It should not react chemically with any of these materials, nor change their physical properties. Moreover, the fluid leaks out from the hydraulic system and encounters other system parts, such as electrical lines, mechanical components, and others, so the fluid must also be compatible with the materials of these parts.

2.2.8 Chemical Stability

Chemical stability is an important property of the hydraulic liquid. It is defined as the ability of the liquid to resist oxidation and deterioration for long periods. All liquids tend to undergo unfavorable changes under severe operating conditions. Some metals, such as zinc, lead, brass, and copper, have undesirable chemical reactions with certain liquids. These reactions result in the formation of sludge, gums, carbon, or other deposits, which clog the openings and cause valves and pistons to stick. As these deposits are formed, certain changes in the physical and chemical properties of the liquid take place. The liquid usually becomes darker, the viscosity increases, and damaging acids are formed. The stability of liquids can be improved by the addition of oxidation inhibitors, which must be compatible with the other required properties of the liquid.

2.2.9 Oxidation Stability

The oxidation stability is the ability of the fluid to resist chemical degradation by reaction with atmospheric oxygen. It is an extremely important criterion concerning the quality of hydraulic fluids, particularly in high-temperature applications. It determines the resistance to aging. The degradation of hydraulic fluids by oxidation can result in significant viscosity increases, development of corrosive organic acids, and lacquering of critical surfaces by resinous oxidation products.

2.2.10 Foaming

All fluids contain dissolved air. The amount of dissolved air depends upon the temperature and pressure. Typically, mineral-based oil can contain up to 10% by volume of dissolved air. As the temperature of the fluid increases or the pressure decreases, the dissolved air is liberated. The liberated air exists in the fluid as discrete bubbles. When the fluid containing the entrained air returns to the reservoir, the air rises to the surface and causes foam. The bad design of a tank and oil return line may also lead to the formation of air bubbles and foam. If the foam builds up, it can cause severe problems in the hydraulic system. Therefore, most hydraulic fluids contain foam-depressant additives that cause the rapid breakdown of the foam.

2.2.11 Cleanliness

Cleanliness in hydraulic systems has received considerable attention. Some hydraulic systems, such as aerospace hydraulic systems, are extremely sensitive to contamination. Fluid cleanliness is of primary importance because the contaminants can cause component malfunction, prevent proper valve seating, cause wear in components, and may increase the response time of servo valves. Air, water, solvent, and other foreign fluids are in the class of fluid contaminants.

Air Contamination

According to Henry's law, *the amount of a given gas dissolved in a given type and volume of liquid is directly proportional to the partial pressure of that gas in equilibrium with that liquid, at a constant temperature*. Hence, the solubility of air in the hydraulic fluid is linearly proportional to the absolute pressure above the liquid surface, and normally decreases with rising temperature.

The solubility is often evaluated by the Bunsen coefficient, defined as *the volume of a gas at 0°C and standard atmospheric pressure which dissolves in a unit volume of a solvent*. Figure 2.21 shows the solubility of air in a typical mineral-based hydraulic fluid. The solubility of air in mineral oil is relatively high compared to the other typical hydraulic media.

The entrained air is of minor significance as long as it remains dissolved in the hydraulic fluid. However, the undissolved air results in serious problems, mainly

- Reduction of the bulk modulus of the oil-air mixture
- Reduction of the density of the fluid by an amount corresponding to the volume fraction of entrained air
- Slight increase in the viscosity of a hydraulic fluid
- Excessive aeration in a hydraulic system results in unreliable operation, high noise levels, and the possible damage of pumps and other components

Free air bubbles passing through a hydraulic pump are subjected to sudden compression under adiabatic conditions, thus raising the temperature of the compressed air bubbles. Assume that the initial volume of the undissolved air is V_i at the standard atmospheric conditions, $T_i = 288.15$ K and $P_i = 100$ kPa (abs). The pressure is increased to P.

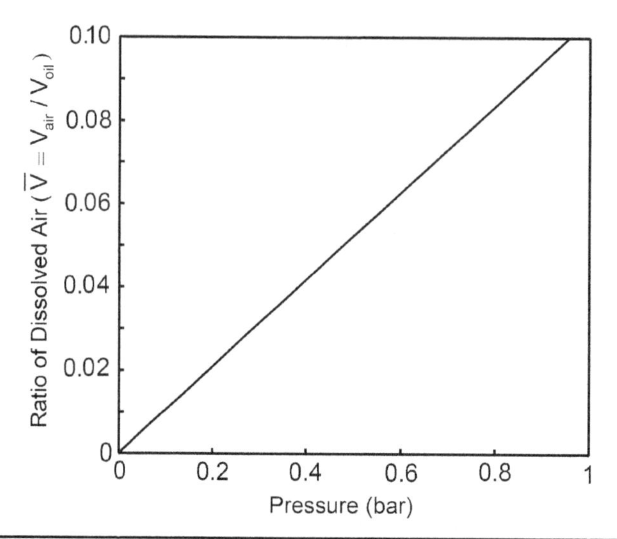

Figure 2.21 Effect of absolute pressure on the air solubility in a typical mineral-based hydraulic oil at constant temperature.

The new volume becomes V and its temperature increases by ΔT. The temperature increment is calculated as follows:

$$P_i V_i^k = P_2 V_2^k = PV^k = \text{constant} \tag{2.86}$$

where k is the adiabatic exponent ($k = 1.4$) and the pressure is absolute.

Then

$$\frac{P}{P_i} = \left(\frac{V_i}{V}\right)^k \quad \text{or} \quad \frac{V}{V_i} = \left(\frac{P_i}{P}\right)^{1/k} \tag{2.87}$$

$$P_i V_i = mRT_i \quad \text{and} \quad P_2 V_2 = mRT_2 \tag{2.88}$$

Then

$$\Delta T = T - T_i = T_i\left(\frac{P}{P_i}\frac{V}{V_i}\right) - T_i = T_i\left\{\left(\frac{P}{P_i}\right)^{\frac{k-1}{k}} - 1\right\} \tag{2.89}$$

Equations (2.87) and (2.89) were used to calculate the volume reduction and temperature increment due to the sudden increase of pressure. The calculation results are plotted in Figs. 2.22 and 2.23. These figures show that by increasing the pressure to 250 bar, the volume decreases to less than 2% of its initial value while its temperature increases to about 1100°C. Consequently, the oil film surrounding the hot bubble is subjected to severe thermal degradation and oxidation.

The undissolved air may exist in the hydraulic oil due to the following reasons:

- Liberation of dissolved air due to local pressure drop
- Air leakages in suction lines, pipe connections, glands, and others

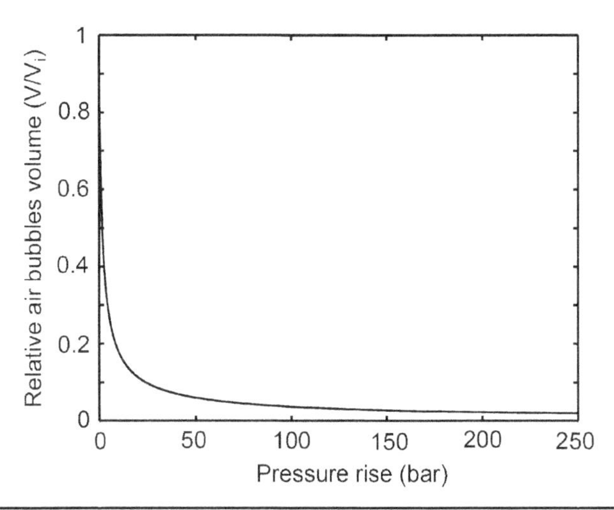

FIGURE 2.22 Variation of the relative volume of air bubbles with the pressure increase, calculated.

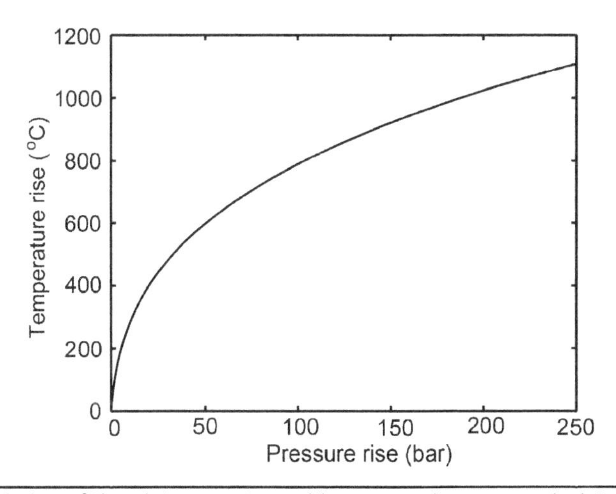

FIGURE 2.23 Variation of the air temperature with pressure increase, calculated.

- Returning fluid, which may contain free air, splashing freely down into the fluid reservoir
- Low fluid level in reservoir, insufficient residence time
- Bad design of tank
- Incorrect maintenance activity

Fluids should be able to absorb and transport a minimal amount of air and release it as quickly as possible. This requirement is fulfilled by using chemical additives.

Water Contamination

It is highly recommended to remove water from the hydraulic systems, except for the water-based fluids. However, this is not easy to attain. The following are the main sources of water contamination:

- Condensation from humid air drawn in via the breather. The ability of a typical mineral hydraulic oil to absorb moisture from humid air is proportional to the air humidity, as shown by Fig. 2.24.
- Leakage from oil-water heat exchangers.
- Filling of moisture-contaminated hydraulic fluid from badly stored containers.

Several harmful effects from moisture contamination may occur, including:

- The formation of rust and subsequent mechanical problems.
- The destructive effect of surface wear. Rust particles can promote the formation of oxidation sludge and deposits that may cause serious malfunctions in system parts.
- Rust particles circulating with the hydraulic fluid may cause abrasive wear.
- Formation or accumulation of rust against dynamic seals results in accelerated wear of the seals and increasing leakage of the hydraulic fluid.
- The free and dissolved water reduce the fatigue strength of ball bearings and stressed alloy steel components.

The following are possible ways to minimize the moisture content:

- Regular draining of undissolved water and sludge from a drain cock fitted at the lowest point of the fluid reservoir. Therefore, a good design of the reservoir and its floor slopes is important.

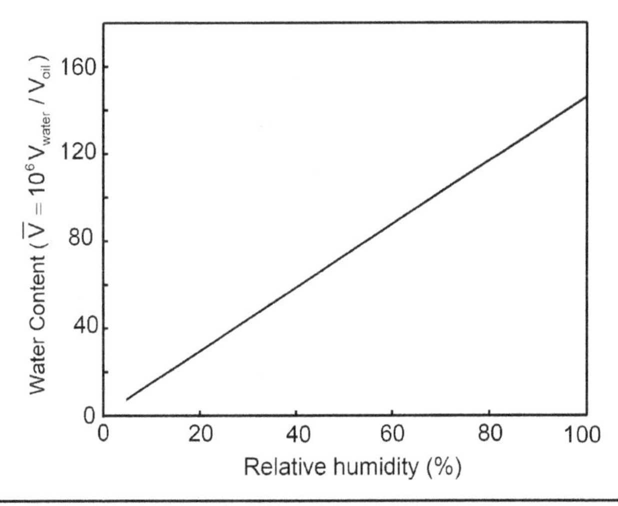

FIGURE 2.24 The effect of the relative humidity of the air in the tank on the concentration of the absorbed water in a typical mineral-based hydraulic oil at constant temperature.

- The hydraulic fluid should have good demulsibility which is the ability of a hydraulic fluid to separate from water. The best demulsibility is achieved by the most highly refined mineral-based oils and certain synthetic oils.
- Using filter elements capable of removing limited amounts of free water from hydraulic fluids. This is not, however, an economic solution for systems subject to continuous contamination by significant quantities of water.

Solvent Contamination

Solvent contamination is a special form of foreign fluid contamination in which the original contaminating substance is a chlorinated solvent. When the chlorinated solvents are allowed to combine with minute amounts of water, they change chemically into hydrochloric acids. These acids attack the internal metallic surfaces in the system and produce severe rust-like corrosion.

Foreign-Fluids Contamination

Hydraulic systems can be contaminated by foreign fluids other than water and chlorinated solvents. This type of contamination is generally a result of lubricating oil, engine fuel, or incorrect hydraulic fluid being introduced by mistake into the system during servicing. The effects of such contamination depend on the contaminant, its amount in the system, and how long it has been present.

2.2.12 Thermal Properties

The *pour point* of a fluid is the temperature 3°C above the temperature at which the fluid ceases to flow. As a general rule, the minimum temperature at which a fluid operates should be at least 10°C above the pour point.

The *flash point* is the temperature at which a liquid gives off vapor of sufficient quantity to ignite momentarily or flash when a flame is applied. A high flash point is desirable for hydraulic liquids because it provides good resistance to combustion and a low degree of evaporation at normal temperatures. The required flash point varies from 150°C for the lightest oils to 265°C for the heaviest oils.

The *fire point* is the temperature at which a substance gives off vapor of a sufficient quantity to ignite and continue to burn when exposed to a spark or a flame. Like the flash point, a high fire point is a basic requirement.

Specific heat capacity is the quantity of heat required to raise the temperature of one unit of mass of a substance by one degree. It is usually expressed in kJ/kg per K or kcal/kg per °C. The specific heat capacity at constant pressure (C_p) is temperature-related, but it is insignificantly affected by the pressure level. For mineral oils and synthetic hydrocarbons, $C_p = 2060$, J/kg per K at 50°C temperature.

Thermal conductivity is the ability to transmit heat, normally expressed in units of W/m per K. For mineral oils and synthetic hydrocarbons, the thermal conductivity is about 0.12 W/m per K.

2.2.13 Acidity

An ideal hydraulic liquid should be free from acids that cause corrosion of the metals in the system. Most liquids cannot be expected to remain completely noncorrosive under severe operating conditions. The degree of acidity of a liquid, when new, may be satisfactory, but after use, the liquid may tend to become corrosive as it begins to deteriorate.

2.2.14 Toxicity

Toxicity is defined as the quality, state, or degree of being toxic or poisonous. Some liquids contain chemicals that are a serious toxic hazard. These toxic or poisonous chemicals may enter the body through inhalation, by absorption through the skin, or through the eyes or the mouth. The result is sickness and, in some cases, death. Manufacturers of hydraulic liquids try to produce suitable liquids that contain no toxic chemicals and, as a result, most hydraulic liquids are free of harmful chemicals. Some fire-resistant liquids are toxic, and suitable protection and care in handling them must be provided.

2.2.15 Environmentally Acceptable Hydraulic Oils

Recent years have been marked by efforts to combat serious threats posed by industrial pollution of our environment. Hydraulic fluids are naturally included in these threats. They are utilized in large amounts by excavating machines, bulldozers, mobile cranes, and other outdoor equipment. An oil leak may result in considerable pollution of the surroundings and ground water. Mineral oils are composed of relatively stable hydrocarbon compounds, and are only very slowly broken down by microorganisms in the environment. Thus, pollution by conventional mineral hydraulic oils can disturb the ecological balance for long periods. This fact led to a growing interest in biodegradable products. Microbiological degradation is the process where by microorganisms, with the help of oxygen, break down organic material and extract nourishment from the decomposed products. The following are the basic specifications of commercially available and environmentally acceptable hydraulic oil. It is a synthetic base oil designed to biodegrade to its natural state when subjected to sunlight, water, and/or microbial activity.

ISO VG	32/46	Viscosity at 100°C,	7.3 cSt
Viscosity at 40°C	40.1 cSt	Viscosity Index (VI)	149
Specific Gravity	0.965	Flash Point	256°C
Fire Point	282°C	Pour Point	–56°C
Biodegradability; percentage biodegradable within 21 days			80%

2.3 Classification of Hydraulic Fluids

2.3.1 Typically Used Hydraulic Fluids

The following are typically used hydraulic fluids:

- Noninhibited refined mineral oil
- Refined mineral oil with improved anti-rust and anti-oxidation properties
- Refined mineral oil with improved anti-rust, anti-oxidation, and anti-wear properties
- Refined mineral oil with improved anti-rust, anti-oxidation, and viscosity-temperature properties
- Refined mineral oil with improved anti-rust, anti-oxidation, anti-wear, and viscosity-temperature properties
- Synthetic fluids with no specific fire-resistant properties
- High water-based fluid (up to 20% combustible materials + min 80% water)

- Chemical solution in water (more than 80% water content)
- Water-in-oil, water droplets in a continuous oil phase (60% oil + 40% water)
- Water/polymer—water glycol (35% water minimum, 80% maximum)
- Pure chemical fluids; water-free
- Phosphate esters
- Chlorinated hydrocarbons
- Mixture of phosphate esters and chlorinated hydrocarbons

2.3.2 Mineral Oils

Mineral-based oils are the most widely used hydraulic fluids. They are relatively inexpensive, widely available, and can be offered in suitable viscosity grades. They are of good lubricity, noncorrosive, and are compatible with most sealing materials with the exception of butyl rubber.

Mineral oils are chemically stable for reasonable operating temperatures. At higher temperatures, however, they suffer chemical breakdown. Premium grade mineral oils contain a package of additives to combat the effects of wear, oxidation, and foam formation, and to improve viscosity index and lubricity. There are, however, certain disadvantages of mineral oils that cannot be remedied by incorporating additives. The two most important are the flammability and the increase in viscosity at high pressures. Fire risk excludes the use of mineral oils in hazardous areas such as injection and plastic molding machines, coal mines, and near furnaces. The viscosity-pressure characteristics limit their use to pressures below 1000 bar (see Fig. 2.3).

2.3.3 Fire-Resistant Fluids

Oil-in-Water Emulsion

This hydraulic fluid consists of tiny droplets of oil dispersed in a continuous water phase. The dilution is normally between 2% and 5% oil in water, and the characteristics of the fluid are more similar to water than oil. It is extremely fire-resistant, is highly incompressible, and has good cooling properties. Its main disadvantages are poor lubricity and low viscosity.

Water-in-Oil Emulsion

The water-in-oil emulsions are the most popular fire-resistant fluids. They have a continuous oil phase in which tiny droplets of water are dispersed. Their lubrication properties are very much reduced. This is partially overcome by running pumps at reduced speeds. Therefore, larger displacement pumps are necessary to obtain the required flow rate. The usual dilution is 60% oil + 40% water. For optimum life, the operating temperatures should not exceed 25°C, but intermittent operation up to 50°C is permissible. At the higher temperature, water content is affected owing to evaporation, which decreases the emulsion's fire-resistance properties. When the system has been idle for long periods, there is a tendency for the oil and water to separate. However, during running, the pump will re-emulsify the fluid.

Water-Glycol Fluids

These fluids were developed primarily for use in aircraft because of their very low flammability characteristics. However, their application is limited since they cannot be used at high temperatures because of their water content. Their lubricating ability is inferior to that of mineral oils; they attack most paints; they are very stable with respect to shear because of the low molecular weight of their constituents; and their good anti-freeze properties make them particularly suitable for low-temperature applications.

Synthetic Oils

Synthetic oils, such as phosphate esters, have remarkably good fire-resistance properties. They are used in industries such as plastic molding and die-casting, where unusually great fire risks occur. Their lubricating ability is similar to that of mineral oil.

Elastomers used in conjunction with phosphate esters must be chosen carefully. Some silicone polymers and butyl rubber are suitable. Certain metals, particularly aluminum, and most paints are susceptible to attack.

Synthetic oils are superior when compared with mineral-oil–based fluids in one or more of the following respects:

- Thermal stability
- Oxidation stability
- Viscosity-temperature properties (VI)
- Low temperature fluidity
- Operational temperature limits
- Fire resistance

However, mineral oil–based fluids may be advantageous regarding:

- Hydrolytic stability
- Corrosion protection
- Toxicity
- Compatibility with elastomers and construction materials
- The solubility of additives
- Frictional characteristics
- Cost and availability

2.4 Additives

The largest class of hydraulic fluids consists of refined hydrocarbon base oils (petroleum oils) and suitable additives to improve the base properties. The main types of additives used in hydraulic fluids are the following.

- *Oxidation inhibitors* improve the ability of liquid to withstand chemical reaction with oxygen/air and avoid subsequent degradation. They are of prime importance when operating at elevated temperatures.

- *Corrosion inhibitors* form molecular layers that are bound to the surface by electrostatic forces and form an effective barrier against penetration by oxygen and water, which is necessary for rust-forming electrochemical reactions.
- *Antifoaming agents*, which form small heterogeneous areas within the bubble walls of surface foam. Due to their low surface tension, these small areas are weak spots and result in bubble bursting.
- *Anti-wear* additives of high thermal stability.
- *Viscosity index improvers*
- *Pour point depressants*
- *Friction modifiers* are necessary under certain conditions to ensure smooth operation, free from juddering (stick/slip).
- *Detergents* are substances providing a cleaning action with respect to the surface deposits.

2.5 Requirements Imposed on the Hydraulic Liquid

The following are the main requirements imposed on hydraulic liquids:

- Satisfactory flow properties throughout the entire range of operating temperatures.
- A high viscosity index that ensures moderate viscosity variation in relation to the temperature fluctuations.
- Good lubricating properties are a prerequisite to reduce the wear and increase the service life of the system.
- Low vapor pressure to avoid cavitation.
- Compatibility with system materials since the fluid should not react chemically with any of the used materials or deteriorate their physical properties.
- Chemical stability is necessary to increase the service life of liquid and avoid performance deterioration.
- Corrosion protection by adding effective corrosion inhibitors.
- Rapid de-aeration and air separation.
- Good thermal conductivity is required to rapidly dissipate the heat generated due to friction between elements and due to hydraulic losses.
- Fire resistance is essential in some applications.
- Electrically insulating properties can be significant in a number of modern designs.
- Environmental acceptability.

Table 2.1 gives the specifications of two typical mineral-based hydraulic liquids.

		VG32	VG68
Density (at 15°C), kg/m³		869	878
Pour point, °C		–45	–36
Flash point, °C		212	252
Minimum startup temperature, °C		–32	–21
Estimated operating range, °C		–14 to 67	0 to 87
Kinematic viscosity, cSt	at 40°C	32.2	67.9
	at 100°C	6.4	10
Viscosity index		156	132

Table 2.1 Specification of Typical All-Season Mineral-Based Hydraulic Oils

2.6 Exercises

1. Derive an expression for the pressure and power losses in a hydraulic transmission line of constant diameter in the case of laminar flow.

2. Derive an expression for the hydraulic resistance, R, of a hydraulic transmission line and find the resultant resistance of two lines connected in series or in parallel.

3. Derive an expression for the hydraulic inertia of a transmission line.

4. Derive an expression for the hydraulic inertia of two hydraulic transmission lines connected in series or in parallel.

5. Define the bulk modulus of oil and derive the equivalent bulk modulus of an oil-air mixture.

6. Derive the expressions for the resultant hydraulic capacitance of two hydraulic lines connected in series or in parallel.

7. Derive an expression for the pressure increment in a volume of liquid trapped in a rigid container when subjected to a temperature increase ΔT.

8. Discuss briefly the effect of the saturated vapor pressure on the function of hydraulic systems.

9. Calculate the percentage of variation in volume of liquid of bulk modulus $B = 1.4$ GPa if its pressure is increased by 10 MPa.

10. A hydraulic pipe line has a diameter D, length L, wall thickness h, material bulk modulus E, and oil bulk modulus B. Prove that the hydraulic capacitance of the line is given by:

(a) $C = \frac{\pi D^2 L}{4B}$, neglecting the pipe wall deformation

(b) $C = \frac{\pi D^2 L}{4}\left(\frac{1}{B} + \frac{D}{Eh}\right)$, considering the radial deformation of pipe material

(c) $C = \frac{\pi D^2 L}{4}\left(\frac{1}{B} + \frac{5D}{4Eh}\right)$, considering both of the axial and radial deformations

11. Calculate the difference between the input and output flow rates of a line if the rate of variation of pressure $dP/dt = 13.4$ MPa/s, given

$D = 10$ mm, $L = 3$ m, $B = 1.3$ GPa, wall thickness $h = 1$ mm.

Modulus of elasticity of wall material $E = 210$ GPa

12. (a) Derive an expression for, and calculate, the viscous friction coefficient for the given spool valve given: $L_1 = L_2 = 10$ mm, $D = 8$ mm, $c = 2$ μm, $\mu = 0.02$ Ns/m^2.

(b) If the spool performs a rotary motion, derive an expression for the viscous friction coefficient and calculate its value.

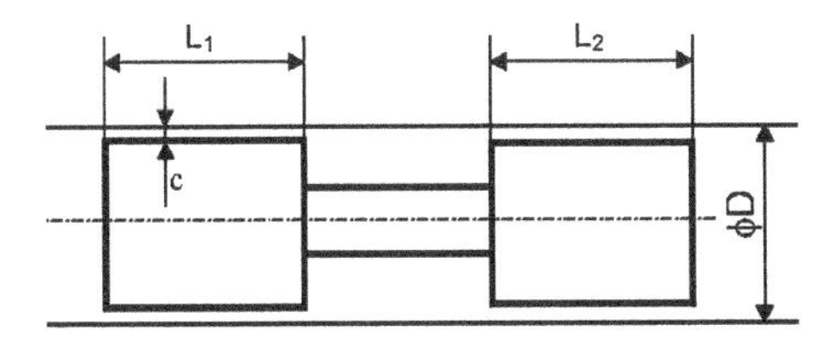

13. Calculate the radial clearance leakage, Q_L, and the resistance to leakage, R_L, in the given spool valve if

$D = 12$ mm, $c = 7$ μm, $L = 20$ mm

$\mu = 0.018$ Ns/m^2, $P_P = 21$ MPa, $P_T = 0$

Recalculate for different values of clearance, up to $c = 100$ μm, and plot the relation $Q_L(c)$.

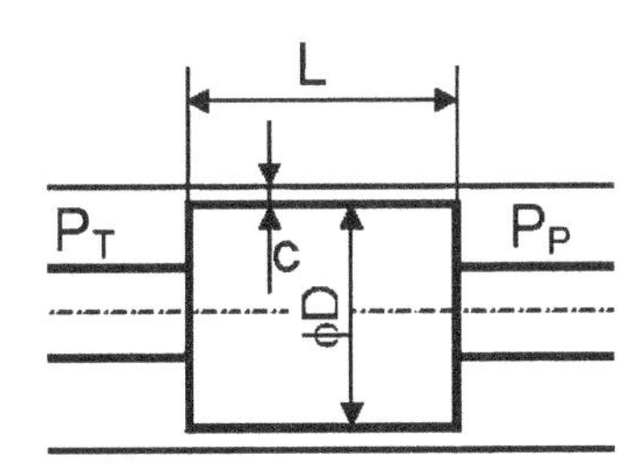

14. (a) Derive the transfer function relating the spool displacement to the applied pressure difference, $X(s)/\Delta P(s)$, where $\Delta P = P_1 - P_2$.

(b) Calculate and plot the transient response if ΔP is a step of magnitude 4 bars, given

Spool and springs reduced mass $m = 0.07$ kg
Springs equivalent stiffness $k = 20$ kN/m
Spool land length $L = 10$ mm
Oil dynamic viscosity $\mu = 0.02$ Ns/m^2
Spool diameter $D = 8$ mm
Radial clearance $c = 4$ μm
Oil density $\rho = 860$ kg/m^3

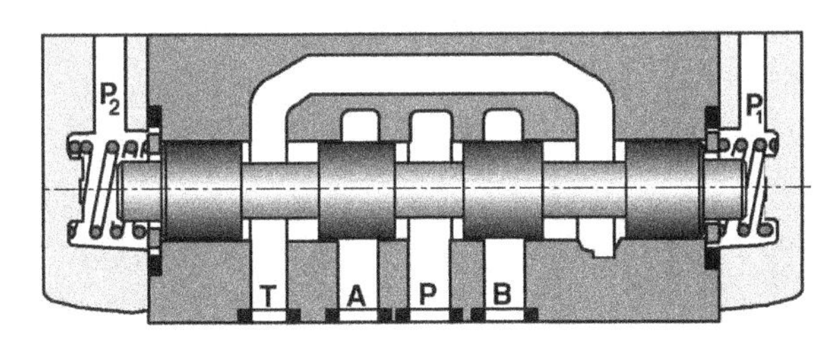

15. Calculate the pressure upstream of the throttle valve, A_t, assume laminar flow in the lines, and then check this assumption, given

Pump flow $Q = 14$ L/min

Dynamic viscosity $\mu = 0.0185$ Ns/m^2

Discharge coefficient $C_D = 0.611$

Oil density $\rho = 866$ kg/m^3

Throttling valve Area $A = 3$ mm^2

Dimensions of lines $D_1 = 14$ mm, $D_2 = 8$ mm, $D_3 = 10$ mm
$L_1 = 1.3$ m, $L_2 = 2$ m, $L_3 = 1.5$ m

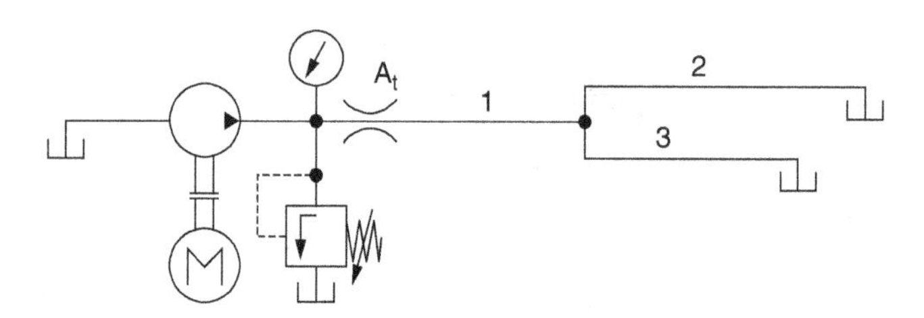

16. Calculate the pressure at the inlet of a pump of 60 L/min discharge if the hydraulic tank is open and the suction line has a 1 m length and a 15 mm diameter, given: oil kinematic viscosity $\nu = 0.2$ cm^2/s and oil density $\rho = 900$ kg/m^3.

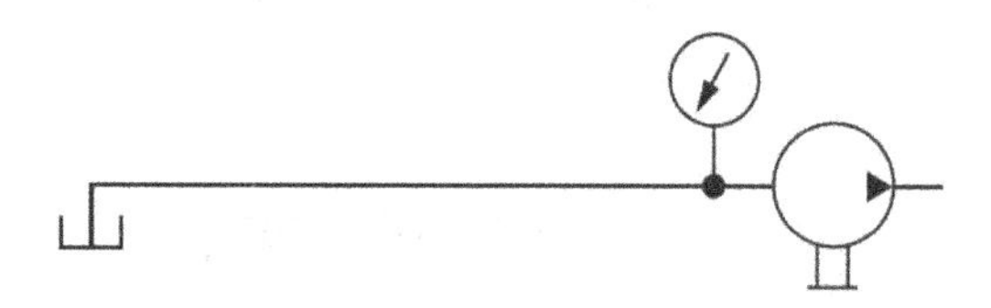

17. Calculate the pressure losses in the given pipe line if

Flow rate $Q = 10$ L/min, oil density $\rho = 850$ kg/m^3, $L_1 = L_2 = 4$ m, $D_1 = 13$ mm, $D_2 = 8$ mm, fluid kinematic viscosity $\nu = 20$ cSt.

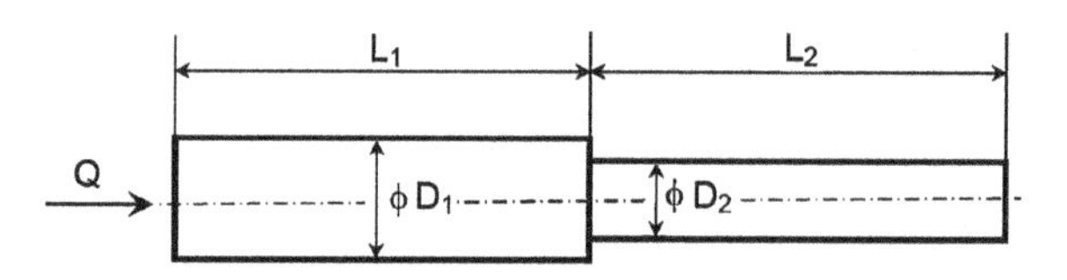

18. Calculate the shift of the piston in the given system if the force F is increased by 100 kN if

Piston area $A_p = .0176$ m^2

Pipe line diameter $d = 13$ mm

Fluid bulk modulus $B = 1.4$ GPa

Initial piston displacement $x_o = 0.5$ m

Pipe wall thickness $h = 1$ mm

Pipe length $L = 5$ m

Cylinder wall thickness $t = 6$ mm

Material bulk modulus $E = 210$ GPa

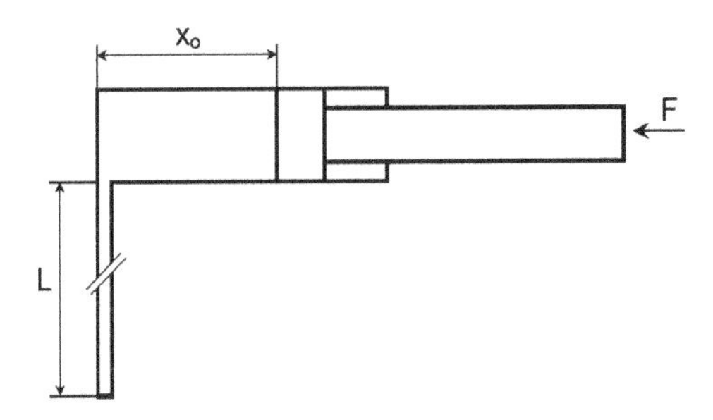

2.7 Nomenclature

A = Line cross-sectional area, m^2
a = Constant, m/s
A_p = Piston area, m^2
A_r = Piston rod-side area, m^2
B = Bulk modulus of oil, Pa
B_e = Equivalent bulk modulus of mixture, Pa
c = Radial clearance, m
C = Hydraulic capacitance, m^5/N
D = Orifice diameter, inner pipe diameter, spool diameter, m
F = Force, N
f_v = Friction coefficient, Ns/m
F_p = Pressure force acting on the fluid element, N
F_τ = Shear force acting on the fluid element, N
k = Spring stiffness, N/m
L = Pipe length, length of spool land, length of leakage path, m
m = Mass of the moving parts, kg
n = Polytropic exponent
P = Pressure, Pa
Q_L = Leakage flow rate, m^3/s
r = Radial distance from the midpoint of the clearance, m
Re = Reynolds number
R_L = Resistance to leakage, Ns/m^5
u = Fluid velocity, m/s
v = Mean fluid velocity m/s
V = Initial oil volume, m^3
V_A = Volume of oil in chamber A, m^3
V_B = Volume of oil in chamber B, m^3
V_T = Total volume of oil-gas mixture, m^3
x = Spool displacement, m
y = Displacement perpendicular to the velocity vector, m
y = Piston displacement, m
α = Ratio of gases volume to mixture volume, at atmospheric pressure
α = Coefficient of volumetric thermal expansion, K^{-1}
ΔP = Pressure difference across the radial clearance, Pa
ΔT = Temperature variation, °C

ΔV_C = Variation of oil volume due to oil compressibility, m^3
ΔV_T = Variation of oil volume due to thermal expansion, m^3
λ = Friction coefficient
μ = Coefficient of dynamic viscosity, Ns/m^2
ρ = Oil density, kg/m^3
τ = Shear stress, N/m^2
ν = Kinematic viscosity, m^2/s

Appendix 2A Transfer Functions

The transfer function of a linear system is defined as the Laplace transform of a system output divided by that of its input when the initial conditions are zero. Conventionally, the symbol $G(s)$ is used for the transfer function. For evaluating the transfer function of a linear system, the differential equation (DE), can be transformed into Laplace domain, just by replacing the term (d/dt) by (s). The resulting transfer function takes the form

$$G(s) = \frac{P(s)}{Q(s)} \text{ where } P(s) \text{ and } Q(s) \text{ are polynomials in } s. \tag{2A.1}$$

For the system to be physically realizable, the order of $P(s)$ should not exceed that of $Q(s)$.

Consider the linear system of input $x(t)$ and output $y(t)$, described by the following differential equation of zero initial conditions:

$$a_4\frac{d^4y}{dt^4} + a_3\frac{d^3y}{dt^3} + a_2\frac{d^2y}{dt^2} + a_1\frac{dy}{dt} + a_0 y = b_2\frac{d^2x}{dt^2} + b_1\frac{dx}{dt} + b_0 x \tag{2A.2}$$

The application of Laplace transform to Eq. (2A.2) yields:

$$\left(a_4s^4 + a_3s^3 + a_2s^2 + a_1s + a_0\right)Y(s) = \left(b_2s^2 + b_1s + b_0\right)X(s) \tag{2A.3}$$

then,

$$G(s) = \frac{Y(s)}{X(s)} = \frac{b_2s^2 + b_1s + b_0}{a_4s^4 + a_3s^3 + a_2s^2 + a_1s + a_0} \tag{2A.4}$$

The transfer function can be deduced as follows:

- Write down the equations governing the system behavior.
- If there are any nonlinear relations, a linearization procedure or simplifying assumptions should be considered to obtain linearized equations.
- By substitution, eliminate the variables that are not of direct interest, leaving the relation between the input and output variables.
- If there are any variables of nonzero initial conditions, $x(t = 0) = x_o$ and $x_o \neq 0$, substitute it by a new variable, $x_1 = x(t) - x_o$. The new variable x_1 is of zero initial value.
- Apply Laplace transform and find the required transfer function.

Appendix 2B Laminar Flow in Pipes

In the laminar flow, the paths of individual particles of fluid do not cross. So, the flow in a pipe may be considered as a series of concentric cylinders sliding over each other. Consider a cylinder of fluid of length dx and radius r, flowing steadily in the center of a pipe.

Consider the cylindrical fluid element illustrated by Fig. 2B.1, moving in a pipe as shown in Fig. 2B.2. In the steady state, the fluid speed is constant; it does not change with time. The forces acting on the fluid element are in equilibrium, thus the shearing forces on the cylinder are equal to the pressure forces.

$$2\pi r dx\,\tau = dpA = dp\pi r^2 \tag{2B.1}$$

or

$$\tau = \frac{r}{2}\frac{dp}{dx} \tag{2B.2}$$

where dp/dx = the pressure gradient. It is a function of x only [$\neq f(r)$].

This equation shows that the shear stress is linearly proportional to the distance r. For Newtonian fluids, the velocity distribution could be calculated as follows:

$$\tau = \mu\, du/dy \tag{2B.3}$$

$$y = R - r \quad \text{then } du/dy = -du/dr \quad \text{and} \quad \tau = -\mu\, du/dr \tag{2B.4}$$

$$\tau = \frac{r}{2}\frac{dp}{dx} = -\mu\frac{du}{dr} \tag{2B.5}$$

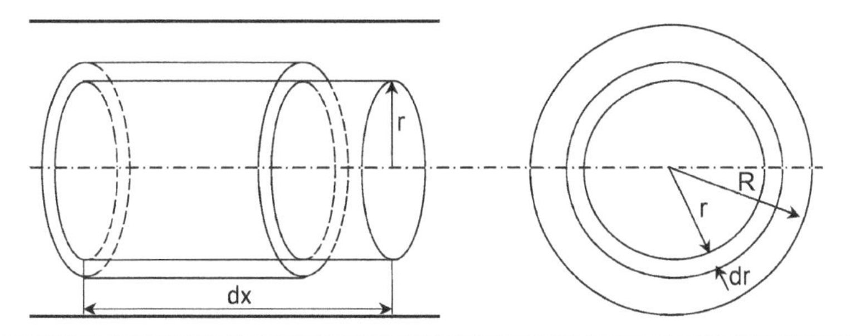

Figure 2B.1 Cylindrical fluid element flowing at the pipe center.

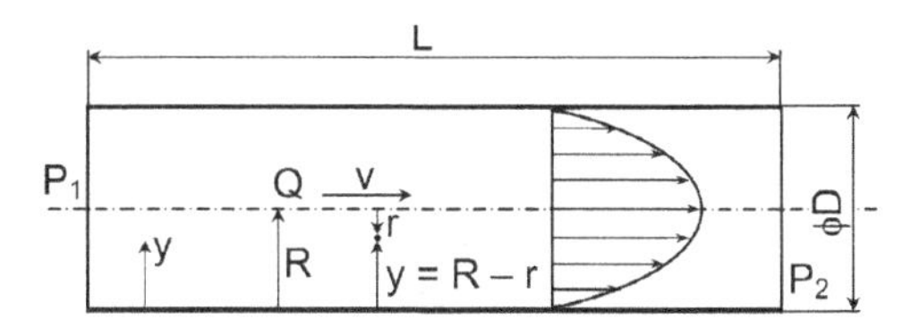

Figure 2B.2 Laminar flow in pipe line.

or
$$\frac{du}{dr} = -\frac{r}{2\mu}\frac{dp}{dx} \tag{2B.6}$$

By integrating Eq. (2B.6), the following expression for the value of the velocity at a point at distance r from the center is obtained;

$$u = -\frac{1}{2\mu}\frac{dp}{dx}\int r\,dr = -\frac{dp}{4\mu dx}r^2 + C \tag{2B.7}$$

At the pipe wall, $r = R$, $u = 0$. The following expression for the integration constant C is obtained by substitution in Eq. (2B.7):

$$C = \frac{R^2}{4\mu}\frac{dp}{dx} \tag{2B.8}$$

The expression for velocity at a point r from the pipe becomes:

$$u = \frac{R^2 - r^2}{4\mu}\frac{dp}{dx} \tag{2B.9}$$

The maximum velocity is at the pipe axis, $r = 0$;

$$u_{\max} = \frac{R^2}{4\mu}\frac{dp}{dx} = \frac{D^2}{16\mu}\frac{dp}{dx} \tag{2B.10}$$

Then, the velocity distribution is a parabolic profile of the form $y = ax^2 + b$ (see Fig. 2B.2). The flow rate in the pipe is calculated as follows:

$$dQ = u(2\pi r dr) = \frac{\pi(R^2 - r^2)}{2\mu}\frac{dp}{dx}r\,dr \tag{2B.11}$$

For a pipe of length L, the pressure gradient can be written as

$$\frac{dp}{dx} = \frac{\Delta p}{L} \tag{2B.12}$$

Then
$$Q = \frac{\pi\Delta p}{2\mu L}\int_0^R (R^2 - r^2)\,r\,dr = \frac{\pi\Delta p}{2\mu L}\left(\frac{R^4}{2} - \frac{R^4}{4}\right) \tag{2B.13}$$

$$Q = \frac{\pi R^4}{8\mu L}\Delta p = \frac{\pi D^4}{128\mu L}\Delta p \tag{2B.14}$$

or
$$\Delta P = \frac{128\mu L}{\pi D^4}Q = RQ \tag{2B.15}$$

The term R expresses the resistance of the hydraulic transmission line.

Equation (2B.15) is the Hagen-Poiseuille equation for laminar flow in a pipe. It shows that the pressure loss is directly proportional to the flow rate when the flow is laminar, (Re < 2300). It has been validated many times by experiments.

The mean velocity $$v = \frac{4}{\pi D^2} Q, \quad \text{or} \quad Q = \frac{\pi}{4} D^2\, v \tag{2B.16}$$

$$\Delta P = \frac{32\mu L}{D^2} v = \frac{64}{\rho v D/\mu} \frac{L}{D} \frac{\rho v^2}{2} \tag{2B.17}$$

Then, $$\Delta P = \lambda \frac{L}{D} \frac{\rho v^2}{2} \tag{2B.18}$$

where, $\lambda = \frac{64}{\text{Re}}$ = friction coefficient
Re = Reynolds number; $\text{Re} = \frac{vD}{\nu} = \frac{\rho v D}{\mu}$

The mean fluid velocity, v, in laminar flow is given by

$$v = \frac{Q}{\pi D^2/4} = \frac{\Delta p}{32\mu L} D^2 = \frac{1}{2} u_{max} \quad \text{or} \quad v = 0.5 u_{max} \tag{2B.19}$$

Thus, in the case of laminar flow in pipe

- The shear stress is linearly proportional to the radius r.
- The velocity distribution is parabolic.
- The velocity is maximum at the pipe axis and zero in the vicinity of the pipe wall.
- The average, mean, velocity is half of the maximum value.

where A = Area, m^2
D = Pipe inner diameter, m
dx = Length of the fluid element, m
P = Pressure, Pa
r = Radius of the cylindrical fluid element, m
R = Pipe radius, m
R = Hydraulic resistance, Pa s/m^5
u = Oil velocity, m/s
v = Mean oil velocity, m/s
τ = Shear stress, N/m^2
μ = Dynamic viscosity, Pa s

CHAPTER 3
Hydraulic Transmission Lines

3.1 Introduction

Hydraulic power systems use hydraulic liquids to transmit hydraulic power. The elements of the system are interconnected by hydraulic transmission lines, through which the hydraulic liquid flows. These lines are either rigid tubing or flexible hoses. The rigid tubing lines connect fixed nonvibrating elements, and the flexible hose lines connect moving parts. In addition, the coaxial joints are employed to transfer liquid between the elements rotating relative to each other.

Regardless of their simple design, the hydraulic conduits have an important effect on the steady state and transient behavior of the system.

Transmission lines affect system performance in the following ways:

- Hydraulic friction losses; hydraulic resistance of lines
- Hydraulic local, or secondary, pressure losses
- Oil compressibility and elasticity of pipe material; hydraulic capacitance of lines
- Oil inertia; the hydraulic inertia of lines

This chapter deals with the construction of tubes and hoses and their fittings, as well as the static and dynamic performance of transmission lines.

3.2 Hydraulic Tubing

The following are the basic factors involved in the selection of the proper tubing:

- Material of tubing
- Fittings
- Tubing size
- Tubing length
- Conveyed fluid compatibility
- Temperature
- Tubing pressure rating
- Installation design

Tube Size ϕ D (mm)	Wall Thickness (mm)	Working Pressure (bar)	Burst Pressure (bar)	Tube Size ϕ D (mm)	Wall Thickness (mm)	Working Pressure (bar)	Burst Pressure (bar)
4	1,0	300	1500	22	2,5	260	890
6	1,0	250	1300	25	2,0	150	600
6	1,5	480	1500	25	2,5	200	780
6	2,0	730	2600	25	3,0	240	940
8	1,0	180	980	25	4,0	340	1200
8	1,5	340	1500	28	2,0	130	570
8	2,0	510	2000	28	2,5	180	700
8	2,5	710	2400	28	3,0	210	840
10	1,0	150	780	30	2,0	120	520
10	1,5	270	1200	30	3,0	200	790
10	2,0	400	1500	30	4,0	280	1000
10	2,5	550	2000	35	2,0	100	450
12	1,5	220	980	35	3,0	170	670
12	2,0	320	1300	38	2,5	120	510
12	2,5	440	1600	38	3,0	170	620
14	1,5	190	840	38	4,0	220	820
14	2,0	270	1100	38	5,0	280	1000
15	1,5	170	790	38	6,0	350	1200
15	2,0	250	1000	42	2,0	84	370
16	1,5	170	730	42	3,0	140	560
16	2,0	230	980	42	4,0	190	740
16	2,5	310	1200	50	3,0	120	420
16	3,0	400	1500	60	3,0	100	340
18	1,5	140	650	75	3,0	76	310
18	2,0	210	870	90	3,5	75	300
20	2,0	180	780	115	4,0	67	270
20	3,0	310	1200	140	4,5	63	250
20	4,0	440	1600	165	5,0	60	230
22	1,5	120	530	220	6,0	54	210
22	2,0	170	710	273	6,0	44	180

TABLE 3.1 Typical Parameters of Common High-Tensile Hydraulic Tubes

High-pressure hydraulic tubes are mostly produced from carbon steel. Manufacturers make a wide range of tube dimensions and wall thicknesses to work with a wide range of flow rates and pressures. Table 3.1 gives typical parameters of commonly used tubes.

The proper tube diameter is determined according to the maximum flow rate and selected fluid speed.

$$d = \sqrt{\frac{4Q_{max}}{\pi v}} \tag{3.1}$$

where d = Tube inner diameter, m
Q_{max} = Maximum flow rate, m^3/s
v = Fluid mean velocity, m/s

The recommended mean fluid velocity in rigid pipes is usually within 2 to 6 m/s for the pressure line and 0.6 to 1.6 m/s for the suction, return, and low pressure lines. The inner diameter of a tube can also be calculated by using the nomogram given in Fig. 3.1, while the pipe wall thickness is determined according to the operating pressure, shown in Table 3.1.

The selection of the proper tube fittings and connectors is essential for the correct operation as well as the safe use of tubing and related equipment. Improper selection of the fittings can result in tube leakage, bursting, and other failures. Figures 3.2 and 3.3 give typical tube end fittings and connectors.

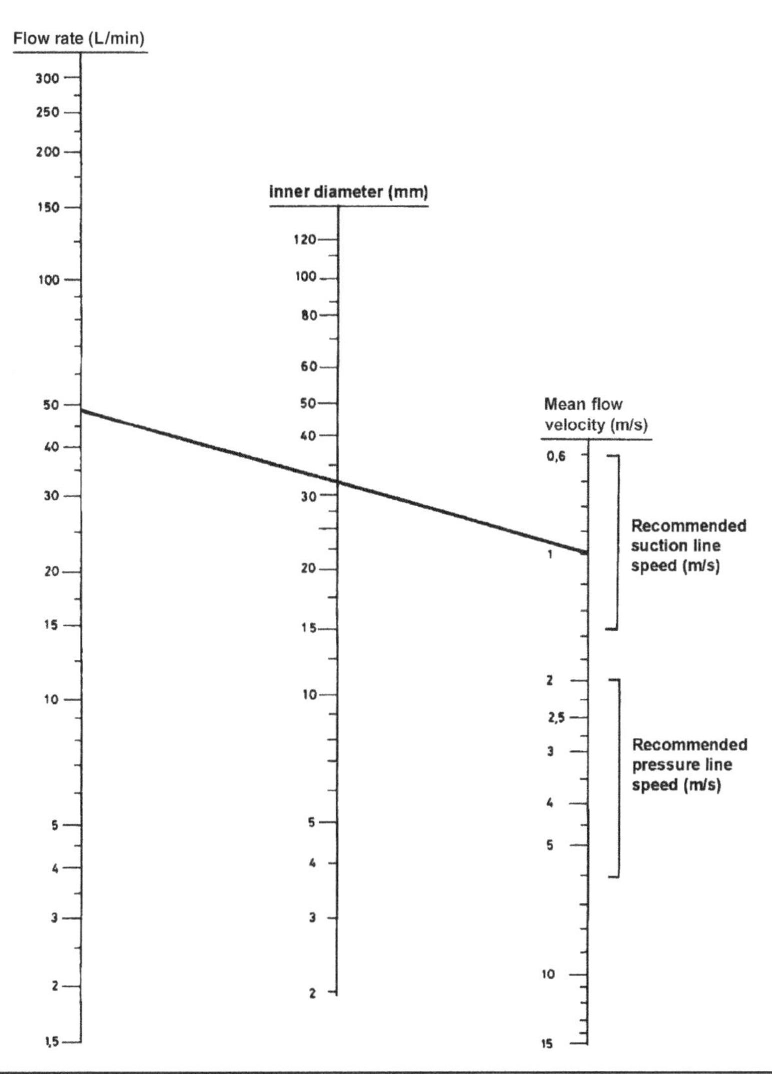

FIGURE 3.1 Nomogram for the calculation of a tube's inner diameter.

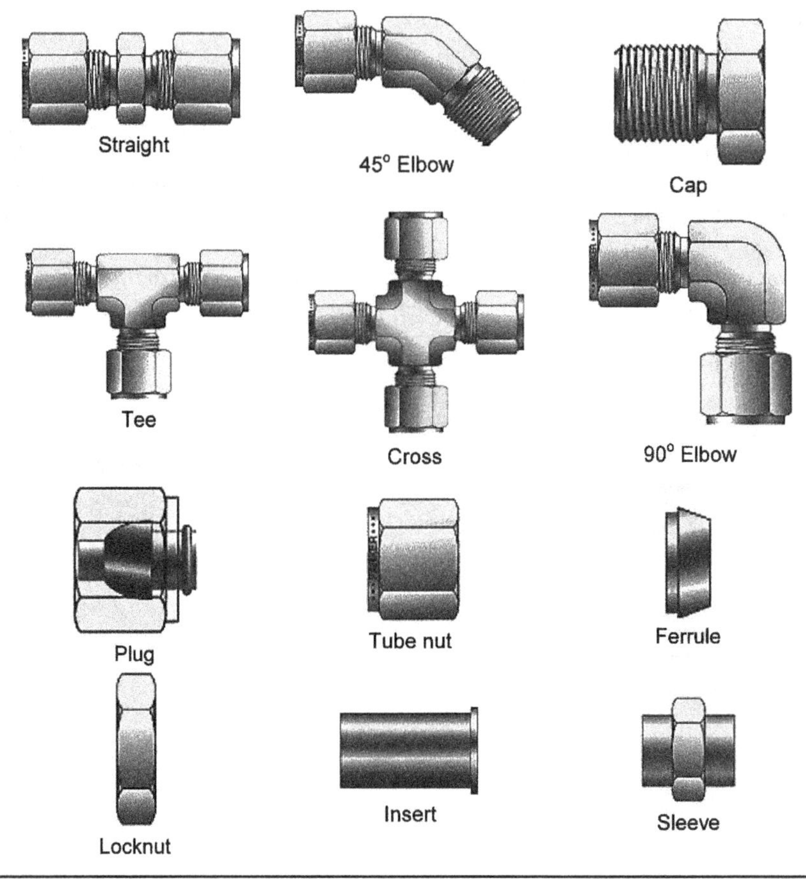

FIGURE 3.2 Commonly used connectors and coupling elements.

The proper installation of the tubing is essential to the correct operation and safe use of the tubing. The following comments should be considered for the mounting of hydraulic tubes, as illustrated by Fig. 3.4:

- Tubing should be bent wherever possible to reduce the number of fittings, and should be bent using proper tube bending equipment (as shown in images 1, 2, and 3 of Fig. 3.4).
- The symmetrical design of piping systems makes them easier to install and present a neat appearance (as shown in image 1).
- Avoid straight-line connections wherever possible (as shown in images 1 and 2 of Fig. 3.4).
- Care should be taken to eliminate stress from tubing lines. Long tubing should be supported by brackets or clips. All parts installed on tubing lines such as heavy fittings, valves, and so on, should be bolted down to eliminate tubing fatigue.
- Tubes should be formed to assemble with true alignment to the center line of the fittings, without distortion or tension (as shown in image 3).

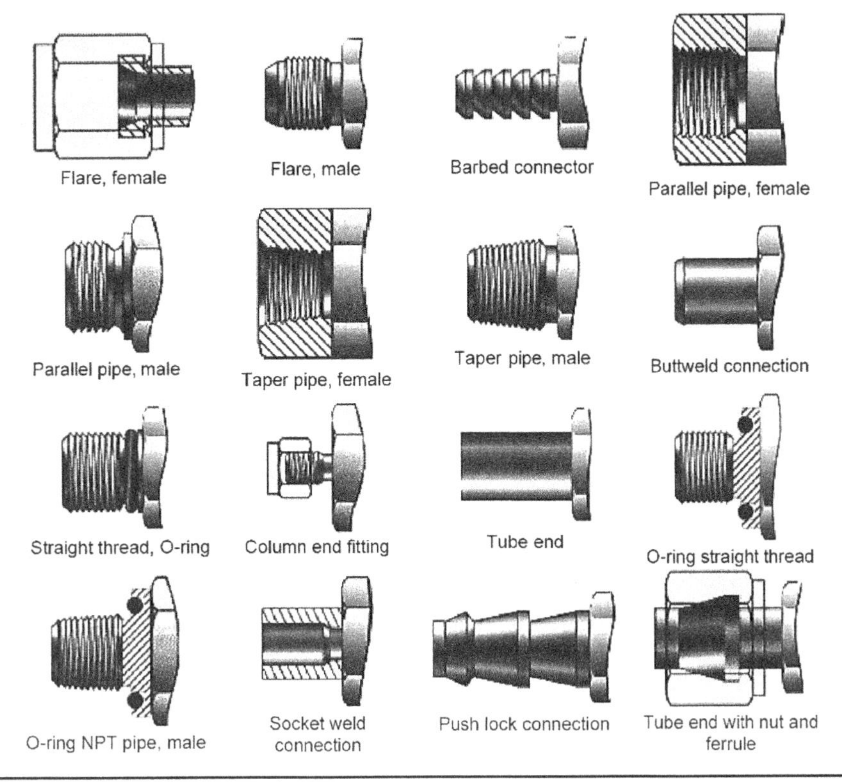

FIGURE 3.3 Commonly used end fittings.

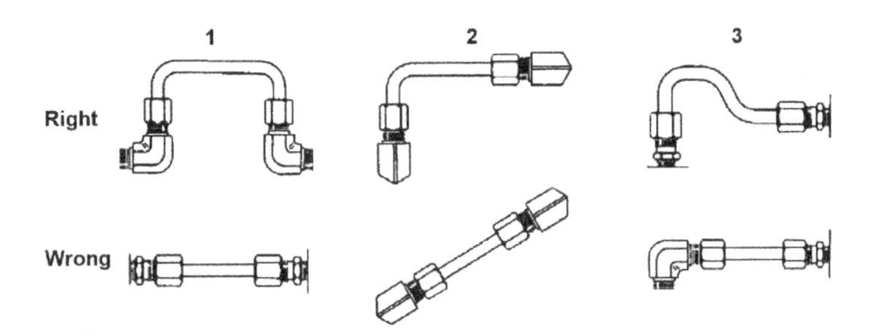

FIGURE 3.4 Proper and improper tube mounting.

3.3 Hoses

Hoses are used to interconnect elements that vibrate or move relative to each other. They insure the required flexibility and can operate at high pressures. The proper hose diameter is determined according to the maximum flow rate and the selected fluid speed.

$$d = \sqrt{\frac{4Q_{\max}}{\pi v}} \tag{3.2}$$

where d = Hose inner diameter, m
Q_{max} = Maximum flow rate, m^3/s
v = Fluid mean velocity, m/s

The recommended mean fluid velocity in flexible hoses is actually within 2.1 to 4.6 m/s for the pressure line and 0.6 to 1.2 m/s for the suction, return, and low pressure lines.

The hose diameter can be also calculated by using the nomogram shown in Fig. 3.5. Figure 3.6 gives some advice and hints about the right and wrong hose mounting.

Table 3.2 shows the typical constructional and operational parameters of high-pressure hydraulic hoses. The hose is built from the following layers, which insure the required stiffness and reliability of operation:

1. Inner synthetic rubber tube, resistant to oils and hydrocarbons. The operating temperature ranges from 40°C to 120°C and the hose may have an internal dressing layer of Viton.
2. High-abrasion-resistant metallic screen layers.
3. Closely braided high tensile steel wire layers separated by antifriction rubber layer.
4. Abrasion resistant synthetic rubber cover.
5. Textile layer.

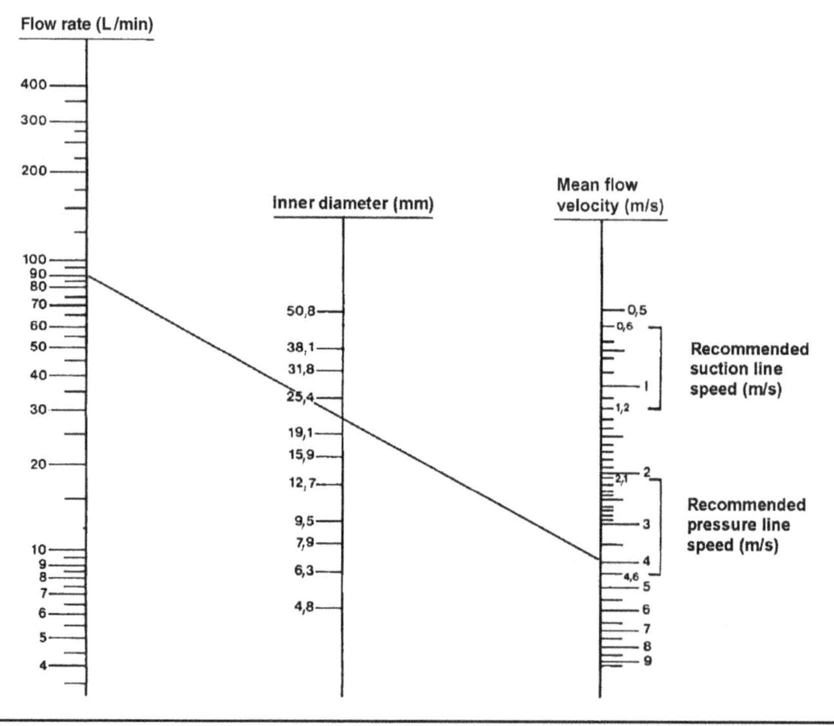

FIGURE 3.5 Nomogram for the calculation of the hose diameter.

1. Provide for length change.

In straight hose installations, allow enough length in the hose line to provide for changes in length that will occur when pressure is applied.

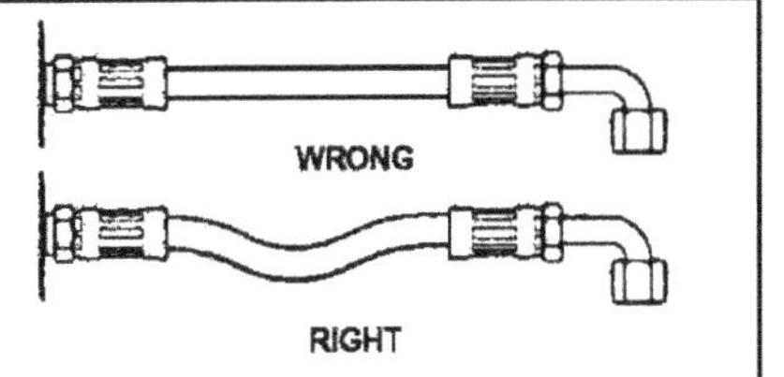

2. Avoid twisting and orient properly.

Do not twist hose during installation. This can be determined by the printed lay-line on the hose. Pressure applied to a twisted hose can cause hose failure.

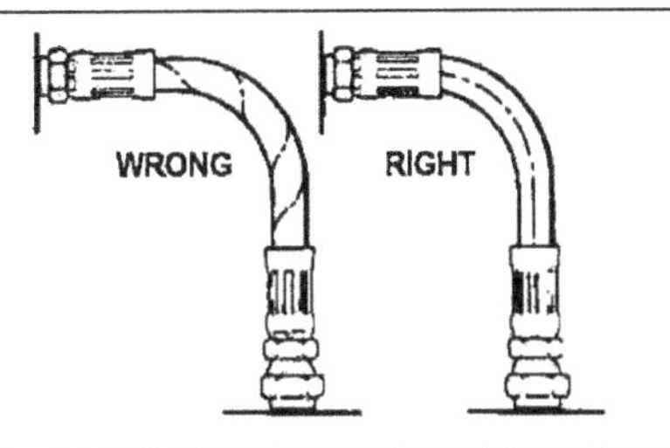

3. Protect from hazardous environment.

Keep hose away from hot parts. High ambient temperature will shorten hose life. It should be routed away from the heat source, and insulated.

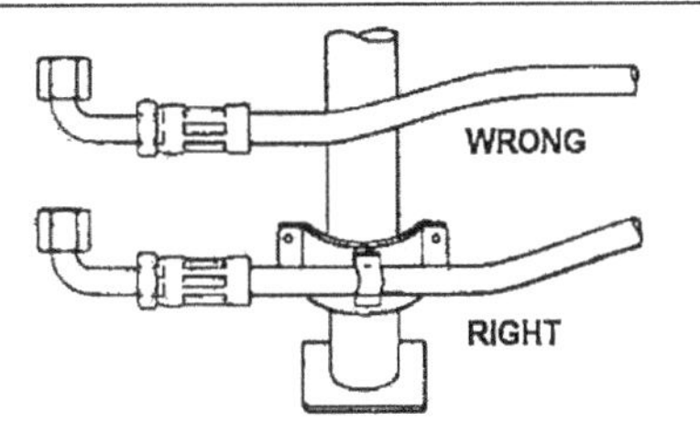

4. Avoid mechanical stress.

Use elbows and adapters in the installation to relieve stress on the assembly.

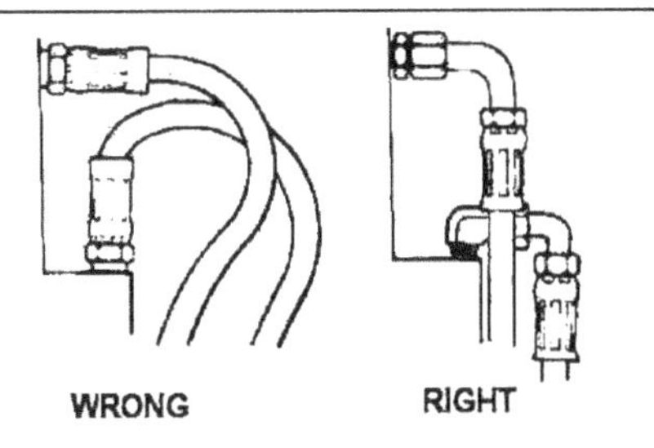

5. Use proper bend radius.

Keep the bend radius of the hose as large as possible to avoid collapsing of the hose and restriction of flow.

Minimum bend radius is measured on the inside bend of the hose as illustrated in the opposite figure.

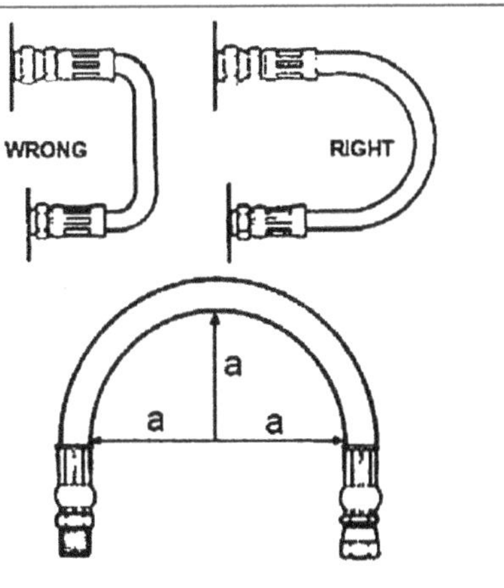

FIGURE 3.6 Comments on hose mounting.

6. Secure for protection. Install hose runs to avoid rubbing or abrasion. Use hose clamps to support long runs of hose or to keep hose away from moving parts. It is important that the clamps not allow the hose to move in order to avoid abrasion and premature hose failure.	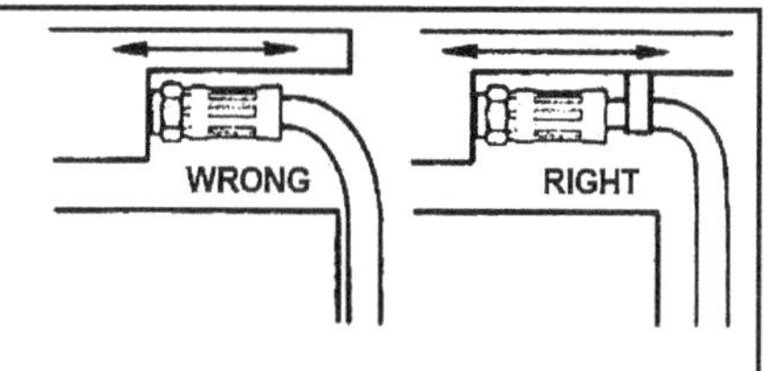
7. Avoid improper hose movement. Make sure the relative motion of the machine components produces bending rather than twisting of the hose. Hose should be routed so the flex is in the same plane as the equipment movement.	

Figure 3.6 *(Continued)*

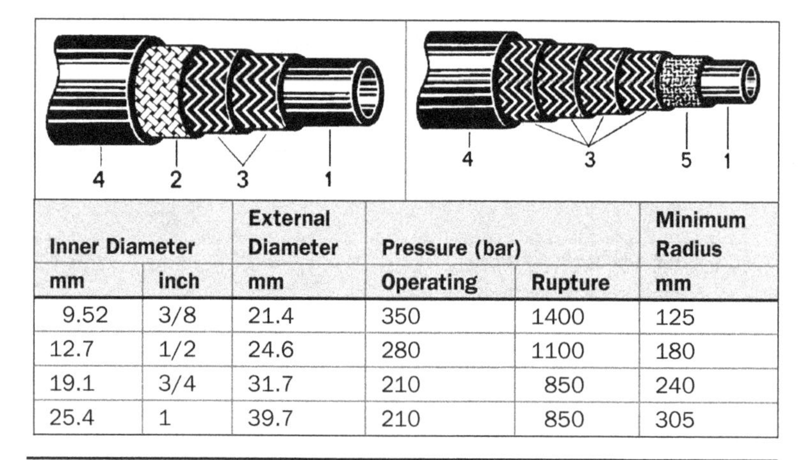

Inner Diameter		External Diameter	Pressure (bar)		Minimum Radius
mm	**inch**	**mm**	**Operating**	**Rupture**	**mm**
9.52	3/8	21.4	350	1400	125
12.7	1/2	24.6	280	1100	180
19.1	3/4	31.7	210	850	240
25.4	1	39.7	210	850	305

Table 3.2 Typical Hose Construction, Dimensions, and Operating Pressure

3.4 Pressure and Power Losses in Hydraulic Conduits

3.4.1 Minor Losses

The minor losses in the hydraulic systems result from the rapid variation of magnitude or direction of the oil velocity. The local pressure losses are calculated by the following formula.

$$\Delta P = \xi \frac{\rho v^2}{2} \tag{3.3}$$

where v = Fluid velocity, m/s
ΔP = Pressure losses, Pa
ξ = Local loss coefficient
ρ = Fluid density, kg/m^3

In laminar flow, the effects of local disturbances are usually insignificant compared with the friction losses. In the case of turbulent flow, the local loss coefficient is determined almost exclusively by the geometry of the local feature. It changes very little with the Reynolds number. The following are some of the local features existing in the hydraulic systems:

- Channel expansion, gradual or abrupt
- Channel contraction, gradual or abrupt
- Channel bend, smooth or sharp (elbow)
- Branching junctions
- Control valves

Table 3.3 gives the values of the local loss coefficient for typical local loss elements.

Figures 3.7 and 3.8 illustrate sudden expansion and sudden contraction elements, respectively. The local pressure losses due to sudden expansion are calculated by the following equation:

$$\Delta P = \xi \frac{\rho v_1^2}{2}, \qquad \xi = \left(1 - \frac{A_1}{A_2}\right)^2 \tag{3.4}$$

or

$$\Delta P = \xi \frac{\rho v_2^2}{2}, \qquad \xi = \left(\frac{A_2}{A_1} - 1\right)^2 \tag{3.5}$$

Local Feature	ξ
Flexible pipe connection	0.3
Standard 90° elbow	1.2–1.3
Tee junction	3.5
Pipe inlet	0.5–1
Pipe outlet	1
Screen filter	1.5–2.5

TABLE 3.3 Local Loss Coefficient of Typical Local Loss Features

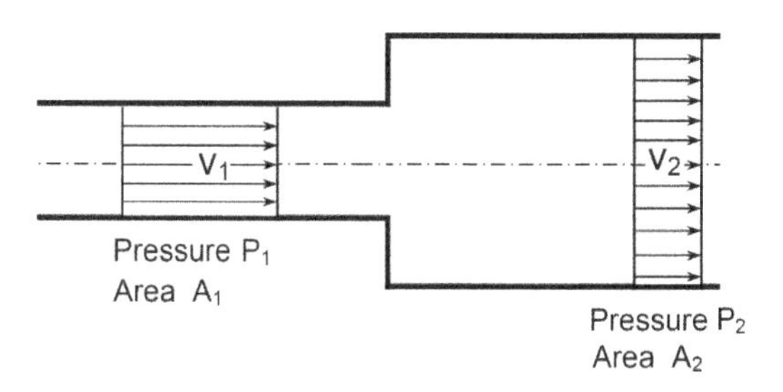

FIGURE 3.7 Sudden expansion.

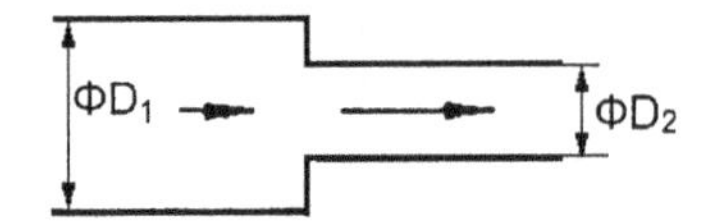

FIGURE 3.8 Sudden contraction.

D_2/D_1	0.2	0.3	0.4	0.5	0.6	0.7	0.8
ξ	0.48	0.455	0.42	0.375	0.32	0.255	0.18

TABLE 3.4 Local Loss Coefficient for Sudden Contraction

The local loss coefficient for sudden contraction is

$$\xi = 0.5\left(1 - \frac{D_2^2}{D_1^2}\right) \quad \text{and} \quad \Delta P = \xi \frac{\rho v_2^2}{2} \tag{3.6}$$

The values of the local loss coefficient ξ for a sudden contraction element are given in Table 3.4.

3.4.2 Friction Losses

The pressure losses in the pipe lines depend mainly on the geometry, surface roughness, fluid properties, and Reynolds number.

The three main types of flow are laminar, turbulent, and transition flow. Laminar flow is a streamlined flow of viscous fluid, where all particles of the fluid move in distinct and separate lines. For Newtonian fluids, the velocity distribution across the cross section of the pipe is parabolic, with zero velocity in the vicinity of the pipe wall and maximum velocity at the pipe center (see App. 2B). Figure 3.9 shows the development of laminar flow in a pipe. The transition length, where the full laminar flow profile is not yet established, is illustrated by this figure. The type of flow is determined by calculating the Reynolds number, Re.

$$\text{Re} = \frac{vD}{\nu} = \frac{\rho v D}{\mu} \tag{3.7}$$

where D = Inner pipe diameter, m
v = Mean fluid velocity = $4Q/\pi D^2$, m/s
μ = Dynamic viscosity, Pa s
ν = Kinematic viscosity, m^2/s

The friction losses in the pipe line are calculated by the following expression:

$$\Delta P = \lambda \frac{L}{D} \frac{\rho v^2}{2} \tag{3.8}$$

The friction coefficient depends mainly of the Reynolds number, Re, as shown in Table 3.5.

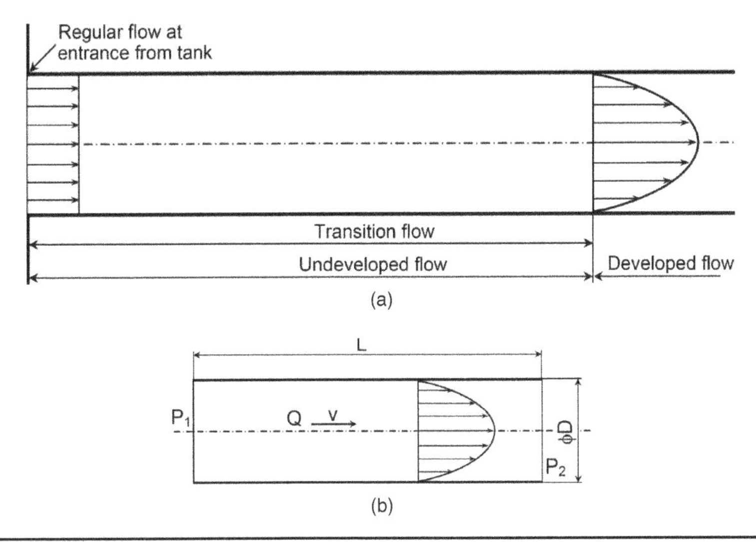

FIGURE 3.9 (a) Development of laminar flow in a pipe line and (b) pipe line parameters.

Laminar flow	$\lambda = \dfrac{64}{\text{Re}}$	Re < 2300	Hagen-Poisseuille's law, 1856
Turbulent flow, smooth pipe	$\lambda = \dfrac{0.3164}{\sqrt[4]{\text{Re}}}$	2300 < Re < 10^5	Blasiu's law, 1915
	$\lambda = 0.0054 + 0.396\,(\text{Re})^{-0.3}$	10^5 < Re < 0.2×10^6	Herman's law, 1930
Turbulent flow, rough pipe	$\dfrac{1}{\sqrt{\lambda}} = -2\,\log\left(\dfrac{\varepsilon/D}{3.7} + \dfrac{2.51}{\text{Re}\sqrt{\lambda}}\right)$	For the whole range of turbulent flow	Colebrook and White, 1939
	Use Moody's diagram (see Fig. 3.10)		

TABLE 3.5 Determination of the Pipe Line Friction Coefficient

The transition from the laminar flow to the turbulent one takes place at a critical value of the Reynolds number. As a rough guide, it is possible to say that flow will be turbulent for Re > 2300. The transition process is a consequence of the instability of laminar flow. The uncertainty of the critical value is due to the fact that the processes, near their stability limits, are easily destabilized even by minute disturbance effects (such as noise of the pump).

In the case of laminar flow, by substituting for v and Re in Eq. (3.8), the following expression was obtained for the pressure losses, ΔP:

$$\Delta P = \frac{128\mu L}{\pi D^4} Q = RQ \tag{3.9}$$

The term R expresses the resistance of the hydraulic transmission line.

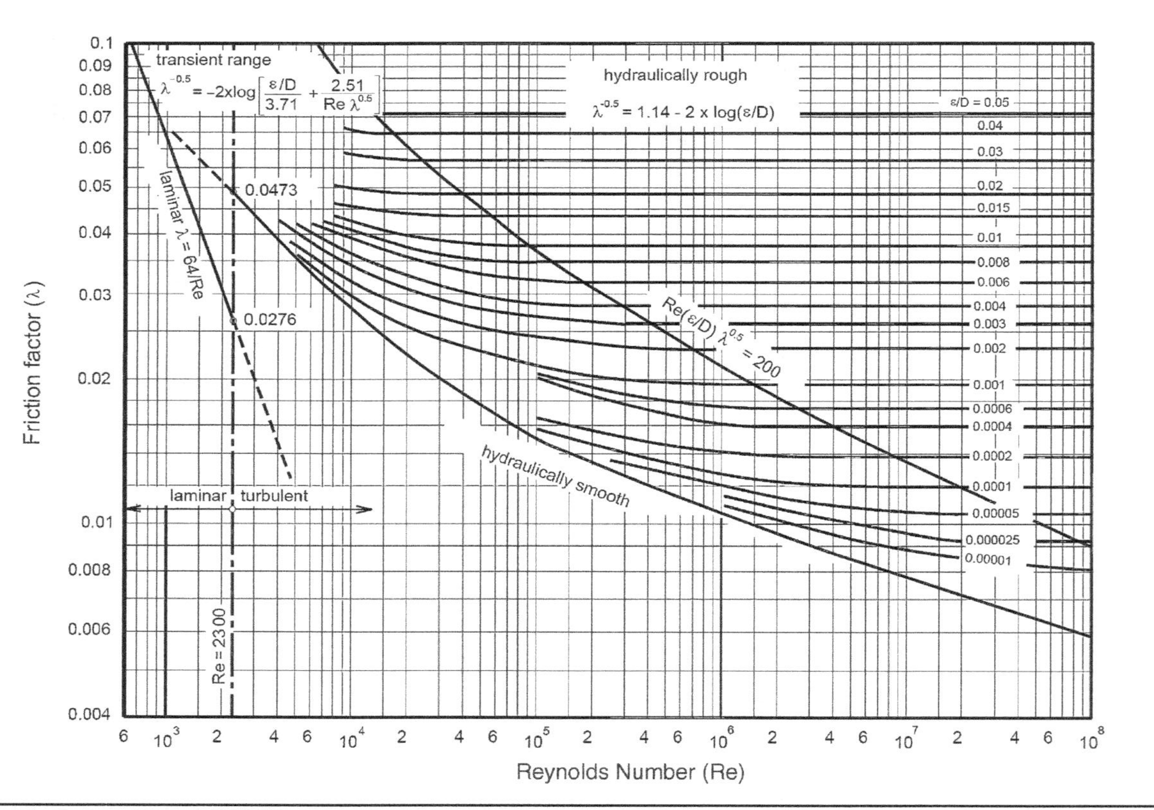

Figure 3.10 Moody's diagram [*Zayed (1999)*].

3.5 Modeling of Hydraulic Transmission Lines

The hydraulic transmission line is actually a distributed parameter system. The motion of the liquid in the transient conditions takes place under the action of the fluid inertia, friction, and compressibility, as well as the driving pressure forces. The oil velocity, pressure, and temperature vary from point to point along the pipe length and pipe radius. The mathematical model of the line becomes too complicated when taking into consideration all the variations of the oil and flow parameters. Therefore, it is necessary to develop a simplified mathematical model, which describes the dynamic behavior of the transmission line with acceptable accuracy. A fairly precise model is the *lumped parameter model*, which can be deduced given the following assumptions:

- The flow is laminar unidirectional.
- The liquid pressure and velocity are looked at as the mean values, and are considered constant along the line cross section.
- The oil moves in the line as one lump (single-lump model) or several lumps (multi-lump model).
- The effect of line resistance, inertia, and capacitance are separate and each of them is localized in one of three separate portions in the line, Fig. 3.11. The effect of the resistance of the whole line is localized in the first portion, the effect of the inertia of the whole line is localized in the second portion, while the effect of the line capacitance takes place in the third portion.

In the first portion, the oil moves as one lump under the action of the friction forces. Therefore, its motion is described by the following equations relating the pressures, P, and flow rates, Q, at both ends of the first portion:

$$P_o - P_1 = RQ_1 \tag{3.10}$$

$$Q_o = Q_1 \tag{3.11}$$

Applying the Laplace transform to these equations, then, after rearrangement, the following equation is obtained:

$$\begin{bmatrix} P_o(s) \\ Q_o(s) \end{bmatrix} = \begin{bmatrix} 1 & R \\ 0 & 1 \end{bmatrix} \begin{bmatrix} P_1(s) \\ Q_1(s) \end{bmatrix} = \mathbf{R} \begin{bmatrix} P_1(s) \\ Q_1(s) \end{bmatrix} \tag{3.12}$$

where R = Whole line resistance $= 128\mu L/\pi D^4$, Ns/m^5
$\mathbf{R}$ = Resistance matrix

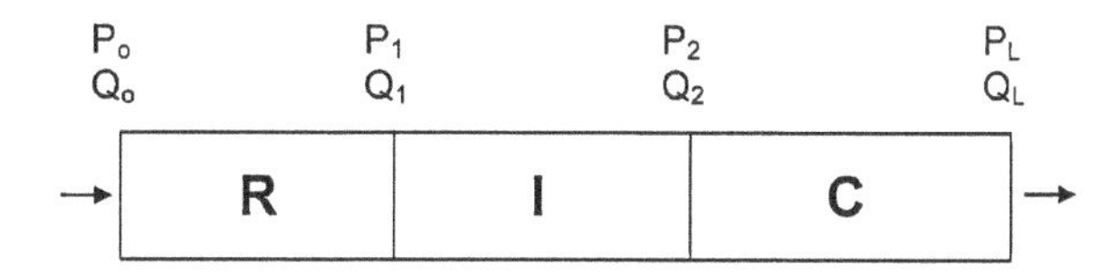

Figure 3.11 Single-lump model.

The following relations describe the motion of the oil lump in the second portion under the action of its inertia, I:

$$P_1 - P_2 = I\frac{dQ_2}{dt} \tag{3.13}$$

$$Q_1 = Q_2 \tag{3.14}$$

Applying Laplace transform to these equations, then, after rearrangement, the following equation is obtained:

$$\begin{bmatrix} P_1(s) \\ Q_1(s) \end{bmatrix} = \begin{bmatrix} 1 & Is \\ 0 & 1 \end{bmatrix} \begin{bmatrix} P_2(s) \\ Q_2(s) \end{bmatrix} = \mathbf{I} \begin{bmatrix} P_2(s) \\ Q_2(s) \end{bmatrix} \tag{3.15}$$

where I = Whole line Inertia = $4\rho L/\pi D^2$, kg/m^4
$\mathbf{I}$ = Inertia matrix

Considering the effect of oil compressibility in the last portion, the following relations can be deduced:

$$Q_2 - Q_L = C\frac{dP_L}{dt} \tag{3.16}$$

$$P_2 = P_L \tag{3.17}$$

Applying Laplace transform to these equations, then, after rearrangement, the following equation is obtained:

$$\begin{bmatrix} P_2(s) \\ Q_2(s) \end{bmatrix} = \begin{bmatrix} 1 & 0 \\ Cs & 1 \end{bmatrix} \begin{bmatrix} P_L(s) \\ Q_L(s) \end{bmatrix} = \mathbf{C} \begin{bmatrix} P_L(s) \\ Q_L(s) \end{bmatrix} \tag{3.18}$$

where C = Whole line capacitance = $\pi D^2 L/4B$, m^3/Pa
$\mathbf{C}$ = Capacitance matrix

The transfer matrix relating the line parameters P_o, Q_o, P_L, and Q_L can be deduced by eliminating the assumed internal variables, P_1, P_2, Q_1, and Q_2.

$$\begin{bmatrix} P_o(s) \\ Q_o(s) \end{bmatrix} = \mathbf{R} \begin{bmatrix} P_1(s) \\ Q_1(s) \end{bmatrix} = \mathbf{RI} \begin{bmatrix} P_2(s) \\ Q_2(s) \end{bmatrix} = \mathbf{RIC} \begin{bmatrix} P_L(s) \\ Q_L(s) \end{bmatrix}$$

or

$$\begin{bmatrix} P_o(s) \\ Q_o(s) \end{bmatrix} = \begin{bmatrix} ICs^2 + RCs + 1 & Is + R \\ Cs & 1 \end{bmatrix} \begin{bmatrix} P_L(s) \\ Q_L(s) \end{bmatrix} \tag{3.19}$$

This equation defines the relation between the pressures and flow rates at both of the line extremities in the transient conditions, assuming a single oil lump.

Example 3.1 Find the transfer function relating the pressures and flow rates at the two extremities of a closed end line.

For a closed end line, $Q_L = 0$.

$$P_o = (ICs^2 + RCs + 1)\,P_L$$

$$Q_o = Cs\,P_L$$

or

$$\frac{P_L}{P_o} = \frac{1}{ICs^2 + RCs + 1} \quad \text{and} \quad \frac{P_L}{Q_o} = \frac{1}{Cs}$$

Example 3.2 Find the transfer function relating the pressures and flow rates at the two extremities of an open end line; the line end is open to atmosphere.

P_o $\quad$ $P_L = 0$

Q_o $\quad$ Q_L

In the case of an open end line, $P_L = 0$

$$Q_o = Q_L \quad \text{and} \quad P_o = (Is + R)Q_L$$

or

$$\frac{Q_L}{P_o} = \frac{1}{Is + R}$$

3.6 Exercises

1. Explain the different types of losses in the hydraulic transmission lines.

2. Explain in detail how to calculate the power losses due to friction in the hydraulic transmission lines.

3. Discuss in detail the modeling of hydraulic transmission lines assuming lumped parameters. State clearly your assumptions and derive the transfer matrix relating the inlet and outlet pressures and flow rates.

4. Equation (3.19) describes the relation between pressures and flow rates in a hydraulic pipe line, assuming lumped parameters. Give the expressions for I, R, and C.

5. (a) Assuming a single oil lump model, derive the transfer function of an open end line, $Q_L/P_o(s)$, and find the expressions for its parameters.

(b) A hydraulic line whose end is open to the atmosphere has the following parameters:

Pipe length $L = 3.5$ m, pipe diameter $D = 8$ mm, oil density $\rho = 850$ kg/m^3, oil viscosity $\mu = 0.018$ Ns/m^2, and oil bulk modulus $B = 1.3$ GPa.

Calculate the coefficients of the transfer function and plot in scale its response to a step input pressure of 0.5 bar.

6. (a) Assuming a single oil lump model, derive the transfer function of a closed end line, $P_L/P_o(s)$, and find the expressions for its parameters.

(b) A hydraulic line with a closed end has the following parameters.

$L = 3$ m, $D = 10$ mm, $\rho = 867$ kg/m^3, $\mu = 0.13$ Ns/m^2, and $B = 1$ GPa.

Calculate the coefficients of the transfer function and plot in scale its response to a step input pressure of 100 bar magnitude.

3.7 Nomenclature

C = Whole line capacitance, m^3/Pa
$\mathbf{C}$ = Capacitance matrix
d, D = Tube inner diameter, m
I = Whole line inertia, kg/m^4
$\mathbf{I}$ = Inertia matrix
Q_{max} = Maximum flow rate, m^3/s
R = Whole line resistance, Ns/m^5
$\mathbf{R}$ = Resistance matrix
v = Mean fluid velocity, m/s
ΔP = Pressure losses, Pa
μ = Dynamic viscosity, Ns/m^2
ξ = Local loss coefficient
ρ = Fluid density, kg/m^3
ν = Kinematic viscosity, m^2/s

Appendix 3A The Laplace Transform

When a differential equation expressed in terms of time, t, is operated on by a Laplace integral, a new equation results, which is expressed in terms of a complex term (s). The Laplace transform translates the time-dependent function from the time domain to the frequency (or Laplace) domain. The transformed equation is in pure algebraic form and may be manipulated algebraically.

The Direct Laplace Transform

The direct Laplace transform is given by the following expression:

$$X(s) = \mathbf{L}\,[x(t)] = \int_0^{\infty} x(t)e^{-st}\,dt \tag{3A.1}$$

The Inverse Laplace Transform

The inverse Laplace transform is an integral operator that enables a transform from the Laplace domain to the time domain.

$$x(t) = \frac{1}{2\pi i}\lim_{R\to\infty}\int_{\sigma_1 - iR}^{\sigma_1 + iR} X(s)e^{st}\,ds \tag{3A.2}$$

Actually, all functions in the time domain have a direct Laplace transform, but some of the functions in the Laplace domain have no inverse Laplace transform.

Properties of the Laplace Transform

The following are the basic properties of the Laplace transform:

1. $\mathbf{L}\ [f_1(t) \pm f_2(t)] = F_1(s) \pm F_2(s)$ (3A.3)
2. $\mathbf{L}\ [af(t)] = aF(s)$ (3A.4)
3. $\mathbf{L}\ [e^{-at} f(t)] = F(s+a)$ (3A.5)
4. Initial value problem $f(0^+) = \lim_{t \to 0} f(t) = \lim_{s \to \infty} sF(s)$ (3A.6)
5. Final value problem $f(\infty) = \lim_{t \to \infty} f(t) = \lim_{s \to 0} sF(s)$ (3A.7)
6. $$\mathbf{L}\left[\frac{d^n f(t)}{dt^n}\right] = s^n F(s) - s^{n-1} f(0^+) - s^{n-2}\frac{df}{dt}(0^+) - s^{n-3}\frac{d^2 f}{dt^2}(0^+) - \cdots - \frac{d^{n-1} f}{dt^{n-1}}(0^+) \quad (3A.8)$$

   $$\mathbf{L}\left[\frac{df(t)}{dt}\right] = sF(s) - f(0^+) \quad (3A.9)$$

   $$\mathbf{L}\left[\frac{d^2 f(t)}{dt^2}\right] = s^2 F(s) - sf(0^+) - \dot{f}(0^+) \quad (3A.10)$$

7. $$\mathbf{L}\left[\int f(t)dt\right] = \frac{1}{s}\left[F(s) + \int f(t)dt\Big|_{0^+}\right] \quad (3A.11)$$

Laplace Transform Tables

The pairs of Laplace transforms are given in math textbook tables. The following are the pairs of Laplace transforms of some key functions:

$F(s)$	$f(t)$
1	$\delta(t)$
$\frac{k}{s}$	k
$\frac{k\,(n!)}{s^{n+1}}$	$k\,t^n$
$\frac{k}{s+a}$	$k\,e^{-at}$
$\frac{k}{(s+a)^2}$	$k\,t\,e^{-at}$
$\frac{k\,(n!)}{(s+a)^{n+1}}$	$k\,t^n\,e^{-at}$
$\frac{k\,\omega}{s^2+\omega^2}$	$k\sin\omega t$
$\frac{k\,s}{s^2+\omega^2}$	$k\cos\omega t$

Appendix 3B Modeling and Simulation of Hydraulic Transmission Lines

This appendix deals with the development of lumped parameter models and the simulation of a hydraulic transmission line. The model takes into consideration the effects of the viscous friction, fluid inertia, fluid compressibility, and elasticity of line material. The developed models are one dimensional. The fluid speed and pressure are thought of as averaged quantities over the cross section of the line. The simplicity of the models results from the assumption of separate effects of the previously mentioned parameters. The validity of the models is evaluated by comparing the step response, calculated by the simulation program, with experimental results.

The lumped parameter model was deduced considering the assumptions given in Sec. 3.5. The effect of line resistance, inertia, and capacitance are assumed to be localized in one of three separate portions in the line, as shown by Fig. 3B.1. The effect of the resistance of the whole line is localized in the first portion, the effect of the inertia of the whole line is localized in the second portion, while the effect of the line capacitance takes place in the third portion.

The Single-Lump Model

The following are the equations describing the single-lump model;

$$P_o - P_1 = RQ_o \tag{3B.1}$$

$$P_1 - P_L = I\frac{dQ_o}{dt} \tag{3B.2}$$

$$Q_o - Q_L = C\frac{dP_L}{dt} \tag{3B.3}$$

Referring to Eq. (3.19), the following is the transfer matrix of the single-lump model, deduced in Section 3.5:

$$\begin{bmatrix} P_o(s) \\ Q_o(s) \end{bmatrix} = \mathrm{RIC}\begin{bmatrix} P_L(s) \\ Q_L(s) \end{bmatrix} = \begin{bmatrix} ICs^2 + RCs + 1 & Is + R \\ Cs & 1 \end{bmatrix}\begin{bmatrix} P_L(s) \\ Q_L(s) \end{bmatrix} \tag{3B.4}$$

The line resistance R, inertia I, and capacitance C are given by Eqs. (2.9), (2.47), and (2.80), respectively.

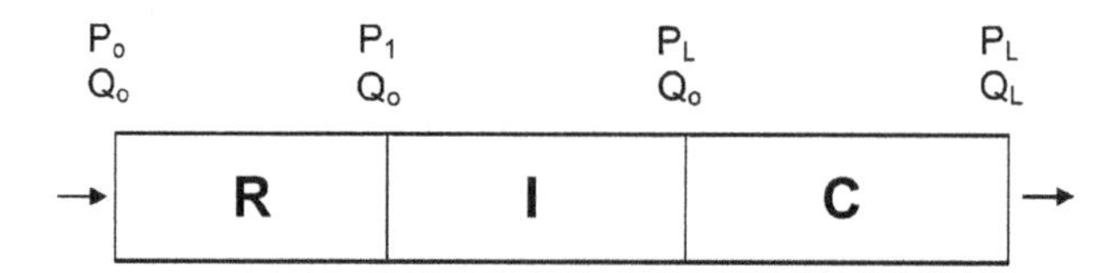

FIGURE 3B.1 The single-lump model.

The Two-Lump Model

Assuming more than one lump shown by Fig. 3B.2, the hydraulic resistance, inertia, and capacitance of each lump in an n-lump model are given by the following expressions:

$$\text{Lump resistance} \quad R_n = \frac{128\mu(L/n)}{\pi D^4} \tag{3B.5}$$

$$\text{Lump inertia} \quad I_n = \frac{4\rho(L/n)}{\pi D^2} \tag{3B.6}$$

$$\text{Lump capacitance} \quad C_n = \frac{\pi D^2(L/n)}{4}\left(\frac{1}{B}+\frac{5D}{4Eh}\right) \tag{3B.7}$$

The following are the equations describing the two-lump model:

$$P_o - P_{11} = R_2 Q_o \tag{3B.8}$$

$$P_{11} - P_{1L} = I_2 \frac{dQ_o}{dt} \tag{3B.9}$$

$$Q_o - Q_{1L} = C_2 \frac{dP_{1L}}{dt} \tag{3B.10}$$

$$P_{1L} - P_{21} = R_2 Q_{1L} \tag{3B.11}$$

$$P_{21} - P_L = I_2 \frac{dQ_{1L}}{dt} \tag{3B.12}$$

$$Q_{1L} - Q_L = C_2 \frac{dP_L}{dt} \tag{3B.13}$$

The transfer matrix of the two-lump model can be deduced as follows:

$$\begin{bmatrix} P_o(s) \\ Q_o(s) \end{bmatrix} = [\mathbf{R_2 I_2 C_2}]^2 \begin{bmatrix} P_L(s) \\ Q_L(s) \end{bmatrix} = \begin{bmatrix} I_2C_2s^2 + R_2C_2s + 1 & I_2s + R_2 \\ C_2s & 1 \end{bmatrix}^2 \begin{bmatrix} P_L(s) \\ Q_L(s) \end{bmatrix}$$

$$= \begin{bmatrix} a_{11} & a_{12} \\ a_{21} & a_{22} \end{bmatrix} \begin{bmatrix} P_L(s) \\ Q_L(s) \end{bmatrix} \tag{3B.14}$$

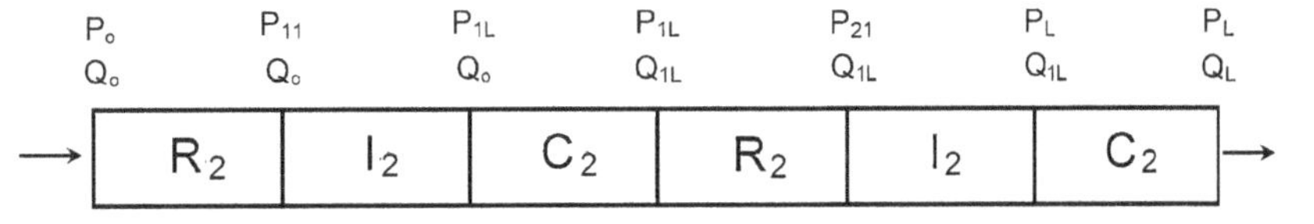

FIGURE 3B.2 The two-lump model.

and

$$\mathbf{R}_2 = \begin{bmatrix} 1 & R_2 \\ 0 & 1 \end{bmatrix}, \quad \mathbf{I}_2 = \begin{bmatrix} 1 & I_2 s \\ 0 & 1 \end{bmatrix} \quad \text{and} \quad \mathbf{C}_2 = \begin{bmatrix} 1 & 0 \\ C_2 s & 1 \end{bmatrix} \tag{3B.15}$$

$$a_{11} = I_2^2 C_2^2 s^4 + 2I_2 R_2 C_2^2 s^3 + (R_2^2 C_2^2 + 3I_2 C_2)s^2 + 3R_2 C_2 s + 1 \tag{3B.16}$$

$$a_{12} = I_2^2 C_2 s^3 + 3I_2 R_2 C_2 s^2 + (R_2^2 C_2 + 2I_2)s + 2R_2 \tag{3B.17}$$

$$a_{21} = I_2 C_2^2 s^3 + R_2 C_2^2 s^2 + 2C_2 s \tag{3B.18}$$

$$a_{22} = I_2 C_2 s^2 + R_2 C_2 s + 1 \tag{3B.19}$$

The Three-Lump Model

The mathematical relations describing the dynamic behavior of the three-lump model can be deduced systematically as those describing the two-lump model. For the three-lump model, the transfer matrix is as follows:

$$\begin{bmatrix} P_o(s) \\ Q_o(s) \end{bmatrix} = [\mathbf{R}_3\, \mathbf{I}_3\, \mathbf{C}_3]^3 \begin{bmatrix} P_L(s) \\ Q_L(s) \end{bmatrix} = \begin{bmatrix} I_3 C_3 s^2 + R_3 C_3 s + 1 & I_3 s + R_3 \\ C_3 s & 1 \end{bmatrix}^3 \begin{bmatrix} P_L(s) \\ Q_L(s) \end{bmatrix}$$

$$= \begin{bmatrix} b_{11} & b_{12} \\ b_{21} & b_{22} \end{bmatrix} \begin{bmatrix} P_L(s) \\ Q_L(s) \end{bmatrix} \tag{3B.20}$$

and $$\mathbf{R}_3 = \begin{bmatrix} 1 & R_3 \\ 0 & 1 \end{bmatrix}, \quad \mathbf{I}_3 = \begin{bmatrix} 1 & I_3 s \\ 0 & 1 \end{bmatrix} \quad \text{and} \quad \mathbf{C}_3 = \begin{bmatrix} 1 & 0 \\ C_3 s & 1 \end{bmatrix} \tag{3B.21}$$

$$\begin{aligned} b_{11} = {} & I_3^3 C_3^3 s^6 + 3I_3^2 R_3 C_3^3 s^5 + (3I_3 R_3^2 C_3^3 + 5I_3^2 C_3^2)s^4 \\ & + (10I_3 R_3 C_3^2 + R_3^3 C_3^3)s^3 + (6I_3 C_3 + 5R_3^2 C_3^2)s^2 \\ & + 6R_3 C_3 s + 1 \end{aligned} \tag{3B.22}$$

The Four-Lump Model

The transfer matrix of the four-lump model can be deduced as follows:

$$\begin{bmatrix} P_o(s) \\ Q_o(s) \end{bmatrix} = [\mathbf{R}_4\, \mathbf{I}_4\, \mathbf{C}_4]^4 \begin{bmatrix} P_L(s) \\ Q_L(s) \end{bmatrix} = \begin{bmatrix} I_4 C_4 s^2 + R_4 C_4 s + 1 & I_4 s + R_4 \\ C_4 s & 1 \end{bmatrix}^4 \begin{bmatrix} P_L(s) \\ Q_L(s) \end{bmatrix}$$

$$= \begin{bmatrix} c_{11} & c_{12} \\ c_{21} & c_{22} \end{bmatrix} \begin{bmatrix} P_L(s) \\ Q_L(s) \end{bmatrix} \tag{3B.23}$$

and

$$\mathbf{R}_4 = \begin{bmatrix} 1 & R_4 \\ 0 & 1 \end{bmatrix}, \quad \mathbf{I}_4 = \begin{bmatrix} 1 & I_4 s \\ 0 & 1 \end{bmatrix} \quad \text{and} \quad \mathbf{C}_4 = \begin{bmatrix} 1 & 0 \\ C_4 s & 1 \end{bmatrix} \tag{3B.24}$$

$$\begin{aligned} c_{11} = {} & I_4^4 C_4^4 s^8 + 4 I_4^3 R_4 C_4^4 s^7 + (6 I_4^2 R_4^2 C_4^4 + 7 I_4^3 C_4^3) s^6 \\ & + (21 I_4^2 R_4 C_4^3 + 4 I_4 R_4^3 C_4^4) s^5 \\ & + (15 I_4^2 C_4^2 + 21 I_4 R_4^2 C_4^3 + R_4^4 C_4^4) s^4 \\ & + (7 R_4^3 C_4^3 + 30 I_4 R_4 C_4^2) s^3 + (15 R_4^2 C_4^2 + 10 I_4 C_4) s^2 \\ & + 10 R_4 C_4 s + 1 \end{aligned} \tag{3B.25}$$

Higher-Order Models

Higher-order models of transmission lines can be deduced by assuming a greater number of fluid lumps. The mathematical models describing these models can be deduced systematically, as shown earlier.

The dynamic behavior of hydraulic transmission lines can be studied by calculating the transient response or the frequency response. These calculations can be carried out by using the detailed mathematical models or by using the transfer functions, deduced from the transfer matrix.

Case Study

The lumped parameter model is used to calculate the transient response of a hydraulic transmission line. The studied line had the following parameters:

Pipe length	$L = 18$ m
Pipe diameter	$D = 0.01$ m
Dynamic viscosity of oil	$\mu = 0.1215$ Ns/m^2
Oil density	$\rho = 868$ kg/m^3
Bulk modulus of oil	$B_e = 1.96 \times 10^9$ N/m^2
Line resistance	$R = \dfrac{128\mu L}{\pi D^4} = 8.912 \times 10^9$ Pa s/m^3
Line inertia	$I = \dfrac{\rho L}{A} = \dfrac{4\rho L}{\pi D^2} = 1.9893 \times 10^8$ kg/m^4
Line capacitance	$C = \dfrac{V}{B_e} = \dfrac{\pi D^2 L}{4}\left(\dfrac{1}{B} + \dfrac{5D}{4Eh}\right) = 7.213 \times 10^{-13}$ m^5/N

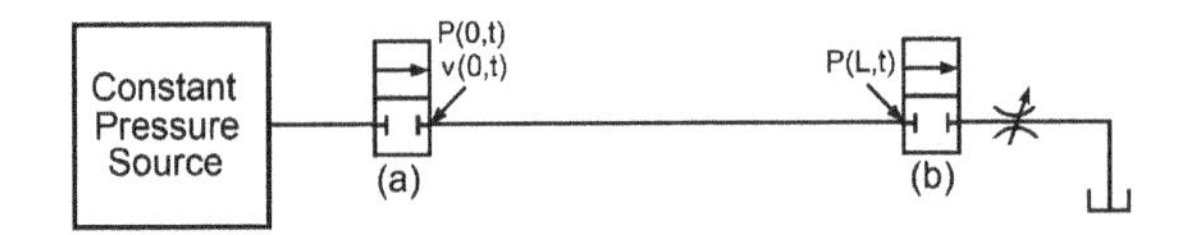

FIGURE 3B.3 Scheme of the line connection.

The line is connected to a constant pressure source and return line by the directional control valves (DCV) (a) and (b), respectively (see Fig. 3B.3). The experiment was conducted using the following sequence:

- The DCV (a) is switched to the closed position.
- The DCV (b) is switched to the opened position, and then to the closed position. The pressure in the tested line is thus equal to the tank pressure.
- The DCV (a) is rapidly switched to the open position to communicate the constant pressure source (447 bar) with the line inlet. The pressure at the line end $P(L)$ is then recorded.

The transient variation of the end pressure $p(L,t)$, was calculated. The calculations were carried out using different lumped parameter models: one-, two-, three-, and four-lump models. The simulation results, calculated using the SIMULINK program, are plotted in Fig. 3B.4. This figure also presents the results of the step response measurements, courtesy of Lallement J, 1976.

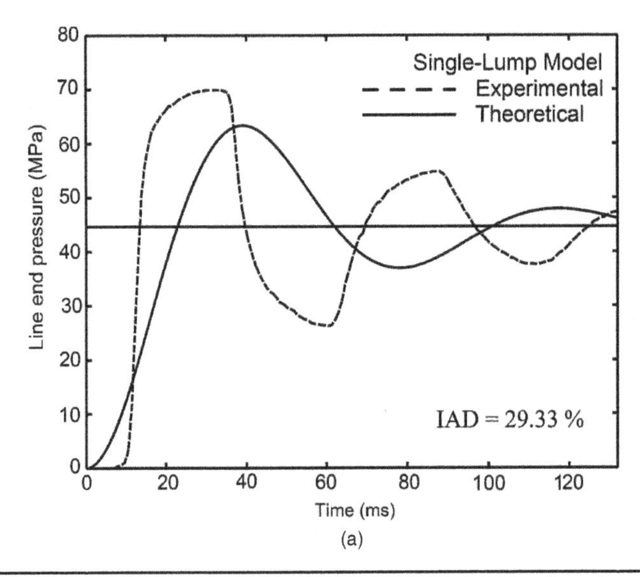

FIGURE 3B.4 (a) Step response of the closed-end hydraulic transmission line, described by a single-lump model, to a step input pressure of 44.7 MPa. (b) Step response of the closed-end hydraulic transmission line, described by a two-lump model, to a step input pressure of 44.7 MPa. (c) Step response of the closed-end hydraulic transmission line, described by a three-lump model, to a step input pressure of 44.7 MPa. (d) Step response of the closed-end hydraulic transmission line, described by a four-lump model, to a step input pressure of 44.7 MPa. (e) Simulation results of the step response of the closed-end hydraulic transmission line, described by a four-lump model, to a step input pressure of 44.7 MPa.

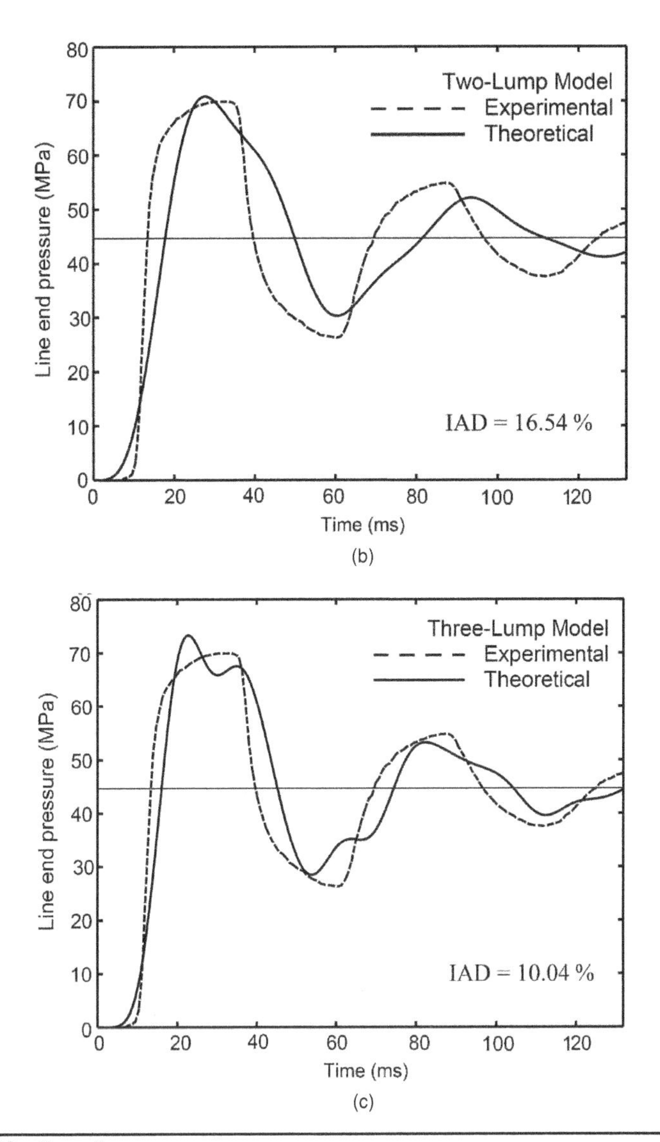

FIGURE 3B.4 *(Continued)*

The precision of model is evaluated by how close is its response to the experimental results. Therefore, the deviation of the theoretical end experimental results is evaluated by calculating the percentage of the integral of absolute difference (IAD), defined as

$$\mathrm{IAD} = \frac{\int_0^T \left| P_{\mathrm{Lth}} - P_{\mathrm{Lexp}} \right| dt}{P_{\mathrm{Lss}} T} \times 100\% \tag{3B.26}$$

where IAD = Integral of absolute difference, %

P_{Lexp} = Experimentally evaluated pressure at the closed end, Pa

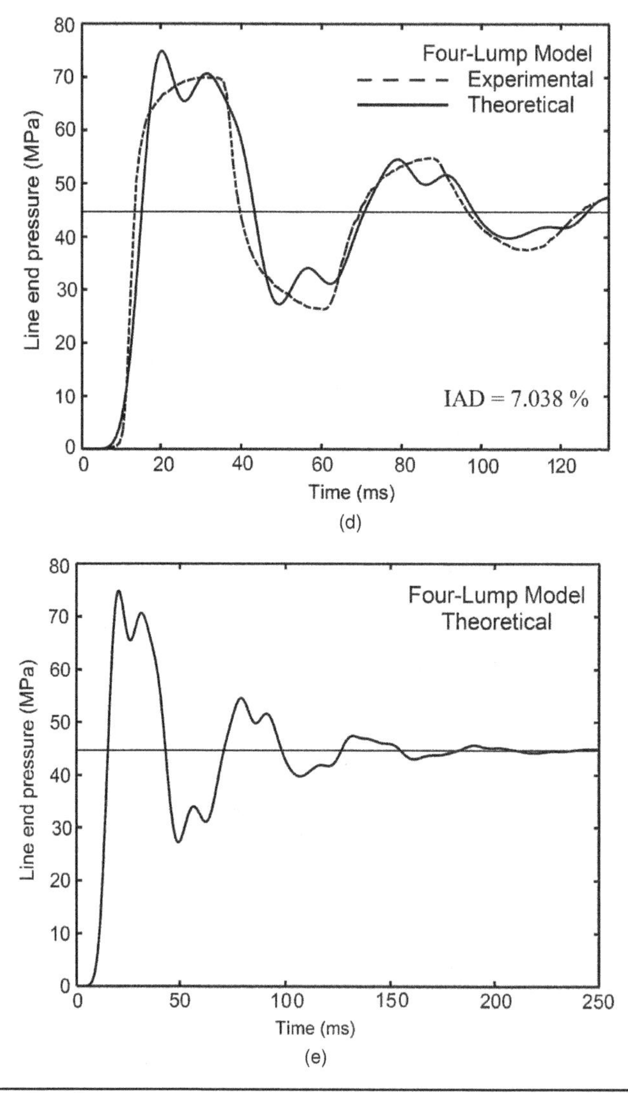

FIGURE 3B.4 *(Continued)*

P_{Lss} = Steady-state pressure at the closed end, Pa
P_{Lth} = Theoretically evaluated pressure at the closed end, Pa
T = Time duration of the response, s

The study of Fig. 3B.4 shows that the three-lump and four-lump models agree satisfactorily with the experimental results. The calculated integral absolute difference decreased rapidly with the increase in the number of lumps.

The frequency response of a closed-end line was calculated assuming a four-lump model for a line having the following parameters: L = 18 m, ρ = 868 kg/m^3,

$B = 1.6$ GPa, $\mu = 0.059$ Ns/m^2, and $\varphi D = 10$ mm. The calculations were based on the deduced transfer function

$$\frac{P_L}{P_o} = \frac{1}{C_{11}} \tag{3B.27}$$

The denominator polynomial C_{11} is given by Eq. (3B.25). The results of the frequency response calculations (magnitude only) are plotted in Fig. 3B.5. The same figure carries the experimental results, published by Lallement J, 1976. The study of these results shows that the four-lump model describes with good precision the first dominant mode of the line resonance from the point of view of the resonance frequency and magnitude.

When dealing with fluid power systems, there is an increased need for the simplest line model to reach a reasonably acceptable simulation program of the whole system. The distributed parameter models are more precise. However, these models, in general, might complicate the total system model in an unacceptable way.

It is a common practice to simplify the hydraulic power system model in one of the following ways:

- The transmission lines behavior is assumed negligible, whenever possible.
- The line inertia is sometimes added to that of the neighboring moving parts.
- The line capacitance (the effect of oil compressibility and wall deformation) is added to that of the attached actuator or neighboring valve cavity.

Therefore, the development of system models, incorporating a lumped parameter description of transmission lines, carries the physical structure of the line, without complicating the mathematical description and the simulation programs.

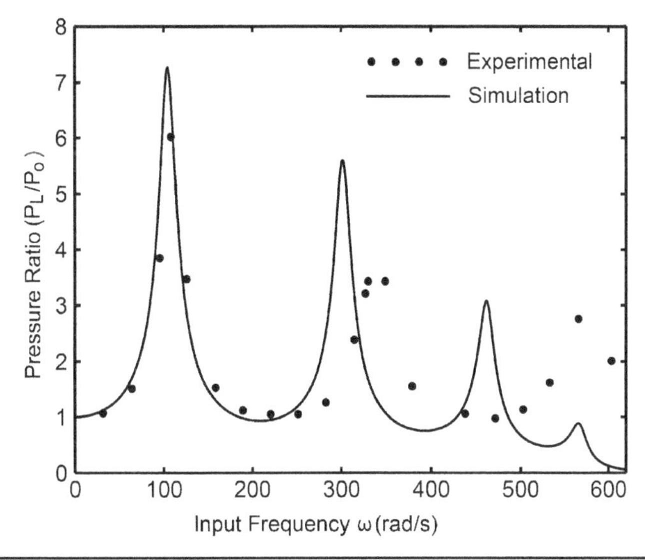

FIGURE 3B.5 Experimental and simulation results of the frequency response of a closed-end hydraulic transmission line.

CHAPTER 4
Hydraulic Pumps

4.1 Introduction

Hydraulic pumps are machines that act to increase the energy of the liquid flowing through them. The three main classes of pumps are displacement, rotodynamic, and special effect pumps. The displacement pumps act to displace the liquid by contracting their oil-filled chambers. In this way, the fluid pressure increases and the fluid is displaced out of the pumping chamber. The rotodynamic pumps increase mainly the kinetic energy of the liquid due to the momentum exchange between the liquid and the rotor. The special effect pumps, such as jet pumps and airlift pumps, operate using different principles.

Rotodynamic pumps derive their name from the fact that a rotating element (rotor) is an essential part of these machines. The mutual dynamic action between the rotor and the working fluid forms their basic principle of operation. The blades, fixed to the rotor, form a series of passages through which a continuous flow of fluid takes place as the rotor rotates. The transfer of energy from the rotor to the fluid occurs by means of rotodynamic action between the rotor and the fluid.

Displacement pumps consist of one or several pumping chambers. These chambers are closed and have nearly perfect sealing. The volume of these chambers changes periodically with the rotation of the pump driving shaft. The fluid is displaced from the suction line to the delivery line by the successive expansion and contraction of the pumping chambers. The displacement pump operation is summarized in the following steps:

1. During its expansion, the pumping chamber is connected to the suction line. The expansion develops an underpressure inside the chamber, forcing the liquid to be sucked in.
2. When the volume of the chamber reaches its maximum value, the chamber is separated from the suction line.
3. During the contraction period, the chamber is linked with the pump delivery line. The fluid is then displaced to the pump exit line and is acted on by the pressure necessary to overcome the exit line resistance.
4. The delivery stroke ends when the volume of the chamber reaches its minimum value. Afterward, the chamber is separated from the delivery line.

This process is repeated continuously as the pump-driving shaft rotates. In addition to the fluid displacement, the pump should act on the fluid by the pressure required to drive the load, or overcome the system resistance.

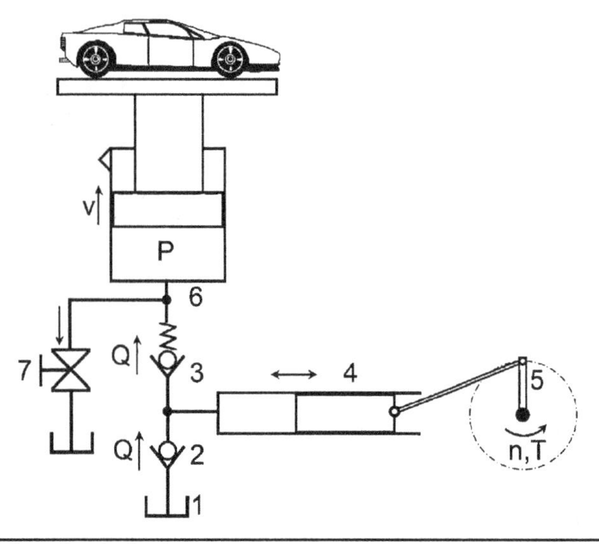

FIGURE 4.1 Operation of a single-piston pump.

The function of the displacement pumps is explained by describing the construction and operation of the single-piston pump, shown in Fig. 4.1. The piston (4), driven by a crank shaft (5), reciprocates between two dead points. During the suction stroke, the piston moves to the right and the oil is sucked from the tank (1) through a check valve (2) of very low cracking pressure. The cracking pressure is the minimum pressure difference needed to open the check valve. Then, during the delivery stroke, the piston moves to the left, displacing the oil to the exit line through the check valve (3). The pump acts on the oil by the pressure, P, needed to drive the load. Therefore, the pump drive should act on the piston by the force needed to produce this pressure, and the crank shaft should be acted on by a torque proportional to this force. The cylinder (6) retracts under the action of the loading force by opening the shut-off valve (7).

4.2 Ideal Pump Analysis

The pump *displacement* is defined as the volume of liquid delivered by the pump per one revolution of its driving shaft, assuming no leakage and neglecting the effect of oil compressibility. It depends on the maximum and minimum values of the pumping chamber volume, the number of pumping chambers, and the number of pumping strokes per one revolution of the driving shaft. This volume depends on the pump geometry; therefore, it is also called the *geometric volume*, V_g. It is given by the following equation:

$$V_g = (V_{max} - V_{min})zi \tag{4.1}$$

where i = Number of pumping strokes per revolution
V_g = Pump displacement (geometric volume), m^3/rev
V_{max} = Maximum chamber volume, m^3
V_{min} = Minimum chamber volume, m^3
z = Number of pumping chambers

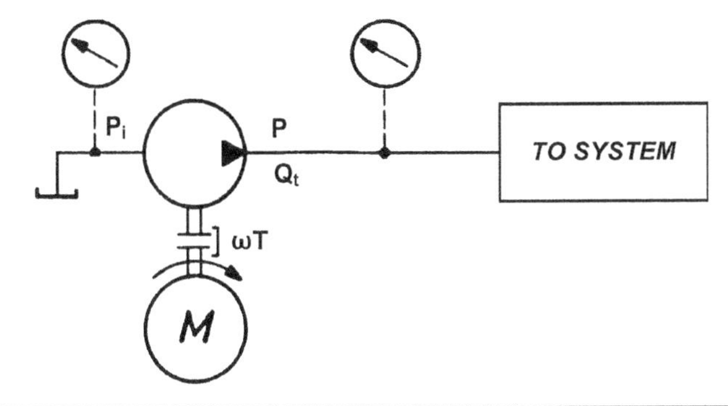

FIGURE 4.2 Typical displacement pump connection in hydraulic power circuits.

Assuming an ideal pump, with no internal leakage, no friction, and no pressure losses, the pump flow rate is given by the following expression:

$$Q_t = V_g n \tag{4.2}$$

where Q_t = Pump theoretical flow rate, m^3/s
n = Pump speed, rev/s

Figure 4.2 shows a typical connection of a displacement pump in the hydraulic power system. Following the assumption of an ideal pump, the input mechanical power is equal to the increase in the fluid power as shown by the following equation:

$$2\pi\, n T_t = Q_t(P - P_i) = V_g n \Delta p \tag{4.3}$$

or

$$T_t = \frac{V_g}{2\pi} \Delta P \tag{4.4}$$

where T_t = Pump theoretical driving torque, Nm
ΔP = Pressure increase due to pump action, Pa

Example 4.1 A gear pump of 12.5 cm^3 geometric volume operated at 1800 rev/min delivers the oil at 16 MPa pressure. Assuming an ideal pump, calculate the pump flow rate, Q_t, the increase in the oil power, ΔN, the hydraulic power at the pump exit line, N_{out}, and the driving torque, T_t, if the inlet pressure is 200 kPa.

$$Q_t = V_g n = 12.5 \times 10^{-6} \times \frac{1800}{60} = 3.75 \times 10^{-4}\ m^3/s = 22.5 \text{ liters/min}$$

$$T_t = \frac{V_g}{2\pi} \Delta P = \frac{12.5 \times 10^{-6}}{2\pi}(16 \times 10^6 - 2 \times 10^5) = 31.4 \text{ Nm}$$

$$\Delta N = Q_t \Delta P = 37.5 \times 10^{-5} \times (16 \times 10^6 - 2 \times 10^5) = 5925 \text{ W}$$

$$N_{out} = Q_t P = 37.5 \times 10^{-5} \times 16 \times 10^6 = 6000 \text{ W}$$

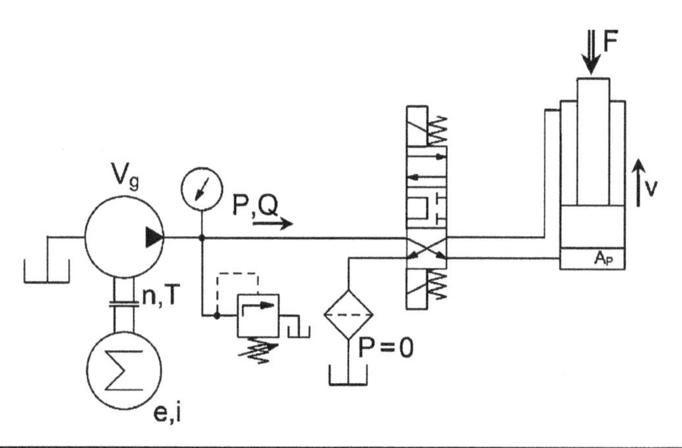

FIGURE 4.3a Operation of a hydraulic system in load lifting mode.

The power transmission and transformation in the hydraulic power systems can be explained through the study of the system shown in Fig. 4.3a. The system is assumed to be an ideal one, with no internal leakage, no friction losses, and no local losses. The prime mover is an electric motor. It converts the electric power into mechanical power. The pump converts the mechanical power ($2\pi nT$) into hydraulic power (PQ). It converts the drive speed into a proportional flow rate ($Q = V_g n$), acting on the liquid by the required pressure. The pressure is determined by the loading conditions. The system is shown in the load lifting operating mode. Assuming an ideal system, the losses in the directional control valve and transmission lines are negligible. When operating at pressure levels less than the cracking pressure of the safety valve, the hydraulic circuit can be redrawn, as shown in Fig. 4.3b. The power transmission and transformation are illustrated in Fig. 4.4. In this figure, the heavy lines with half-arrows indicate the direction of power transmission, while the dot-dash lines indicate the causality relation. The arrows go from the cause to the effect.

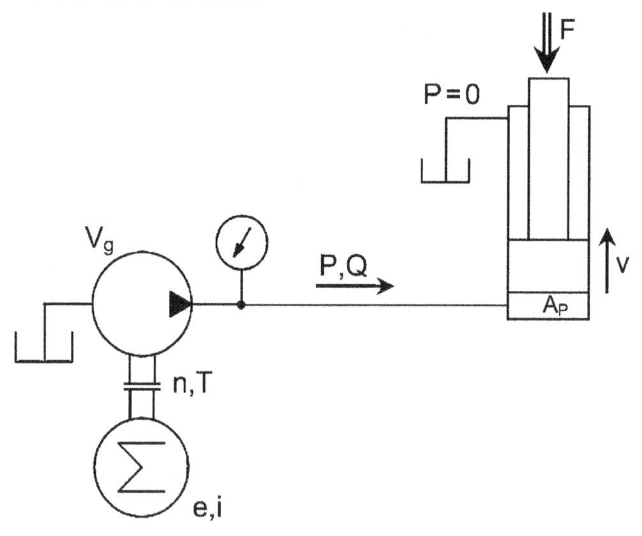

FIGURE 4.3b Load lifting mode, neglecting the losses in the valves and piping.

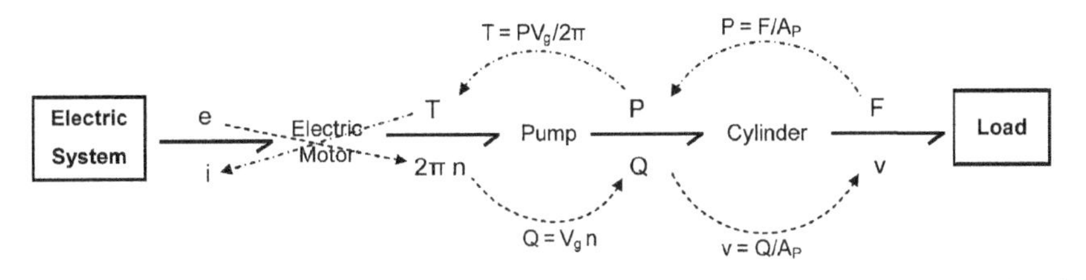

FIGURE 4.4 Power transmission in the hydraulic power system and causality relations.

4.3 Real Pump Analysis

The hydraulic power delivered to the fluid by the real pumps is less than the input mechanical power due to the volumetric, friction, and hydraulic losses. The actual pump flow rate, Q, is less than the theoretical flow, Q_t, mainly due to:

- Internal leakage
- Pump cavitation and aeration
- Fluid compressibility
- Partial filling of the pump due to fluid inertia

The first source of power losses is the internal leakage. Actually, when operating under the correct design conditions, the flow losses are mainly due to internal leakage, Q_L. The leakage flow through the narrow clearances is practically laminar and changes linearly with the pressure difference (see Fig. 4.5). The resistance to internal leakage, R_L, is proportional to oil viscosity, μ, and inversely proportional to the cube of the mean clearance, c. (See Sec. 2.2.1.)

$$Q_L = P/R_L \tag{4.5}$$

$$Q = Q_t - Q_L \tag{4.6}$$

where $R_L = K\mu/c^3$

For high-pressure levels and increased radial clearances, the leakage flow rate increases and the leakage flow becomes turbulent. Figure 4.5 shows the typical variation of pump flow rates with pressure.

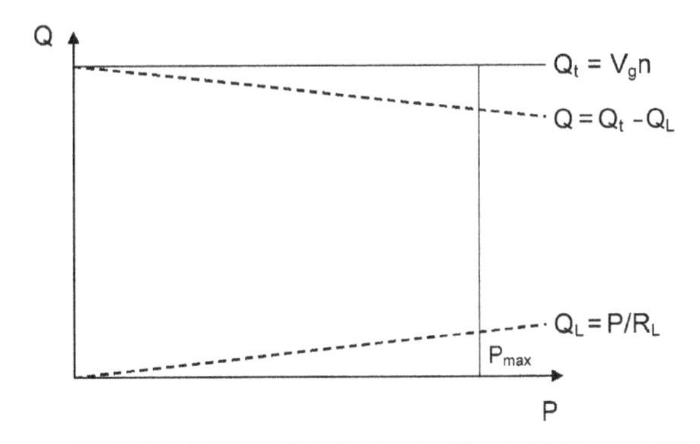

FIGURE 4.5 Pump flow characteristics.

The effect of leakage is expressed by the volumetric efficiency, η_v, defined as follows:

$$\eta_v = \frac{Q}{Q_t} = \frac{Q_t - Q_L}{Q_t} = 1 - \frac{Q_L}{Q_t} = 1 - \frac{P}{R_L V_g n} \tag{4.7}$$

The volumetric efficiency of displacement (geometric) pumps ranges from 0.8 to 0.99. Piston pumps are of high volumetric efficiency, while vane and gear pumps are, in general, of lower volumetric efficiency.

The friction is the second source of power losses. The viscous friction and the mechanical friction between the pump elements dissipate energy. A part of the driving torque is consumed to overcome the friction forces. This part is the friction torque, T_F. It depends on the pump speed, delivery pressure, and oil viscosity. Therefore, to build the required pressure, a higher torque should be applied. The friction losses in the pump are evaluated by the mechanical efficiency, η_m, defined as follows:

$$\eta_m = \frac{\omega(T - T_F)}{\omega T} = \frac{T - T_F}{T} \tag{4.8}$$

where T = Actual pump driving torque, Nm
T_F = Friction torque, Nm
$T - T_F$ = Torque converted to pressure, Nm
ω = Pump speed, rad/s

The third source of power losses in the pump is the pressure losses in the pump's inner passages. The pressure, built inside the pumping chamber, P_C, is greater than the pump exit pressure, P. These losses are caused mainly by the local losses. The hydraulic losses are of negligible value for pumps running at speeds less than 50 rev/s, and mean oil speeds less than 5 m/s. For greater speeds of oil, the pressure losses are proportional to the square of the flow rate. These pressure losses are evaluated by the hydraulic efficiency, η_h.

$$\eta_h = \frac{QP}{QP_C} = \frac{P}{P_C} \tag{4.9}$$

where P_C = Pressure inside the pumping chamber, Pa
P = Pump exit pressure, Pa

An expression for the total pump efficiency, η_T, is deduced as follows:

$$\eta_T = \frac{QP}{\omega T} = \frac{Q}{Q_t}\frac{T - T_F}{T}\frac{P}{P_C}\frac{Q_t P_C}{\omega(T - T_F)} = \eta_v\,\eta_m\,\eta_h\,\frac{Q_t P_C}{\omega(T - T_F)} \tag{4.10}$$

The mechanical power $\omega(T - T_F)$ is converted into equal hydraulic power, $Q_t P_C$, then

$$\eta_T = \eta_v\,\eta_m\,\eta_h \tag{4.11}$$

In the steady-state operation, the real displacement pump is described by the following relations:

$$Q = V_g n \eta_v \tag{4.12}$$

$$N_h = N_m \eta_T \quad \text{or} \quad Q\Delta P = 2\pi n T \eta_T \tag{4.13}$$

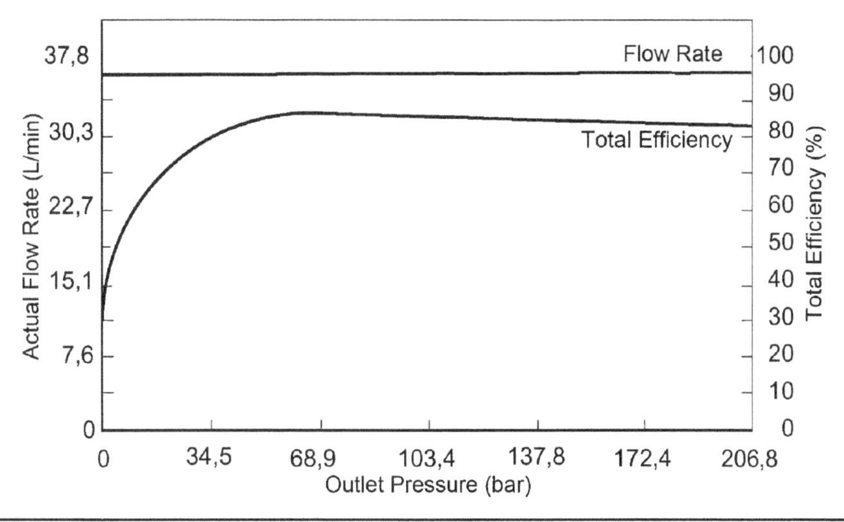

FIGURE 4.6 Typical flow and efficiency characteristics of an axial piston pump.

Then

$$T = \frac{V_g}{2\pi\eta_m\eta_h}\Delta P \tag{4.14}$$

where N_h = Hydraulic power, W
N_m = Mechanical power, W
ΔP = Difference between the pump output and input pressures, $\Delta P = P - P_i$, Pa

If the pump input pressure, P_i, is too small compared with the delivery pressure, P, then it may be neglected, and the pressure difference, ΔP, equals the pump exit pressure, P. If so, then

$$T = \frac{V_g}{2\pi\eta_m\eta_h}P \tag{4.15}$$

Figure 4.6 shows the typical characteristics of an axial piston pump.

4.4 Cavitation in Displacement Pumps

The cavitation characteristics of a pump describe the effect of input pressure on the pump flow rate. The reduction of the pump inlet pressure to values less than the vapor pressure leads to the evaporation or boiling of oil. The fluid flow to the pump inlet becomes a mixture of liquid, liberated gases, and vapors. At zero or very low exit pressure, when the pump is bypassed for example, the vapors do not condensate and the vapor cavities do not collapse. But during normal operating conditions, the pump is loaded by great load pressures. The vapor cavities collapse due to the rapid condensation of vapors when transmitted to the high-pressure zone. Therefore, the net flow rate of the pump decreases. Generally, a 1% increase in the vapor volume in the oil-vapor flow reduces the pump volumetric efficiency by about 1%. Figure 4.7 shows the typical effect of pump inlet pressure on the flow rate at constant exit pressure, for different pump speeds.

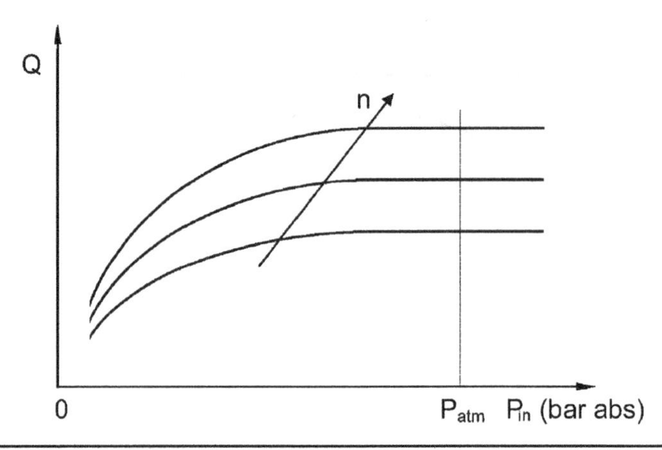

FIGURE 4.7 Typical cavitation characteristics of displacement pumps.

In addition to the reduction of the volumetric efficiency, the pump elements are subjected to great impact pressures resulting from the fluid rushing to fill the space of collapsed vapor cavities. The impact pressure reaches very high values, up to 7000 bar. When subjected to cavitation, the pump noise level increases and a very loud sharp noise is heard. The surfaces of the inner pump elements are damaged due to the pitting resulting from the impact pressure forces. Therefore, the pump inlet pressure should be higher than the saturated vapor pressure of oil at the maximum operating temperature by a convenient value. This value is called cavitation reserve and ranges from 0.3 to 0.4 bar (see Fig. 4.8).

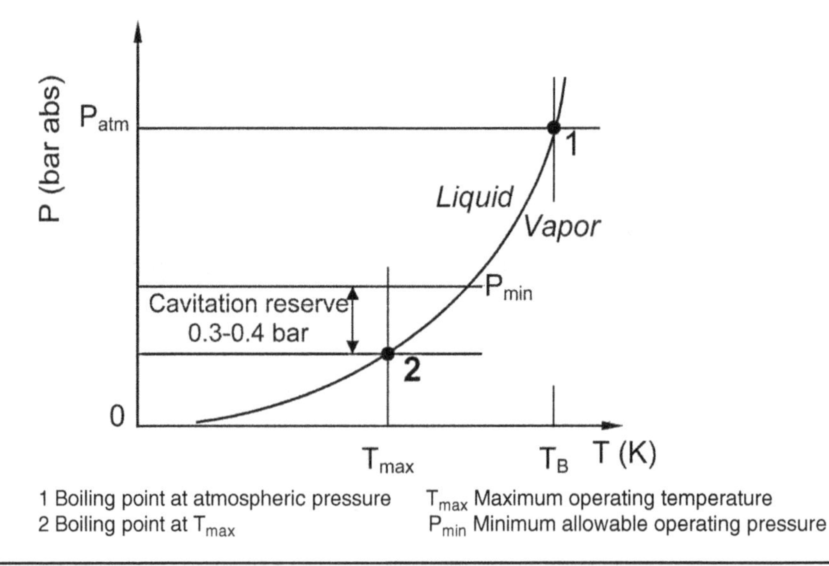

FIGURE 4.8 Determination of the minimum allowable operating pressure.

This undesirable phenomenon can be avoided by taking the following actions, whenever possible:

- Reduce the pressure losses in the pump inlet line by increasing the suction line diameter and decreasing its length.
- Avoid using the inlet line filter and other local-loss elements.
- Increase the pump suction pressure by doing one of the following:
  - Use a booster pump.
  - Use a closed pressurized tank.
  - Mount the pump below the tank by a convenient distance.

4.5 Pulsation of Flow of Displacement Pumps

Theoretically, the pump flow rate is calculated as $Q = V_g n$. This expression gives the average or mean pump flow rate. Actually, the pump flow rate is not constant. Each pumping chamber delivers a flow rate that equals the rate of reduction of its volume. The net pump flow rate at a certain instant is the summation of the flow rates delivered by the chambers connected to the delivery port at that instant.

The flow rate delivered by the pumping chamber starts at a zero value at the beginning of the delivery stroke. It increases progressively until it reaches its maximum value at the midpoint of the delivery stroke. Then, it decreases progressively until it becomes null at the end of the delivery stroke. Therefore, the net pump flow is pulsating, as illustrated in Fig. 4.9. The magnitude of flow pulsation is evaluated by the pulsation coefficient and is defined as

$$\sigma_Q = \frac{Q_{max} - Q_{min}}{Q_m} \times 100\% \tag{4.16}$$

where σ_Q = Flow pulsation coefficient
Q_{min} = Minimum value of pump flow rate, m^3/s
Q_{max} = Maximum value of pump flow rate, m^3/s
$Q_m = V_g n$ = mean flow rate, m^3/s

The pulsation of flow results in pressure oscillation and the non-uniform motion of hydraulic cylinders and motors.

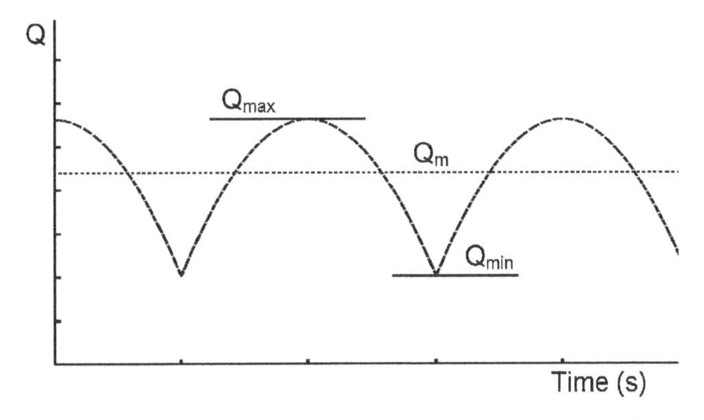

Figure 4.9 The pulsation of flow in displacement pumps.

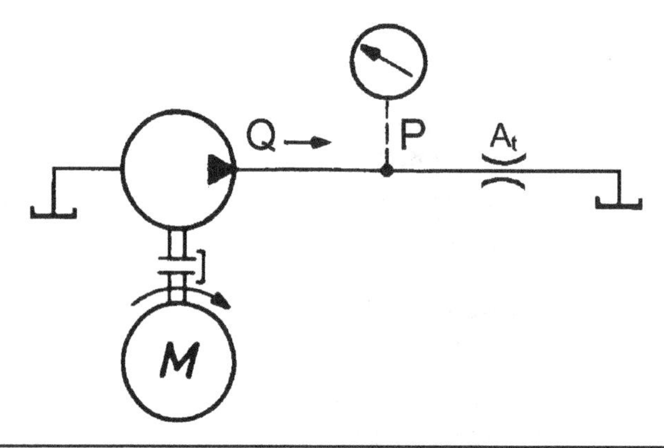

FIGURE 4.10 Pump loaded by a fixed area throttle.

Considering the case of a throttled pump exit line (see Fig. 4.10) and neglecting the fluid compressibility, the pressure at the pump exit is given by

$$P = \frac{\rho}{2C_d^2 A_t^2} Q^2 \tag{4.17}$$

$$P_m = \frac{\rho}{2C_d^2 A_t^2} Q_m^2 \tag{4.18}$$

$$P_{\max} = \frac{\rho}{2C_d^2 A_t^2} Q_{\max}^2 \tag{4.19}$$

$$P_{\min} = \frac{\rho}{2C_d^2 A_t^2} Q_{\min}^2 \tag{4.20}$$

$$\sigma_P = \frac{P_{\max} - P_{\min}}{P_m} \times 100\% \tag{4.21}$$

or

$$\sigma_P = \frac{Q_{\max}^2 - Q_{\min}^2}{Q_m^2} \times 100\% \tag{4.22}$$

where σ_p = Pressure pulsation coefficient
$P_{\min}$ = Minimum value of pump exit pressure, Pa
$P_{\max}$ = Maximum value of pump exit pressure, Pa
P_m = Mean exit pressure, Pa

If the flow rate oscillates between $0.9Q_m$ and $1.04Q_m$, then σ_Q = 14% and σ_p = 27.16%. Actually, considering the effect of oil compressibility, the pressure oscillation decreases especially for the increased volume of the exit line.

4.6 Classification of Pumps

Figure 4.11 shows the classification of the hydraulic pumps, focusing on the most commonly used displacement pumps. The following sections deal with their construction, operation, and special features.

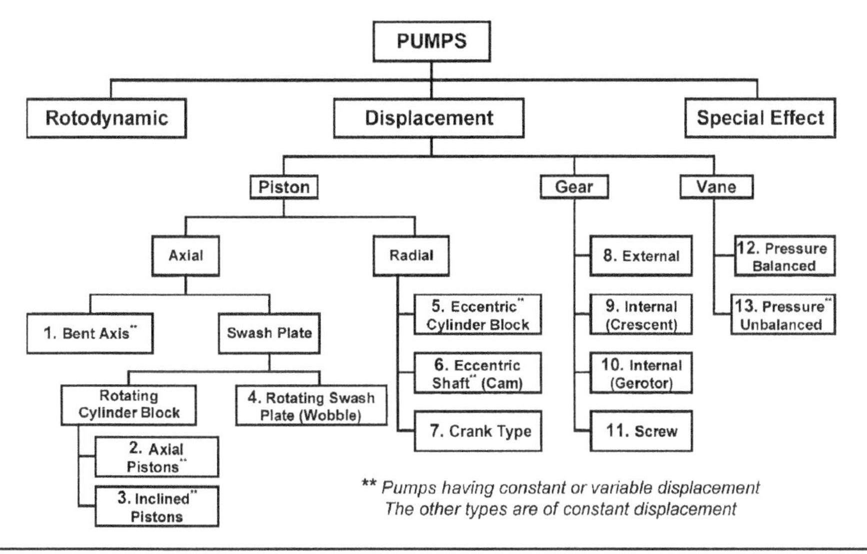

Figure 4.11 Classifications of hydraulic pumps.

4.6.1 Bent Axis Axial Piston Pumps

Construction and Operation

Figure 4.12 shows a typical construction of the bent axis axial piston pump. The pump consists of a drive shaft (1), cylinder block (3), pistons (4), and a port plate (5). The spherical ends of the pistons are attached to the disk (2), coupled to the driving shaft. As the drive shaft is rotated, the cylinder block also rotates. The cylinder block slides on the port plate, which includes two kidney-shaped control openings (see Fig. 4.13).

The driving shaft rotates around a horizontal axis while the axis of rotation of the cylinder block is inclined by an angle, α. The cylinder block inclination forces the pistons, which rotate with the cylinder block, to reciprocate with respect to this cylinder block. Therefore, each of the pistons performs a reciprocating motion between its upper and lower dead points. The piston movement from the lower dead point to the upper dead point produces a suction stroke. The fluid is sucked via the control opening on the suction side of the port plate into the cylinder block bore. As the drive shaft is further rotated and the piston moves from the upper dead point to the lower dead point, the fluid is displaced out through the other control opening (pressure side). During the delivery stroke, the driving shaft acts on the disk by the torque needed to produce the forces that drive the pistons against the load pressure.

The pump geometric volume is given by the following expression:

$$h = D \sin \alpha, \qquad V_g = zAh \tag{4.23}$$

or

$$V_g = \frac{\pi}{4} d^2 D z \sin \alpha \tag{4.24}$$

where A = Piston area, m^2
D = Pitch circle diameter, m
d = Piston diameter, m

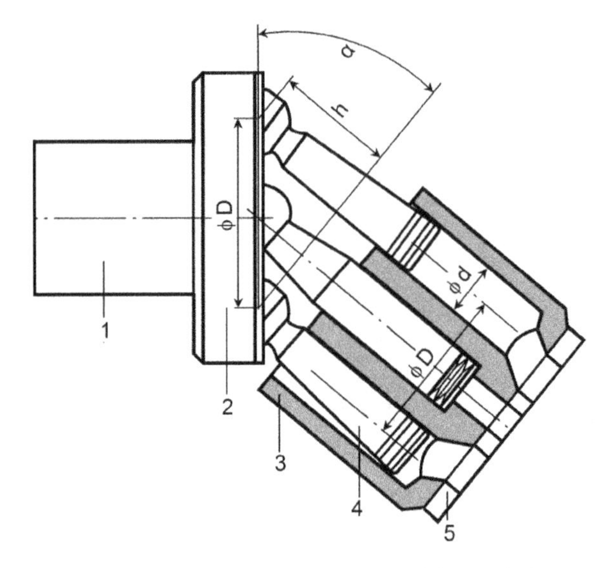

FIGURE 4.12 Illustration of a bent axis axial piston pump of conical pistons.

h = Piston stroke, m
z = Number of pistons
α = Inclination angle, rad

Pulsation of Flow of Axial Piston Pumps

In the case of axial piston pumps, the pistons perform simple harmonic motions, following the sinusoidal law. The flow rate delivered by each piston equals its speed multiplied by the piston area. Neglecting the effects of internal leakage, fluid inertia, and compressibility, the resulting flow rate from each piston is also sinusoidal. Figure 4.14

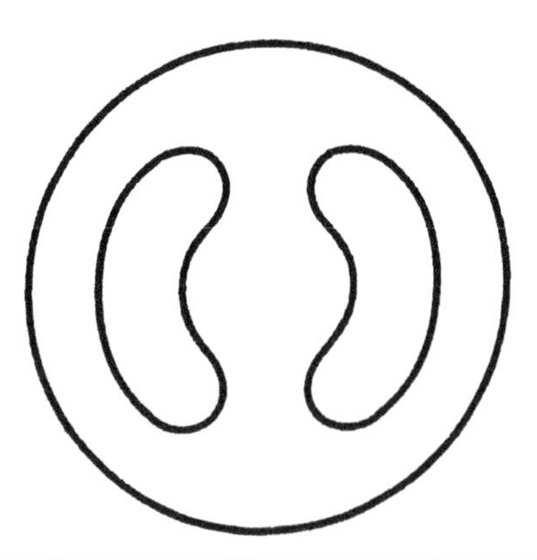

FIGURE 4.13 Layout of the port plate.

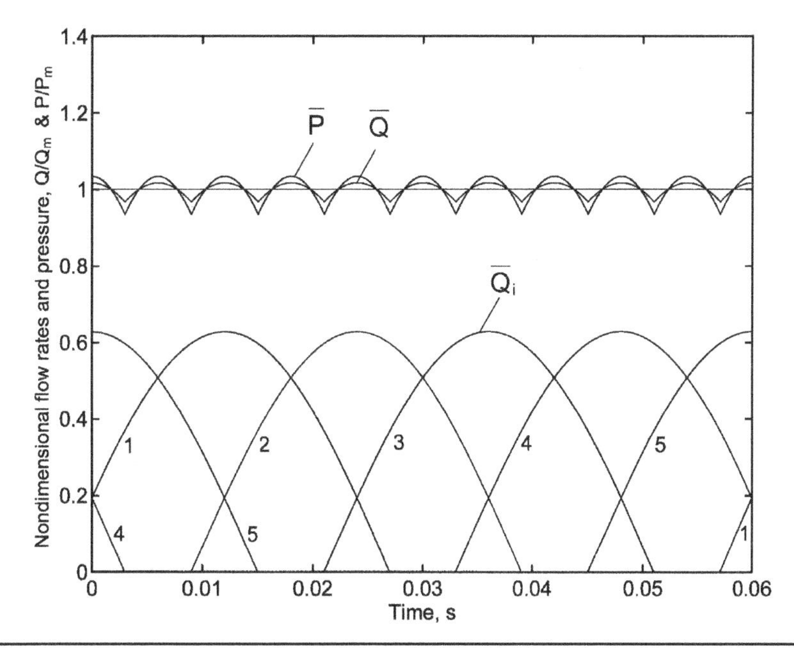

FIGURE 4.14 Flow rate and pressure pulsation of a five-piston axial piston pump, calculated neglecting the effects of internal leakage, fluid inertia, and compressibility.

shows the flow rate delivered by the individual pistons of a five-piston pump. The pump delivery is the sum of the flow rates delivered by all of the pistons in connection with the delivery port. There are either two or three pistons delivering at the same time and the total pump flow rate is shown in Fig. 4.14. The pump is loaded by a throttle valve (see Fig. 4.10 earlier). The pump exit pressure is calculated and also plotted (see Fig. 4.14). The pressures and flow rates are plotted in nondimensional form relative to the mean values P_m and Q_m; $\bar{Q} = Q/Q_m$, $\bar{P} = P/P_m$, $\bar{Q}_i = Q_i/Q_m$.

The flow rate oscillations were calculated for axial piston pumps of a different number of pistons. The calculation results are plotted in Fig. 4.15. The associated pressure oscillations were also calculated neglecting the oil compressibility and assuming a constant-area-loading sharp-edged orifice. The flow and pressure pulsation coefficients were obtained from the simulation results and shown in the table in Fig. 4.15. The results show that the flow pulsation is minimized in the case of pumps having a greater odd number of pistons. Therefore, axial piston pumps with pistons moving in simple harmonic motion, are of odd number of pistons.

4.6.2 Swash Plate Pumps with Axial Pistons

Figure 4.16 shows the construction and operation of a swash plate pump. The drive shaft (1) rotates and drives the cylinder block (5). Both the driving shaft and the cylinder block have the same axis of rotation. The cylinder block (5) and its pistons (6), rotate with the drive shaft. Each of the pistons is attached to a slipper pad (3). The pistons and their slipper pads are inserted in the holes of the retaining plate (4). Therefore, the retaining plate rotates with the pistons and the cylinder block. It is guided to rotate in a plane parallel to the swash plate (2) by a fixed guide (8). The trajectory of the slipper

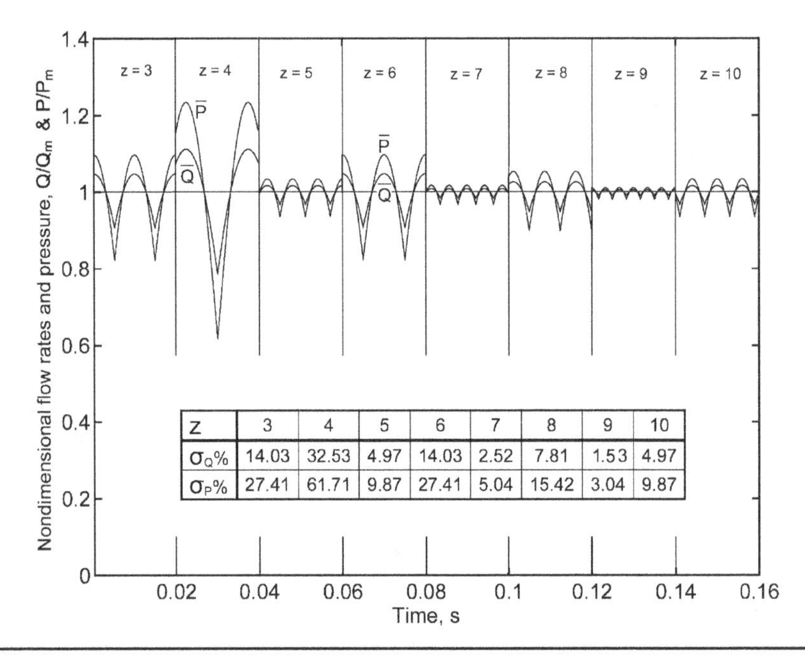

z	3	4	5	6	7	8	9	10
σ_Q%	14.03	32.53	4.97	14.03	2.52	7.81	1.53	4.97
σ_P%	27.41	61.71	9.87	27.41	5.04	15.42	3.04	9.87

FIGURE 4.15 Effect of the number of pistons on the flow and pressure pulsation, calculated neglecting the effects of internal leakage, fluid inertia, and compressibility.

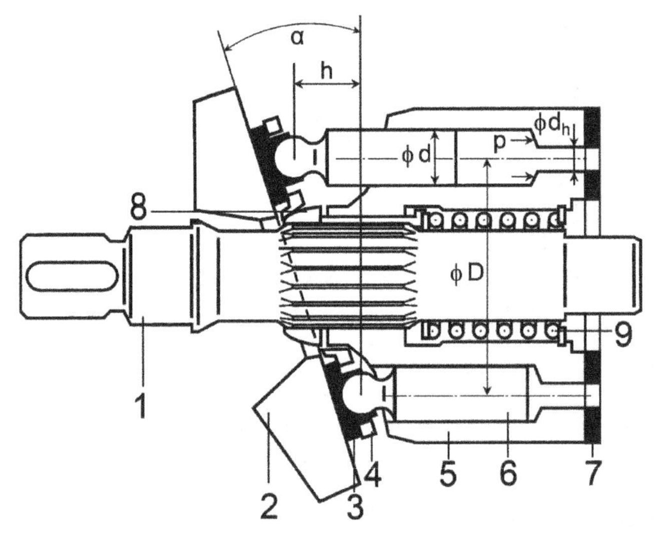

1. Drive shaft, 2. Swash plate, 3. Slipper pad, 4. Retaining plate,
5. Cylinder block, 6. Piston, 7. Port plate, 8. Fixed guide of the retaining plate,
9. Cylinder block loading spring

FIGURE 4.16 Illustration of a swash plate axial piston pump.

pad is determined by the swash plate and the retaining plate. During rotation, each piston performs a reciprocating motion. During this process, a volume of fluid, corresponding to the piston area and stroke, is sucked or delivered via both control openings in the port plate (7).

The cylinder block is pushed against the port plate by means of a spring (9), which minimizes the leakage through the clearance separating them at the beginning of the pump operation. When the pressure builds up, it acts on the cylinder block with a tightening force given by $\{0.25\pi\,(d^2 - d_h^2)P\}$. This force acts to the right, against the repulsion force due to the pressure distribution in the clearance between the cylinder block and the port plate. The resultant force acts to reduce this clearance and minimize the leakage through it. The pump geometric volume is given by the expression

$$V_g = \frac{\pi}{4} d^2 Dz \tan\alpha \tag{4.25}$$

where α = Swash plate inclination angle, rad.

4.6.3 Swash Plate Pumps with Inclined Pistons

The swash plate pump with inclined pistons, also called a semi-axial piston pump, is produced with cylinder holes inclined to its axis (see Fig. 4.17). In this case, the pistons reciprocate in a direction inclined to the axis of rotation by an angle, φ. The swash plate also inclines by an angle, α. This design increases the piston stroke and the pump geometric volume. In addition, the centrifugal force acting on the pistons has a component in the direction of the pistons' axis, F_x, which assists the suction stroke. The following expressions for the piston stroke and the pump geometric volume are systematically deduced.

$$L_1 = \frac{D\sin(\alpha)}{2\cos(\varphi - \alpha)} \quad \text{and} \quad L_2 = \frac{D\sin(\alpha)}{2\cos(\varphi + \alpha)} \tag{4.26}$$

$$V_g = \frac{\pi}{8} d^2 Dz \sin(\alpha)\left\{\frac{1}{\cos(\varphi - \alpha)} + \frac{1}{\cos(\varphi + \alpha)}\right\} \tag{4.27}$$

4.6.4 Axial Piston Pumps with Rotating Swash Plate-Wobble Plate

In axial piston pumps with rotating swash plates (see Fig. 4.18), the swash plate is rotated by the driving shaft. The cylinder block is fixed. The pistons are displaced inwards by the rotating swash plate, while the piston displacement in the opposite direction is insured by a spring. During the pump operation, the pumping chamber is connected with the inlet and exit lines through the check valves. The inlet line check valves should be of low cracking pressure to avoid pump cavitation. The pump displacement is calculated by Eq. (4.25).

4.6.5 Radial Piston Pumps with Eccentric Cam Ring

Radial piston pumps are usually used in applications requiring high pressures, above 400 bar. In presses, for example, operating pressures are required to be up to 700 bar. The radial piston pumps can operate reliably at such high pressures. The construction and operation of a radial piston pump with an eccentric cam ring are illustrated by

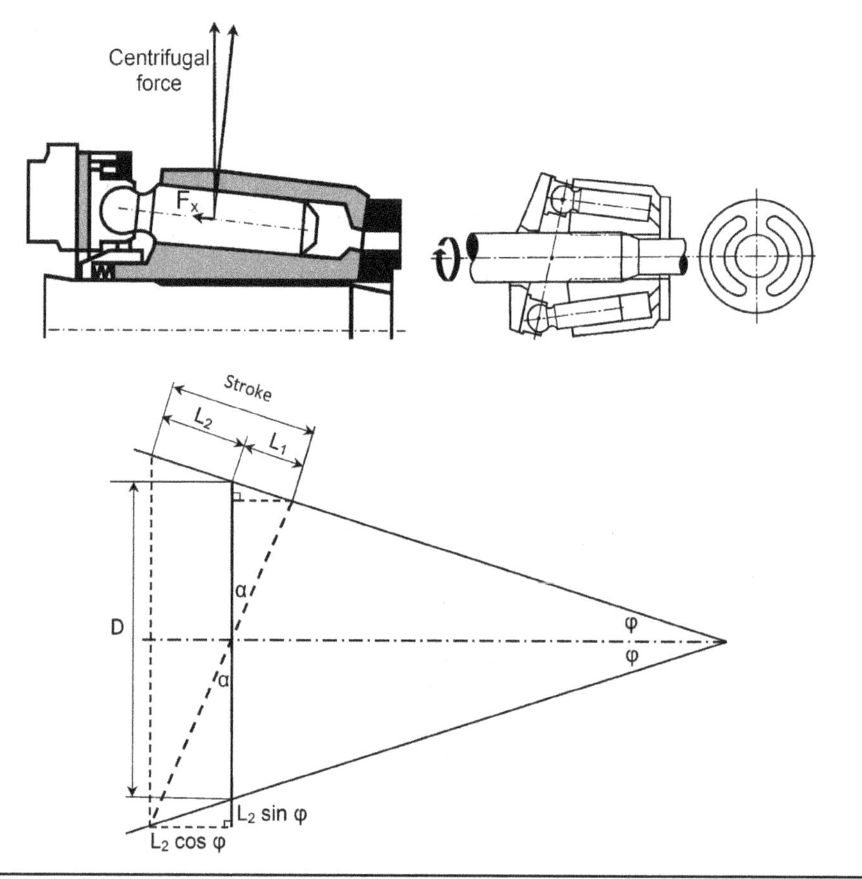

FIGURE 4.17 The inclined pistons swash plate pump.

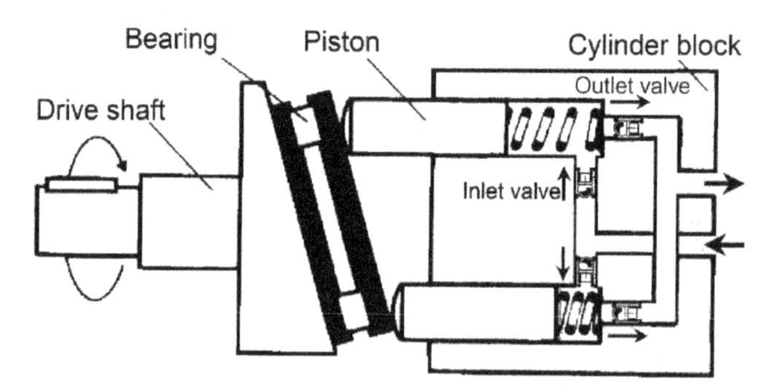

FIGURE 4.18 Typical design of an axial piston pump with a rotating swash plate (wobble plate).

Fig. 4.19. This example shows a variable displacement pump, where the pistons are arranged radially in the cylinder block. They are held in contact with the inner surface of a cam ring by means of a retainer ring and slipper pads. The pistons and slipper pads are connected to each other by means of ball-and-socket joints. The stroke of pistons

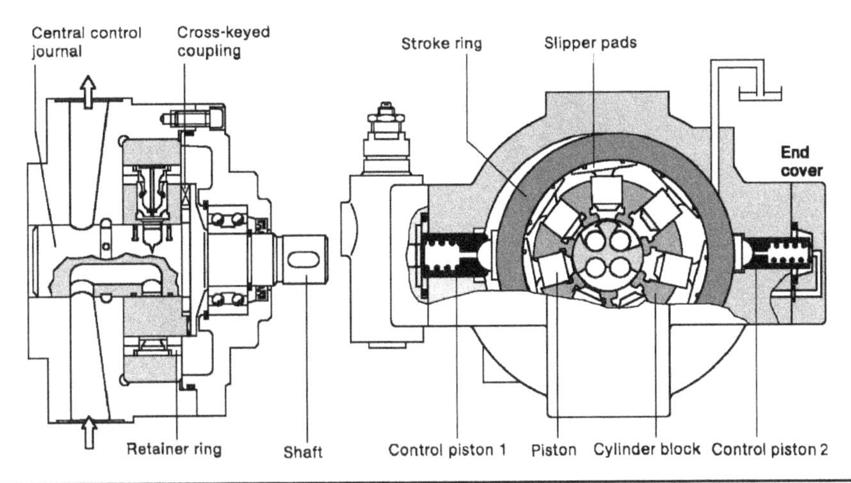

Figure 4.19 Radial piston pump with an eccentric cam ring. (*Gotz, 1984, courtesy of Bosch Rexroth AG.*)

and consequently the pump geometric volume are controlled by adjusting the eccentricity of the cam ring (stroke ring) by means of two control pistons. The cylinder block is rotated by the driving shaft. During the suction stroke, the pistons' motion is governed by the retainer ring, while during the delivery stroke the pistons are displaced by the cam ring. The piston stroke, h, is twice the eccentricity, e. The pump geometric volume is given by

$$V_g = \frac{\pi}{2} d^2 e z \tag{4.28}$$

where e = Eccentricity, m.

4.6.6 Radial Piston Pumps with Eccentric Shafts

Figure 4.20 shows the construction of a radial piston pump with an eccentric shaft (cam). The pistons reciprocate in the radial direction under the action of the eccentric cam. The cam (11) is eccentric to the pump driving shaft (2). The pump consists of the pistons (6), cylinder sleeve (7), pivot (9), compression spring (8), suction valve (4), and exit valve (5). The pivot is screwed into the housing (1). The piston is positioned with the slipper pad (6) on the cam. The compression spring causes the slipper pad to lie on the cam, and the cylinder sleeve is supported by the pivot. The pumping process in this class of pumps takes place in the following four phases:

Phase 1: The piston is at the upper dead point and the volume of the pumping chamber (10) is minimum. The suction valve (4) and exit valve (5) are closed.

Phase 2: As the shaft rotates, the piston moves towards the axis of the cam. The volume of the pumping chamber increases and the suction valve opens due to the underpressure produced. The fluid flows via a groove in the cam surface to the bore of the piston into the pumping chamber.

Phase 3: The piston is at the lower dead point. The pumping chamber is completely filled (maximum volume). The suction valve and exit check valve are closed.

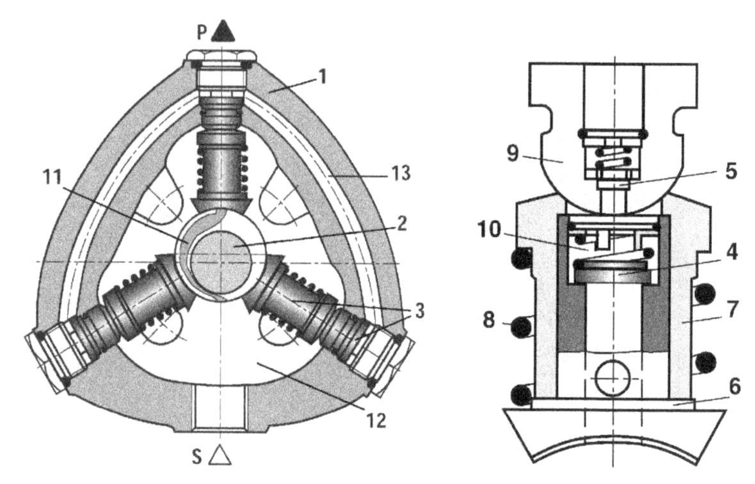

1. Housing, 2. Driving shaft, 3. Piston assembly, 4. Suction valve, 5. Exit valve, 6. Piston, 7. Sleeve, 8. Spring, 9. Pivot, 10. Pumping chamber, 11. Cam, 12. Case inner cavity, 13. Ring channel.

FIGURE 4.20 Radial piston pump with eccentric shaft (cam). (*Courtesy of Bosch Rexroth AG.*)

Phase 4: As the cam rotates, the piston is moved outwards in the radial direction. The fluid is compressed in the displacement chamber. The increased fluid pressure opens the exit check valve, and the fluid flows into the ring channel (13), which connects the pumping elements.

4.6.7 Radial Piston Pumps of Crank Type

The typical construction of this class of pumps is illustrated by Fig. 4.21. The pump consists of a fixed housing incorporating the pistons and crank shaft assembly.

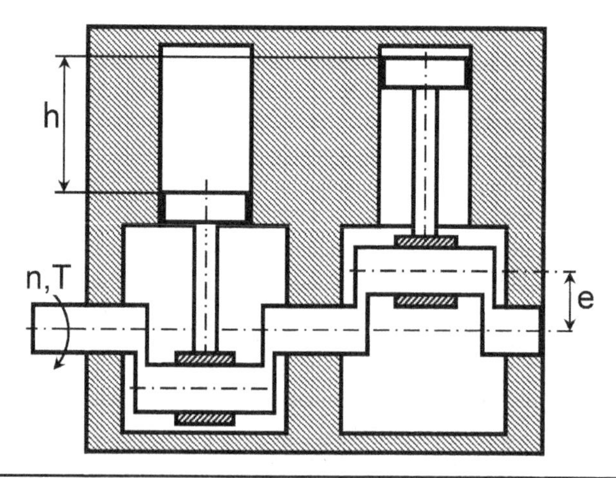

FIGURE 4.21 A radial piston pump of crank type.

The pistons are driven by means of a crank shaft. The pumping chambers are connected to the suction and delivery ports through two check valves (not illustrated). The pump displacement (geometric volume) is given by the following expression:

$$V_g = \frac{\pi}{4} d^2 h z = \frac{\pi}{2} d^2 e z \tag{4.29}$$

where h = Piston stroke = $2e$, m.

4.6.8 External Gear Pumps

Construction and Operation

Gear pumps are of the multirotor displacement type. The four main types of gear pumps are external gear pumps, internal gear pumps, screw pumps, and gerotors. External gear pumps (see Fig. 4.22) consist of: the housing (1), mounting flange (2), drive shaft (3), two side plates (4), bearing bush (5), two gears (9 and 10), and disc (6). The driving gear (9) is connected to the driving shaft (3). The pumping chamber is formed by the surfaces of two adjacent teeth, the inner surface of the housing, and the two side plates. During the rotational movement of the gears, the un-meshing gears release the pumping chambers. The resulting underpressure, together with the pressure in the suction line, forces the fluid to flow to the pump inlet port (7). This fluid fills the pumping chambers, and then is moved with the rotating gear from the suction side to the pressure side. Here, the gears mesh once more and displace the fluid out of the pumping chambers and prevent its return to the suction zone.

In the case of an external gear pump with two spur gears, the pump geometric volume is given by the following relation:

$$V_g = 2\pi b m^2 (z + \sin^2 \gamma) \tag{4.30}$$

where b = Tooth length, m
m = Module of tooth, m
z = Number of teeth per gear
γ = Pressure angle of tooth, rad

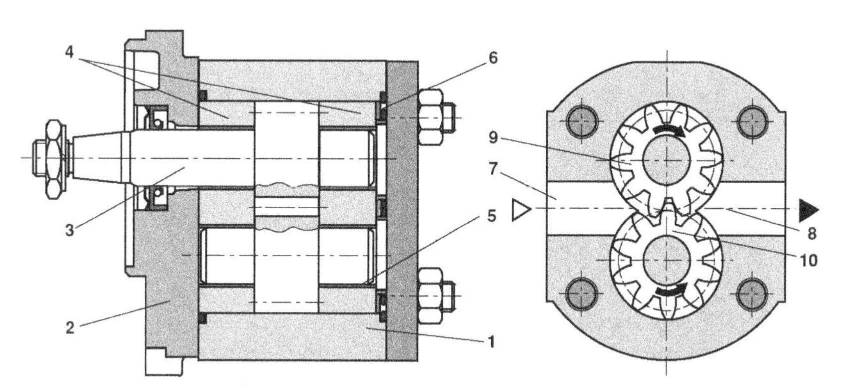

1. Housing, 2. Mounting flange, 3. Drive shaft, 4. Two bearing blocks, side plates, 5. Bearing bush, 6. Discs, 7 and 8. Inlet and exit ports, 9. Driving gear, 10. Driven gear

FIGURE 4.22 An external gear pump. (*Courtesy of Bosch Rexroth AG.*)

Internal Leakage in External Gear Pumps

The internal leakage in gear pumps takes place through two main paths:

- Over the tip of the tooth, tip clearance leakage
- Between the sides of the gears and the side plates, side clearance leakage.

The tip clearance leakage is affected by the tip clearance, the number of teeth, and the pump exit pressure. An increase in the number of teeth increases both the local losses and resistance to internal leakage. The excessive wear of the pump casing increases the tip clearance and consequently the tip clearance leakage.

The side clearance leakage takes place through the clearance between the gears' sides and the side plates. In the case of pumps operating at low pressure levels, this leakage is not so high; therefore, the side plates are fixed. Then, the wear on the side plates increases this leakage. However, the pumps operating at high pressures present higher leakage through this path. Therefore, they should include an arrangement for the hydrostatic compensation of the side clearance. The side plates are pushed towards the gears under the action of a pressure force (see Fig. 4.22). The pump exit pressure is communicated to act on a part of the side plate's area (6). This area is well calculated to generate the force necessary to produce the required tightness without too much increase in the friction torque. In this way, the side clearance is automatically adjusted according to the system pressure. At low-pressure levels, the leakage is reduced, and a smaller tightening force acts on the wear plates. In addition, the wear of the side plate has no significant effect on the side clearance leakage since it is constantly pressed against the gear side.

The Pulsation of Flow in Gear Pumps

The flow at the pump exit is pulsating due to the variable rate of delivery from the pump chambers. The following relation gives the frequency of pulsation:

$$f = 2zn \tag{4.31}$$

where f = Flow pulsation frequency, Hz
n = Pump speed, rev/s
z = Number of teeth per gear

For the gear pump, the pulsation coefficient is calculated by the following expression:

$$\sigma = \frac{\pi^2 \cos^2 \gamma}{4(z+1)} \times 100\% \tag{4.32}$$

A gear pump with ten teeth per gear and a pressure angle $\gamma = 25°$ has a flow pulsation coefficient $\sigma = \pi^2 \cos^2 25/\{4(10+1)\} = 0.184 = 18.4\%$.

Oil Trapping and Squeezing in Gear Pumps

During the normal operation of the pump, as the tooth comes to the meshing point, a volume of oil becomes trapped in the space between two successive teeth. The oil trapping takes place where the gears come in contact at two contact lines simultaneously. The further rotation of gears reduces the volume of the trapped oil and its pressure increases to very great values (see Fig. 4.23). A 1% reduction of the oil volume results in

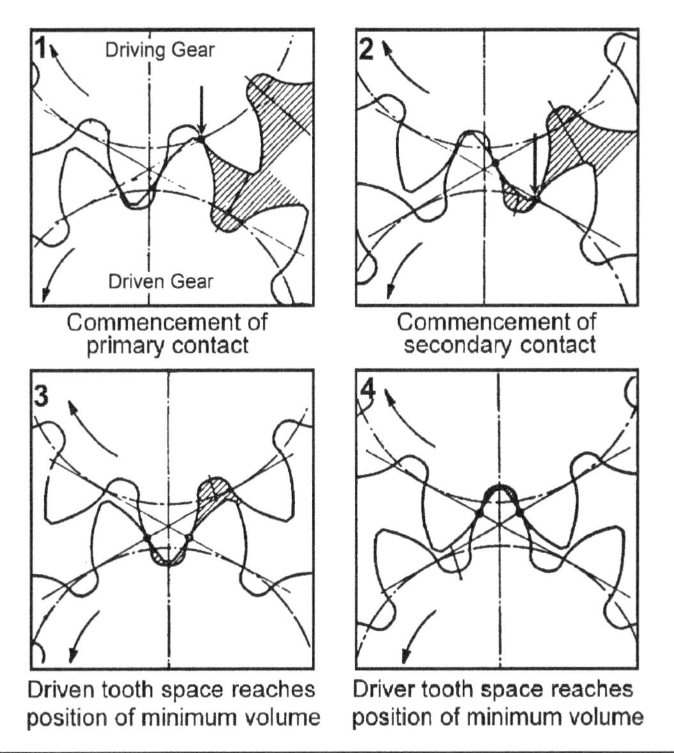

FIGURE 4.23 Steps of oil trapping in gear pumps having gears without backlash.

a pressure rise of 100 to 200 bar. The excessive pressure rise of the trapped oil can be avoided by using one of the following techniques:

- By cutting grooves in the side plates to communicate the inter-teeth space with the pressure side
- By designing gears with a small number of teeth running with a definite backlash of 0.4 to 0.5 mm
- By using a helical gear train

Limitations of Gear Pump Speeds

In gear pumps, the oil enters the pumping chambers along the gear circumference. On entering the pump, the fluid starts to rotate with the gears and is subjected to centrifugal forces. These forces tend to push it away and out of the pumping chamber. Therefore, the maximum pump speed should be limited and the inlet pressure should be high enough to avoid this phenomenon. An expression for the maximum speed is deduced in the following:

Considering the pressure and centrifugal forces acting on an element of fluid (see Fig. 4.24), the following relations are deduced neglecting the term ($drd\xi$) compared with ($rd\xi$):

$$(P + dP)brd\xi = Pbrd\xi + F_r \quad \text{or} \quad F_r = brd\xi dP \tag{4.33}$$

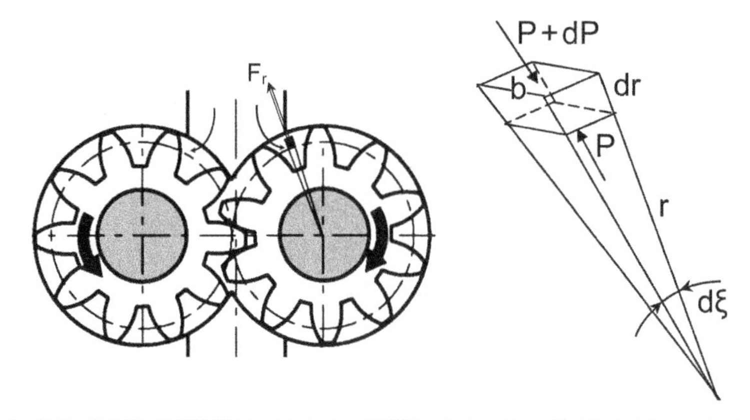

FIGURE 4.24 Centrifugal force acting on a fluid element.

The centrifugal force F_r is given by

$$F_r = mr\omega^2 = \rho r\, d\xi\, b\, dr\, r\, \omega^2 \tag{4.34}$$

Then,

$$dP = \rho r\omega^2 dr \tag{4.35}$$

$$\int_0^{Pc} dP = \int_0^r \rho\omega^2 r\, dr \tag{4.36}$$

or

$$P_C = \rho r^2 \frac{\omega^2}{2} \tag{4.37}$$

The pump input pressure P_i should be greater than the centrifugal forces pressure P_C. Therefore, the maximum pump speed should be limited as follows:

$$\omega = 2\pi n \tag{4.38}$$

$$P_i > P_C \tag{4.39}$$

or

$$n_{\max} < \frac{1}{\pi r}\sqrt{P_i / 2\rho} \tag{4.40}$$

where P_i = Pump inlet pressure, Pa
r = Gear addendum radius, m
ρ = Oil density, kg/m³

The bearing lubrication is another factor imposing minimum recommended speed in some pumps. The gears are loaded by the pressure forces, F_p, and gear contact forces, F_C, as shown in Fig. 4.25. The rotors are nonpressure compensated and the pressure forces are unbalanced. These forces are transmitted to the shaft bearing. Therefore, in the case of sliding bearing lubricated hydrodynamically, a minimum pump speed, $n_{\min}$, is recommended to insure the required bearing lubrication.

4.6.9 Internal Gear Pumps

Figure 4.26 shows the construction of an internal gear pump of the leak gap compensated type. The shown pump consists of: a housing (1), bearing cover (1.1), blanking

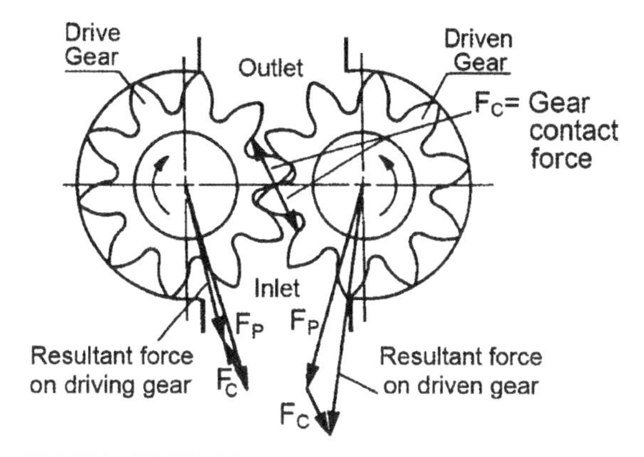

FIGURE 4.25 Pressure and gear contact forces acting on the gears.

plate (1.2), internal gear (2), driving shaft with external gear (pinion) (3), shaft bearing (4), side, wear plates (5), and a stop pin (6), as well as the segment assembly (7), which is comprised of the segment (7.1), segment carrier (7.2), and the sealing rolls (7.3).

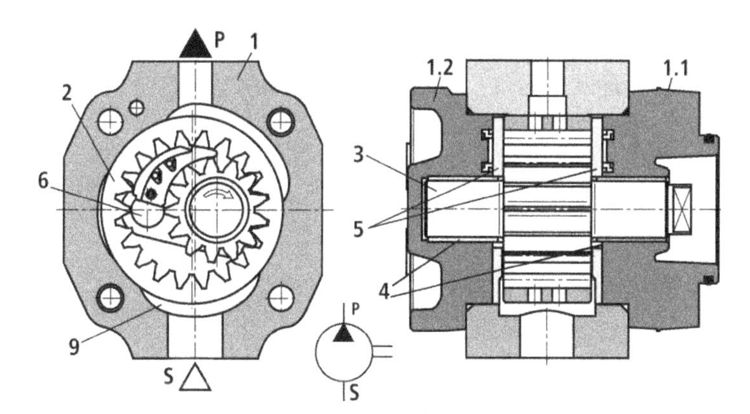

1. Housing, 1.1. Bearing cover, 1.2. Blanking plate, 2. Internal gear, 3. Pinion shaft, 4. Shaft bearings, 5. Side plates, 6. Stop pin, Sealing rolls, 9. Hydrostatic bearing

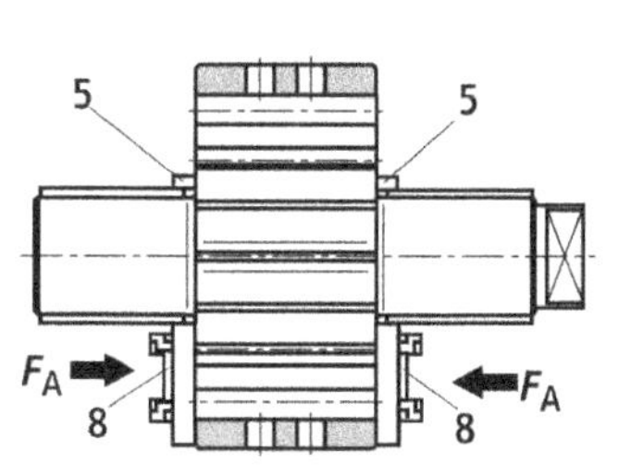

5. Wear plates, 8. Pressure field on the side plate

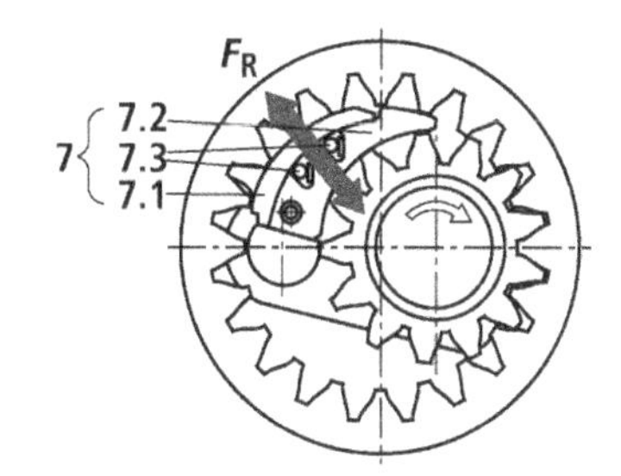

7. Segment assembly, 7.1. Segment, 7.2. Segment carrier, 7.3. Sealing rolls

FIGURE 4.26 Construction of an internal gear pump. (*Courtesy of Bosch Rexroth AG.*)

The Suction and Displacement Process

The pinion (3), which is carried in the hydrodynamic bearings, drives the internal toothed ring gear (2). During the rotation, a negative pressure is produced in the un-meshing zone and the fluid flows into the pump through a hole in the side plate. The segment assembly (7) separates the suction chamber from the delivery chamber. The two gears mesh in the delivery zone, displacing the oil out of the pump.

Axial Compensation Forces

The axial compensation force, F_A, acts within the pressure chamber areas, located on the outer side of the wear plates (5). It is generated by the pressure field (8). The side clearances, between the nonrotating and the rotating components, are extremely minimized, which insures minimum side clearance leakage.

Radial Compensation Forces

The radial compensation force, F_R, acts on the segment (7.1) and the segment carrier (7.2). The area ratios and the position of the sealing rolls (7.3) between the segments and the segment carrier are designed to provide an almost leak-free seal between the internal gear, the segment assembly (7), and the external gear. The spring elements under the sealing rolls (7.3) ensure a sufficient contact force at startup.

Figure 4.27 shows the typical flow characteristics in modern designs of internal gear pumps with different geometric volumes (sizes).

4.6.10 Gerotor Pumps

The term "gerotor" means *ge*nerated *rotor*, and is the trade name for a popular internal gear element. The cross-sectional illustration (see Fig. 4.28) shows a typical Gerotor element, which consists of a pair of gear-shaped parts. The internal gear (or rotor) (1), drives the outer gear (2), in the same direction of rotation. This is similar to the internal gear pump with a crescent seal.

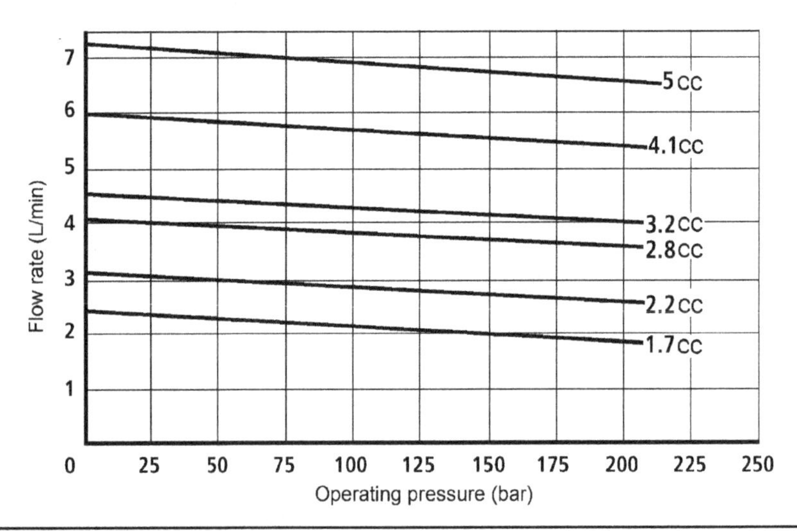

FIGURE 4.27 Flow characteristics of typical internal gear pumps of different sizes at 1450 rpm speed. (*Courtesy of Bosch Rexroth AG.*)

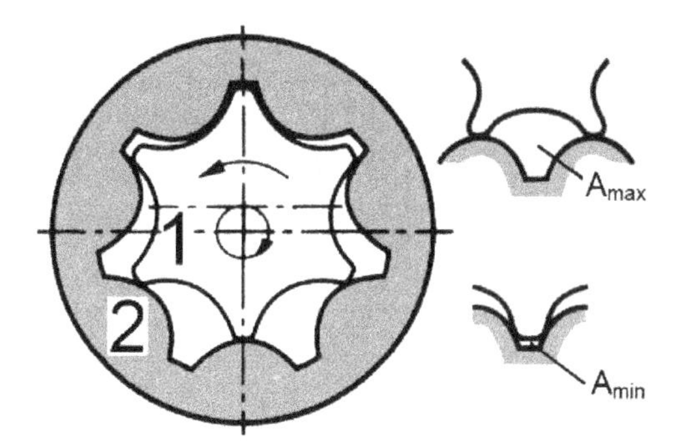

FIGURE 4.28 Construction of a Gerotor pump. (*Courtesy of Bosch Rexroth AG.*)

The drive gear has one less tooth than the outer, driven, element. The pumping chambers are formed by the adjacent pairs of teeth, which are constantly in contact with the outer element, except for clearance. As the rotor is turned, its gear tips are accurately machined so they precisely follow the inner surface of the outer element. The expanding chambers are created as the gear teeth withdraw. The chamber reaches its maximum size when the female tooth of the outer rotor reaches the top dead center. During the second half of the revolution, the spaces collapse, displacing the fluid to the outlet port, formed at the side plate. The geometric volume of the Gerotor pump is given as follows:

$$V_g = bz(A_{\text{max}} - A_{\text{min}}) \tag{4.41}$$

where b = Tooth height, m
z = Number of rotor teeth

4.6.11 Screw Pumps

In screw pumps (see Fig. 4.29), the displacement chamber is formed between the threads and housing. The following is an expression for the geometric volume of a twin-gear screw pump:

$$V_g = \frac{\pi}{4}(D^2 - d^2)s - D^2\left(\frac{\alpha}{2} - \frac{\sin 2\alpha}{2}\right)s \tag{4.42}$$

with

$$\cos(\alpha) = \frac{D+d}{2D} \tag{4.43}$$

4.6.12 Vane Pumps

Construction and Operation

Figure 4.30 shows the construction of a fixed displacement vane pump. The rotor (2) is driven by the shaft (1) and rotates inside the stator ring (cam ring) (3). The vanes (4) are mounted into radial slots in the rotor. They are pressed radially outwards to be in contact with the inner surface of the cam ring under the action of the centrifugal force. As the pressure builds in the delivery line, the roots of some vanes are pressurized to insure the required sealing in the proper positions. The pumping chamber is bounded by two

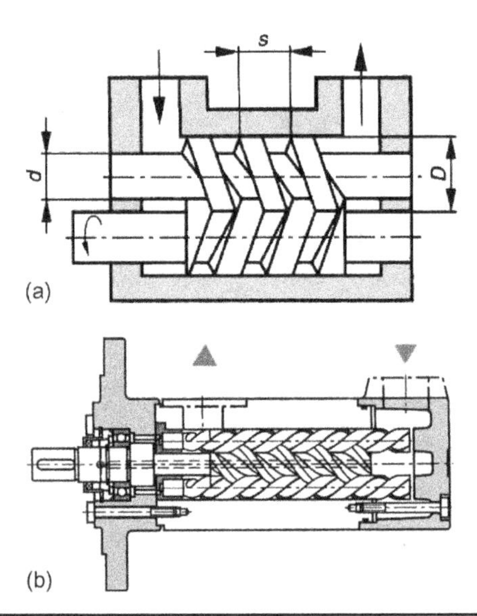

FIGURE 4.29 (a) A twin-gear screw pump and (b) a treble-gear screw pump. (*Courtesy of Bosch Rexroth AG.*)

successive blades (vanes), the inner surface of the cam ring, the outer surface of the rotor, and the two surfaces of the side plates (5). The volume of the pumping chamber changes (increases and decreases) during rotation due to the ovality of the cam ring. During its expansion, the chamber is connected to the suction line through a hole in the side plate. Therefore, the oil is sucked into the chamber. Then, during its contraction, the pumping chamber displaces the oil to the outlet line through another hole in the side plate. In this way, each chamber performs two pumping strokes per revolution.

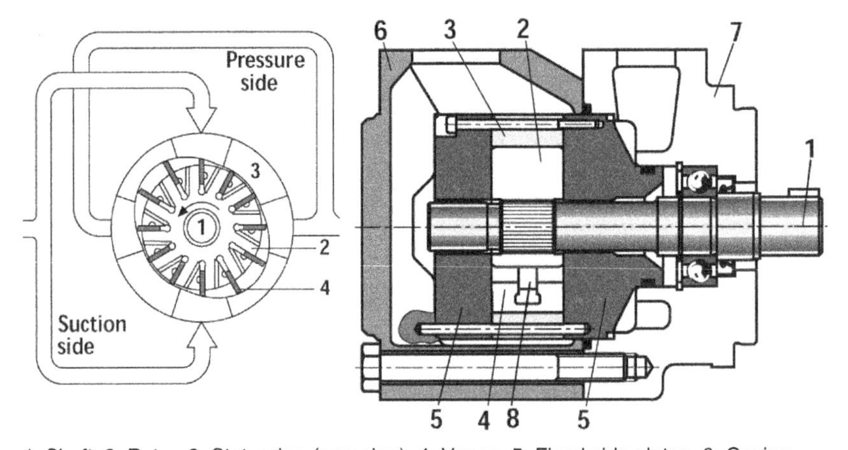

1. Shaft, 2. Rotor, 3. Stator ring (cam ring), 4. Vanes, 5. Fixed side plates, 6. Casing, 7. Bearing mount, 8. Intra-vane

FIGURE 4.30 Fixed displacement vane pump with fixed side plates. (*Courtesy of Bosch Rexroth AG.*)

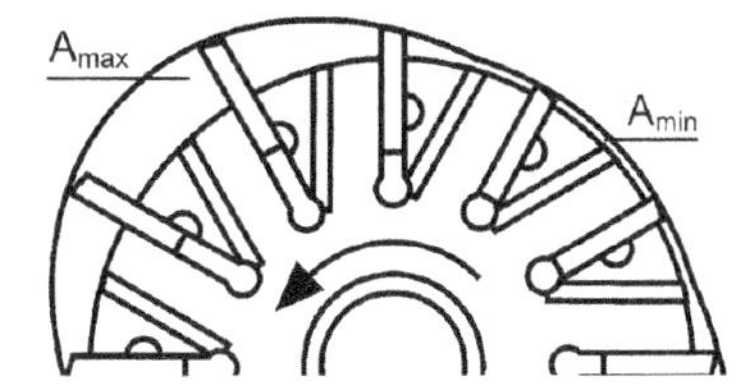

Figure 4.31 Location of maximum and minimum chamber volume.

The pump displacement is given by the following expression (see Fig. 4.31):

$$V_g = 2bz(A_{max} - A_{min}) \quad (4.44)$$

where b = Rotor height, m.

Due to the oval shape, the pump construction is axi-symmetrical. There are two pressure and two suction chambers opposite each other. Therefore, the rotor is pressure balanced. It is loaded mainly by the pump driving torque. In this way, the bearing load is minimized, which reduces the friction and wear and increases the pump efficiency and service life.

Side Clearance Leakage

The internal leakage in vane pumps takes place, mainly, through the rotor side clearance and blade tip clearance. In the case of fixed side plates, the side clearance leakage is uncontrollable. The wear of the fixed side plates is not compensated for, which increases the internal leakage. By using movable side plates, it is possible to:

- Minimize the side clearance leakage
- Compensate for the thermal expansion of the rotor
- Act against sudden pressure changes

Figure 4.32 illustrates the construction of a vane pump with two movable side plates (1). Two counter pressure chambers (2) are created and communicated to the pump delivery line. The side plate area, subjected to the pressure, is designed to create the required balance against the pressures inside the pumping chambers. Because of this, an optimum clearance between the rotor and the side plates is guaranteed, and the side clearance leakage is minimized.

Tip Clearance Leakage

The tip clearance leakage takes place through the clearance between the blade tip and the inner surface of the cam ring. Therefore, the blade root should be pressurized wherever the blade separates two chambers of different pressure (see Fig. 4.34c and 4.34d). Otherwise, wherever the blade separates two chambers of equal pressure, the blade root should have the same pressure as that in these chambers (see Fig. 4.34a and 4.34b). In this way, the blade is pressure balanced and is pushed radially outwards under the action of the centrifugal force only. This zone of blade displacement has minimal friction and reduced wear.

Most of the pressure-balanced vane pumps have an intra-vane feature. This design helps keep the vane in continuous contact with the cam ring during the rises and falls

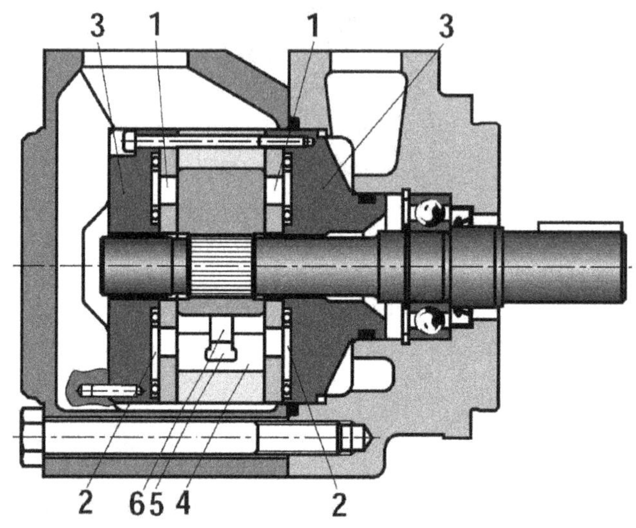

1. Movable side plates, 2. Counter pressure chambers,
3. Fixed side plates, 4. Blade (vane), 5. Intra-vane chamber,
6. Intra vane

FIGURE 4.32 A fixed displacement vane pump with movable side plate. (*Courtesy of Bosch Rexroth AG.*)

while minimizing the vane-tip contact force. The construction of the intra-vane structure is illustrated in Fig. 4.32. Instead of a single vane, there are two vanes of the same thickness. A smaller one (intra vane 6) is installed at the inner side of the main vane (4), which creates an inner chamber (intra-vane chamber) between them (5).

The operation of the intra-vane structure is explained by Figs. 4.33 and 4.34. Figure 4.33 shows: the rotor (1), the main vane (2), the cam ring (the stator) (3), the intra-vane pressurizing hole (4), and the vane root pressurizing hole (5). This hole connects the vane root with the vane trailing chamber. The intra-vane inner chamber is pressurized through the hole (4) and the pressurizing grooves on the side plate. The side plate has

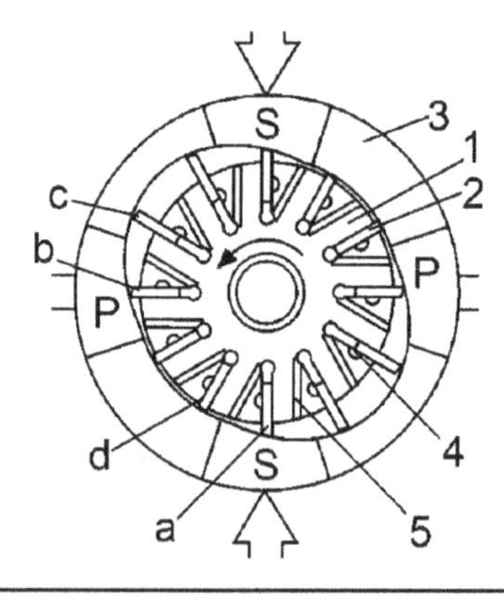

FIGURE 4.33 Function of the intra-vane structure. (Note: a, b, c and d indicate the locations of blade loading cases illustrated in Fig. 4.34.)

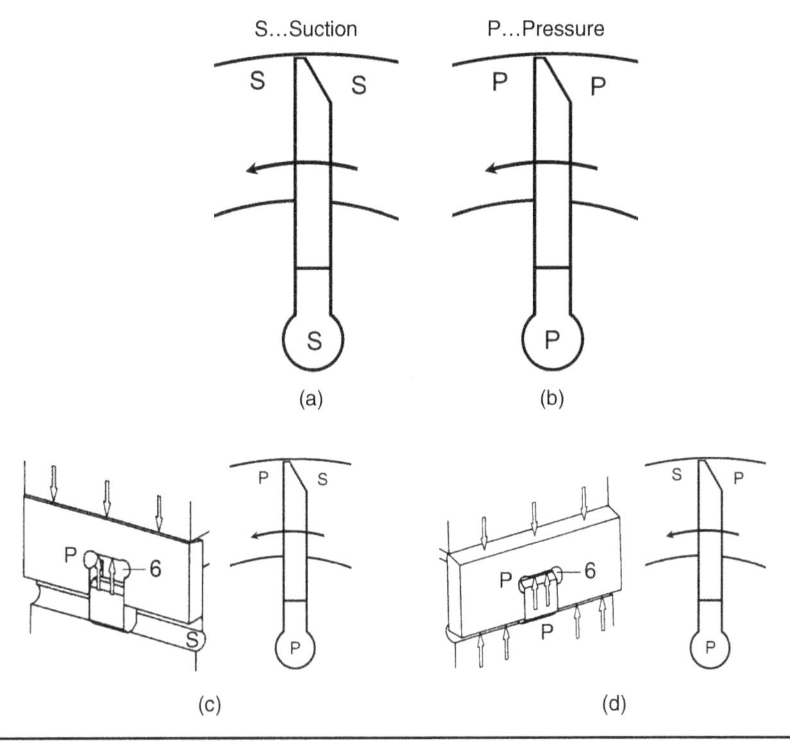

FIGURE 4.34 Pressurization of vains' roots and intra vane.

four grooves: two are in connection with the suction line, while the other two are connected to the pressure line. These grooves serve to communicate the intra vane with either the pressure or the suction lines. The four grooves are arranged on a circle, facing the pressurizing grooves (4). The vane has four main positions:

- The vane separates two chambers in connection with the suction line. Both the vane root and the intra-vane chamber are not pressurized and the vane is pressure balanced (see Fig. 4.34a).
- The vane separates two chambers in connection with the pressure line, Fig. 4.34b. At this position, the vane root and intra vane are pressurized, which makes the vane pressure balanced and pushed radially by the centrifugal force. If the blade root is not pressurized, the vane retracts under the action of the pressure force on its tip area. Then, the vane will be fired when the root is pressurized, which takes place four times per revolution, to insure the required tightness. The vane firing will cause severe damage to the cam's inner surface.
- The vane separates two chambers of different pressures, with the high pressure facing the vane leading side (see Fig. 4.34c). In the case of single-vane design, the root should be pressurized to insure the required tightness. The vane extension force is much greater than the vane retraction force since the pressure in the leading chamber acts on a small portion of the vane tip area. The pumps having this design suffer from severe wear at these zones.

Here comes the main advantage of the intra-vane structure. At this position, the blade root is not pressurized since it is connected with the trailing nonpressurized chamber. Only the intra vane is pressurized. The net forces at the vane can be kept small at all times; thus, the wear during normal operation is extremely reduced.

- The vane separates two chambers of different pressures, with high pressure facing the vane trailing side (see Fig. 4.34d). At this position, most of the blade tip area is subjected to the pressure in the trailing chamber. Therefore, both the vane root and intra vane are pressurized to create the required tightening force.

4.7 Variable Displacement Pumps

4.7.1 General

Variable displacement pumps are much more expensive and of a more complicated design compared to fixed displacement pumps. However, the designers use the variable displacement pumps for two reasons: economy and control.

Economic Reasons

The hydraulic systems designers try to minimize the hydraulic power generation (*PQ*) whenever it is not needed. Several solutions satisfy this requirement, for example, by using convenient valves and an accumulator assembly. The use of variable displacement pumps with pressure compensators is among the most widely used solutions. The flow characteristics of this class are illustrated in Fig. 4.38.

Another important issue is that hydraulic systems operate within a wide range of pressures and flow rates. Therefore, these systems need prime movers of very high power. Thus, the use of displacement pumps with constant power controllers offers a reasonable solution to this problem. This text presents examples for these controllers.

Control Reasons

A wide range of controllers of displacement pumps are available, which are intended to control the flow rates and direction of flow to the hydraulic cylinders and motors, aiming to control the magnitude and direction of their velocity. Among these controllers are:

- Hydraulic proportional controllers, for open-loop control of the pump displacement
- Hydraulic servo controllers, for closed-loop control of the pump displacement
- Electrohydraulic proportional controllers, for open-loop control of the pump displacement
- Electrohydraulic servo controllers, for open- and closed-loop control of the pump displacement

4.7.2 Pressure-Compensated Vane Pumps

Construction and Operation

The construction of a variable displacement vane pump with a pressure-unbalanced rotor is shown in Figs. 4.35 through 4.37. The pump consists of: the housing (1), rotor (2),

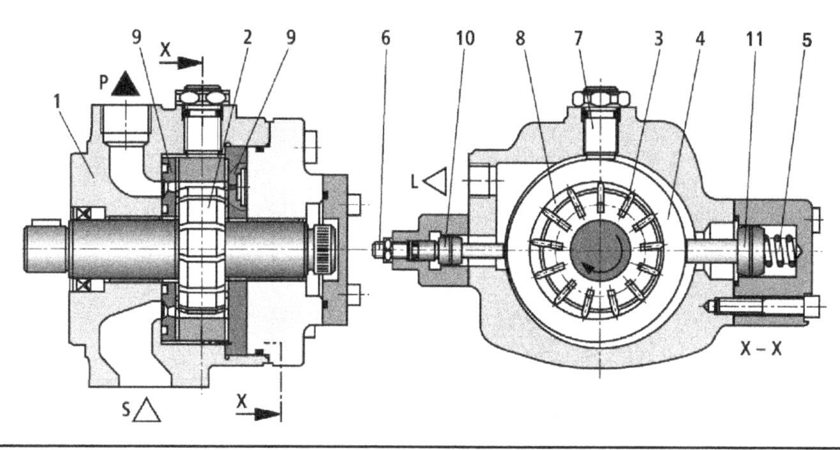

Figure 4.35 A vane pump with a pressure unbalanced rotor. (*Courtesy Bosch Rexroth AG.*)

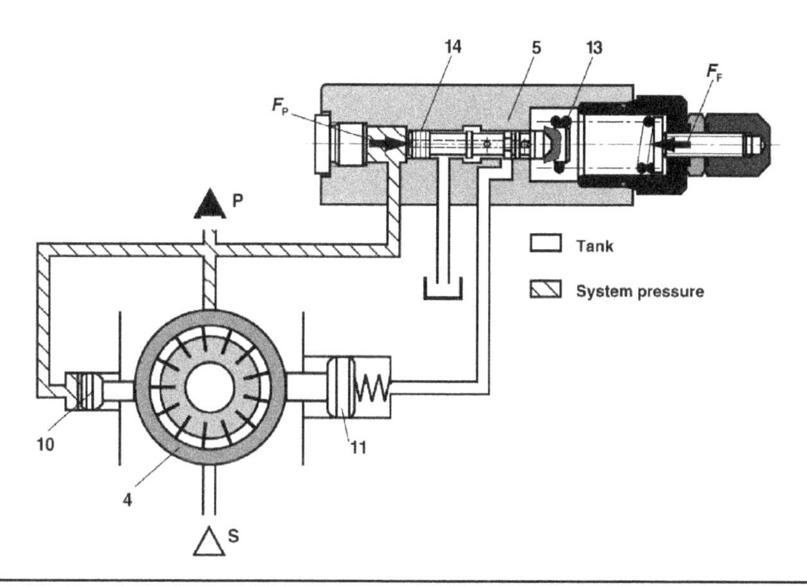

Figure 4.36 A variable displacement vane pump with a pressure-unbalanced rotor in off-stroke mode. (*Courtesy of Bosch Rexroth AG.*)

vanes (3), stator ring (cam) (4), and an adjustment screw (6). The position of the cam (stator) ring (4) is controlled horizontally by means of two pistons: a small control piston (10) and a large one (11). The vertical position of the cam is determined by the adjustment screw (7). The pumping chambers (8) are formed by the vanes (3), the rotor (2), the stator ring (4), and the side (control) plates (9). In order to ensure the pump function during start-up, the stator ring (4) is held in the eccentric position (displacement position) by the spring (5). The stator is eccentrically mounted with respect to the rotor.

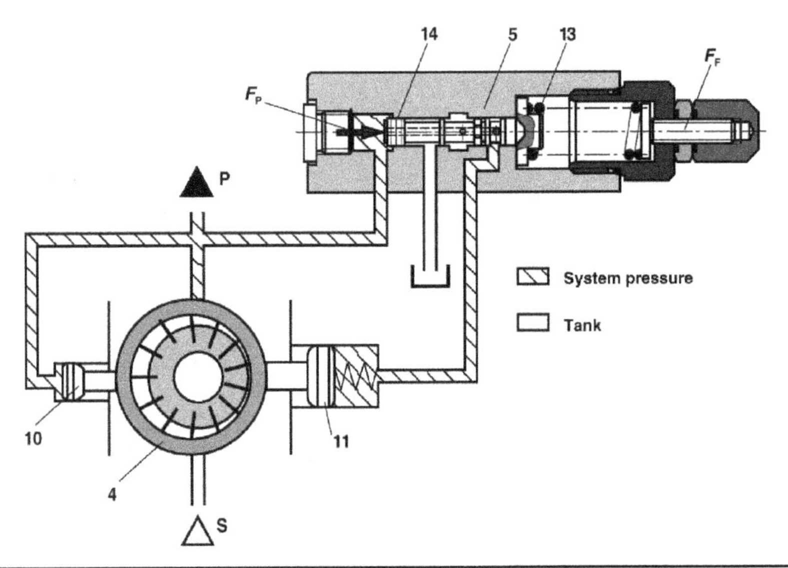

FIGURE 4.37 A variable displacement vane pump with a pressure-unbalanced rotor in on-stroke mode. (*Courtesy of Bosch Rexroth AG.*)

Then, due to the rotation of the rotor (2), the volume of the chambers (8) increases and fills with fluid via a suction channel (S). When the maximum chamber volume is reached, the chambers (8) are disconnected from the suction side. Then, as the rotor (2) continues to rotate, the chambers' volume decreases as they are connected with the pressure side. The oil is then displaced to the pump exit line via the pressure channel (P). The small control piston (10) is in permanent connection with the pump pressure line. The large piston is loaded by a spring (5), which puts the cam in its extreme eccentric position, at the pump's startup. Then, when the pump operates and the pump exit pressure builds up, the pressurization of the spring chamber determines the position of the stator (cam) and consequently the pump displacement.

Pump Controls

Off-Stroke Mode If the pressure force, F_P, exceeds the counter force of the spring, F_F, the controller piston (14) is moved against the spring (13). (See Fig. 4.36.) In this way, the chamber behind the large control piston (11) is connected to the tank and is thus unloaded. The small control piston (10), which is constantly under system pressure, moves the stator ring (4) towards the center position. Its end position is determined by a mechanical displacement limiter. The pumping chamber's volume decreases to its minimum value. The pump delivers a minimal flow rate, which compensates for the internal leakage. The pump's real delivery is null. The power loss and heating of the fluid are kept at a low level. The pump flow characteristics, *Q(P)*, is shown in Fig. 4.38. The commencement of the pump displacement control is set by the spring (13).

On-Stroke Mode When the pressure in the system falls below the set zero stroke pressure, the spring (13) moves the controller spool (14) back to its initial position, the extreme left position (see Fig. 4.37). The large control piston (11) is subjected to the pressure and moves

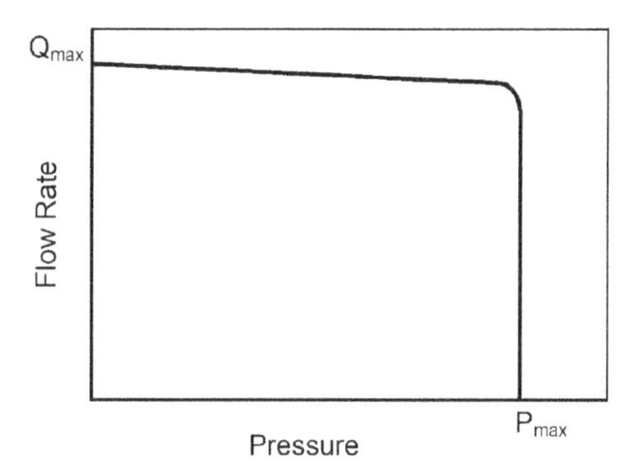

FIGURE 4.38 Typical flow characteristics of a variable displacement vane pump with a pressure compensating controller.

the stator ring (4) into the eccentric position, the extreme left position. The pump is in its maximum displacement position, delivering maximum flow rate.

4.7.3 Bent Axis Axial Piston Pumps with Power Control

Pump Description and Operation

Figure 4.39 shows a schematic of a variable displacement bent axis axial piston pump with power controller. This pump consists of two main groups. The first group is the pumping mechanism group, which contains the shaft, port plate, and seven pistons within a common cylindrical block. The cylinder block is held tightly against a port plate using the compression force of the holding spring and the pressure forces. During the operation, the port plate with the cylinder block moves along the sliding surface.

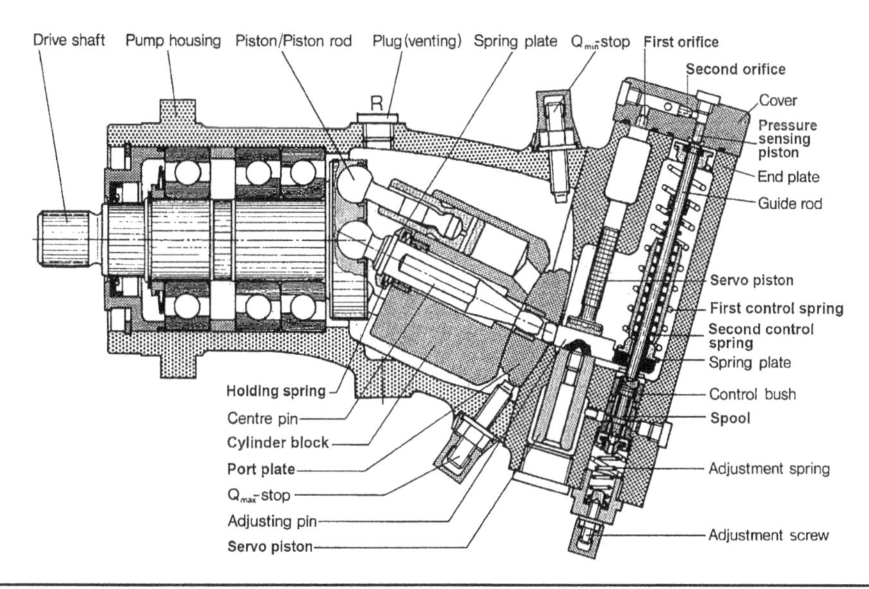

FIGURE 4.39 Bent axis piston pump with power controller. (*Courtesy of Bosch Rexroth AG.*)

The second group is the pump controller, which contains the pressure-sensing piston, the servo piston, the guide rod, the spool valve, the adjustment spring, and two control springs (the first and the second springs). The two main groups are connected together by means of a pin. The upper smaller area of the servo piston is in permanent connection with the pump exit pressure line, through the first damping orifice. The lower servo piston chamber is connected to either the pressure or the tank, means of the spool valve. The pressure-sensing piston is connected to the pressure line through the second damping orifice. This piston displaces the spool downwards, through the guide rod, while the adjusting spring acts on the other direction. In the steady state, the pump has four operating modes, as described in the following:

First Mode The pressure is less than the control commencement pressure. The value of this pressure is set by the adjusting spring. The pump exit pressure is less than the value pre-set by the adjustment spring, and the spool is shifted upwards and connects the lower servo piston chamber with the tank. The servo piston is at its lower position and the pump operates at its maximum displacement (lines 1–2 in Fig. 4.40).

Second Mode The operating pressure exceeds the value pre-set at the adjustment spring; the control piston, guide-rod, and spool are moved against this spring. The high-pressure oil is thus allowed to flow via the spool valve to the lower servo piston chamber. Should the resultant pressure force be greater than the force of the first control spring, the servo piston is moved upwards, compressing the first spring until the balance of these forces is restored. The second control spring is still out of action (lines 2–3 in Fig. 4.40).

Third Mode The third mode is similar to the second one, except that the pressure is higher, such that the equilibrium of the servo piston and attached elements is attained when the second control spring comes into action (lines 3–4 in Fig. 4.40).

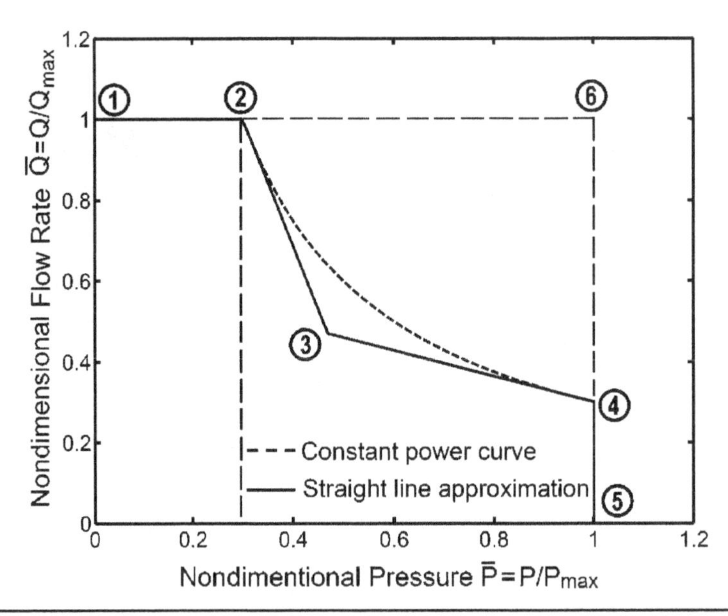

FIGURE 4.40 Steady-state characteristics of the pump constant power controller.

Fourth Mode The pump displacement is minimal, as limited by the adjusting screw. The pump delivers the fluid by the flow rate and pressure corresponding to point (4). The maximum pump exit pressure is determined by the relief valve (lines 4–5 in Fig. 4.40).

The study of Fig. 4.40 shows that:

- When using a fixed displacement pump, the power required to drive the pump is too high: $N_6 = P_5Q_1$.
- In the case of a pump equipped with a constant power controller, the needed power is determined by the pressure and flow rates of point 2: $N_2 = P_2Q_2$.
- Point 2 can be displaced horizontally by means of the adjustment spring setting, and displaced vertically by using the maximum inclination angle limiter.
- The slope of line 2–3 is imposed by the stiffness of the first control spring, while that of line 3–4 is determined by the summation of stiffness of both of the first and second control springs.
- The operation in the vicinity of point 3 does not consume the whole available power of the prime mover due to the straight line approximation of the constant power curve. But the simplicity and relative lower cost give good appreciation to this design. However, other power controllers are commercially available, which better agree with the constant power curve.

4.8 Rotodynamic Pumps

Rotodynamic pumps, such as centrifugal pumps, are machines that transfer energy to the liquid by increasing its momentum. This class of pumps is used widely wherever high flow rates are required under low or medium heads. They offer advantages such as simple construction, low price, easy maintenance and repair, and the ability to work with liquids of low lubricity.

The main components of a typical centrifugal pump are shown in Fig. 4.41. It consists of:

- Rotor, which is known as the impeller (1). The function of the impeller is to convert the mechanical energy delivered by the pump shaft to hydraulic energy of flowing liquid.

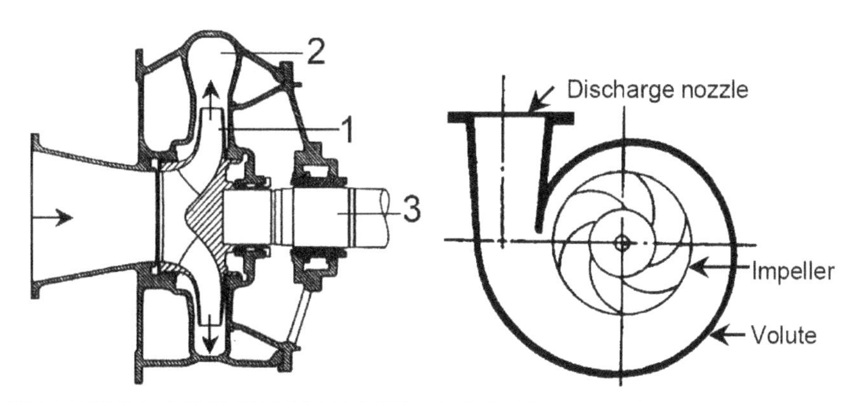

Figure 4.41 Construction of a typical rotodynamic pump of the centrifugal type.

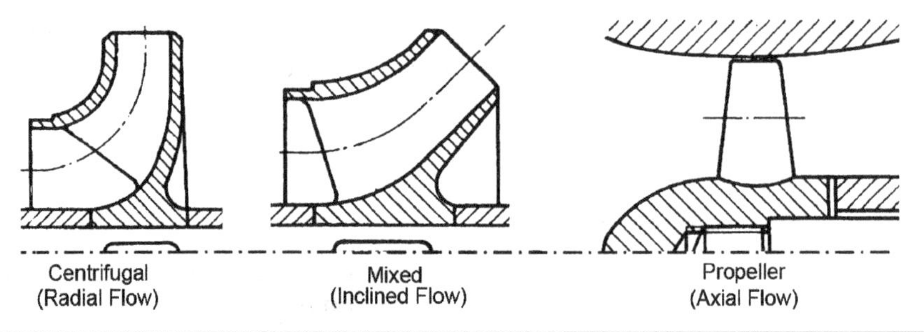

FIGURE 4.42 Impeller types of rotodynamic pumps.

- Stator which may include guide vanes and volute or spiral casing (2). The function of the casing is to accumulate the liquid flowing out of the impeller, as well as recuperating the major part of the kinetic energy of the liquid by converting it to pressure energy.
- Drive shaft with convenient sealing.
- Mechanical parts such as bearings, couplings, and so on.

Different shapes of impellers can be found to adapt different design requirements (see Fig. 4.42). Impellers are classified according to the major direction of flow in reference to the axis of rotation. Thus, the centrifugal pumps may have radial flow impellers, mixed flow impellers, or axial flow impellers. Impellers are further classified as single suction with a single outlet on one side, and double suction with liquid flowing to the impeller symmetrically from both sides.

Figure 4.41 shows an example of a single-stage centrifugal pump. As the impeller rotates, an underpressure is created at its inlet. The liquid enters the pump through the inlet nozzle to the inlet of the impeller. Inside the impeller, and due to the hydrodynamic effect (centrifugal force), the liquid momentum increases as it moves radially out of the impeller to the casing with high kinetic energy. The kinetic energy of the delivered flow is recuperated (converted) partially as pressure energy in the guide vanes, volute casing, and the outlet diffuser. The liquid then flows with higher energy out from the diffuser to the pump discharge nozzle.

Rotodynamic pumps can deliver continuous flow with high flow rates. These pumps are sensitive to air and gases. Nevertheless, they are of low sensitivity to solids due to their wide flow passages. They operate at relatively low-pressure levels. Most of commercial pumps operate at pressures less than 6 bar for a single-stage pump. Therefore, they are not used as main pumps in the hydraulic power systems. However, they are used in some systems to increase the pressure level at the inlet of the main pump as a way of protecting against cavitation. Figure 4.43 shows the head and efficiency characteristics of a typical single-stage centrifugal pump.

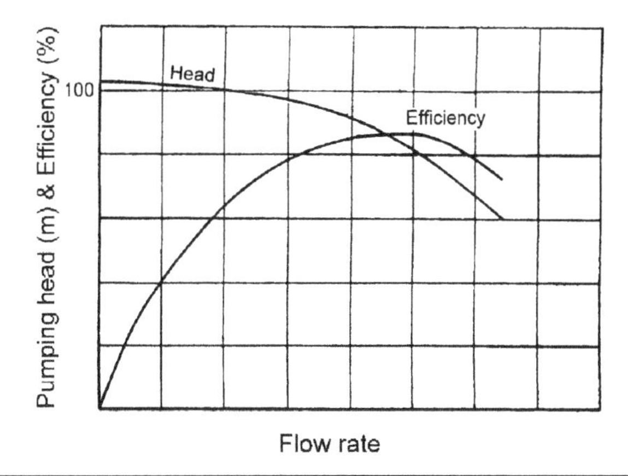

Figure 4.43 The pressure head and efficiency of a typical single-stage centrifugal pump.

4.9 Pump Summary

The following are the common features of the displacement pumps (most of the illustrations in this section appear courtesy of Bosch Rexroth AG):

No.	Pump Type and Typical Data	Illustration
1	**Axial piston pump** (bent axis) Depending on the inclination angle, the pistons move within the cylinder bores when the shaft rotates. V_g = 2–3600 cc n = 500–5000 rpm p_{max} = 400 bar $V_g = \frac{\pi}{4} d^2 D z \sin\alpha$	

2	**Axial piston pump** (swash plate type) The pistons rotate with the cylinder block, supported by the swash plate. The swash plate angle determines the piston stroke. V_g = 2–600 cc n = 500–4500 rpm p_{max} = 400 bar $V_g = \frac{\pi}{4} d^2 Dz \tan\alpha$	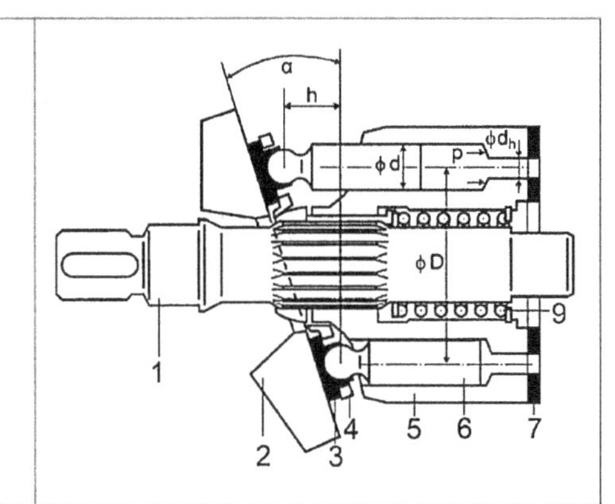
3	**Axial piston pump** (swash plate type, inclined pistons) V_g = 8–300 cc n = 500–3500 rpm p_{max} = 350 bar $V_g = \frac{\pi}{8} d^2 D z \sin(\alpha) \times \left\{\frac{1}{\cos(\varphi-\alpha)} + \frac{1}{\cos(\varphi+\alpha)}\right\}$	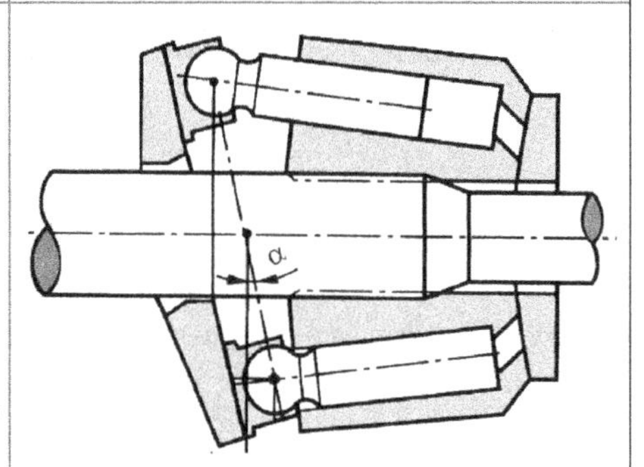
4	**Axial piston pump** (rotating swash plate type, wobble plate) p_{max} = 630 bar $V_g = \frac{\pi}{4} d^2 Dz \tan\alpha$	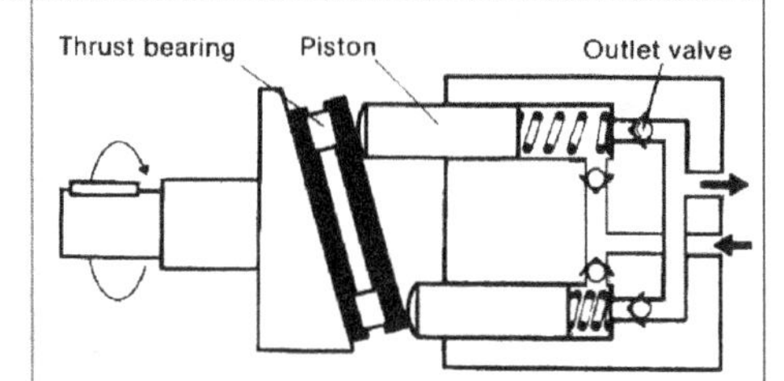
5	**Radial piston pump** (eccentric cam ring) The pistons rotate within the rigid cam ring. The eccentricity, e, causes the radial motion of the pistons. V_g = 0.5–20 cc n = 500–2000 rpm p_{max} = 400 bar $V_g = \frac{\pi}{2} d_k^2 ez$	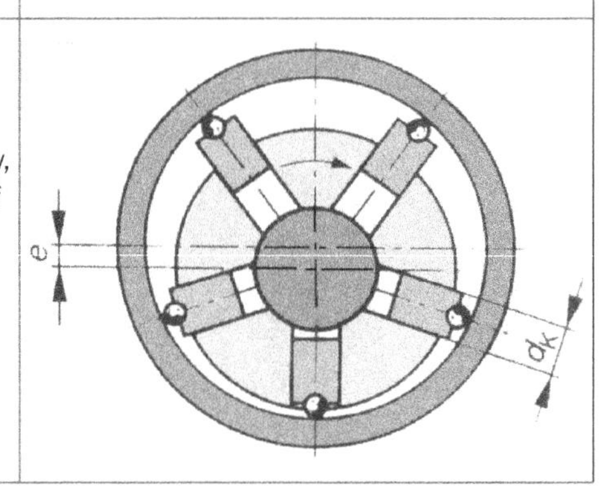

(Continued)

6	**Radial piston pump** (eccentric shaft, valve type) The rotating eccentric shaft produces a radial motion of the pistons. $V_g = 0.5\text{–}200$ cc $n = 500\text{–}2000$ rpm $p_{max} = 780$ bar $V_g = \frac{\pi}{2} d_k^2 ez$	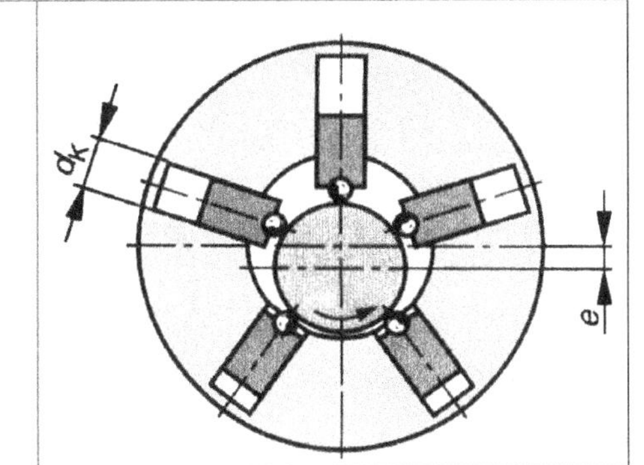
7	**Radial piston pump** (with crank shaft) $V_g = 500\text{–}4000$ cc $n = 500\text{–}2000$ rpm $p_{max} = 1200$ bar $V_g = \frac{\pi}{4} d^2 hz$ $h = 2e$	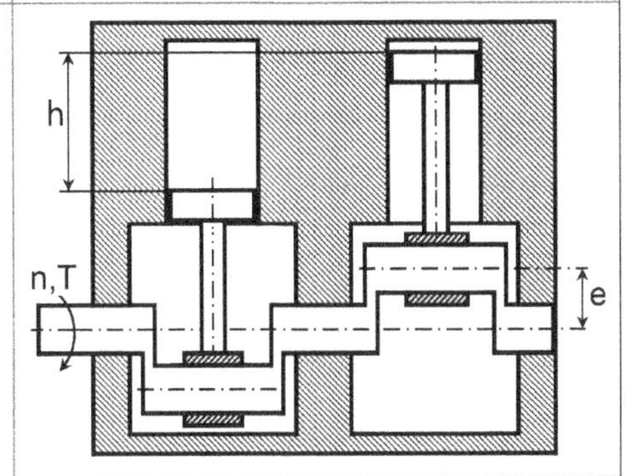
8	**External gear pump** Volume is created between gears and housing. $V_g = 0.4\text{–}1200$ cc $n = 300\text{–}4000$ rpm $p_{max} = 250$ bar $V_g = 2\pi bm^2(z + \sin^2 \gamma)$	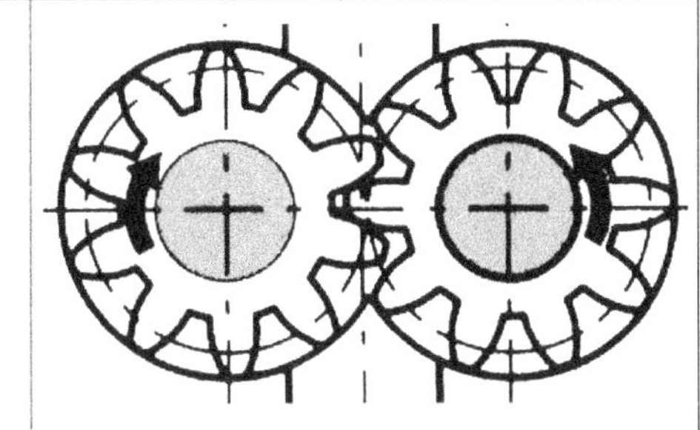
9	**Internal gear pump** Volume is created between gears, the crescent sealing element, and housing. $V_g = 0.4\text{–}350$ cc $n = 500\text{–}2000$ rpm $p_{max} = 350$ bar	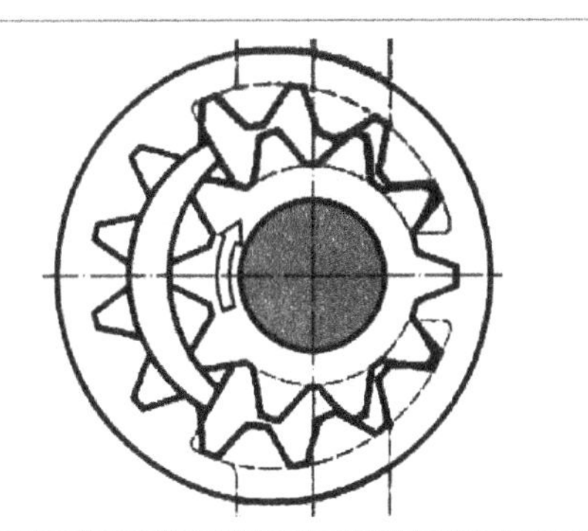

(Continued)

10	**Gerotor** The inner rotor has one less tooth than the outer (internally toothed) element. The rotor moves in a planetary motion. $V_g = 50\text{–}350$ cc $n = 500\text{–}2000$ rpm $p_{max} = 100$ bar $V_g = bz(A_{max} - A_{min})$	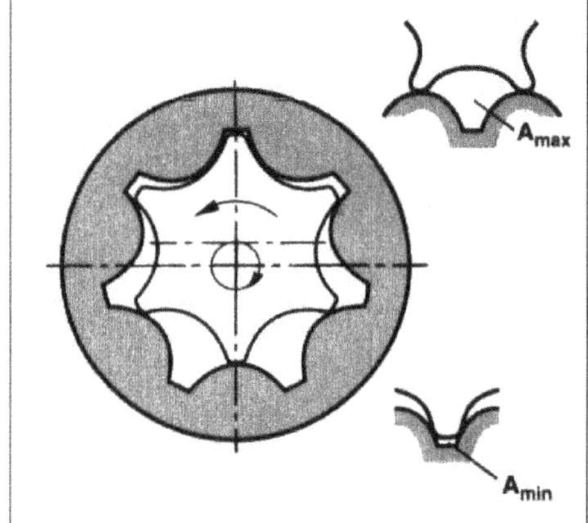
11	**Screw pump** The displacement chamber is formed between the threads and housing. $V_g = \frac{\pi}{4}(D^2 - d^2)s - D^2 \times \left(\frac{\alpha}{2} - \frac{\sin 2\alpha}{2}\right)s$ with $\cos(\alpha) = (D + d)/2D$ $V_g = 2\text{–}1500$ cc $n = 500\text{–}3500$ rpm $p_{max} = 200$ bar	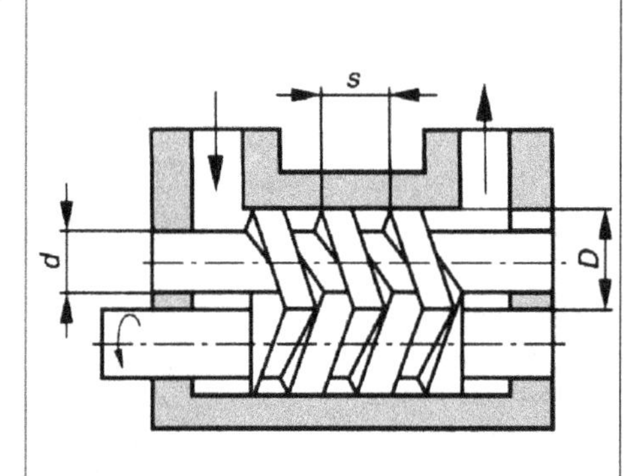
12	**Vane pump** (two strokes per revolution, pressure balanced) Volume is created between the rotor, stator (cam ring), and vanes. $V_g = 10\text{–}500$ cc $n = 500\text{–}2500$ rpm $p_{max} = 200$ bar $V_g = 2(V_{max} - V_{min})z$	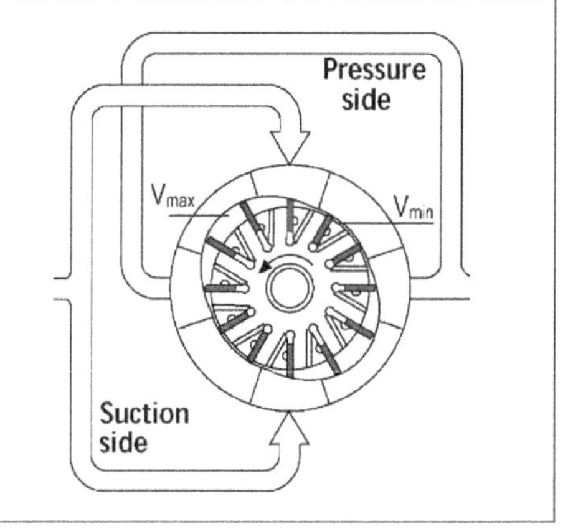

(Continued)

13	**Vane pump** (single stroke per revolution, pressure un-balanced) Volume is created between the rotor, stator (cam ring), and vanes. V_g = 10–150 cc n = 1000–3500 rpm p_{max} = 100 bar $V_g = z(V_{max} - V_{min})$	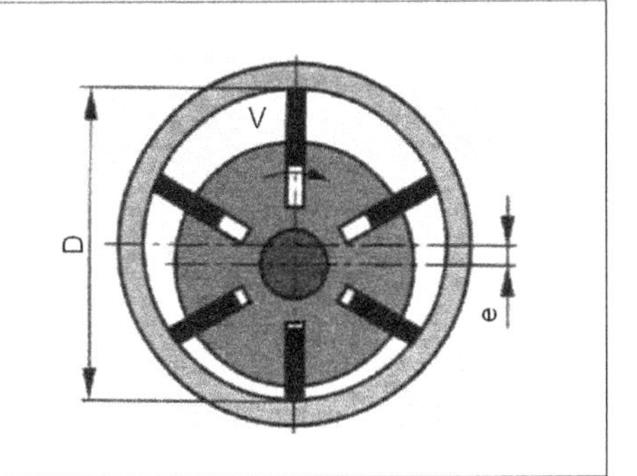

4.10 Pump Specification

The following list shows the basic specifications that should be available to specify the pump precisely:

- Size (displacement)
- Speed (maximum and minimum speeds)
- Maximum operating pressure (continuous/intermittent)
- For open/closed circuit
- Direction of rotation (viewed to shaft end; clockwise [R], counterclockwise [L])
- Controller (for variable displacement pumps)
- Seals (oil)
- Drive shafts
- Port connections
- Mounting type
- External dimensions
- Installation position
- Operating temperature range
- Further details in clear text

4.11 Exercises

1. Explain briefly the principal of operation of displacement pumps.
2. Give the expressions describing ideal and real pumps.
3. A displacement pump of 23 cm^3 geometric volume operated at 1450 rev/min delivers the oil at 20 MPa pressure. Assuming an ideal pump, calculate the pump flow rate, Q_t, the increase in the oil power, ΔN, the pump outlet power, N_{out}, and the driving torque, T_t, if the inlet pressure is 100 kPa.

4. The figure shown here explains the power transmission in a hydraulic power system, as well as the causality relation of the system variables. Discuss it in detail.

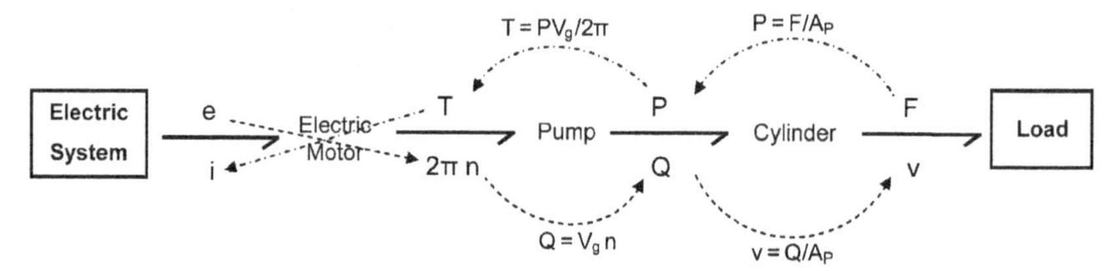

5. Discuss the different types of power losses in displacement pumps.

6. Discuss the cavitation phenomenon in displacement pumps.

7. Give and discuss briefly the classification of hydraulic pumps.

8. Draw a schematic of a bent axis axial piston pump. Explain briefly its function and give an expression for its geometric volume.

9. Deal with the pulsation of flow of axial piston pumps and the associated pressure oscillation for a pump loaded by a loading orifice.

10. Draw a schematic of a swash plate axial piston pump. Explain briefly its function and give an expression for its geometric volume.

11. Draw a schematic of a swash plate axial piston pump with inclined pistons. Explain briefly its function and give an expression for its geometric volume.

12. Discuss briefly the need for variable displacement pumps.

13. Explain briefly the principal of operation of wobble plate pumps.

14. Classify and explain briefly the principal of operation of radial piston pumps. Use the illustrations in Figs. 4.19, 4.20, and 4.21.

15. Draw a schematic of an external gear pump. Explain briefly its function and give an expression for its geometric volume.

16. Discuss the speed limitations in gear pumps and derive the necessary relations.

17. Discuss briefly the pulsation of flow in external gear pumps.

18. Deal with the problem of oil squeezing in external gear pumps.

19. Explain briefly the principal of operation of internal gear pumps and discuss their flow and efficiency characteristics. (Use Figs. 4.26 and 4.27.)

20. Explain briefly the principal of operation of gerotors (Fig. 4.28) and screw pumps (Fig. 4.29).

21. Explain briefly the construction and operation of pressure-balanced vane pumps. (Use Fig. 4.30.)

22. Explain briefly the function of an intra-vane structure in pressure-balanced vane pumps. (Use Figs. 4.33 and 4.34.)

23. Explain briefly the construction, operation, and control of variable displacement vane pump. (Use Figs. 4.35 through 4.38.)

24. Explain briefly the construction, operation and control of variable displacement axial piston pumps with constant power controllers. (Use Figs. 4.39 and 4.40.)

25. Calculate the displacement volume, delivery pulsation coefficient, input power leakage flow rate, resistance to internal leakage, and the driving torque of a gear pump with the following parameters:

pump speed	= 1450 rpm	number of teeth	= 12
tooth module	= 3.5 mm	tooth width	= 20 mm
pressure angle	= 20°	inlet pressure	= 0.2 MPa
exit pressure	= 15 MPa	mechanical efficiency	= 0.85
volumetric efficiency	= 0.9		

Calculate the volumetric efficiency if the pressure is increased to 220 bar.

26. Calculate the maximum allowable speed of a gear pump connected as shown in the following figure, given:

number of teeth	= 12	tooth module	= 3.5 mm
tooth width	= 20 mm	pressure angle	= 20°
pipe length	= 1 m	pipe diameter	= 13 mm
pressure head H	= 0.3 m	tank over pressure	= 0.13 MPa
oil density	= 870 kg/m^3	pipe friction coefficient λ	= 0.035

Neglect the local losses in the suction line.

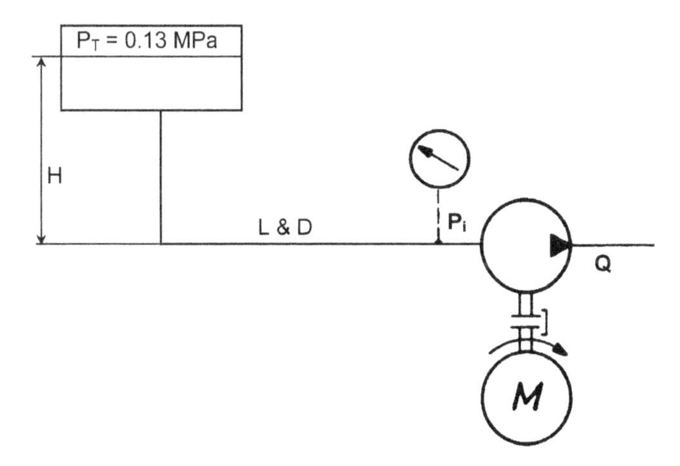

27. A bent axis pump has the following parameters:

number of pistons z	= 9	piston diameter d	= 9.3 mm
pitch circle diameter D	= 33 mm	driving speed n	= 4000 rpm
inlet pressure P_i	= 0.3 MPa	exit pressure P	= 18 MPa
volumetric efficiency	= 0.94	total efficiency	= 0.89
hydraulic efficiency	= 1	inclination angle of cylinder block	= 20°

(a) Calculate the pump theoretical flow, real flow, input mechanical power and driving torque.

(b) Calculate the leakage flow rate and resistance to leakage.

(c) Calculate the pump real flow and driving torque if the exit pressure is increased to 30 MPa, keeping the resistance to leakage and mechanical efficiency constant.

28. A swash plate axial piston pump has the following parameters:

$z = 7$, $d = 10$ mm, $D = 35$ mm, $\gamma = 20°$, $n = 3000$ rpm, $\eta_m = 0.9$, $\eta_h = 0.99$, $P_i = 0$ and resistance to internal leakage $R_L = 258$ GNs/m^5.

(a) Calculate the geometric volume of the pump and plot in scale the relation between the real pump flow and exit pressure in the range from 0 to 30 MPa.

(b) Calculate the total pump efficiency at an exit pressure of 10 MPa.

(c) Calculate or find graphically the maximum pressure in the delivery line if it is completely closed, in the absence of any relief valves.

4.12 Nomenclature

A = Piston area, m^2
b = Tooth length, m
d = Piston diameter, m
d_h = Hole diameter on the cylinder block, m
D = Pitch circle diameter, m
e = Eccentricity, m
f = Flow pulsation frequency, Hz
h = Piston stroke, m
i = Number of pumping strokes per revolution
L = Inclined piston stroke, m
m = Module of tooth, m
n = Pump speed, rev/s
P = Pump exit pressure, Pa
P_C = Pressure inside the pumping chamber, Pa
P_i = Pump inlet pressure, Pa
P_m = Mean exit pressure, Pa
P_{max} = Maximum pressure, Pa
P_{min} = Minimum pressure, Pa
$\bar{P}$ = Nondimensional pump pressure
$\bar{Q}$ = Nondimensional pump flow
Q_m = Mean flow rate, m^3/s
Q_{max} = Maximum pump flow rate, m^3/s
Q_{min} = Minimum pump flow rate, m^3/s
Q_t = Pump theoretical flow rate, m^3/s
r = Gear radius, m
R_L = Pump resistance to leakage, Ns/m^5
T = Actual pump driving torque, Nm
T_F = Friction torque, Nm
T_t = Pump theoretical driving torque, Nm

V_g = Pump displacement (geometric volume), m^3/rev
V_{max} = Maximum chamber volume, m^3
V_{min} = Minimum chamber volume, m^3
z = Number of pumping chambers, pistons, and teeth per gear
α = Swivel angle, swash plate inclination angle, rad
η_h = Hydraulic efficiency
η_m = Mechanical efficiency
η_T = Total efficiency
η_v = Volumetric efficiency
φ = Inclination angle of the inclined pistons, rad
σ_p = Pressure pulsation coefficient
σ_Q = Flow pulsation coefficient
ω = Pump speed, rad/s
γ = Pressure angle of tooth, rad
ρ = Oil density, kg/m^3

CHAPTER 5
Hydraulic Control Valves

5.1 Introduction

The control of hydraulic power in hydraulic power systems is carried out by means of control valves. The control requirements are imposed by the function of the system. The parameters of the mechanical power delivered to the load are managed hydraulically by controlling the pressure, flow rate, or the direction of flow. The control valves are classified into the following main categories:

- Ordinary switching valves
- Proportional valves
- Servovalves
- Digital valves

This chapter is dedicated to the study of the following ordinary valves:

- Pressure control valves (PCVs)
  - Relief valves (direct- and pilot-operated)
  - Pressure-reducing valves (direct- and pilot-operated)
  - Sequence valves (direct- and pilot-operated)
  - Accumulator charging valves
- Directional control valves (DCVs; direct- and pilot-operated)
- Flow control valves (FCVs)
  - Throttle valve
  - Series pressure-compensated FCV
  - Parallel pressure-compensated flow control valves
  - Flow dividers
- Check valves
  - Direct-operated check valves
  - Pilot-operated check valves (hydraulically or mechanically piloted)

The three basic control elements used in hydraulic control valves are poppet valves, sliding spool valves, and rotating spool valves. The construction, advantages, and disadvantages of these parts are presented in Table 5.1.

Type	Illustration
Poppet valve The valve consists of a head (called a poppet), a spring, and a seat. The poppet may be spherical, conical, plane-like, or any other shape. During operation, the valve is either closed (poppet seated) or opened (poppet unseated). This valve has the advantages of low cost, simple construction, repair, and maintenance, and negligible leakage. The main disadvantage is the limitation of the number of ports and poor controllability.	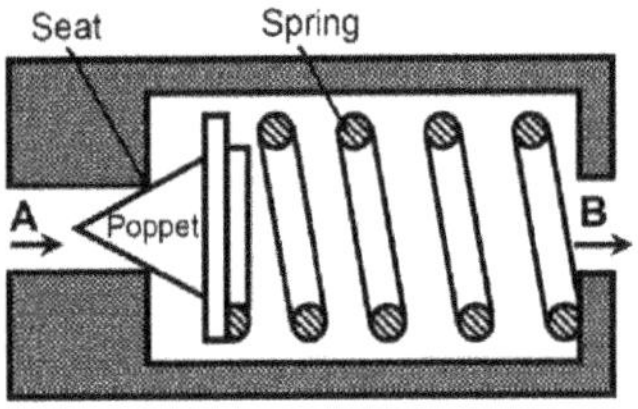 Seat, Spring, A, Poppet, B
Spool valve (sliding spool) The valve consists of a spool mounted in a sleeve. The valve is usually ax-symmetrical and its spool slides axially. The main advantages of this valve are the increased number of control ports and its greater controllability. The basic disadvantage is the increased leakage and high initial and current cost.	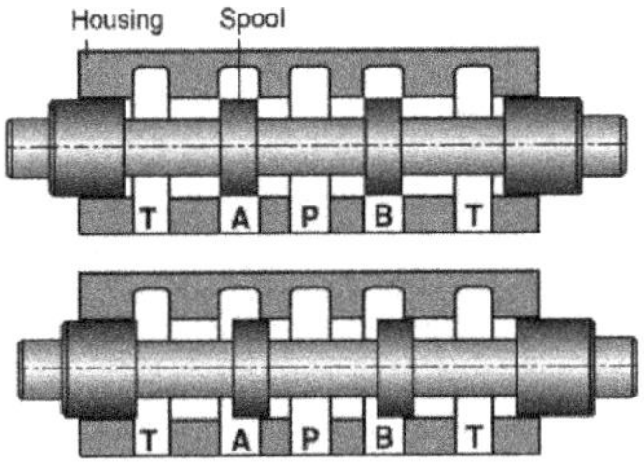 Housing, Spool, T, A, P, B, T; T, A, P, B, T
Spool valve (rotating spool) The valve consists of a spool mounted in a sleeve. The spool rotates inside the sleeve. This class of valves is used in the steering systems of some vehicles.	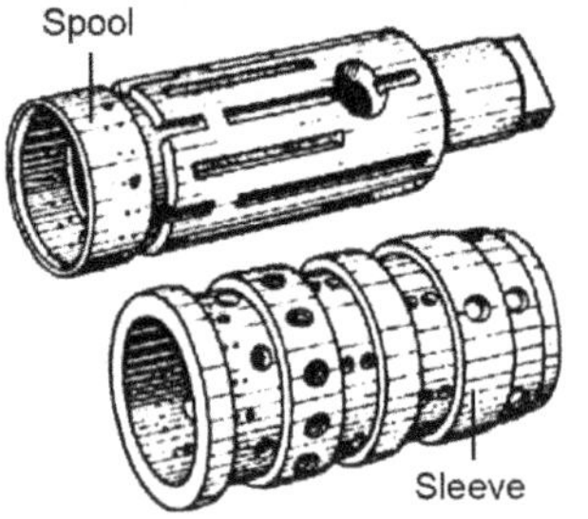 Spool, Sleeve

TABLE 5.1 Basic Constructional Elements of Hydraulic Control Valves

5.2 Pressure-Control Valves

5.2.1 Direct-Operated Relief Valves

Relief valves are connected with high-pressure and return low-pressure lines. They are used to limit the maximum operating pressure in the high-pressure lines. The relief valve consists mainly of a poppet, loaded by a spring (see Fig. 5.1). The poppet is pushed by the spring to rest against its seat in the valve housing. The spring pre-compression force is adjusted by a spring seat screw or by inserting distance rings.

The poppet is subjected to both the spring and pressure forces. The poppet rests against its seat as long as the pressure force, $F_p = PA_p$, is less than the spring force, $F_x = kx_o$.

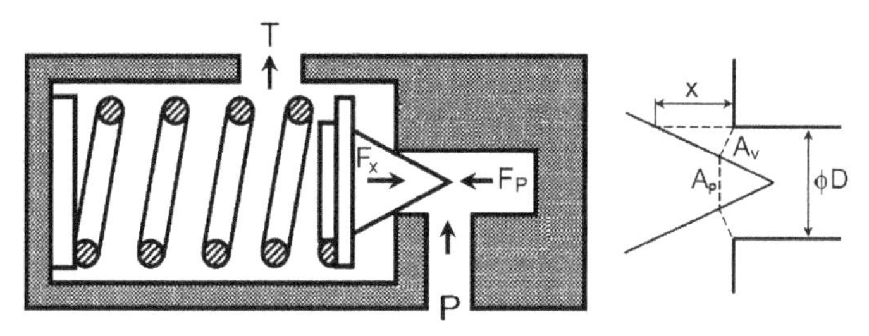

FIGURE 5.1 A simple direct-operated relief valve.

The two forces are equal when the pressure reaches the cracking pressure, P_r. For further increases of pressure, the poppet is displaced and the oil flows from the high-pressure line, P, to the return line, T.

$$A_P P_r = k x_o \quad \text{or} \quad P_r = \frac{k}{A_p} x_o \tag{5.1}$$

where k = Spring stiffness, N/m
x_o = Spring pre-compression distance, m
P_r = Cracking pressure, Pa
A_p = Poppet area subjected to pressure; for a seated poppet, $A_p = \pi D^2/4$, m²
D = Seat diameter, m

The valve opening (throttling) area, A_v, and poppet area subjected to pressure, A_p, change nonlinearly with the poppet displacement in this example (see App. 5A). However, in other cases, the area subjected to pressure is constant, as shown in Fig. 5.2. This figure illustrates a direct operated relief valve consisting of spool (1), housing (2), spring (3), spring seat (4), and adjusting knob (5).

Considering the valve shown in Fig. 5.2, the relation between the valve flow rate and system pressure is deduced in the following:

$$P_r = \frac{k}{A_p}(x_o + x_r) \tag{5.2}$$

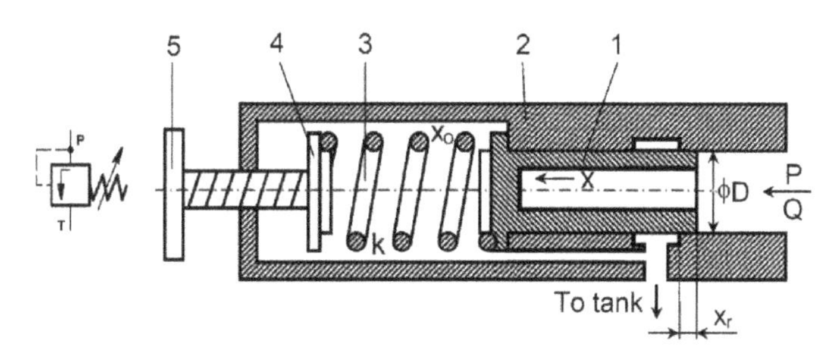

FIGURE 5.2 A direct-operated relief valve of the guided spool type.

Then $$x_r = P_r \frac{A_P}{k} - x_o \quad \text{or} \quad x_o + x_r = \frac{A_P}{k} P_r \tag{5.3}$$

$$A_P = \pi D^2/4 \tag{5.4}$$

In the steady state, the valve spool (1) reaches equilibrium under the action of the pressure forces, spring force, and jet reaction forces. Neglecting the radial clearance leakage and jet reaction forces and assuming a zero return line pressure, when the pressure increases such that the spool displaces by a distance x, then

$$Q = \begin{cases} 0 & \text{for } x \le x_r, \quad A_v = 0 \\ C_d A_v \sqrt{2P/\rho} & \text{for } x > x_r, \quad A_v = \omega(x - x_r) \end{cases} \tag{5.5}$$

$$PA_P = k(x_o + x) \quad \text{Then} \quad x = P\frac{A_P}{k} - x_o \tag{5.6}$$

$$A_v = \omega\left(P\frac{A_P}{k} - x_o - x_r\right) \quad \text{or} \quad A_v = \omega\frac{A_P}{k}(P - P_r) \tag{5.7}$$

Then $$Q = C_d \omega \frac{A_P}{k}(P - P_r)\sqrt{2P/\rho} = K(P - P_r)\sqrt{P} \tag{5.8}$$

$$K = C_d \omega \frac{A_P}{k}\sqrt{2/\rho} \tag{5.9}$$

where A_v = Valve throttling area, m²
C_d = Discharge coefficient
P = Valve input pressure, Pa
x = Spool displacement, m
x_r = Spool overlap, m
ω = Valve throttling area proportionality coefficient, m

The steady-state characteristics of the valve are described by the pressure-flow rate relation, as illustrated in Fig. 5.3. This figure shows that the maximum pressure, P, corresponds to the relieved flow rate, Q_r. The pressure difference $(P - P_r)$ is called the override pressure. At a pressure $P = P_r$, the poppet is in equilibrium under the action of the pressure and spring pre-compression forces. The valve is closed and its flow rate is zero, assuming no leakage. In order to allow the oil to flow, the poppet should displace, which takes place at pressures higher than P_r. This increase in pressure is the override pressure. For higher flow rates, the override pressure is higher.

The slope of the $Q(P)$ curve depends on the proportionality coefficient, K [see Eq. (5.9)]. The valves operating at low flow rates and high-pressure levels are of small dimensions, ω and A, with a great stiffness spring, k. The constant K has a small value and the $Q(P)$ characteristic curve inclines more toward the P axis. In this case, greater

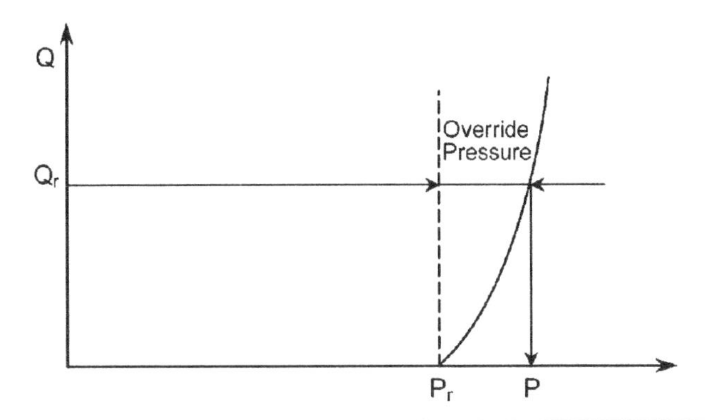

FIGURE 5.3 Steady-state characteristics of a relief valve.

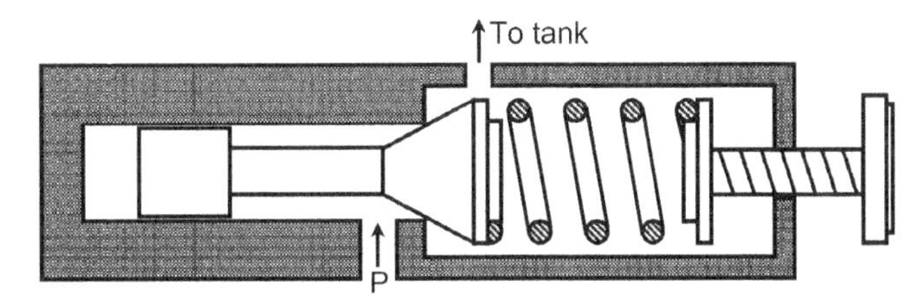

FIGURE 5.4 A direct-operated relief valve with damping spool.

override pressures are reached when the valve operates at greater flow rates. Thus, the maximum system pressure is much affected by the flow rate. Therefore, the valve should be designed to have a nearly vertical $Q(P)$ characteristic curve. In this way, the maximum pressure is not much affected by the variation of the relieved flow rate. For this reason, the value of coefficient K should be maximized by increasing the valve dimensions, A and ω, and decreasing the spring stiffness, k. But these requirements contradict one another since the valve should close under the action of the spring force (kx_o) against the large pressure forces ($\pi D^2 P/4$). This contradiction is solved by using the pilot-operated relief valves.

In the case of a direct-operated relief valve, the poppet is actually a spring-supported mass. This mass-spring system is subjected to very low viscous friction and spring material structural-damping forces. Then, in the steady state, this valve may suffer from sustained oscillations of the poppet, which results in observable pressure oscillations (see Fig. 5B.4). Therefore, it might be necessary to add a damping element to the valve. Figure 5.4 shows a direct-operated relief valve with a damping spool. Appendix 5B presents a detailed analysis of this valve.

5.2.2 Pilot-Operated Relief Valves

The problem of increased override pressure in the direct-operated relief valves is solved by using the pilot-operated valve design (see Fig. 5.5). The pilot-operated relief valve consists of a main valve (1) loaded by a spring (2). The valve is designed with a relatively

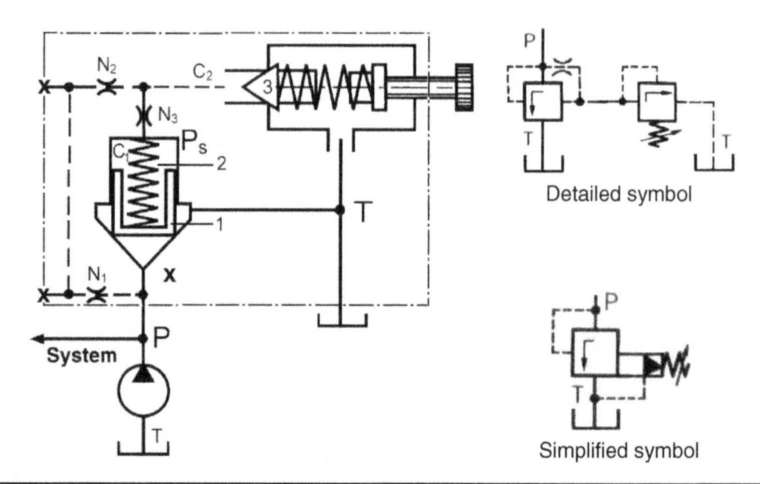

FIGURE 5.5 Functional scheme and symbols of a pilot-operated relief valve. (*Courtesy of Bosch Rexroth AG.*)

large diameter of the main poppet (1) and a small stiffness of the main valve spring (2), which decreases the override pressure. The spring is pre-compressed. But the pre-compression force can be overcome by a pressure difference of about 4–10 bar between the input pressure, P, and the spring chamber pressure, P_s.

The operation of the main valve is controlled by installing a pilot stage (3). The pilot valve is a direct-operated relief valve, connected to the input high-pressure line through the two nozzles, N_1 and N_2. The diameters of these nozzles are usually less than 1 mm. The flow rate passing through them is too small. The direct-operated relief valve (3) is used to impose an upper limit to the spring chamber pressure (2). The pilot stage has small dimensions and a very stiff spring. However, the override pressure is of negligible value in this valve, due to the very small flow rate.

When the supply pressure is less than the cracking pressure of the pilot stage, its poppet is seated and the pressures in the valve input line and spring chamber, C_1, are equal. At this condition, the pressure forces acting on the main poppet (1) are compensated. The spring (2) acts to close the main valve. When the pressure, P, becomes greater than the pilot valve cracking pressure, the pilot poppet valve opens, and the pressure in chambers C_1 and C_2 becomes equal to the relief pressure of the pilot stage, P_r. As the pilot valve flow rate increases, the dynamic depression in nozzles N_1 and N_2 results in a sufficient pressure difference $(P - P_s)$ between the inlet chamber and the chamber C_1. The main valve poppet moves to allow the fluid to flow to the return line. The pilot-operated relief valves allow for great flow rate with smaller override pressure, compared with the direct-operated relief valves.

Figure 5.6a illustrates the static characteristics of direct- and pilot-operated relief valves. Note that, in commercial catalogues, these characteristics are plotted with the flow rate on the horizontal axis and pressure at the vertical one (see Fig. 5.6b).

Figure 5.7 shows a pilot-operated, sandwich-type, relief valve. It consists of a housing (7) and a relief cartridge. The system pressure is set by the adjustment element (4). The pressure at port P acts upon the main poppet (1). Simultaneously, the high-pressure oil passes through the orifice (2) to the spring-loaded side of the main poppet (1), and

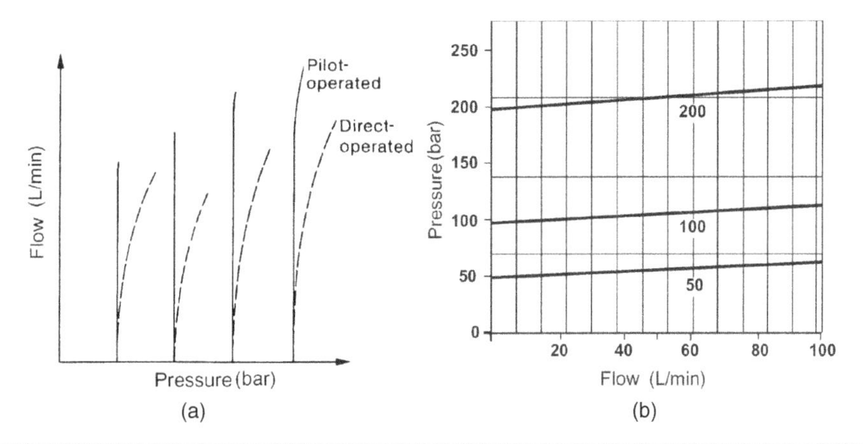

Figure 5.6 Typical flow characteristics of the relief valves. (a) Flow-pressure characteristics of direct- and pilot-operated relief values. (b) Typical characteristics of a commercial pilot-operated relief valve, as presented in the commercial documents.

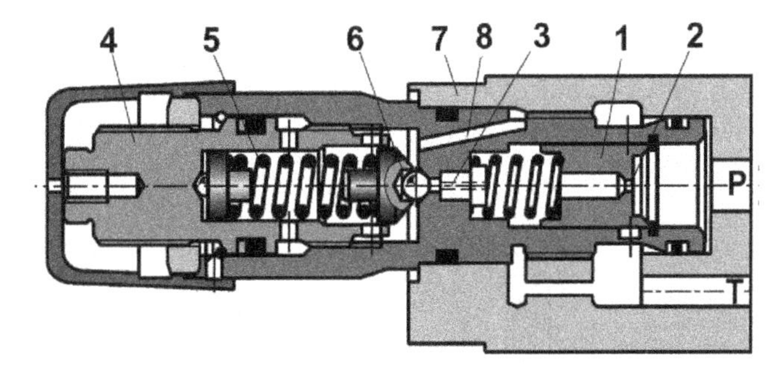

Figure 5.7 Pilot-operated relief valve, sandwich construction. (*Courtesy of Bosch Rexroth AG.*)

via the orifice (3), on to the pilot valve poppet (6). If the pressure at port P exceeds the value set on the spring (5), the pilot poppet (6) opens. The fluid flows from the spring-loaded side of the main poppet (1), through the orifice (3), and then the channel (8) to the return line (T). The created pressure drop across the orifice (2) acts on the main poppet (1). The main valve opens as this pressure drop exceeds its cracking pressure. Figure 5.8 shows another design of the pilot-operated relief valve with two separate cartridges for the main and pilot stages.

In addition to the override pressure reduction, the pilot-operated relief valves can be used for system unloading. It can also be used to produce multipressure limiting values. The symbols of these two applications of the pilot-operated relief valves are given in Fig. 5.9.

5.2.3 Pressure-Reducing Valves

Pressure reducers are used when a subsystem operates at a pressure lower than that of the main system. Generally, the pressure reduction and control is carried out by means of throttling elements. Figure 5.10 illustrates the principle of operation of hydraulic pressure reducers. Two throttles are used to connect the reduced-pressure line to the

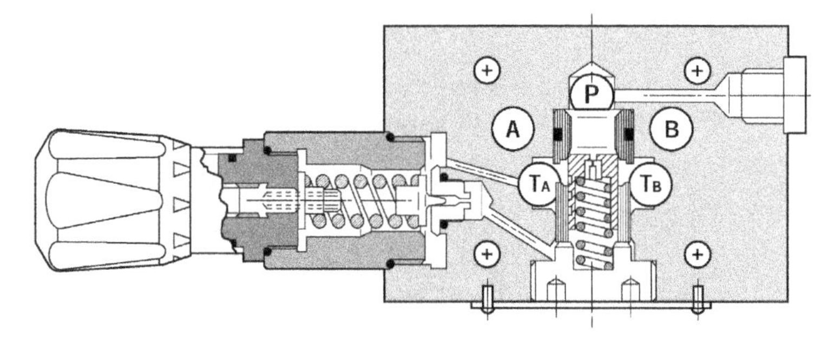

Figure 5.8 Construction of a pilot-operated relief valve. (*Courtesy of Bosch Rexroth AG.*)

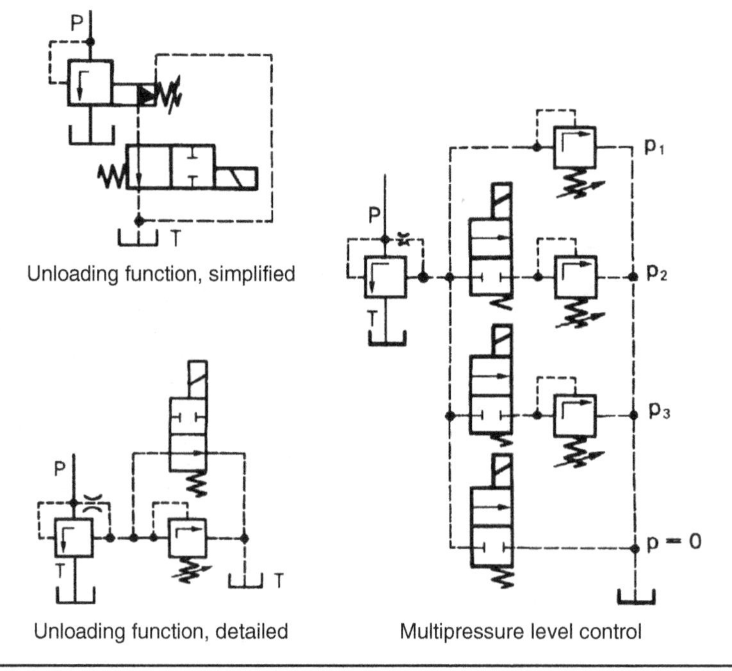

Figure 5.9 Typical applications of a pilot-operated relief valve.

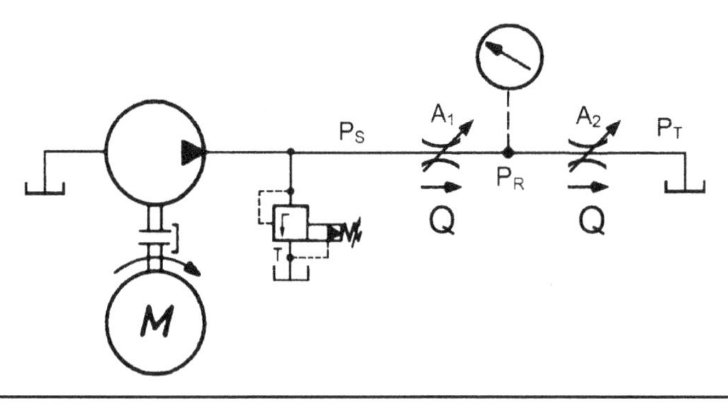

Figure 5.10 A hydraulic circuit illustrating the pressure reduction principle.

high-pressure line and return (tank) lines. The reduced pressure, P_R, is increased by increasing the area, A_1, or decreasing the area, A_2, and vice versa. An expression for the reduced pressure is deduced in the following:

$$Q = C_d A_1 \sqrt{2(P_S - P_R)/\rho} = C_d A_2 \sqrt{2(P_R - P_T)/\rho} \quad (5.10)$$

Assuming that the tank line pressure is zero, $P_T = 0$, then, for equal discharge coefficients,

$$P_R = \frac{A_1^2}{A_1^2 + A_2^2} P_S \quad (5.11)$$

The reduced pressure can be controlled by adjusting the throttle areas A_1 and A_2. When the first valve is fully closed, $A_1 = 0$, and the second one is open, $A_2 > 0$, the reduced pressure equals the return pressure, $P_R = P_T$, while for $A_2 = 0$ and $A_1 > 0$, the reduced pressure equals the supply pressure, $P_R = P_S$. Consequently, the reduced pressure can have any value between those of the supply and return line pressures.

Figure 5.11 shows a direct-operated pressure reducer. It consists of a spool (2) loaded by a spring (3). The pressure in the exit port (A) is connected to the control chamber, at the right hand side of the spool, via the line (6). It acts on the spool, against the spring (3). If the pressure in the exit port (A) is less than the value corresponding to the spring pre-compression force, the spool shifts to its extreme right-hand position. The pressure line (P) is then connected to the exit port (A). The pressure in the port (A) increases and the force acting on the spool increases. When this force overcomes the spring force, the spool moves to the left and the connection (P-A) is throttled. At the final position, the spool lands separate the line (A) from both the pressure and tank lines, except for the radial clearance. If the pressure increases to values greater than that preset at the spring, the spool moves further to the left. This spool displacement connects the line (A) to the tank, which decreases the pressure in the line (A). The spool displacement reaches a final steady-state value when the pressure and spring forces reach equilibrium. Thus, the value of the reduced pressure is simply adjusted by controlling the spring pre-compression. The reduced pressure is precisely adjusted at no-flow conditions. During the valve opening, the fluid flows from line (P) to line (A). The pressure in line (A) is determined by the loading conditions of the downstream subsystem. For each pressure level, the spool takes a corresponding position; $x = (P_A A_s - kx_o)/k$, and a corresponding flow rate is developed. In this case, the reduced pressure decreases with the increase of the valve flow rate.

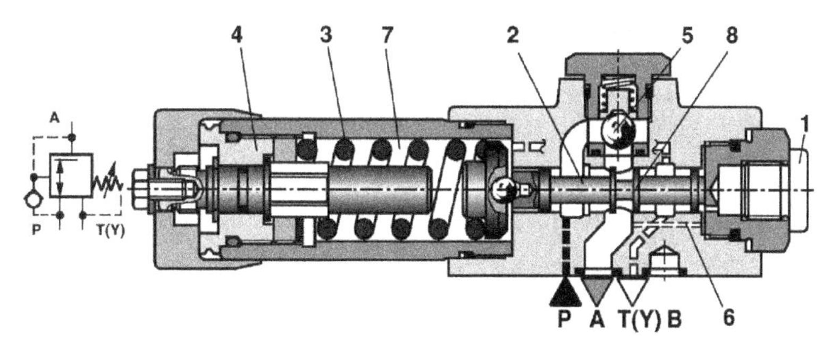

Figure 5.11 A direct-operated pressure reducer. (*Courtesy of Bosch Rexroth AG.*)

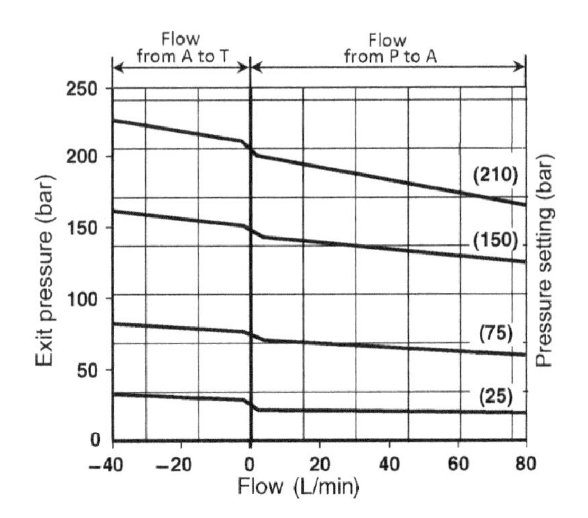

FIGURE 5.12 Static characteristics of a typical direct-operated pressure reducer. (*Courtesy of Bosch Rexroth AG.*)

Figure 5.12 gives the flow characteristics of a pressure reducer for different preset pressure levels. If the exit pressure, P_A, is increased, the spool moves to the left, allowing the fluid to flow from port (A) to the tank line (T), creating a negative flow rate. In this way, the valve acts as a pressure reducer and as a relief valve for line (A). A built-in check valve (2) allows for the free flow from line (A) to line (P) whenever needed. The different operating modes of a typical direct-operated pressure reducer are presented in Fig. 5.13. A hydraulic circuit illustrating the application of a direct-operated pressure reducer is given in Fig. 5.14.

In the case of increased flow rates, the pilot-operated pressure reducers are used. Figure 5.15 shows the construction of a pilot-operated pressure reducer. This reducing valve consists of the main housing (1), pilot valve (2), and main cartridge assembly (3). The pilot valve is a simple direct-operated relief valve. It limits the pressure behind the main valve to a maximum value, P_L. The flow rate through this valve is too small; therefore, it operates with a negligible override pressure. The spring chamber of pilot valve (14) is drained externally through passage (15). The main valve (13) is normally open, permitting flow from the high-pressure port (P) to the exit port (A). Port A is connected to the spring chamber in two ways, through the orifice (4) and through the orifices (7 and 10).

If the reduced pressure at port (A) is less than the limiting value P_L, the pilot stage is closed and the pressure in the chamber (13) equals the input pressure, and the pressure forces acting on the main spool are in equilibrium. The spring in the chamber (12) holds the main spool (13) open. When the pressure in port (A) becomes greater than (P_L), the pilot poppet (6) opens, allowing the orifices (4, 7, and 10) to maintain the flow over the pilot poppet (6). When pressure difference across the main spool exceeds the value needed to overcome the spring force in the chamber (12), the main spool (13) begins to close. The fluid flow from (P) to (A) throttles gradually. Thus, the main spool (13) reaches a final steady state when the resultant force acting on it is null. Or

$$P_A = P_L + k(x_o + x)/A_s \tag{5.12}$$

where A_s is the spool area.

The spool displacement x is actually too small compared with the spring precompression distance x_o.

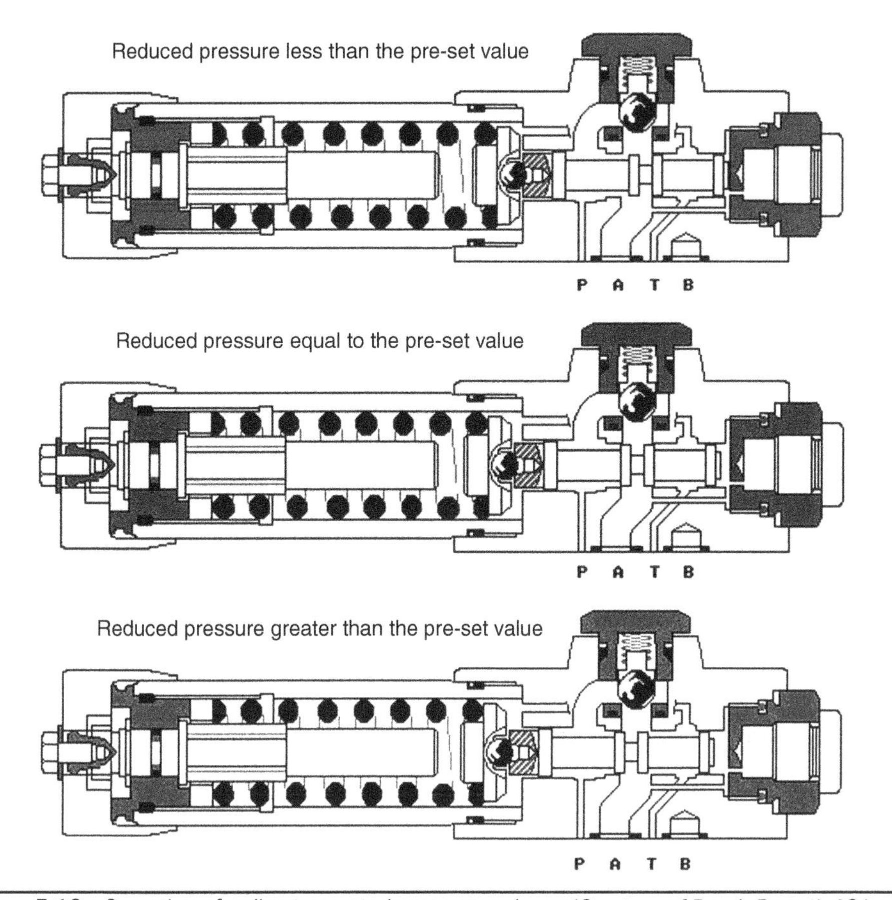

FIGURE 5.13 Operation of a direct-operated pressure reducer. (*Courtesy of Bosch Rexroth AG.*)

5.2.4 Sequence Valves

The sequence valves are used to create a certain sequence of operations according to the pressure level in the system. Figure 5.16 shows a direct-operated sequence valve consisting of a spool (2) loaded by a spring (3). The pressure in the inlet port (P) is connected to the control chamber at the right side of spool, through the passage (6). The pressure in this chamber acts on the spool against the spring force. If the pressure forces overcome the spring force, the spool displaces to the left connecting line (P) to (A). The valve can be externally controlled through port (B). In this case, the connection of port (P) with the control chamber should be blocked. Optionally, the valve is equipped with a check valve to allow for free reverse flow. The open and closed operating modes of a direct-operated sequence valve are shown in Fig. 5.17.

An expression for the valve steady-state flow rate is deduced in the following, neglecting the return line pressure:

$$F_S = k(x + x_o) \qquad F_P = A_S P_P \quad \text{and} \quad F_P = F_S \tag{5.13}$$

Then

$$x = P_P \frac{A_S}{k} - x_o \tag{5.14}$$

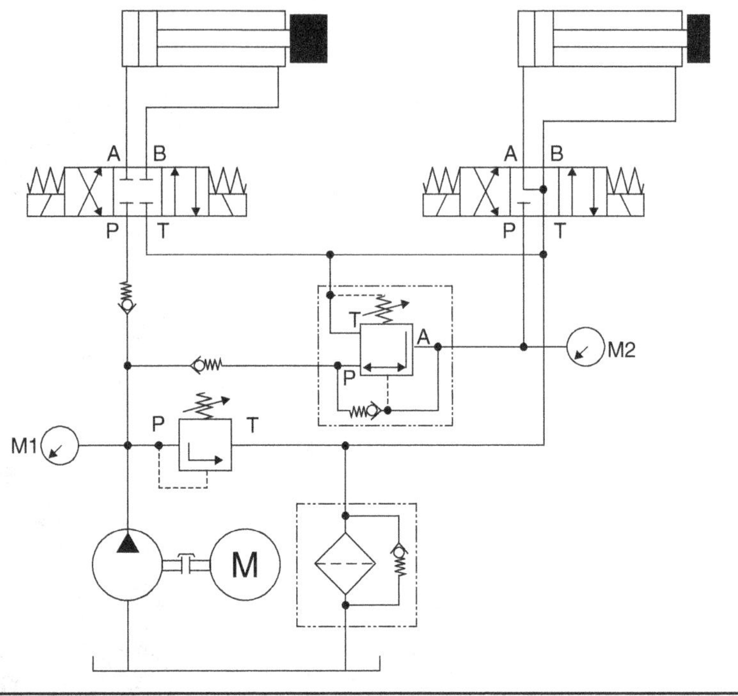

FIGURE 5.14 Typical application of the pressure reducer.

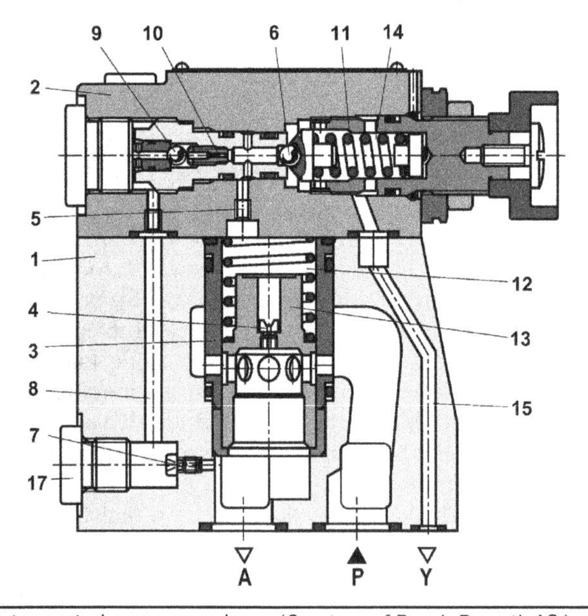

FIGURE 5.15 A pilot-operated pressure reducer. (*Courtesy of Bosch Rexroth AG.*)

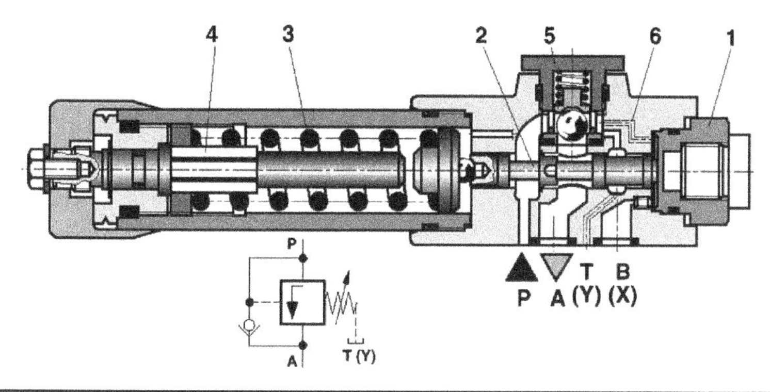

FIGURE 5.16 A direct-operated sequence valve. (*Courtesy of Bosch Rexroth AG.*)

The flow rate, Q, flowing from the input port (P) to exit port (A) in Fig. 5.17 is

$$Q = \begin{cases} 0 & x \le x_r \\ C_d \omega \left(P_P \dfrac{A_S}{k} - x_o - x_r \right) \sqrt{\dfrac{2}{\rho}(P_P - P_A)} & x \ge x_r \end{cases} \tag{5.15}$$

For a certain flow rate (Q), the pressure drop (ΔP) across the sequence valve is

$$\Delta P = P_P - P_A = \frac{\rho Q^2}{2C_d^2 \omega^2 \left(P_P \dfrac{A_S}{k} - x_o - x_r \right)^2} \tag{5.16}$$

where A_S = Spool area, m^2
P_P = Inlet line pressure, Pa
P_A = Exit line pressure, Pa
x_r = Spool overlap, m

Equation (5.16) shows that the pressure drop (ΔP) across the sequence valve is inversely proportional to the input pressure (P_p). The increase in the input pressure, above the cracking pressure, decreases the pressure and power losses in the valve. The pressure losses in a typical direct-operated sequence valve as well as its flow characteristics are shown by Fig. 5.18. These characteristics show that the pressure losses in the sequence valve, path (P-A) are of the same order of magnitude as those in the check valve.

The sequence valves are also called multifunction valves and are used in various configurations to control sequencing, braking, unloading, load counter balancing, or other functions. Figure 5.19 shows the hydraulic circuit of a typical application of the sequence valve.

Pilot-operated sequence valves (shown in Fig. 5.20) are used for applications requiring increased flow rates. They consist of a main housing with a cartridge assembly (1), a pilot valve (2), and an optional reverse free-flow check valve (3). The function of this valve varies depending on the pilot line (internal or external) and the drain line (internal or external).

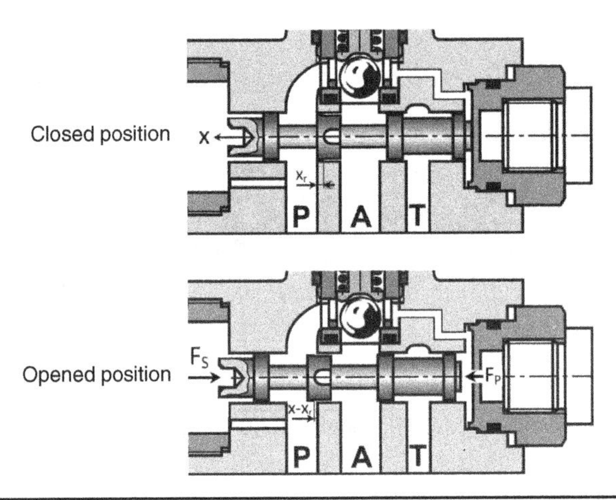

FIGURE 5.17 Operation of a direct-operated sequence valve.

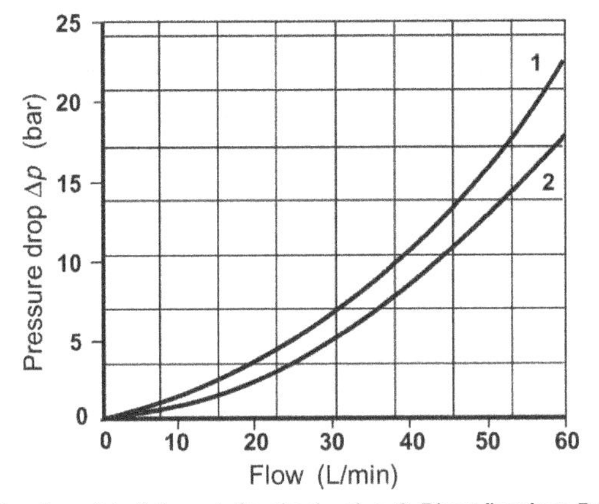

FIGURE 5.18 Static characteristics of a typical direct-operated sequence valve.

The pilot stage is a 2/2 direct-operated directional control valve of spool type (see Sec. 5.3.3). The spool is loaded by the adjustable force of the spring (8). For internal piloting, the plug (4.1) is removed and the plug (4.2) is installed. The control pressure acts on the spool through the plunger (5). For an externally piloted valve, the line (X) is active; the plug (4.1) is installed while the plug (4.2) is removed.

If the pilot pressure force is less than the spring force, the spool is shifted to the left. The spool land (10) closes the spring chamber of the main valve (7). The two sides of the main valve are interconnected by the orifice (6). The pressures at the input line and the spring chamber are equal and the spring acts to keep the main valve closed. On the other hand, if the control pressure force overcomes the force of the pilot-valve spring (8), the spool shifts to the right. The liquid flows from the input port (A) to the output port (B) through the orifices (6 and 9), the spool valve (10) and the passages (11 and 12).

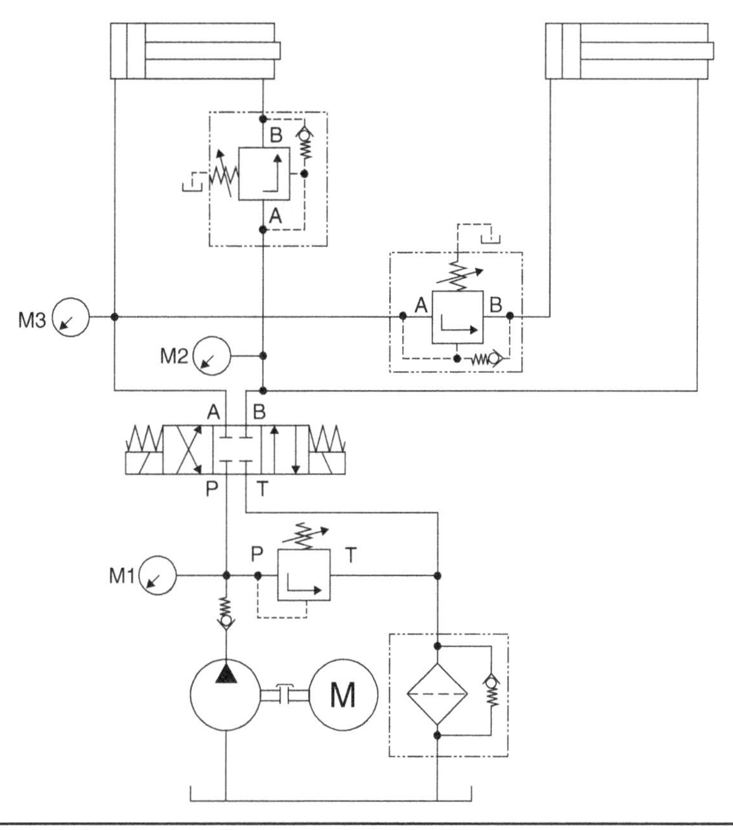

FIGURE 5.19 Typical application of a sequence valve.

The pressure difference developed across the orifice (6) acts to displace the main valve poppet upward. The main valve opens and connects the inlet port (A) with the exit port (B). The orifice (9) acts as a damping element.

5.2.5 Accumulator Charging Valve

The hydraulic accumulators are used to store hydraulic energy. (See Sec. 6.2.2.) They are installed in the hydraulic power systems for different reasons: energy storage, protection against hydraulic shock, pump unloading, and others. When using the hydraulic accumulator for pumps unloading, it is considered the main source of hydraulic energy in the system. The pump serves mainly to charge the hydraulic accumulator. In this case, the system operates between two operating pressure levels: maximum pressure P_2 and minimum pressure P_1. Figure 5.21 illustrates a construction and typical application of an accumulator charging valve. It operates to keep the pressure in the accumulator within the prescribed limits.

The accumulator charging valve consists of a pressure sensor (1) and a pilot-operated check valve (see Fig. 5.21a). The spool of the pressure sensor displaces downward as the accumulator pressure increases. The pilot-operated check valve

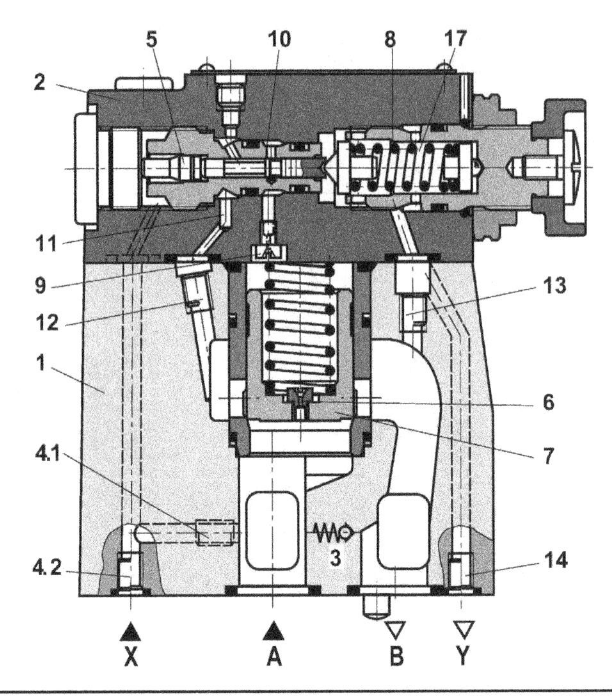

Figure 5.20 A pilot-operated sequence valve. (*Courtesy of Bosch Rexroth AG.*)

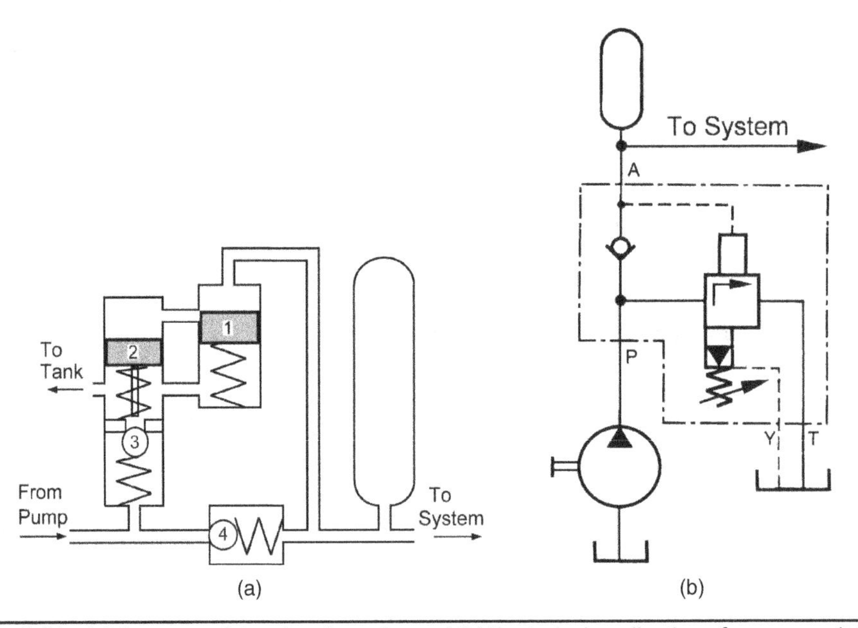

Figure 5.21 (a) An accumulator charging valve and (b) the typical application of an accumulator charging valve.

consists of a check valve (3) and a pilot piston (2). The check valve is forced to open if the pilot piston is displaced downward; otherwise, the poppet rests against its seat. When the pressure in the accumulator reaches the maximum value P_2, the spool of the pressure sensor (1) displaces downward to connect the high pressure line to the pilot piston (2). This piston displaces downward to open the check valve (3). The pump delivery line is then drained to the tank through the opened check valve and the pump operates unloaded. The check valve (4) prevents the pressurized oil in the accumulator from returning to the tank. During this operating mode, the accumulator supplies the system with the required high-pressure oil, which decreases the oil volume in the accumulator and decreases its pressure. Then, the piston (1) shifts upward. The oil in the upper chamber of the pilot piston is trapped. When the accumulator pressure reaches its minimum value, P_1, the pressure sensor spool reaches its upper position, and the pilot piston chamber is drained to the tank. The pilot piston moves upward under the action of its spring and the check valve (3) closes the pump bypass line. The pump flow is then redirected to recharge the accumulator.

5.3 Directional Control Valves

5.3.1 Introduction

Directional control valves (DCVs) are used to start, stop, or change the direction of fluid flow. These valves are specified by the number of connected lines (ways) and the number of control positions. The control positions determine the way in which the lines are interconnected, and consequently the directions of fluid flow. A 4/3 DCV has four ways and three positions. The application of a DCV in controlling the direction of motion of hydraulic cylinders is illustrated in Fig. 5.22. A 4/3 directional control valve is connected to the pressure line (P), return line (T), and cylinder lines (A and B). In its neutral position, the valve closes all of the four lines and the cylinder is stopped. By switching the valve to any of the other positions, the cylinder moves in the corresponding direction.

5.3.2 Poppet-Type DCVs

Poppet-type directional control valves are of two positions and three or four ways. The construction and operation of this class of valves is illustrated in Figs. 5.23 and 5.24. Generally, the direct-operated DCV of the poppet type operates at pressure levels up to 630 bar and flow rates up to 40 L/min.

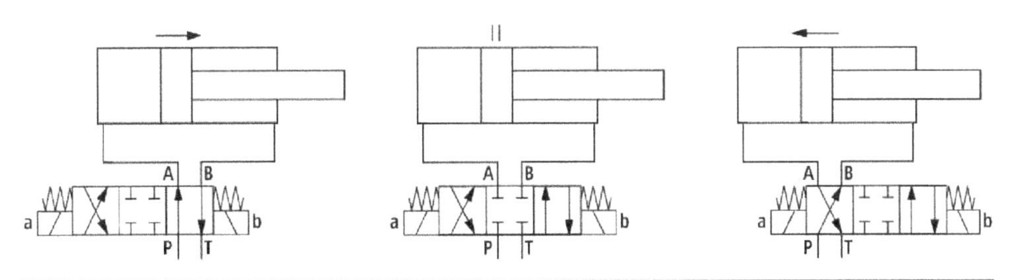

FIGURE 5.22 A hydraulic cylinder controlled by a 4/3 DCV.

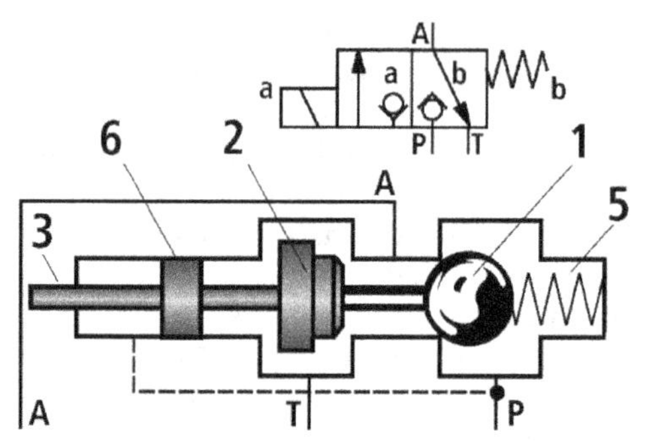

FIGURE 5.23 A 3/2 poppet directional control valve. (*Courtesy of Bosch Rexroth AG.*)

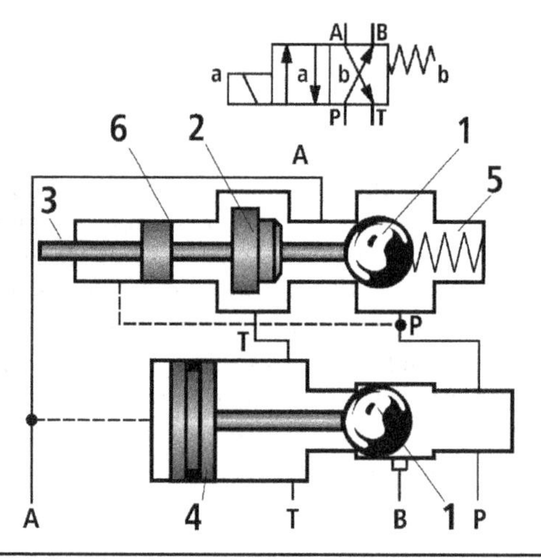

FIGURE 5.24 A 4/2 poppet directional control valve. (*Courtesy of Bosch Rexroth AG.*)

Figure 5.23 illustrates the construction of a 3/2 DCV of poppet type. When the plunger (3) is not depressed, the spherical poppet is seated under the action of the spring. The pressure line (P) is closed and the main poppet is in a mid position, connecting the exit port (A) with the tank (T). When the plunger is displaced, usually by an electric solenoid, it displaces the main and spherical poppets to the right. The tank port closes and the port (A) connects with the pressure line (P). The valve is equipped with an arrangement for compensating the pressure forces, acting on the moving parts. The 3/2 DCV can be converted to a 4/2 DCV, by adding a sandwich plate, including a control piston (4) and a spherical poppet (see Fig. 5.24).

5.3.3 Spool-Type DCVs

The spool valves are widely used in directional controls. They allow designing valves of two, three, four, five, or six ways, and even more, in addition to a wide variety of control positions. Figures 5.25 through 5.29 illustrate the construction and operation of

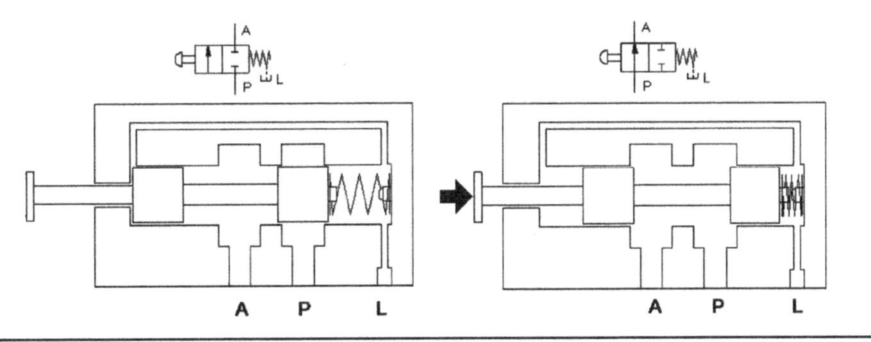

FIGURE 5.25 A 2/2 DCV of spool type.

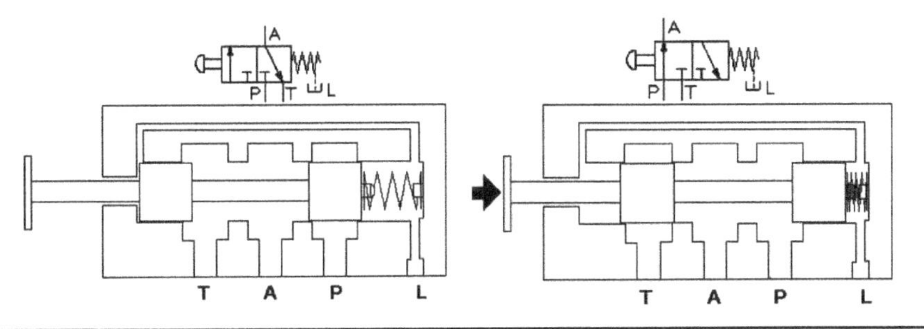

FIGURE 5.26 A 3/2 DCV of spool type.

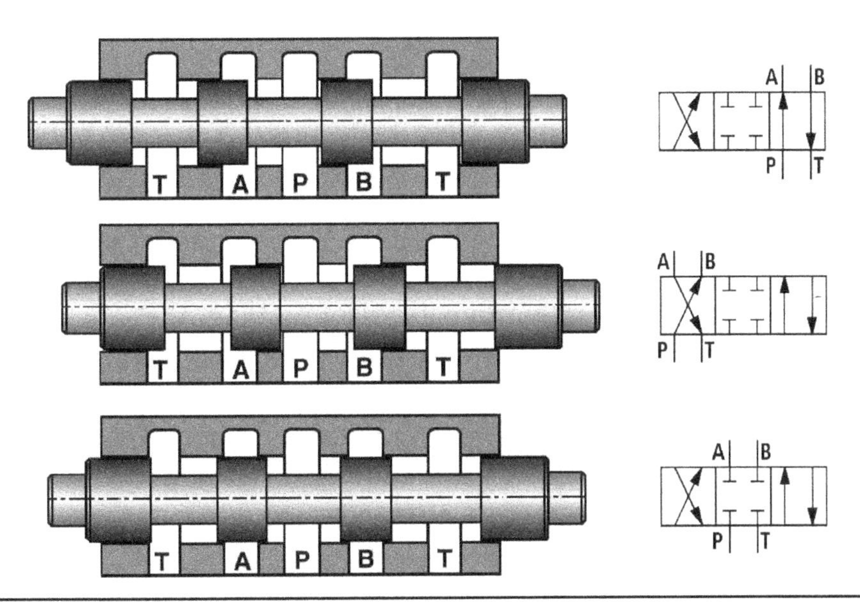

FIGURE 5.27 Operation of a 4/3 DCV of spool type (closed center).

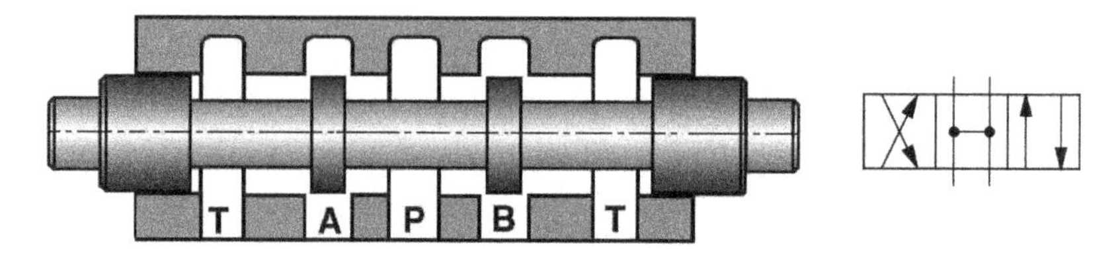

Figure 5.28 A 4/3 DCV (open center).

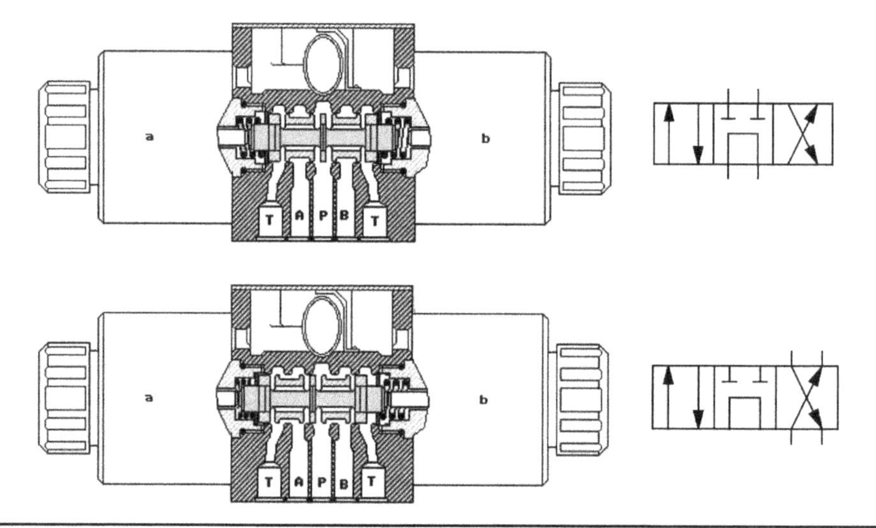

Figure 5.29 A 4/3 DCV with a bypass at the neutral position. (*Courtesy of Bosch Rexroth AG.*)

typical DCVs of spool type. These examples show that the spool can be designed to give the required way of connection of valve ports. Figure 5.30 shows the symbols of the most widely used industrial directional control valves.

5.3.4 Control of the Directional Control Valves

Basic Control Devices

Controlling a DCV means to switch the valve from one position to another. The main control devices for the DCV are shown in Table 5.2, while the standard symbols of the different controllers are illustrated by Fig. 5.31.

Electric Solenoids

The electric solenoid actuators are widely used for the control of a wide variety of hydraulic valves, and are available for many voltages in both AC and DC versions. Both air gap solenoids and wet pin solenoids are also used.

Basic Solenoid Operation Whenever electric current flows through a wire, it creates a magnetic field around that wire. If the wire is wound into a coil, the magnetic field is

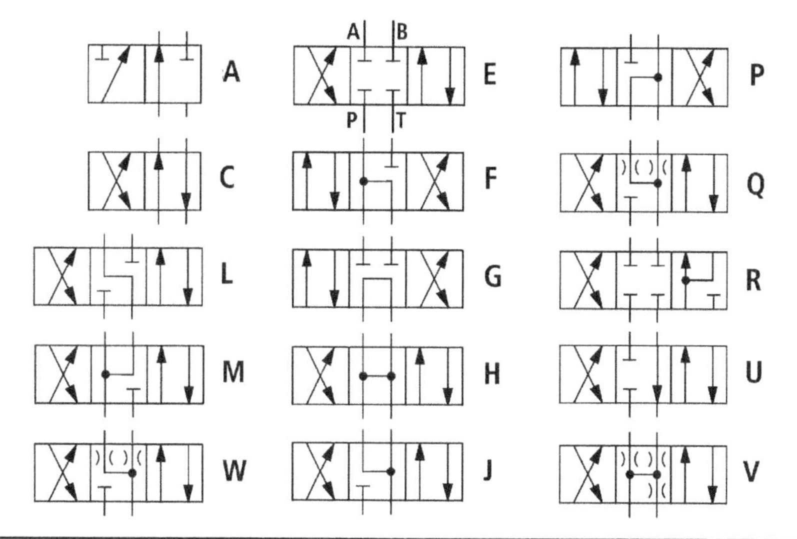

FIGURE 5.30 Symbols of the most used configurations of industrial DCVs.

Control Device	Illustration
Mechanical control by hand lever	
Mechanical control by cam and roller	
Mechanical control by rotary knob	

Courtesy of Bosch Rexroth AG.

TABLE 5.2 Control Devices of Directional Control Valves

Control Device	Illustration
Hydraulic control	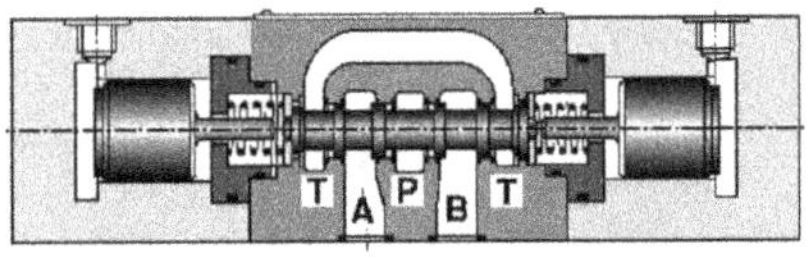
Pneumatic control	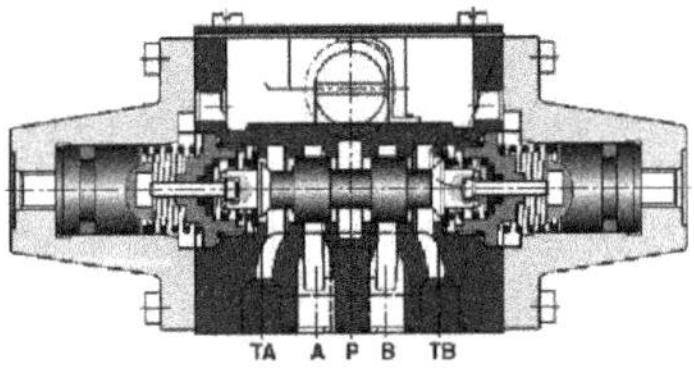
Electric control by direct current solenoids	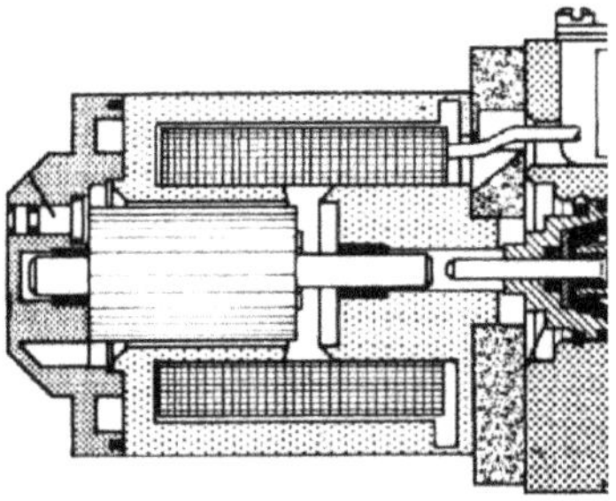
Electric control by alternating current solenoids	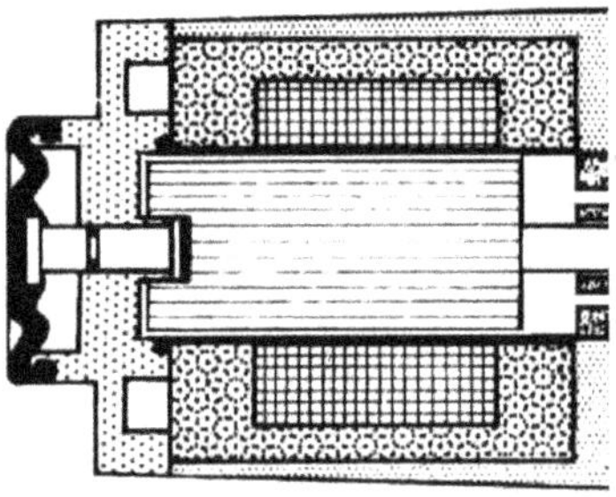

TABLE 5.2 Control Devices of Directional Control Valves (*Continued*)

generated around the coil windings (see Fig. 5.32). The magneto-motive force, λ, is proportional to the number of turns of coil, N, and electric current, i.

$$\lambda = Ni \tag{5.17}$$

The magnetic field builds up more readily in soft magnetic material such as iron or steel than it does through air. Therefore, the magnetic field can be concentrated by adding a "C-frame" of iron around the outside the coil (see Fig. 5.33). Then, if a movable iron core is placed inside the coil, the magnetic field will be more intense when the core is in such a position that the C-frame and core are totally within the magnetic field.

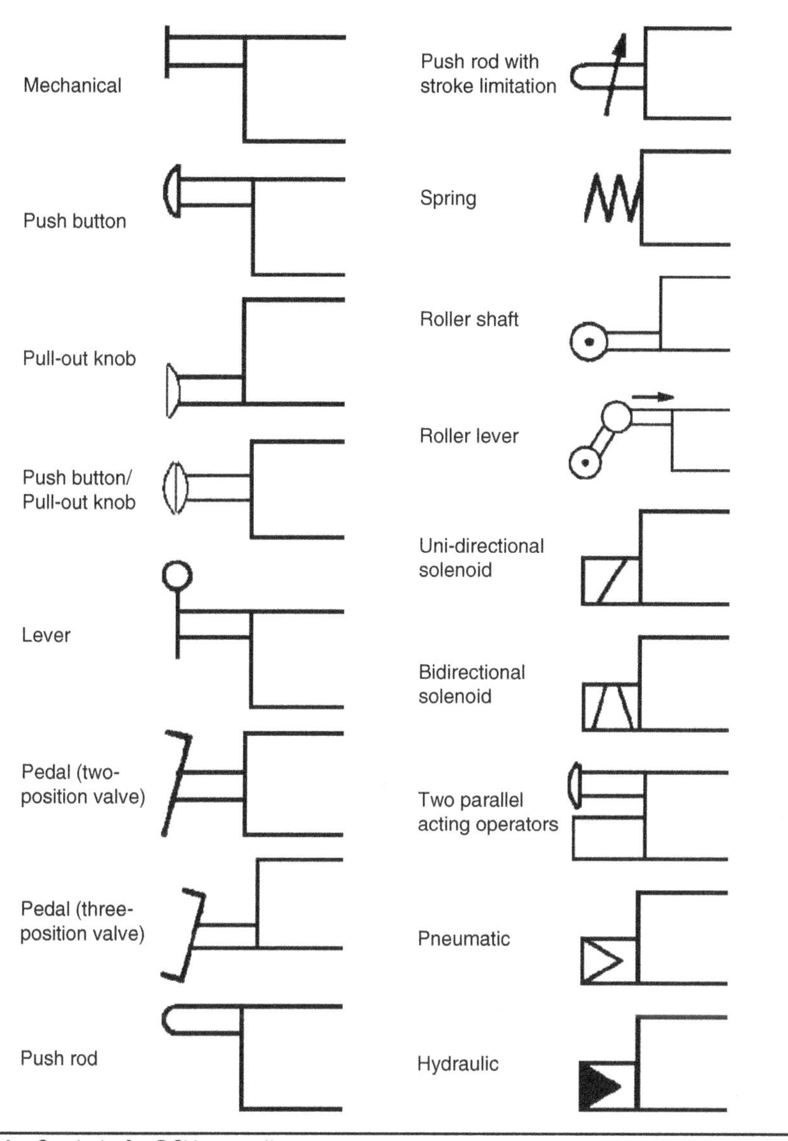

FIGURE 5.31 Symbols for DCV controllers.

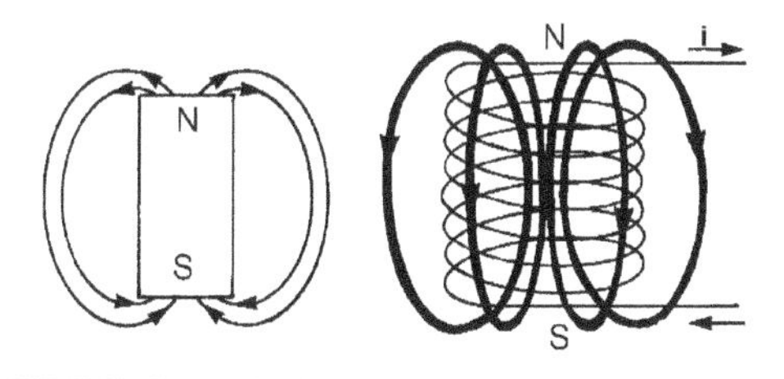

FIGURE 5.32 Magnetic field due to permanent- and electro-magnets.

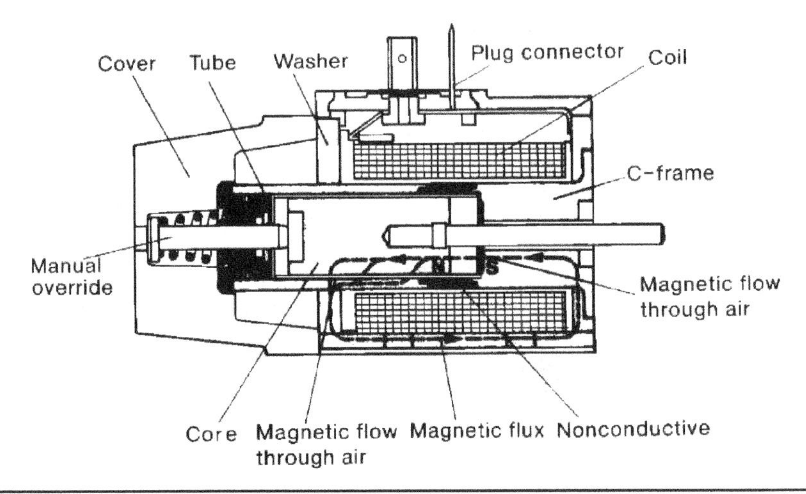

FIGURE 5.33 Construction of a DC solenoid. (*Courtesy of Bosch Rexroth AG.*)

The solenoid force will develop when the coil is energized. This force will pull the core to its point of equilibrium at the mid-position of the C-frame.

DC Solenoids In the case of direct current solenoids, the current develops a magnetic field of fixed polarity. Both the C-frame and the core will be magnetized with definite north and south poles. When a current is applied to a DC solenoid, the north and south poles of the C-frame attract the south and north poles of the core, respectively. The resultant attraction force is the solenoid force.

The DC solenoids are practically safe from burning out if the correct voltage is applied. The solenoid force depends not only on the solenoid design and current but also on the core position as shown by Fig. 5.34. It decreases as the core approaches its point of equilibrium. Therefore, the core displacement should be limited (see Fig. 5.33), otherwise the solenoid force drops below the required value. The direction of this force reverses if the core is shifted over its point of equilibrium, but it does not change if the current polarity is reversed.

The available commercial solenoids produce a force within 60 to 70 N. For a greater force, the number of turns of coil or current should be increased, which increases the solenoid volume to inconvenient values.

AC Solenoids Although the AC solenoid functions in the same manner as the DC model, its magnetic field is influenced by the alternating current. This has the net result of changing the polarity of the magnetic field at the same rate. The magnetic force is high only when the AC current is at its positive or negative peak. The change in magnitude of electric current induces an electromotive force and eddy currents in the metallic parts of the magnetic circuit, mainly the C-frame and core. Therefore, these parts are produced from electrically isolated laminates of metals (see Fig. 5.35). In this way, the eddy currents and associated heating problems are reduced. The problem of overheating in AC solenoids is treated by using wet pin solenoids, which are also used in some DC solenoids because of the following advantages:

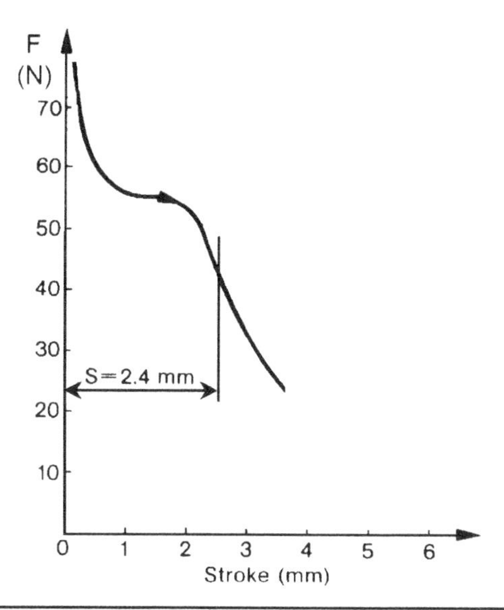

Figure 5.34 A typical solenoid force-stroke relation.

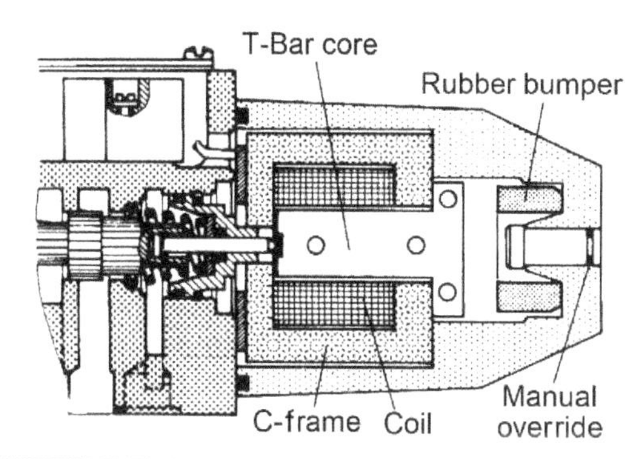

Figure 5.35 An AC solenoid. (*Courtesy of Bosch Rexroth AG.*)

- Better heat dissipation.
- A dynamic pushpin seal is not needed.
- Quiet operation.
- The moisture problem is eliminated.

As the current changes from positive to negative, or vice versa, it must pass through a neutral point where there is no current. During this short period of time, a point is reached where no magnetic force exists. Without this magnetic attraction, the load can push the core slightly out of equilibrium. Then, as the current builds up, the magnetism increases and pulls the core back. This movement of the core in and out at a high cycle rate creates noise, which is commonly referred to as buzz.

Property	DC	AC
Switching time	50–60 ms	Within 20 ms
Service-life expectations	20 to 50 million cycles	10 to 20 million cycles
Max. switching frequency	Up to 4 cycles/second	Up to 2 cycles/second
Continuous operating period	Practically unlimited	15–20 min for dry solenoids 60–80 min for wet solenoids
Costs (relative)	1	1.2
Occurrence rate	10	2

TABLE 5.3 Comparison of DC and AC Solenoids

To eliminate "buzz" and to increase the solenoid's holding power, most AC solenoids incorporate a shading coil. A magnetic field passing through a coil of wire induces an electric current in the same way electric current passing through a coil of wire creates a magnetic field. The flow of current in the shading coil creates its own magnetic field. The current produced in the shading coil lags behind the applied current to the coil inductance. When the applied current passes through the zero value in its change from one polarity to another, the current, and thus the magnetic field of the shading coil, are at their maximum value. When the solenoid is used within its force rating, the magnetic field of the shading coil is sufficient in strength to keep the core in position, thus eliminating the buzz. Table 5.3 gives a brief comparison of DC and AC solenoids.

5.3.5 Flow Characteristics of Spool Valves

The spool valves are classified into three types according to spool land length: overlapping (positive), zero-lapping (ideal), and under-lapping (negative). (See Fig. 5.36.)

The valve is said to be of overlapping type if the spool land length is greater than the valve opening width (see Fig. 5.37). Assuming that the valve throttle area is linearly proportional to the valve opening, and neglecting the radial clearance leakage, the flow rate through the spool valve is given by the following relations:

$$\text{For} \quad |x| < \varepsilon, \quad A_v = 0, \quad Q = 0 \tag{5.18}$$

$$\text{For} \quad \varepsilon < |x| < (\varepsilon + a), \quad A_v = \omega(|x| - \varepsilon), \quad Q = C_d \omega(|x| - \varepsilon)\sqrt{2\Delta P/\rho} \tag{5.19}$$

$$\text{For} \quad |x| \geq \varepsilon + a, \quad A_v = \omega a, \quad Q = C_d \omega a \sqrt{2\Delta P/\rho} \tag{5.20}$$

where A_v = Throttling area, m^2
x = Spool displacement, m
ε = Overlapping length, m
ω = Width of the valve port, for annulus area $\omega = \pi D$, m

The flow characteristics of the three basic types of spool valves are better illustrated by assuming a constant pressure difference across the valve restriction.

$$\Delta P = P_P - P_A = P_A - P_T = \text{Const.} \tag{5.21}$$

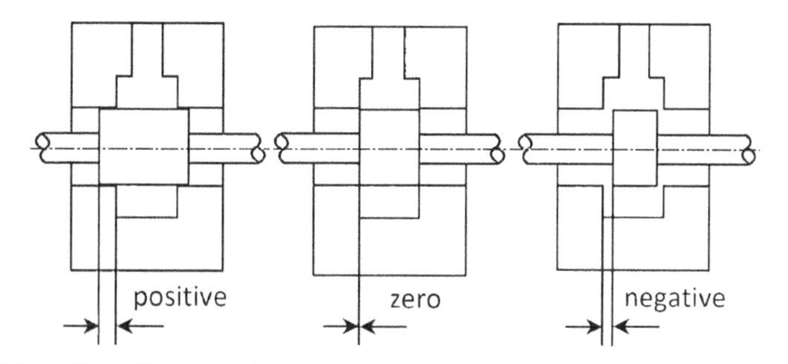

FIGURE 5.36 Spool valve classification according to spool land length.

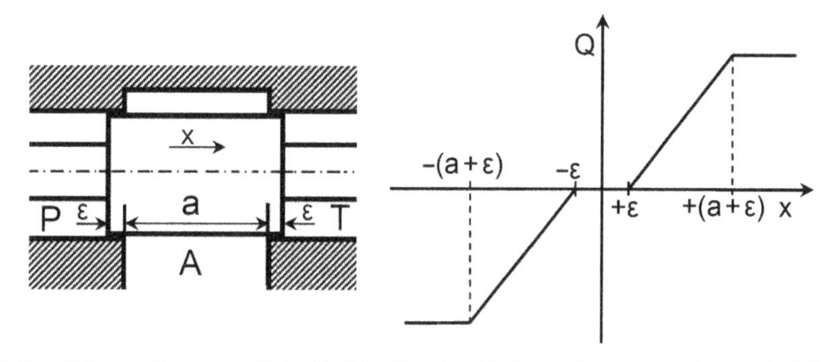

FIGURE 5.37 Flow characteristics of the over-lapping spool valve.

The flow characteristics of the overlapping valve are plotted in Fig. 5.37. The flow from (P) to (A) is positive and from (A) to (T) is negative. The valve has a dead zone of width 2ε. The valve flow rate reaches the saturation conditions when the valve port is completely cleared out; $|x| \geq \varepsilon + a$. This class of spool valves is convenient for the directional control valves. The overlap reduces the internal leakage and production cost. Meanwhile, nonlinearity of flow characteristics does not affect the DCV operation since the spool is displaced to its extreme position without stopping within the dead zone.

The ideal zero-lapping valve has a spool land length equal to the width of the valve port (see Fig. 5.38). The flow rate through the valve port is given by the following equations:

$$\text{For} \quad |x| < a, \quad A_v = \omega\,|x|, \quad Q = C_d \omega\,|x| \sqrt{2\Delta P/\rho} \tag{5.22}$$

$$\text{For} \quad |x| \geq a, \quad A_v = \omega a, \quad Q = C_d \omega a \sqrt{2\Delta P/\rho} \tag{5.23}$$

The flow-displacement relation of the zero-lapping valve is plotted in Fig. 5.38. The valve shows a linear flow-displacement relation for ($-a < x < a$). On the other hand, the zero-lap increases the internal leakage and the production cost. This class of spool valves is convenient for the servovalves due to their linear characteristics.

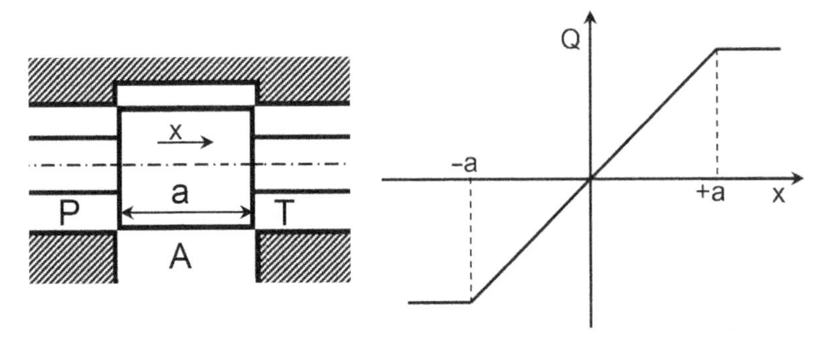

FIGURE 5.38 Flow characteristics of a zero-lapping spool valve.

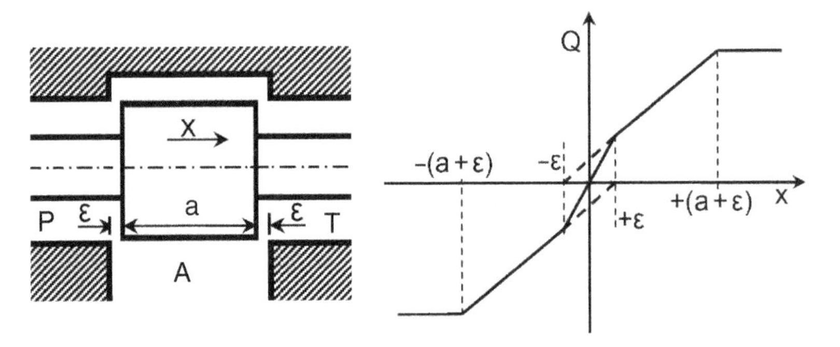

FIGURE 5.39 Flow characteristics of an under-lapping spool valve.

The construction and flow characteristics of the under-lapping valve are shown in Fig. 5.39. This valve design gives increased internal leakage and higher flow gain for spool displacements within the under-lapping zone. This class of spool valves is used in ordinary directional control valves and sometimes in proportional and servo-valves.

5.3.6 Pressure and Power Losses in the Spool Valves

Flow rates through the spool valve restrictions, Q, and the pressure drop across it, ΔP, are related by the following expressions:

$$Q = C_d A_v \sqrt{2\,\Delta P/\rho} \quad \text{or} \quad \Delta P = \frac{\rho}{2C_d^2 A_v^2} Q^2 \tag{5.24}$$

The loss of power, ΔN, due to oil flow through the valve restriction is given by the following relations:

$$\Delta N = Q\,\Delta P = C_d A_v \Delta P \sqrt{2\,\Delta P/\rho} \quad \text{or} \quad \Delta N = \frac{\rho}{2C_d^2 A_v^2} Q^3 \tag{5.25}$$

It is important to reduce the pressure and power losses in the valves. Therefore, the restriction area, A_v, should be increased as much as possible. The manufacturers of DCVs offer the characteristic chart, shown in Fig. 5.40. This chart gives the pressure-flow relation for the directional control valves, size 6, illustrated by Figs. 5.27 through 5.30.

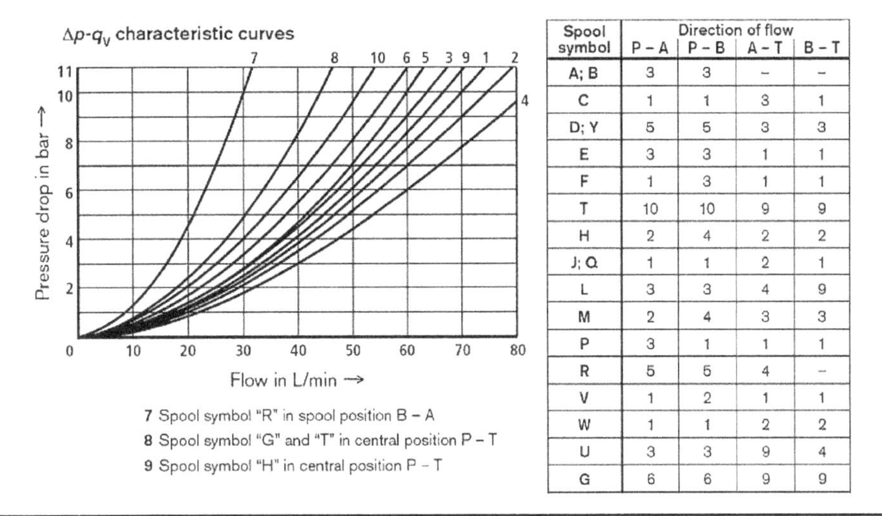

Spool symbol	Direction of flow P – A	P – B	A – T	B – T
A; B	3	3	–	–
C	1	1	3	1
D; Y	5	5	3	3
E	3	3	1	1
F	1	3	1	1
T	10	10	9	9
H	2	4	2	2
J; Q	1	1	2	1
L	3	3	4	9
M	2	4	3	3
P	3	1	1	1
R	5	5	4	–
V	1	2	1	1
W	1	1	2	2
U	3	3	9	4
G	6	6	9	9

FIGURE 5.40 Typical pressure drops in a size 6 DCV for different inner paths. (*Courtesy of Bosch Rexroth AG.*)

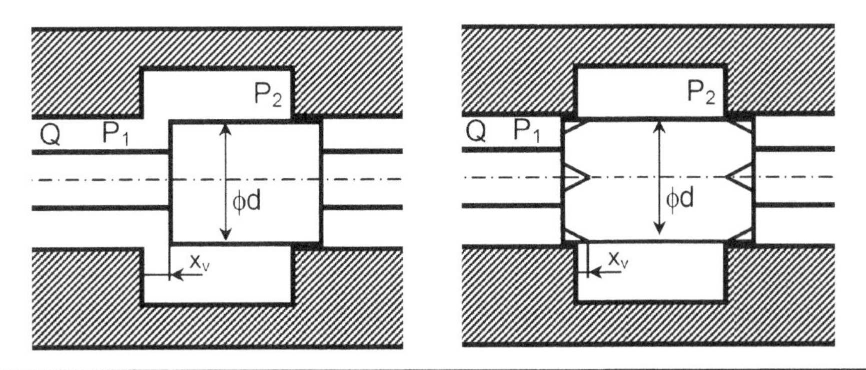

FIGURE 5.41 A spool with and without control edges.

Generally, the throttle area, A_v, changes with the spool valve opening, x_v, depending on the spool geometry, as illustrated by Fig. 5.41 and discussed in App. 5A. For the annulus throttling area of a spool of diameter d, the throttling area is $A_v = \pi d x_v = \omega x_v$. Therefore, the pressure losses are reduced by increasing the spool diameter and stroke. These requirements are in contradiction to the requirements of minimum dimensions and weight. Moreover, the increase in the diameter increases the mass, decreases the valve natural frequency, and increases the response time. Therefore, for small flow rates, the spool dimensions may be minimized, while the increase in flow rate imposes the necessity for dimensions to increase.

Sometimes, the spools are machined with control notches or grooves of different shapes at the edge of the spool land. These grooves are designed to change the valve throttle area in relation with the spool displacement according to certain laws. This may be intended for gradual compression or decompression of oil.

5.3.7 Flow Forces Acting on the Spool

The fluid flow results in a steady-state force equal to the rate of change of the momentum of the fluid. Considering a machine bounded by a control volume (Fig. 5.42), where the inlet fluid velocity is v_1 the exit velocity is v_2, the rate of fluid flow is Q, and the fluid density ρ is constant, the machine is subjected to a momentum force, F_j, given by the following formula:

$$\vec{F_j} = \rho Q(\vec{v_1} - \vec{v_2}) \tag{5.26}$$

Considering the throttling of oil in the valve restriction, the oil velocity in the valve chamber is much smaller than that in the throttle section. In the case of fluid flow out of the valve (Fig. 5.43a), the exit oil velocity is given by

$$v_2 = \frac{Q}{C_c A_v} \tag{5.27}$$

where C_c = Contraction coefficient. Then, for $v_1 << v_2$,

$$\vec{F_j} = -\rho Q \vec{v_2} \tag{5.28}$$

The negative sign means that the momentum force acts in the direction opposite to that of the exit oil velocity. The magnitude of the momentum force is deduced as follows:

$$F_j = \frac{\rho Q^2}{C_c A_v} \tag{5.29}$$

$$Q = C_d A_v \sqrt{2\Delta P/\rho} \tag{5.30}$$

$$F_j = \frac{2C_d^2 A_v}{C_c}\Delta P \tag{5.31}$$

If the radial clearance is negligible compared to the valve opening distance, the exit flow and jet forces are in a direction inclined to the spool axis by an angle $\vartheta = 69°$. This force is decomposed into two components: axial F_x and radial F_r.

$$F_x = F_j \cos\vartheta \quad \text{and} \quad F_r = F_j \sin\vartheta \tag{5.32}$$

The radial component of the jet forces increases the friction and wear of spools. Therefore, the inlet and exit ports of the valve should be distributed ax-symmetrically to compensate for the radial components of the momentum forces. However, the axial

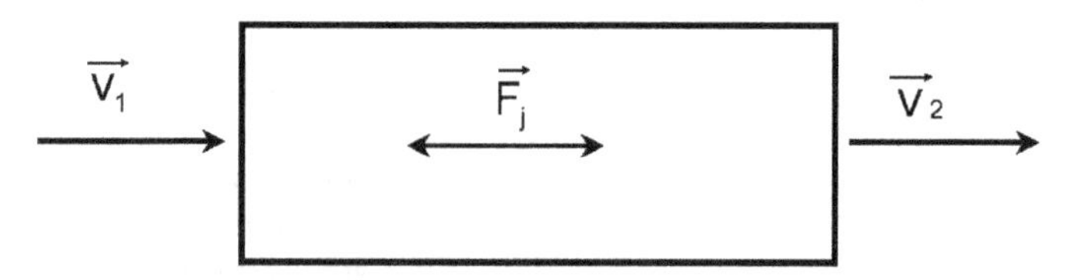

FIGURE 5.42 Fluid flow across system boundaries.

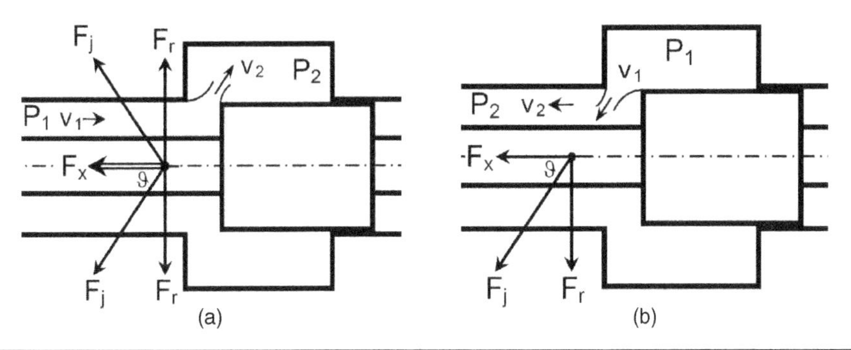

Figure 5.43 Momentum: jet reaction, forces.

components are added together and produce an axial resultant force. This force acts in the direction of valve restriction closure. Thus, the valve controller should act on the spool by the force necessary to displace the spool or poppet against the momentum force in addition to the other loading forces.

$$F_x = \frac{2C_d^2 A_v}{C_c} \cos\vartheta \, \Delta P \tag{5.33}$$

For $C_d = 0.611$, $C_c = 0.62$ and $\vartheta = 69^o$, $A_v = \omega x$

Then $$F_x = 0.428\omega \Delta P x = Kx \tag{5.34}$$

If the fluid flow is inlet to the spool valve (Fig. 5.43b), then $v_1 >> v_2$ and the momentum force is

$$\vec{F_j} = \rho Q \vec{v_1} \tag{5.35}$$

Thus, the momentum force acts all the time in the direction to close the valve restrictions, regardless of the direction of flow. It acts as a centering spring.

5.3.8 Direct-Operated Directional Control Valves

In the case of valves operating for low flow rates, the valve restriction areas are small and the flow forces are of negligible value. The controller can act directly on the spool by the force required to overcome the spring compression, friction, momentum, and inertia forces. These valves are called direct-operated valves (see Fig. 5.44). The increase in the valve flow rate increases the momentum forces and imposes the need for greater dimensions of the valve. Therefore, the valve controller should act by greater force to drive the spool. In the case of valves controlled by electric solenoids, the solenoid mass and volume increase rapidly as the force increases. This is due to the need for a coil with a greater number of turns and greater wire diameter in order to support the higher electric current. Therefore, the electric solenoids are used for the direct-operated DCV of flow rates within 100 L/min. For greater flow rates, the direct-operated directional control valves become inconvenient due to the limitation of the solenoid force. In this case, the pilot-operated DCVs are used. However, direct operated DCVs, controlled mechanically, can operate at higher flow rates.

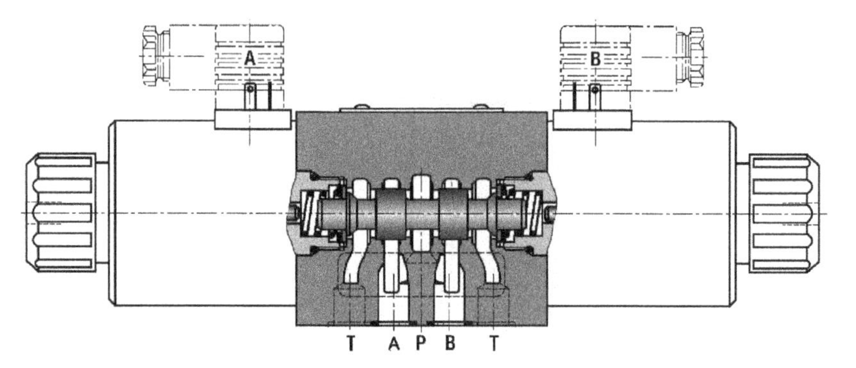

Figure 5.44 A 4/3 direct-operated directional control valve. (*Courtesy of Bosch Rexroth AG.*)

5.3.9 Pilot-Operated Directional Control Valves

The pilot-operated directional control valve consists of two stages: a pilot valve and a main valve. Figures 5.45 through 5.47 show functional schematics and symbols of this class of valves. The pilot valve is a direct-operated directional control valve, controlled electrically in this example. The valve is supplied by the high-pressure oil from the main valve high-pressure port (P). The oil is also drained through the main valve return line (T). With the pilot valve spool in the neutral position, the control chambers (C and D) are drained to the tank and the main spool is put in its neutral position. When the solenoid (a) is energized, the pilot valve spool moves to the right. The high-pressure oil reaches the chamber (D) and the main spool moves to the left.

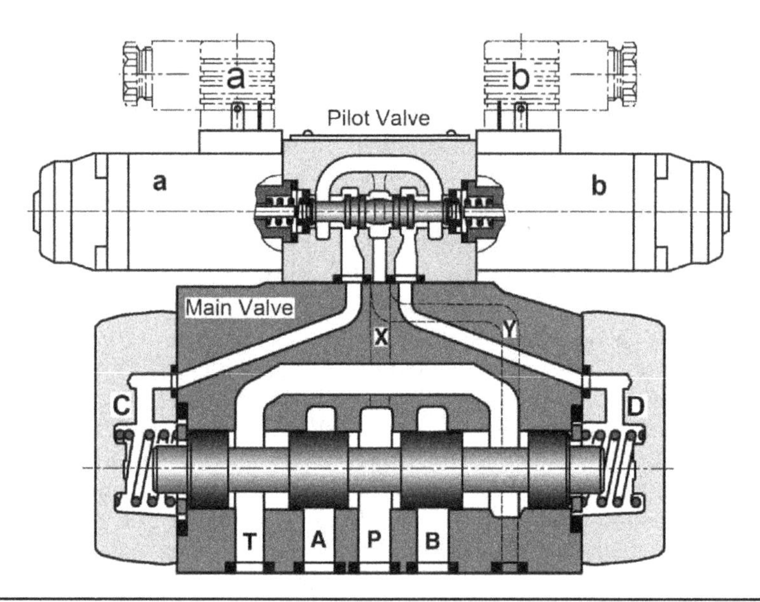

Figure 5.45 The functional schematic of a pilot-operated DCV.

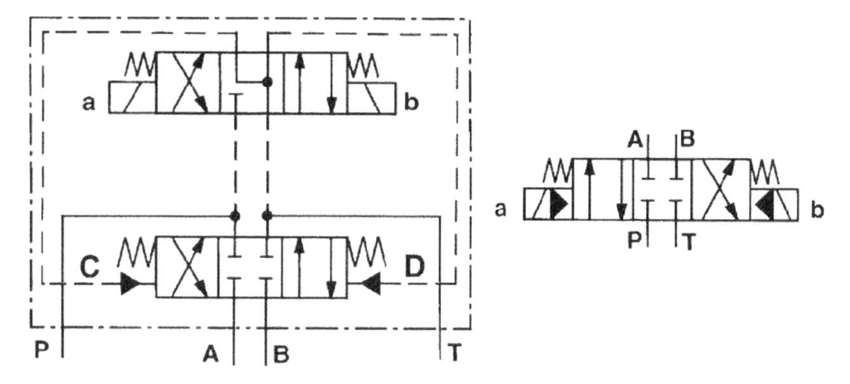

Figure 5.46 The detailed and standard symbol of a pilot-operated DCV with an internal pilot supply and return.

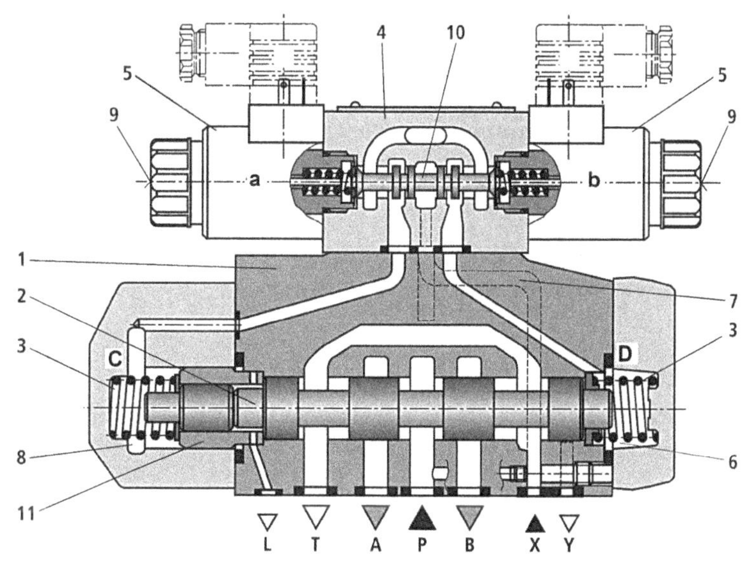

Figure 5.47 A 4/3-way pilot operated directional valve with pressure centering of the main control spool. (*Courtesy of Bosch Rexroth AG.*)

In the same way, the connection of electric power to the solenoid (b) results in a motion of the main spool to the right.

The supply and return lines of the pilot stage are mostly connected to the main stage supply and return lines. However, optionally, the valve may include an arrangement for switching any of them to external ports.

5.4 Check Valves

The check valves are generally used to allow for free flow in one direction, and prevent (obstruct) the fluid flow in the opposite direction. Figure 5.48 shows a classification of the check valves.

5.4.1 Spring-Loaded Direct-Operated Check Valves

The direct-operated check valves consist of a simple poppet valve with a poppet loaded by a spring (see Fig. 5.49). The poppet rests against its seat, obstructing the direction from (B) to (A). It allows the fluid flow in the direction (A) to (B) if the pressure difference $(P_A - P_B)$ is greater than the cracking pressure P_r, defined as the pressure diference which produces a pressure force equal to the spring force. The cracking pressure is usually less than 10 bar for the check valves.

$$(P_A - P_B)A_P = kx_o \tag{5.36}$$

$$P_r = P_A - P_B = kx_o/A_p \tag{5.37}$$

where A_p = Poppet area subjected to pressure difference, m^2
P_r = Cracking pressure, Pa
x_o = Spring pre-compression distance, m

5.4.2 Direct-Operated Check Valves Without Springs

Some applications require a very low cracking pressure. In this case, the check valve is designed with springs of very low stiffness or even without springs (see Fig. 5.50). These valves operate with a cracking pressure less than 0.2 bar. Their symbol is drawn without the spring.

5.4.3 Pilot-Operated Check Valves Without External Drain Ports

Some applications, such as the hydraulic locking of hydraulic cylinders, require the installation of check valves. In certain operating modes of these systems, it is recommended to

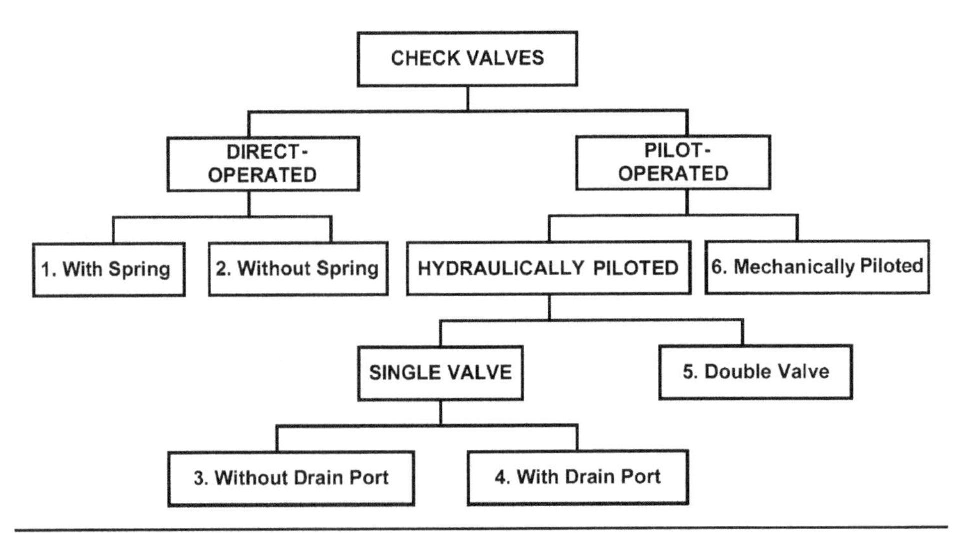

FIGURE 5.48 Classification of check valves.

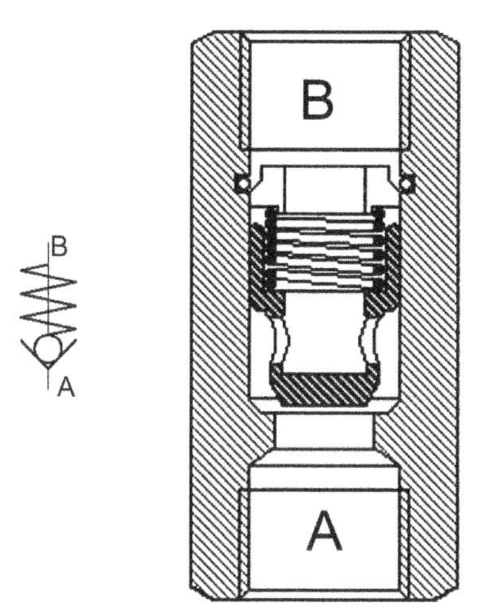

FIGURE 5.49 Direct-operated check valve, spring loaded. (*Courtesy of Bosch Rexroth AG.*)

open the check valve to allow free fluid flow in both directions. The pilot-operated check valves are designed to fulfill this requirement (see Figs. 5.51 and 5.53).

These valves allow the fluid to flow in one direction (A to B) and are piloted to allow for the reverse flow (from B to A). These pilot-operated check valves consist of valve housing (1), main poppet (2), spring (3), pilot piston (4), and an optional decompression poppet assembly (5). In the checked direction (B to A), the main poppet (2) and the decompression poppet (5) are seated by the spring (3) and by the pressure in port (B). When the pilot pressure is applied to the port (X), the pilot piston (4) moves to the right. The decompression poppet (5) opens first, followed by the main poppet (2). This design permits the rapid and smooth decompression of the fluid.

Figure 5.52 shows an example of the application of a pilot-operated check valve for the hydraulic locking of a cylinder position.

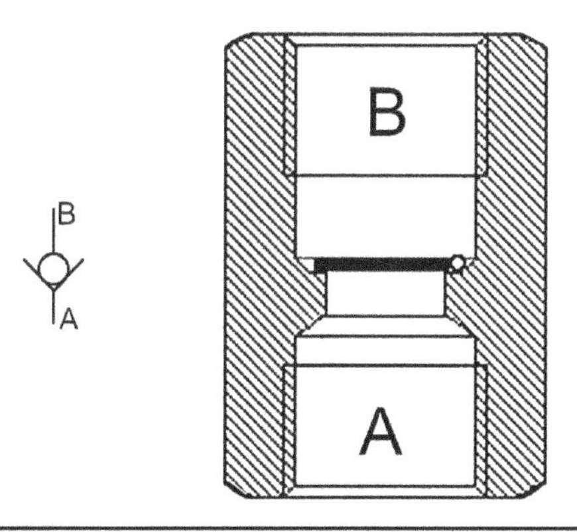

FIGURE 5.50 Direct-operated check valve without spring.

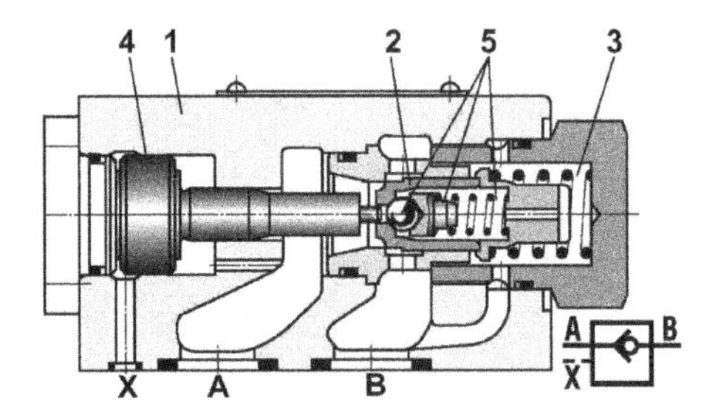

FIGURE 5.51 Pilot-operated check valves, with internal drains. (*Courtesy of Bosch Rexroth AG.*)

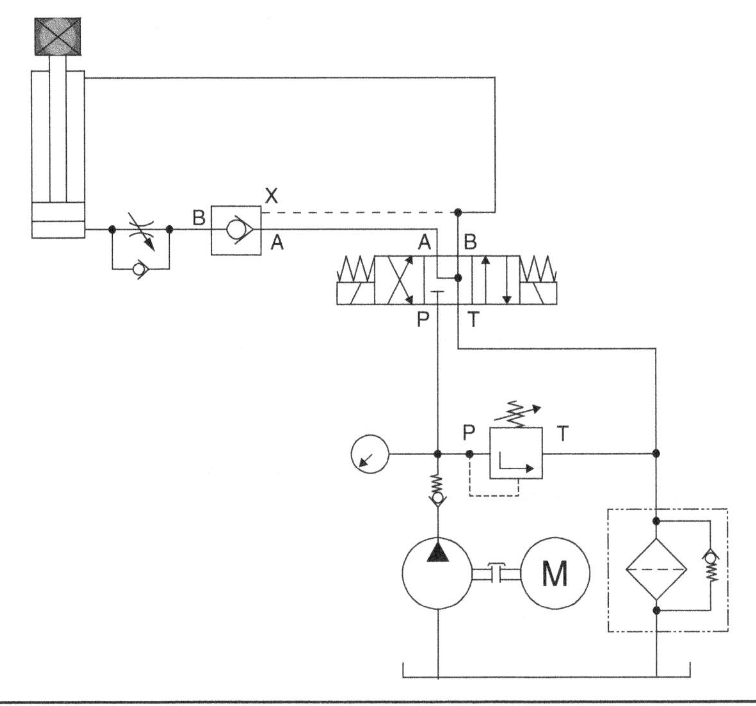

FIGURE 5.52 Example of a pilot-operated check valve application for hydraulic cylinder position locking.

5.4.4 Pilot-Operated Check Valves with External Drain Ports

An additional external drain port (Y), allows the annulus area of the pilot piston (4) to be drained separately (see Fig. 5.53). The pressure at the port (A) only acts on the rod area of pilot piston (9), which reduces the influence of downstream pressure.

5.4.5 Double Pilot-Operated Check Valves

Figure 5.54 shows a double pilot-operated check valve of sandwich plate design. This valve provides leak-free closure of the two actuator ports (A2 and B2) during the idle periods. Free flow (from A1 to A2 or B1 to B2) is permitted, while the flow in the opposite direction is not allowed. The flow (from A1 to A2 or B1 to B2) applies a pressure force to spool (1), which moves to the left (or the right), unseating the opposite

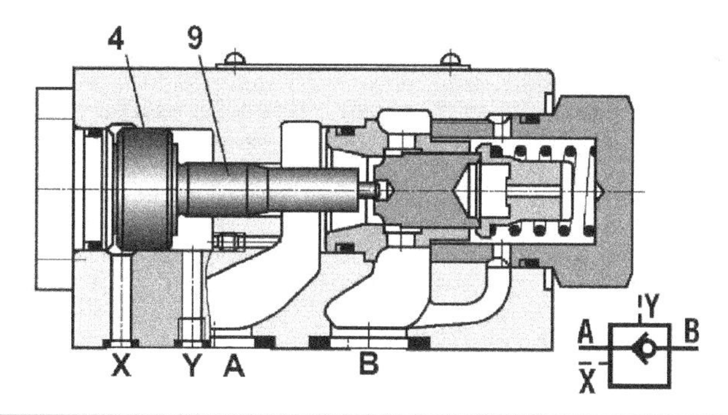

FIGURE 5.53 Pilot-operated check valve, with external drains. (*Courtesy of Bosch Rexroth AG.*)

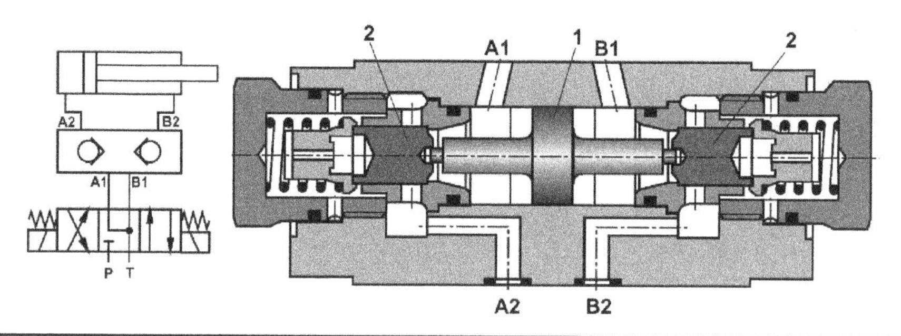

FIGURE 5.54 A double pilot-operated check valve. (*Courtesy of Bosch Rexroth AG.*)

poppet (2). The oil now flows from (B2 to B1 or A2 to A1). To ensure that the poppet valves seat correctly, the ports of the DCV (A1 and B1) should be drained. They should be connected to the tank when the DCV is put in its neutral position.

5.4.6 Mechanically Piloted Pilot-Operated Check Valves

Another version of the pilot-operated check valves is piloted mechanically (see Fig. 5.55). When depressed by an external force, the pin displaces the poppet. Then, the valve opens and allows the fluid to flow in both directions. This valve is commonly used as a means for creating a sequence of operating hydraulic actuators.

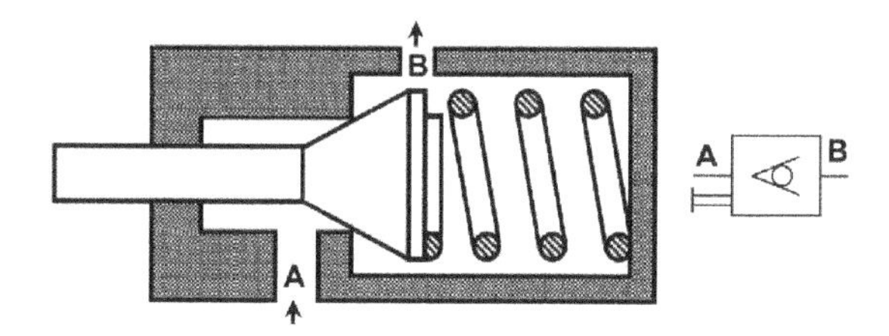

FIGURE 5.55 A mechanically piloted check valve.

5.5 Flow Control Valves

The fluid flow rate is controlled by using throttling elements. The flow rate through a throttling element is governed by the following equation (see Orifice Flow in Sec. 2.2.2).

$$Q = C_d A_o \sqrt{2(P_1 - P_2)/\rho} \tag{5.38}$$

Sharp-edged restrictors are viscosity independent. They have a practically constant discharge coefficient ($C_d = 0.611$). Short tube and other shaped orifices have variable discharge coefficients, depending on the Reynolds number and orifice geometry. They are temperature- and viscosity-dependent. Moreover, all of the throttle elements are pressure-dependent. The flow rate through them changes with the variation of the pressure difference. Therefore, the simple throttle valves do not control precisely the fluid flow rate. A pressure compensator should be added to have a pressure-independent flow control valve. Figure 5.56 gives a classification of the flow control valves.

5.5.1 Throttle Valves

The throttle valves are used to restrict the fluid flow in both directions while the throttle-check valves restrict the flow in one direction only. Figure 5.57 shows examples of the throttle and throttle-check valves for threaded line mounting. They consist of an adjustment sleeve (1) and an inner housing (2). The throttle valve (see Fig. 5.57a) restricts the flow in both directions. The fluid flows through the radial drillings (3) to the throttling area (4), which is defined by the inner housing (2) and the adjustment sleeve (1). The valve restriction area is controlled by turning the adjustment sleeve (1).

Figure 5.57b shows a throttle-check valve. This valve restricts the flow in one direction and allows for free flow in the opposite direction. The fluid passes through the radial drillings and throttling area (4). The throttling is achieved in one direction. In the reverse direction, the pressure acts on the check valve poppet (5). When the pressure difference exceeds the cracking pressure, the poppet opens, allowing reverse flow. In parallel, the fluid also passes through the throttle area (4).

5.5.2 Sharp-Edged Throttle Valves

The flow rate through sharp-edged throttle valves is independent of viscosity. Figure 5.58 shows the construction of a fine throttle valve of sharp edges. It is comprised of a housing (1), an adjusting element (2), and an orifice (3). The fluid flow (from A to B) is throttled at

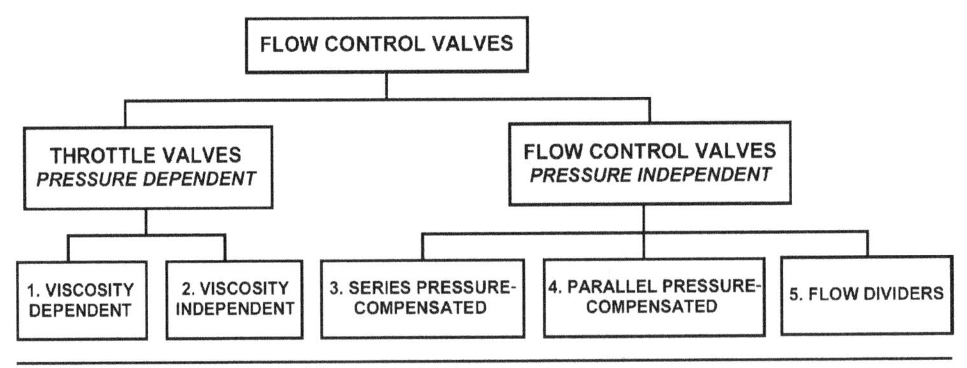

FIGURE 5.56 Classifications of flow control valves.

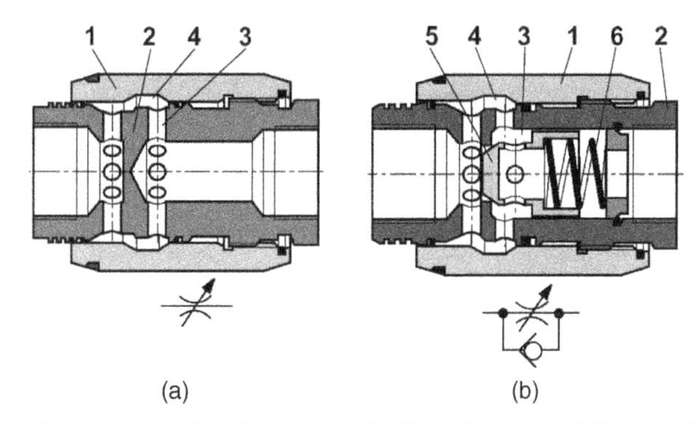

Figure 5.57 (a) A throttle valve and (b) a throttle-check valve for line mounting. (*Courtesy of Bosch Rexroth AG.*)

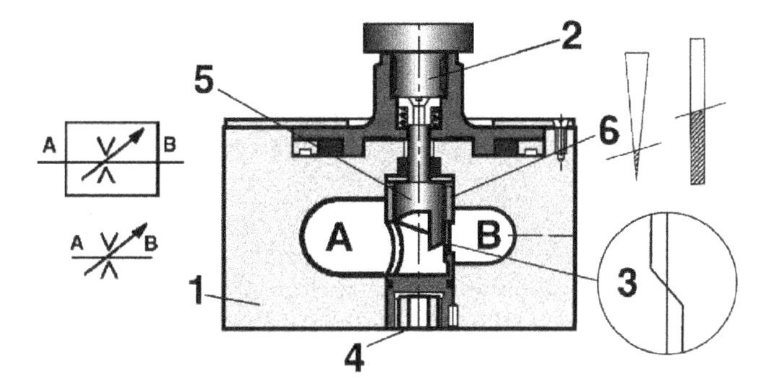

Figure 5.58 A fine throttle valve. (*Courtesy of Bosch Rexroth AG.*)

an orifice window. The throttle opening is adjusted by rotating the core (5), the lower end of which is a lip of helical shape. The preferred direction of flow is from (A) to (B). The area (3) of the orifice revealed by the core pin is controlled by positioning the sleeve (6), using the adjustment screw (4). The throttling area is indicated by means of an adjustment scale at the top surface of the housing (1). During operation, the orifice with the adjustment screw is supported on the valve mounting face. The variation of the throttle area with the core rotational angle may be linear or nonlinear, depending on the shape of the orifice on the sleeve (6).

5.5.3 Series Pressure-Compensated Flow Control Valves

Figure 5.59 gives the hydraulic circuit of a system incorporating a series pressure-compensated flow control valve FCV, also called two-way FCV. The valve consists of a sharp-edged throttle and a pressure compensator connected in series. The pressure compensator is installed downstream of the throttle. It consists of a spool valve loaded by a spring. The pressure difference across the main throttle ($P_1 - P_2$) acts on the spool by the force $F_p = A_c(P_1 - P_2)$, against the spring force F_x. The compensator keeps a constant pressure drop, ΔP_t, across the main throttle. Typically, the value of the pressure difference is selected in the range 4 to 10 bar. In the steady state, this pressure difference produces a force equal to the spring force. The two-way flow control valve operates as follows:

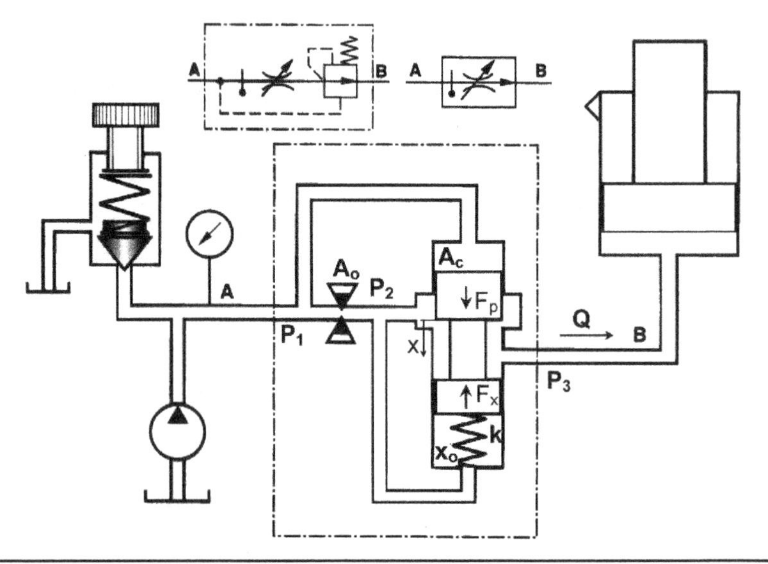

Figure 5.59 A series pressure-compensated FCV with a pressure compensator mounted downstream of the main orifice. (*Courtesy of Bosch Rexroth AG.*)

In the steady state, the pressure difference across the main throttle reaches its required value, ΔP_t. The pressure and spring forces are in equilibrium and the spool gets in its steady-state position. The flow rate reaches the required value given by

$$Q = C_d A_o \sqrt{2(P_1 - P_2)/\rho} \tag{5.39}$$

For $\rho = 900$ kg/m^3, $C_d = 0.611$, and $\Delta P = \Delta P_t = 5$ bar,

$$Q = 20.37\, A_o \tag{5.40}$$

If the pressure difference is increased, $P_1 - P_2 > \Delta P_t$, the flow rate increases. Simultaneously, the spool moves downward, against the spring, to decrease the area of the spool valve restriction. The flow rate through the main throttle decreases and so does the pressure difference across the main restriction. Afterward, the valve again reaches the steady state, where $P_1 - P_2 = \Delta P_t$.

If the pressure difference is decreased, $P_1 - P_2 < \Delta P_t$, the flow rate decreases. The pressure force acting on the compensator spool becomes less than the spring force. The spring pushes the spool upward, increasing the restriction area of the spool valve. The flow rate through the main throttle and the pressure difference $(P_1 - P_2)$ increase until they reach the required steady-state values.

The pressure compensator acts constantly to compensate for the effect of the variation of supply and load pressures. In the steady state, the spool of the pressure compensator is in equilibrium

$$(P_1 - P_2)A_c + F_j = k(x_o + x) \tag{5.41}$$

Actually, the jet reaction force, F_j, is negligible compared with the spring force. Moreover, considering the real valve operation, the spool displacement x is too small

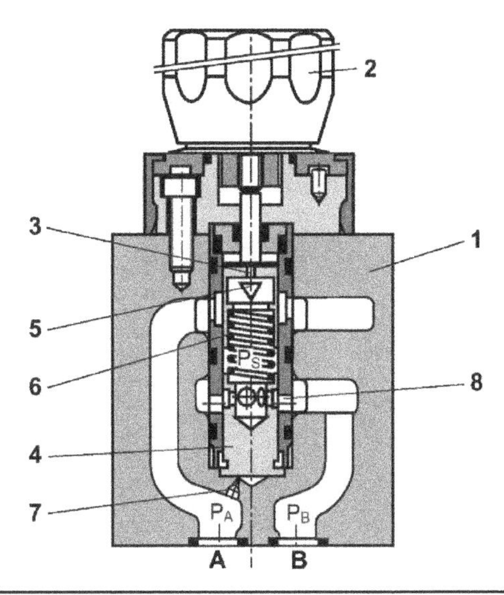

FIGURE 5.60 A two-way FCV. (*Courtesy of Bosch Rexroth AG.*)

compared with the spring pre-compression distance, $x << x_o$. Then, the pressure difference $(P_1 - P_2)$ reaches a steady-state value $\Delta P = kx_o/A_c = \Delta P_t$.

Figure 5.60 shows a commercially available flow control valve. It is a series pressure-compensated flow control valve. It controls the flow rate, independent of the changes in fluid viscosity or pressure drop across the valve as long as it is greater than ΔP_t. It consists of a housing (1), a hand knob with scale (2), and a downstream pressure compensator (4). The flow (from port A to B) passes through the orifice window (5) on the side wall of the hollow spool (3). The rotation of the knob (2) adjusts the vertical position of the spool (3) and thus determines the area of the throttle (5).

The pressure compensation is achieved by the downstream compensator, which consists of a spool (4) of area A_S, and an orifice (8) connected in series with the main orifice (5). The displacement of the spool (4) upward decreases the throttling area of orifice (8) and vice versa. The spool is subjected to the pressure and the spring forces. The input chamber is connected to the lower side of the spool (4) through the damping orifice (7). In the steady state, the pressure force acting on the spool (4) is $F_p = (P_A - P_S)A_S$.

The spool (4) is subjected to the pressure forces and the spring force. In the steady state, the pressure force equals the spring force. Considering the actual operating conditions, the spool displacement is too small compared with the spring pre-compression. Therefore, the pressure difference $(P_A - P_S)$ and the valve flow rate are kept constant in the steady state. This class of valves regulates the fluid flow and produces practically constant flow rate if the pressure difference across the whole valve is greater than the preset value ΔP_t, as shown in Fig. 5.61. Figure 5.62 shows a typical application of the two-way flow control valve.

The speed of hydraulic actuators is controlled by managing either the inlet flow rate, meter-in control, or the outlet flow rate, meter-out control. The meter-in control is preferred when moving against the load, while the meter-out control is obligatory when moving in the same direction of the load. However, if the load direction is not determined or reversible, the meter-out control is the best choice.

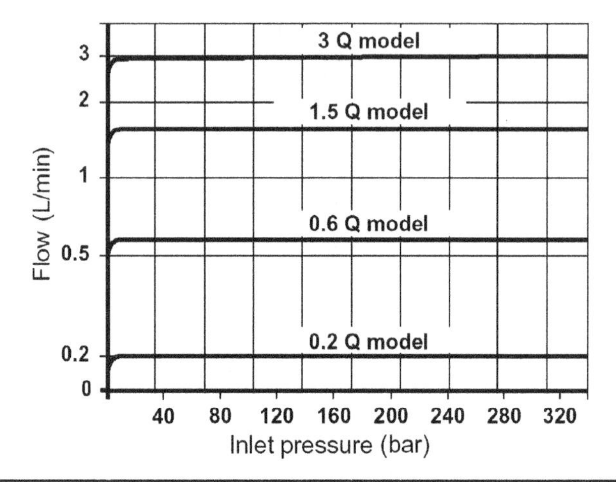

FIGURE 5.61 Flow characteristics of series pressure-compensated flow control valves. (*Courtesy of Bosch Rexroth AG.*)

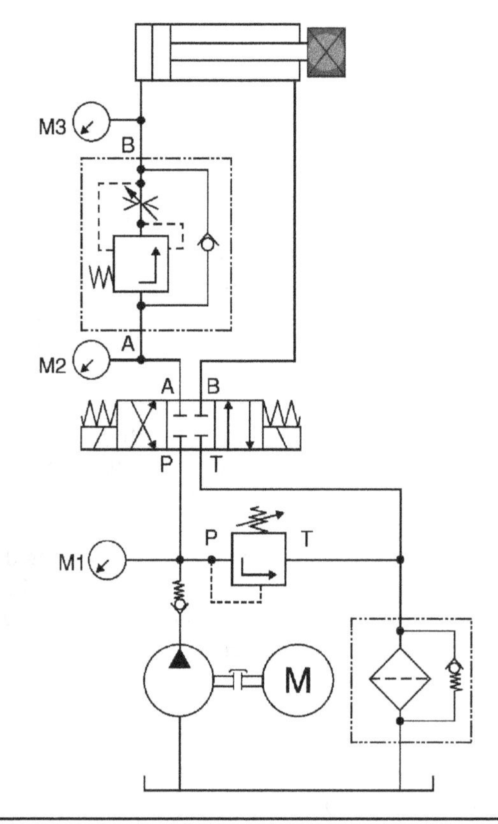

FIGURE 5.62 Typical application of a series pressure-compensated flow control valve.

5.5.4 Parallel Pressure-Compensated Flow Control Valves—Three-Way FCVs

Figure 5.63 shows a hydraulic circuit incorporating a parallel pressure-compensated flow control valve—a three-way flow control valve. This class of valves regulates the fluid flow and produces practically constant flow rate if the pressure difference across the whole valve is greater than the preset value ΔP_t. The pump loading pressure is slightly higher than the load pressure, $P_1 = P_2 + \Delta P_t$, where $\Delta P_t = 4$ to 10 bar.

5.5.5 Flow Dividers

The flow dividers are used to divide the fluid flow rate into two or more parts: either equal parts or by a certain division ratio. The two main classes of flow dividers are displacement and spool types. Displacement flow dividers consist of two or more hydraulic motors mounted on the same shaft, rotating at the same speed. Figure 5.64 illustrates the idea of dividing the inlet flow rate, Q, into three parts. Neglecting the internal leakage, the flow rates through the three motors are

$$Q_1 = V_{g1}\, n, \qquad Q_2 = V_{g2}\, n \quad \text{and} \quad Q_3 = V_{g3}\, n \tag{5.42}$$

Then:

$$Q_1 : Q_2 : Q_3 = V_{g1} : V_{g2} : V_{g3} \tag{5.43}$$

and

$$Q_1 + Q_2 + Q_3 = Q \tag{5.44}$$

where n = Motor speed, rps
Q = Motor flow rate, m^3/s
V_g = Motor displacement, m^3/rev

By using the displacement flow dividers, the flow can be divided into two or more parts, with the required division ratio.

This class of flow dividers can be used as a pressure intensifier (see Fig. 5.65). The second motor is connected to the tank. It drives the first motor, which operates as a

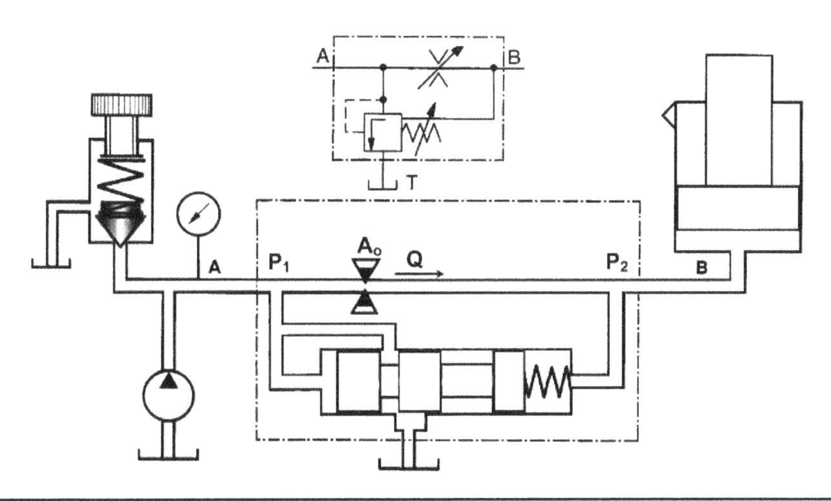

Figure 5.63 A three-way FCV. (*Courtesy of Bosch Rexroth AG.*)

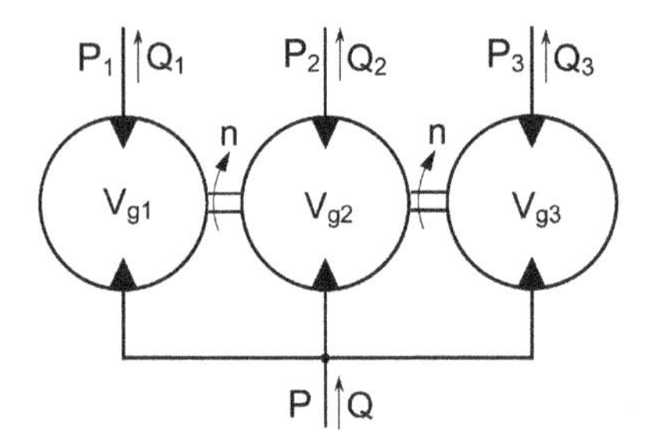

FIGURE 5.64 Displacement-type flow divider.

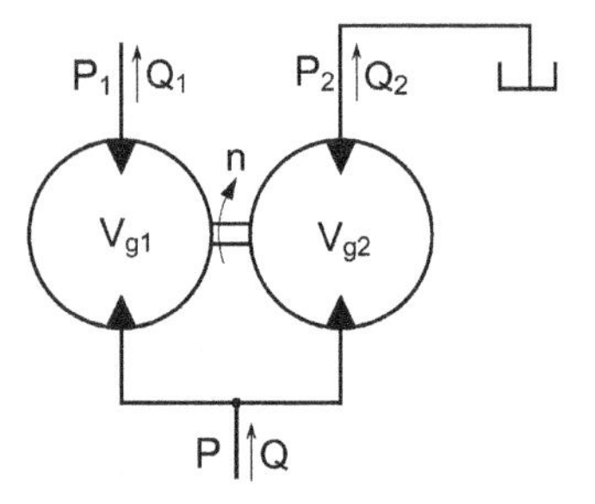

FIGURE 5.65 Application of a displacement flow divider as a pressure intensifier.

pump with input pressure, P. Assuming ideal machines, the input and output hydraulic powers are equal. Then, if $P_2 = 0$,

$$Q_1 P_1 + Q_2 P_2 = QP \tag{5.45}$$

$$Q_1 = V_1\, n, \qquad Q_2 = V_2\, n \quad \text{and} \quad Q_1 + Q_2 = Q \tag{5.46}$$

Then

$$P_1 = P(V_1 + V_2)/V_1 \tag{5.47}$$

The principle of operation of the flow divider of spool type is illustrated by the system shown in Fig. 5.66. This system includes two symmetrical cylinders connected in parallel. The displacement of the cylinders should be synchronized during their extension. Therefore, a flow divider valve is installed. The flow divider acts to divide the pump flow equally between the two cylinders in the extension stroke. The oil flows into the mid-chamber of the spool valve, then through the spool valve restrictions (A_A and A_B) to the cylinders lines (A and B). If the cylinders are equally loaded, then the right and left sides are symmetrical, having equal hydraulic resistance and the main spool is centered. At these conditions, equal flow rates will flow to the cylinders. If the loading force, F_A, is increased, for example, the pressure P_A increases, the pressure difference $(P_P - P_A)$ decreases compared with $(P_P - P_B)$, which decreases the flow rate, Q_A. Simultaneously, the spool displaces to the left, increasing the throttle area A_A and decreasing the area A_B. In this way, the hydraulic resistance of path (P-A) decreases and that of path (P-B) increases. Thus, the increase of load of one side is compensated by the reduction of its resistance and the increase of resistance of the other side. In the steady state, equal flow rates are practically reached.

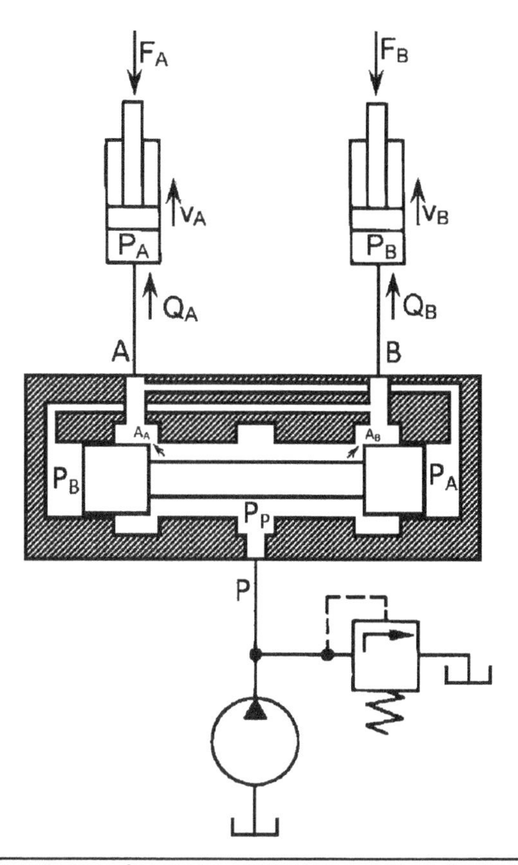

FIGURE 5.66 The synchronization of hydraulic cylinders by spool-type flow divider.

The construction of a commercially available spool-type flow divider is shown in Fig. 5.67. The valve consists mainly of the housing (1), a control spool (2), and three springs (3). The fluid flow from the port (P) is divided into two equal partial flows.

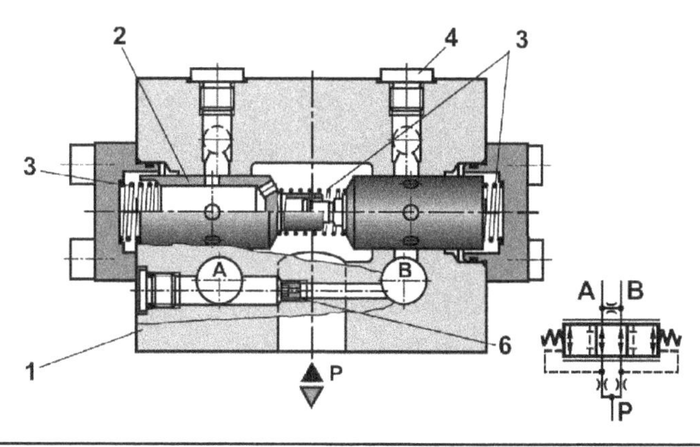

FIGURE 5.67 A spool-type flow divider. (*Courtesy of Bosch Rexroth AG.*)

The centering springs as well as the pressure forces keep the control spool in its mid-position. Any imbalanced flow rate consumption from any of the two exit ports (A or B) results in a pressure difference. The resulting pressure difference displaces the spool to throttle the branch of excessive flow and opens the other branch. As a result, in the steady state, the flow rates to both outlets remain practically equal.

5.6 Exercises

1. Discuss briefly the control of power in the hydraulic power systems.

2. Explain briefly the construction and compare the spool valves and poppet valves.

3. Deal with the function of the pressure control valves in the hydraulic power systems.

4. Discuss in detail the static characteristics of a direct-operated relief valve, and explain how to reduce the override pressure. Derive the needed mathematical relations. (See Fig. 5.2.)

5. Draw schematically a pilot-operated relief valve, and explain its function.

6. Explain the function of the pilot-operated relief valve illustrated in Figs. 5.7 and 5.8.

7. Discuss the applications of the pilot-operated relief valve shown in Fig. 5.9.

8. Discuss the principle of pressure reduction in a hydraulic power system.

9. Explain the function of the direct-operated hydraulic pressure reducer illustrated in Figs. 5.11 and 5.13.

10. Discuss the application of the hydraulic pressure reducer shown in Fig. 5.14.

11. Discuss briefly the operation of the pilot-operated pressure reducer illustrated in Fig. 5.15.

12. Explain the function of the direct-operated sequence valve shown in Figs. 5.16 and 5.17.

13. What are the differences between the relief valves and the sequence valves?

14. Discuss the application of the hydraulic sequence valve shown in Fig. 5.19.

15. Discuss briefly the construction and operation of the pilot-operated sequence valve illustrated in Fig. 5.20.

16. Draw a schematic of an accumulator-charging valve, and explain its function.

17. Explain briefly the function of the directional control valves.

18. Draw a schematic of a 4/3 DCV that is direct-operated electrically, and briefly explain its function.

19. Discuss in detail the flow displacement relation of the ideal overlapping and under-lapping spool valves, giving their possible applications in the hydraulic control valves.

20. Discuss in detail the flow forces acting on the spool valves, and derive an expression for these forces.

21. Draw schematically and explain the function of a 4/3 pilot-operated directional control valve.

22. State the different ways of control of directional control valves.

23. Explain briefly the construction and operation of the electric solenoids, and compare the DC and AC solenoids.

24. Cite the classifications of check valves, and explain the function of a pilot-operated check valve, giving the necessary drawings.

25. Discuss the application of the pilot-operated check valve shown in Fig. 5.52.

26. Cite and briefly discuss the various classifications of flow control valves.

27. Explain the operation of the throttle and throttle check valves given by Fig. 5.57.

28. Explain the operation of the series pressure-compensated flow control valve illustrated in Fig. 5.59.

29. Explain the construction and operation of the series pressure-compensated flow control valve illustrated in Fig. 5.60.

30. Discuss the application of the two-way flow control valve shown in Fig. 5.62.

31. Explain the construction and operation of the parallel pressure-compensated flow control shown in Fig. 5.63.

32. Discuss briefly the function of a displacement-type flow divider.

33. Discuss briefly the function of the spool-type flow divider as illustrated in Figs. 5.66 and 5.67.

34. The speed of a hydraulic cylinder is controlled by means of a series pressure-compensated flow control valve, as shown in the given circuit. *Given:*

Pump:

$V_g = 25\ \text{cm}^3/\text{rev}$

$n = 1000$ rpm

$\eta_v = 0.95$

$\eta_m = 0.93$

$\eta_h = 1$

Directional control valve:

$Q = 6 \times 10^{-7}\sqrt{\Delta P}$

Relief valve:

Cracking pressure = 6 MPa

Override pressure = 0

Hydraulic cylinder:

$f = 2000$ Ns/m

$v = 0.1$ m/s

$F = 9000$ N

Piston diameter = 60 mm

Rod diameter = 25 mm

No inner leakage

Calculate:

$P_1, P_2, P_3, P_P, Q_1, Q_2$

Power losses in the DCV

Power losses in the FCV

Pump real flow rate

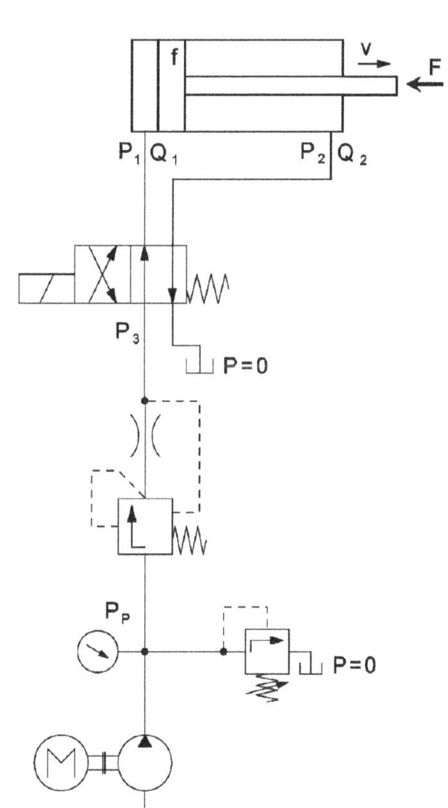

5.7 Nomenclature

A_o = Orifice area, m^2
A_p = Poppet area subjected to pressure, m^2
A_c & A_s = Spool area, m^2
A_v = Valve throttling area, m^2
C_c = Contraction coefficient
C_d = Discharge coefficient
F_j = Jet reaction, momentum, force, N
k = Spring stiffness, N/m
n = Rotational speed, rps
P = Pressure, valve input pressure, Pa
P_A = Inlet line pressure, Pa
P_B = Exit line pressure, Pa
P_r = Cracking pressure, Pa
P_R = Reduced pressure, Pa
Q = Flow rate, m^3/s
V_g = Displacement, geometric volumes, m^3/rev
v = Oil velocity, m/s
x = Displacement, m
x_r = Overlap, m
x_o = Spring pre-compression, m
ρ = Oil density, kg/m^3
ω = Valve throttling area proportionality coefficient, m

Appendix 5A Control Valve Pressures and Throttle Areas

Conical Poppet Valves

The conical poppet is one of the frequently used designs of poppet valves. When the poppet rests against its seat, the contact line is a circle of diameter d (see Fig. 5A.1). The poppet is subjected to a pressure force, F_P, where

$$F_P = \frac{\pi}{4} d^2 (P_1 - P_2) \tag{5A.1}$$

If the poppet is displaced by distance x, the fluid is allowed to flow through the opened area. This area consists of the side area of the truncated cone (abce) whose left

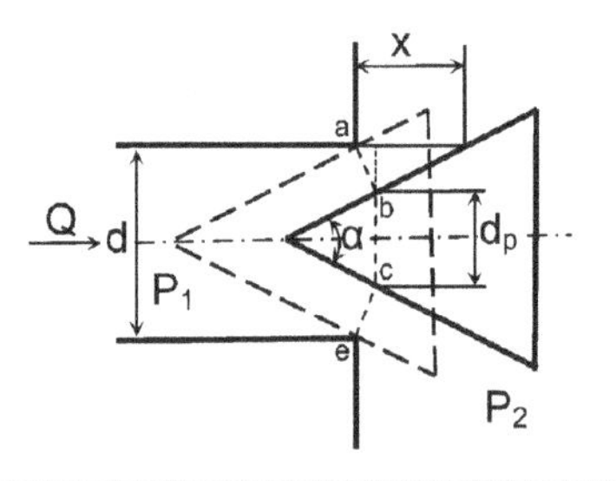

FIGURE 5A.1 A conical poppet valve.

base is of diameter ϕd and whose right base is of diameter ϕd_p. The area A_t is calculated as follows:

$$\overline{ab} = x \sin(\alpha/2) \tag{5A.2}$$

$$d_p = d - 2x \sin(\alpha/2)\cos(\alpha/2) \tag{5A.3}$$

$$A_t = \frac{\pi}{2}(d + d_p)\overline{ab} = \pi x \sin(\alpha/2)\{d - x\sin(\alpha/2)\cos(\alpha/2)\} \tag{5A.4}$$

The area subjected to the pressure force is of diameter ϕd_p. It is calculated by the following formula:

$$A_p = \frac{\pi}{4}\{d - 2x\sin(\alpha/2)\cos(\alpha/2)\}^2 \tag{5A.5}$$

where A_t = Throttle area, m^2
A_p = Poppet area subjected to the pressure, m^2
d = Diameter of the circular poppet seat, m
d_p = Diameter of the circular poppet area subjected to the pressure, m
x = Poppet displacement, m
α = Cone vertex angle, rad

The area subjected to pressure, A_p, and throttle area, A_t, were calculated for a conical poppet of vertex angle 60° and inlet diameter $d = 4$ mm. The calculation results are plotted in Fig. 5A.2. The areas were calculated in nondimensional form, relative to the inlet area A_i and plotted versus the relative poppet displacement (x/d). The throttle area, A_t, being the minimum cross-sectional area of the stream tube, should be less than or equal to the inlet area, $A_i = 0.25\pi\, d^2$. For practical applications, the poppet displacement is within 0.2 d.

Cylindrical Poppets with Conical Seats

The cylindrical poppet with a conical seat (see Fig. 5A.3) is frequently used as a main stage of the pilot-operated pressure control valves. In this design, when the poppet rests

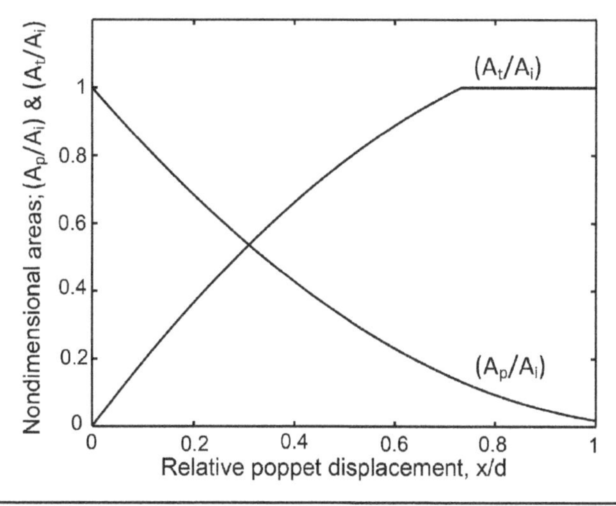

FIGURE 5A.2 A nondimensional throttle and pressure areas of a conical poppet valve.

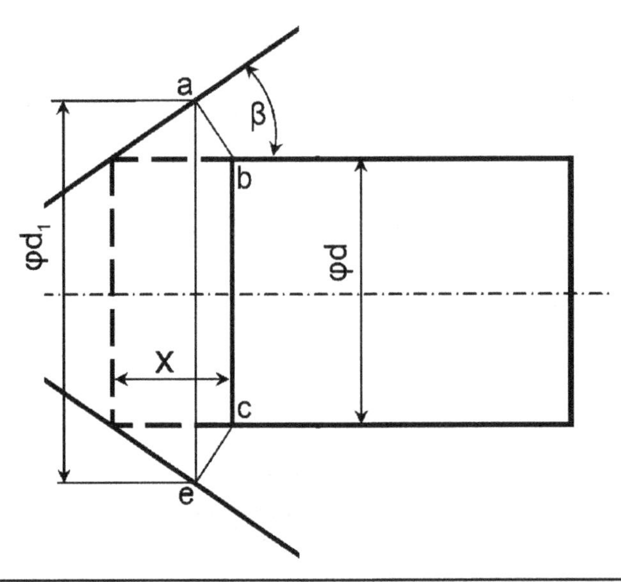

Figure 5A.3 A poppet valve with a cylindrical poppet and conical seat.

against its seat, the contact line is a circle of diameter d. The poppet area subjected to pressure force is A_p. It is constant, and independent of the poppet displacement, where

$$A_P = \frac{\pi}{4} d^2 \tag{5A.6}$$

If the poppet is displaced by a distance x, the fluid is allowed to flow through the opened area, A_t. This area consists of the side area of the truncated cone, (abce). The left base area of this cone is of diameter d_1, while its right base area is of diameter d.

The area A_t is calculated as follows:

$$\overline{ab} = x \sin(\beta) \tag{5A.7}$$

$$d_1 = d + 2x \sin(\beta)\cos(\beta) \tag{5A.8}$$

$$A_t = \frac{\pi}{2}(d + d_1)\overline{ab} = \pi x \sin(\beta)\{d + x \sin(\beta)\cos(\beta)\} \tag{5A.9}$$

The throttle area was calculated for a seat angle $\beta = 30°$. The calculated area is plotted in nondimensional form (A_t/A_p) versus the relative poppet displacement (x/d) on Fig. 5A.4. This plot shows a quasi-linear displacement relation for $(x/d < 0.3)$.

Spherical Poppet Valves

Poppets of spherical shape are widely used. They are of simple and economic design. When the poppet rests against its seat, the contact line is a circle of diameter d (Fig. 5A.5), and the poppet area subjected to the pressure force is A_p, where

$$A_P = \frac{\pi}{4} d^2 \tag{5A.10}$$

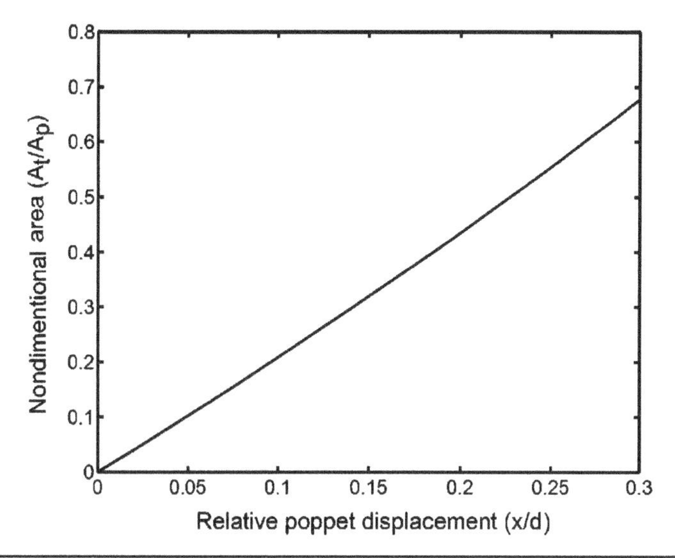

Figure 5A.4 The nondimensional throttle area of a cylindrical poppet valve with a conical seat.

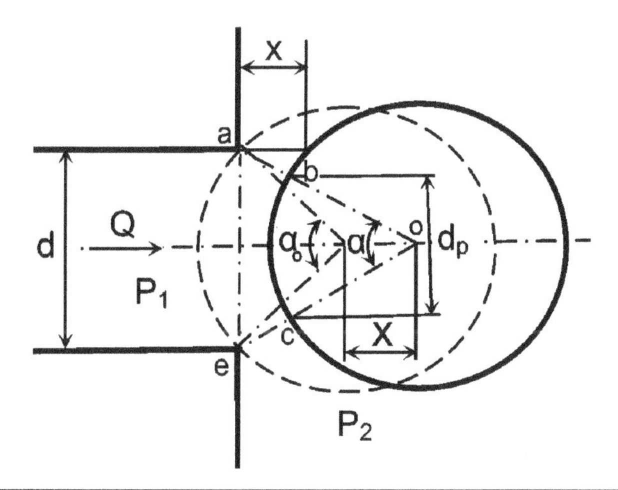

Figure 5A.5 A spherical poppet valve.

If the poppet is displaced by a distance x, the fluid is allowed to flow through the opened area. This area consists of the side area of the truncated cone (abce), whose left base area is of diameter ϕd, and whose right base area is of diameter ϕd_P. The area A_t is calculated as follows:

$$\alpha_o = 2\sin^{-1}\left(\frac{d}{2r}\right) \tag{5A.11}$$

where r is the spherical poppet radius, m.

$$\alpha = 2\tan^{-1}\left[\frac{d}{2\{r\cos(\alpha_o/2)+x\}}\right] \tag{5A.12}$$

$$\overline{ab} = \frac{d}{2\sin(\alpha/2)} - r \tag{5A.13}$$

$$d_P = 2r\sin(\alpha/2) \tag{5A.14}$$

$$A_t = \frac{\pi}{2}(d+d_P)\overline{ab} = \frac{\pi}{2}[d+2r\sin(\alpha/2)]\left\{\frac{d}{2\sin(\alpha/2)} - r\right\} \tag{5A.15}$$

The area subjected to the pressure force is of diameter ϕd_P. It is calculated by the following formula:

$$A_P = \frac{\pi}{4}d_p^2 = \pi r^2 \sin^2(\alpha/2) \tag{5A.16}$$

The area subjected to the pressure A_p and throttle area A_t were calculated for a poppet of radius $r = 4$ mm and inlet diameter $d = 5$ mm. The calculation results are plotted in Fig. 5A.6. The throttle area, A_t, being the minimum cross-sectional area of the stream tube, should be less than or equal to the inlet area, $A_i = 0.25\pi d^2$. The areas were calculated in nondimensional form, relative to the inlet area A_i and plotted versus the relative poppet displacement (x/d).

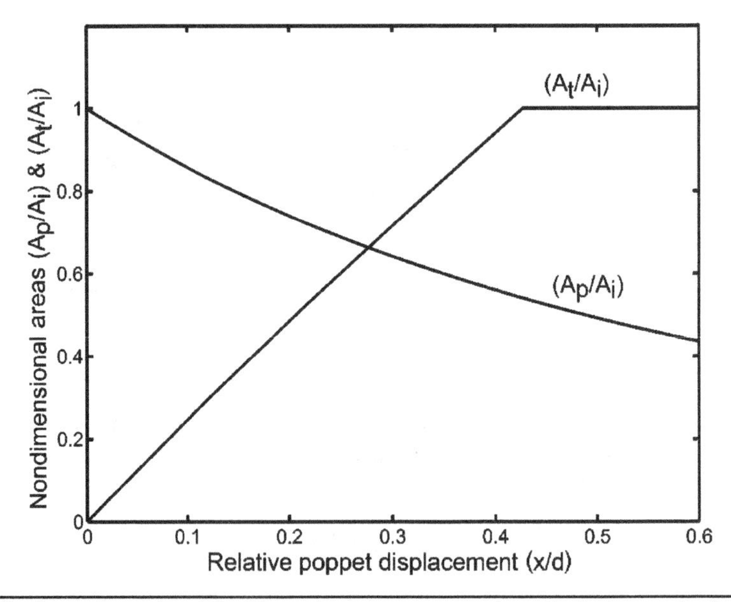

FIGURE 5A.6 A nondimensional throttle and pressure areas of a spherical poppet valve.

Circular Throttling Area

Circular opening areas are among the widely used throttling elements in the spool valves due to the simplicity of production (see Fig. 5A.7).

The area cleared due to the spool displacement, x, can be calculated as follows:

$$\text{For} \quad 0 < x < r \quad \begin{cases} \alpha = 2\cos^{-1}\dfrac{r-x}{r} \\ \overline{ab} = 2r\sin(\alpha/2) \\ A_t = 0.5r^2\alpha - 0.5\overline{ab}(r-x) \end{cases} \tag{5A.17}$$

$$\text{For} \quad r < x < 2r \quad \begin{cases} \alpha = 2\cos^{-1}\dfrac{x-r}{r} \\ \overline{ab} = 2r\sin(\alpha/2) \\ A_t = \pi r^2 - \left\{0.5r^2\alpha - 0.5\overline{ab}(x-r)\right\} \end{cases} \tag{5A.18}$$

$$\text{For} \quad x > 2r; \quad A_t = \pi r^2 \tag{5A.19}$$

The throttling area cleared by the spool is calculated and plotted (see Fig. 5A.8). Being a part of the circle, the throttle area, A_t, is limited by the circle area: $A_C = \pi r^2$, where A_c = Area of the circular hole, m^2; r = Circular hole radius, m; and x = Height of the cleared circle segment, m. The opened area was calculated in nondimensional form, relative to the area A_C.

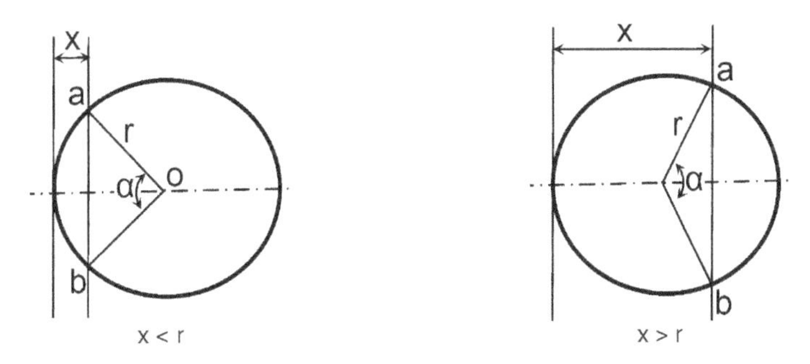

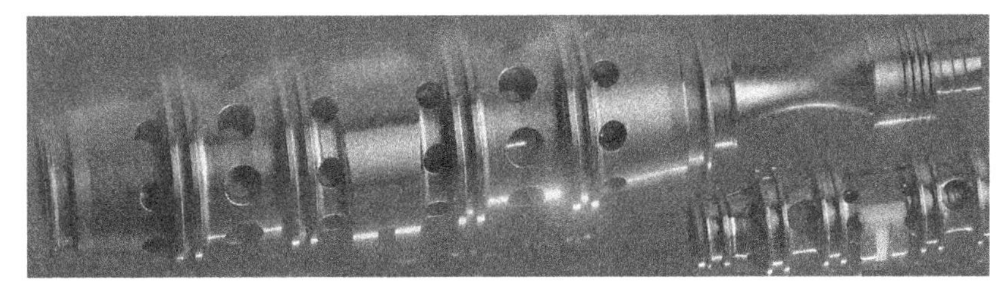

Figure 5A.7 Spool valves with circular throttling areas. (*Courtesy of Moog Inc.*)

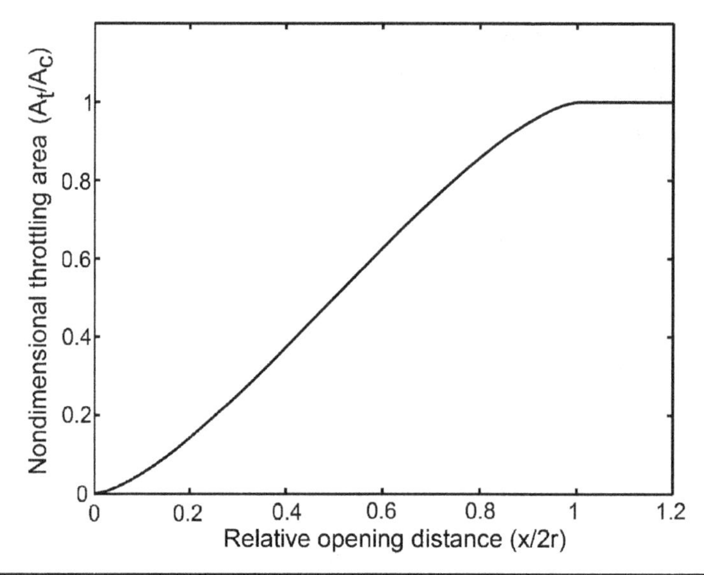

FIGURE 5A.8 A nondimensional plot of the opened area of a circular hole.

Triangular Throttling Area

Some spools are machined with triangular cuts (see Fig. 5A.9). The cut is of depth h, and the area cleared by the spool valve A_{TR} is of triangular shape (Fig. 5A.9). The fluid flows through this area, then through a rectangular area A_{RE} perpendicular to it.

where
$$b = 2x\tan(\alpha/2)$$

$$A_{TR} = bx/2 = x^2\tan(\alpha/2)$$

$$A_{RE} = bh = 2hx\tan(\alpha/2)$$

The triangular and rectangular areas are equal at $x = 2h$ and the throttling area is calculated by the following expression:

$$A_t = \begin{cases} x^2\tan(\alpha/2) & \text{For } x \le 2h \\ 2hx\tan(\alpha/2) & \text{For } x \ge 2h \end{cases} \tag{5A.20}$$

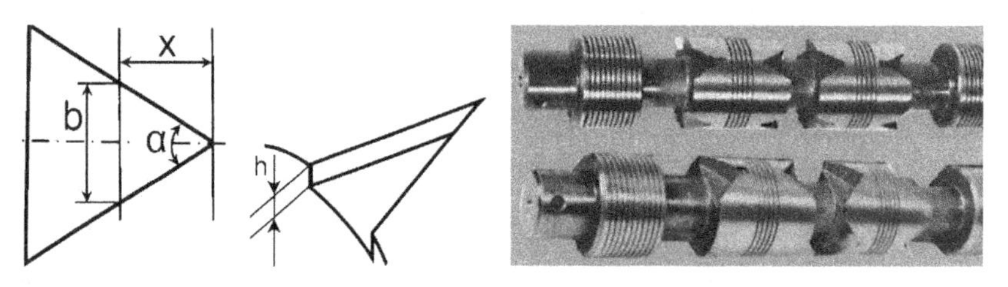

FIGURE 5A.9 Triangular throttling area. (*Photo courtesy of Bosh Rexroth AG.*)

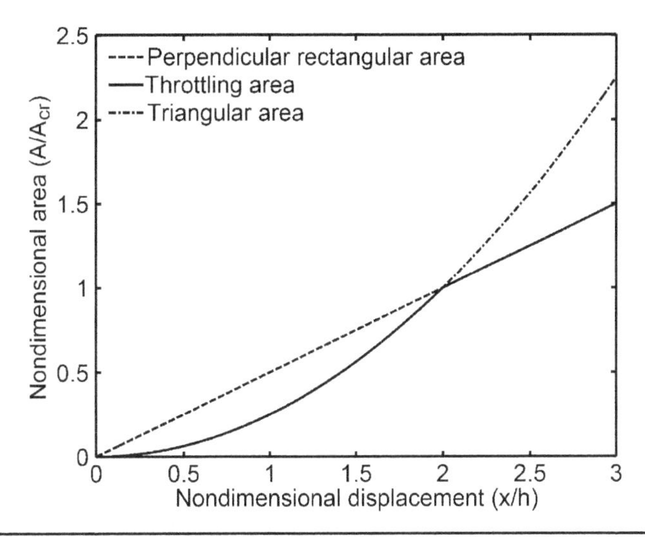

Figure 5A.10 The nondimensional throttling area of triangular hole.

The critical area is at $x = 2h$; $A_{cr} = 4\ h^2 \tan(\alpha/2)$.

The opened throttling area is calculated in nondimensional form for a vertex angle of 90°, and plotted in Fig. 5A.10.

Appendix 5B Modeling and Simulation of a Direct-Operated Relief Valve

This appendix presents a study of a direct-operated relief valve. The study includes the following:

- Construction and operation of the valve
- Mathematical modeling
- Computer simulation
- A brief discussion of the static and dynamic characteristics

Construction and Operation of the Valve

Figure 5B.1 shows a schematic of the studied direct-operated relief valve. The valve consists of a poppet valve (2), rigidly attached to a damping spool (1). The poppet is loaded by a spring of stiffness k_{rv}. The spring is pre-compressed by an adjustable pre-compression distance, x_o. When the valve is not operating, the spring pre-compression force ($k_{rv}\, x_o$) pushes the poppet against its seat. The seat produces an equal seat reaction force F_{SR}. When the input pressure P is increased, the liquid flows to the damping chamber through the radial clearance of the damping spool, Q_d. The pressure, P_d, increases and acts on the damping spool. When the valve is closed, the pressure P at the valve inlet chamber does not produce any axial force on the moving parts. But, as the poppet valve opens, the poppet area subjected to inlet pressure A_p (see Figs. 5A.1 and 5A.2) becomes less than the damping spool area A_d. The pressure P acts on the area difference ($A_d - A_p$) to the left.

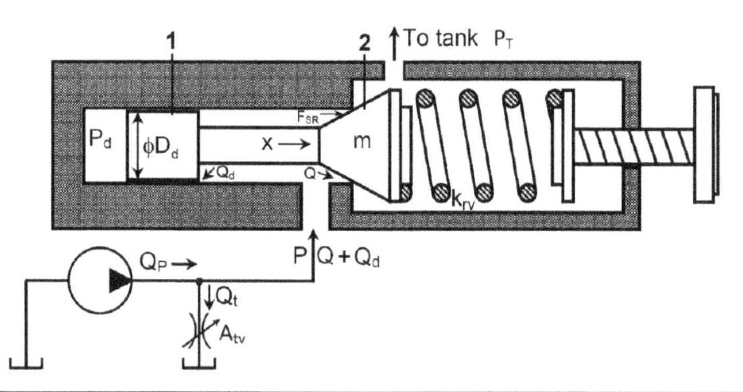

FIGURE 5B.1 A direct-operated relief valve connected to a pump and a bypass valve.

Neglecting the jet reaction forces, the motion of the damping spool and poppet is governed by the spring force, the seat reaction force, and the pressure forces. When the pressure force ($P_d A_d$) exceeds the spring force ($k_{rv} x_o$), the poppet displaces, opening the path from the inlet port to the drain port, with a pressure P_T. The variation of the pressure P_d is resisted by the radial clearance, which throttles the connection of the inlet port with the damping spool chamber.

The relief valve is connected to the delivery line of a fixed displacement pump rotated by a constant speed. A bypass valve is connected to the pump delivery line to control the loading pressure.

Mathematical Modeling

The dynamic behavior of the valve is described by the following set of mathematical relations. The effect of the transmission lines was neglected and the return pressure is $P_T = 0$.

The Poppet Valve Throttling Area

The following mathematical expressions for the poppet area, A_P, subjected to the pressure P and poppet valve throttle area A_t, were deduced in App. 5A.1.

$$A_t = \pi x \sin(\alpha/2)\{D_d - x\sin(\alpha/2)\cos(\alpha/2)\} \tag{5B.1}$$

$$A_P = \frac{\pi}{4}\{D_d - 2x\sin(\alpha/2)\cos(\alpha/2)\}^2 \tag{5B.2}$$

Equation of Motion of the Poppet

$$P_d A_d + F_{SR} - P(A_d - A_P) = m\frac{d^2x}{dt^2} + f\frac{dx}{dt} + k_{rv}(x_o + x) \tag{5B.3}$$

Seat Reaction Force

The poppet displacement in the closure direction is limited mechanically. When reaching its seat, a seat reaction force takes place due to the action of the seat stiffness and structural damping of the seat material. These two effects are introduced by the equivalent seat stiffness k_s and damping coefficient R_s.

$$F_{SR} = \begin{cases} 0 & x > 0 \\ k_S|x| - R_s \dfrac{dx}{dt} & x < 0 \end{cases} \tag{5B.4}$$

Flow Rate Through the Radial Clearance of the Damping Spool

$$Q_d = \frac{\pi D_d c^3}{12\mu L}(P - P_d) \tag{5B.5}$$

Flow Rate Through the Poppet Valve

$$Q = C_d A_t \sqrt{2P/\rho} \tag{5B.6}$$

Continuity Equation Applied to the Damping Spool Chamber

$$Q_d - A_d \frac{dx}{dt} = \frac{V_o + A_d x}{B} \frac{dP_d}{dt} \tag{5B.7}$$

Pump Flow Rate

$$Q_P = Q_{th} - P/R_L \tag{5B.8}$$

Flow Rate Through the Bypass Valve

$$Q_t = C_d A_{tv} \sqrt{2P/\rho} \tag{5B.9}$$

Continuity Equation Applied to the Pump Exit Line

$$Q_P - Q - Q_d - Q_t = \frac{V_P}{B} \frac{dP}{dt} \tag{5B.10}$$

Computer Simulation

Equations (5B.1) through (5B.10) describe the static and dynamic behavior of the studied system. These equations were used to develop a computer simulation program (see Fig. 5B.2), which was employed to calculate and plot the valve's static and dynamic characteristics.

Steady-State Characteristics

The steady-state characteristic of the relief valve is the steady-state relation between the input valve pressure, P, and the relief flow rate, Q. The valve flow characteristics were calculated for different values of the cracking pressure and plotted in Fig. 5B.3. The simulation results show that for pressures less than the cracking pressure, the flow rate is zero. If the pressure exceeds the cracking pressure, the valve opens and the fluid flow increases as the override pressure increases. This figure also carries the pump flow

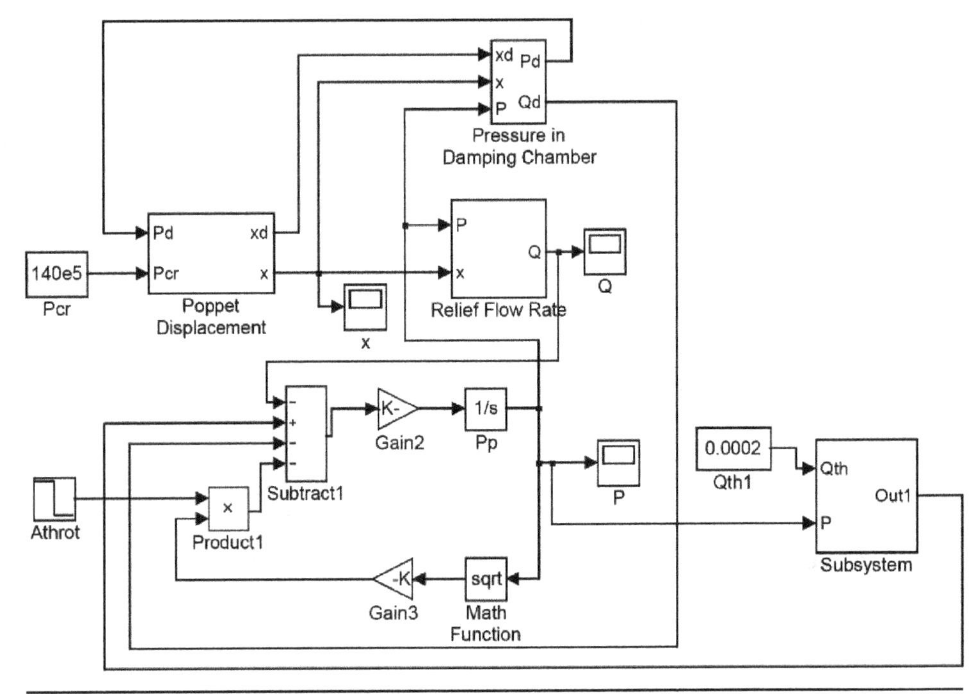

FIGURE 5B.2 The block diagram of the relief valve simulation program (SIMULINK model).

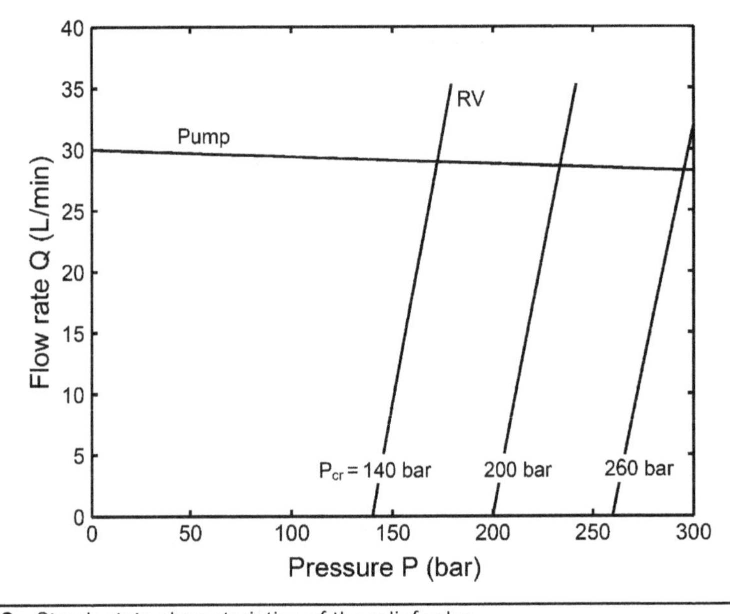

FIGURE 5B.3 Steady-state characteristics of the relief valve.

characteristic curve. The intersection of this curve with the relief valve characteristic lines gives the operating points for the considered cracking pressures.

In the commercial documents, the static characteristics of relief valves are plotted with the flow rate at the horizontal axis, *P*(*Q*). This type of causality assumes that the independent variable is the flow rate.

Transient Response

The transient response of the valve was calculated for step closure of the throttling area, A_{tv}, of the bypass valve. The calculations were repeated for different values of the damping spool radial clearance. The transient response of valve input pressure was calculated and plotted (see Fig. 5B.4). The simulation results show that the radial clearance has a significant effect on the valve response. For smaller radial clearances, the flow rate into the damping spool chamber is throttled and the pressure building in this chamber is delayed. The poppet takes a longer time to open, which results in greater pressure overshoot. For great radial clearances, the damping effect weakens and sustained pressure oscillations appear in the steady-state part of the response.

An optimum value of the radial clearance was estimated by calculating the integral of time absolute error (ITAE) defined by Eq. (5B.11). This error criterion is most suitable for systems presenting steady-state error or sustained oscillations.

$$\text{ITAE} = \int_0^T t|e(t)|dt = \int_0^T t(P - P_{SS})dt \tag{5B.11}$$

The variation of the integral of time absolute error (ITAE) with the radial clearance is plotted in Fig. 5B.5. This figure shows that a radial clearance of 35 to 60 μm produces the minimum ITAE. The transient response of the valve pressure to step closure of the bypass valve, for 55-μm radial clearance, was calculated and plotted in Fig. 5B.6. This figure shows that the settling time is within 15 ms and the maximum percentage overshoot is considerably reduced.

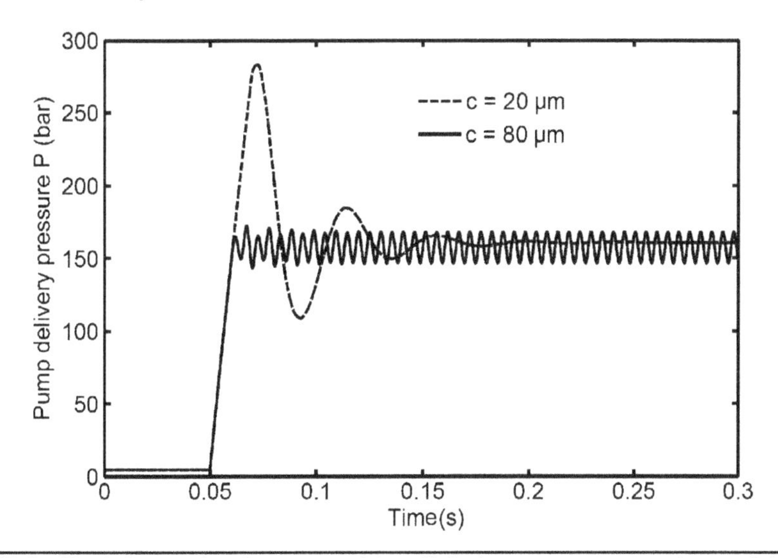

Figure 5B.4 Simulation results of response of the relief valve pressure to step closure of the pump bypass valve for damping spool radial clearances of 20 and 80 μm.

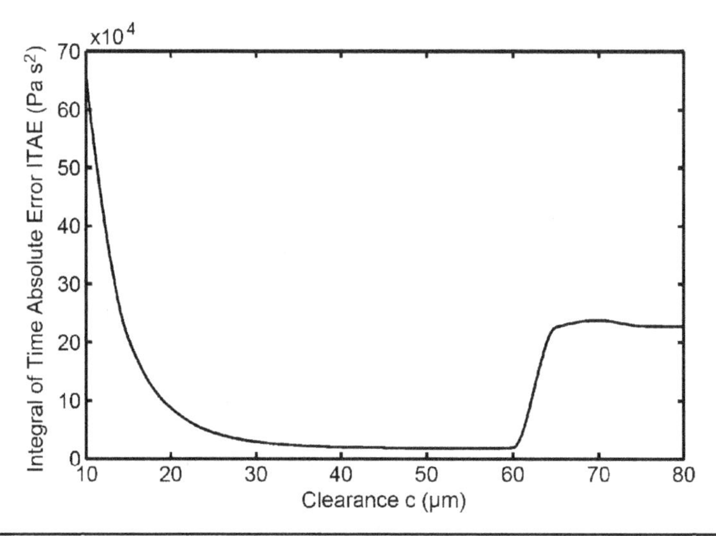

Figure 5B.5 Effect of the damping spool radial clearance on the ITAE.

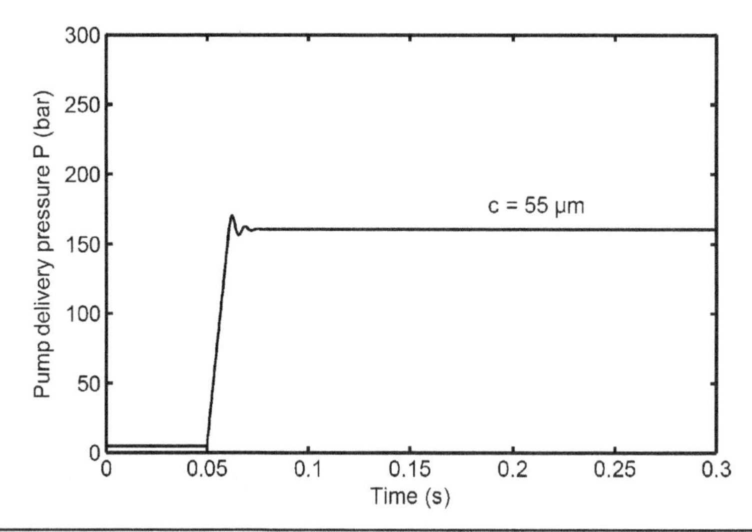

Figure 5B.6 Transient response of the relief valve pressure to step closure of the pump bypass valve for damping spool radial clearances of 55 μm, calculated.

Nomenclature

A_d = Damping spool area, m^2
A_p = Poppet area subjected to pressure, m^2
A_t = Throttle area, m^2
A_{tv} = Throttle valve area of the bypass valve, m^2
B = Bulk modulus of oil, Pa
c = Radial clearance of the damping spool, m
C_d = Discharge coefficient
D_d = Damping spool diameter, m
f = Equivalent spring structural damping and piston friction coefficient, Ns/m
F_S = Seat reaction force, N
k_{rv} = Spring stiffness, N/m
k_s = Equivalent seat material stiffness, N/m
L = Damping spool length, m
m = Reduced mass of the moving parts, kg
P = Valve inlet pressure, Pa
P_d = Pressure in the damping chamber, Pa
P_T = Return pressure, Pa
Q = Relief valve flow rate, m^3/s
Q_{SS} = Steady-state value of the flow rate, m^3/s
Q_d = Flow rate in the radial clearance of the damping spool, m^3/s
Q_p = Pump flow rate, m^3/s
Q_t = Flow rate through the throttle valve, m^3/s
Q_{th} = Pump theoretical flow rate, m^3/s
R_L = Pump resistance to internal leakage, Pa s/m^3
R_s = Equivalent seat material damping, Ns/m
V_o = Initial volume of the damping spool chamber, m^3
T = Time limit for the transient response calculation, s
x = Poppet displacement, m
x_o = Spring pre-compression distance, m
α = Poppet cone vertex angle, rad
μ = Dynamic viscosity, Pa s
ρ = Oil density, kg/m^3

CHAPTER 6
Accessories

6.1 Introduction

Hydraulic power systems are used to transmit and control mechanical power. They consist of the following:

- Elements converting mechanical power into hydraulic power: *pumps*
- Elements transmitting hydraulic power: *pipelines*
- Elements controlling hydraulic power: *valves*
- Elements converting controlled hydraulic power into the required mechanical power needed to drive the load: *hydraulic cylinders and motors*

In addition to the basic elements, other elements are needed for the proper operation of hydraulic systems. These elements do not take part in the power transmission, transformation, or control. However, they are important for the reliable operation of the system, such as:

- Hydraulic tanks: needed to locate the required volume of hydraulic liquid.
- Coolers and heaters: needed to keep the oil temperature within the required operating range.
- Hydraulic filters: required to control the amount and size distribution of contaminants in the hydraulic oil.
- Monitoring elements: pressure gauges, thermometers, flow meters, and other elements needed for monitoring system operation or even switching the system off in some cases.
- Energy storage elements: needed for various operating modes, such as accumulators.

This chapter deals with descriptions of the construction and operation of hydraulic accumulators, pressure switches, and filters.

6.2 Hydraulic Accumulators

6.2.1 Classification and Operation

In the case of pneumatic systems, using highly compressible fluids, the energy transmission medium is directly used to store energy. Compressed air reservoirs and high-pressure bottles are commonly used elements in pneumatic systems. But in the case of hydraulic systems, the used energy transmission medium has a very poor compressibility. Commonly used hydraulic oils have a bulk modulus within 1 to 2 GPa; therefore, these oils are very poor at storing energy. For comparison, one liter of oil at 15 MPa pressure stores about 80 J, while one liter of compressed air at the same pressure stores 28 kJ (see Sec. 11.2.1). Therefore, in the hydraulic circuits requiring energy storage, the hydraulic accumulators are used. The three main types of hydraulic accumulators are weight-loaded, spring-loaded, and gas-charged (see Table 6.1).

Weight-loaded accumulators store the energy in the form of potential energy in the mass of the piston and load. It is charged by pumping the oil into the lower chamber, and then displacing the piston and load upward. The pressure variation due to piston displacement is of negligible value; therefore, this type delivers the oil at a constant pressure.

In the spring-type accumulator, the energy is stored as elastic energy of the spring. The spring is compressed by pumping oil into the accumulator. This class of accumulators delivers the oil at varying pressures. The delivery pressure of this type decreases with the spring relaxation due to the decrease of the spring force. The delivery pressure is proportional to the volume of oil in the oil chamber of the accumulator.

The following expression can be deduced systematically for the pressure of oil in the spring-loaded accumulator:

$$P = k\left(\frac{x_o}{A} + \frac{V}{A^2}\right) \tag{6.1}$$

where A = Piston area, m^2
k = Spring stiffness, N/m
P = Pressure, Pa
V = Oil volume in the accumulator, m^3
x_o = Spring pre-compression, m

Weight-loaded	Spring-loaded	Gas-charged (with Separating Element)		
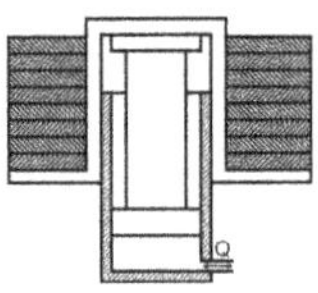	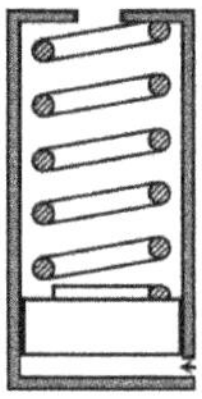	Piston Type 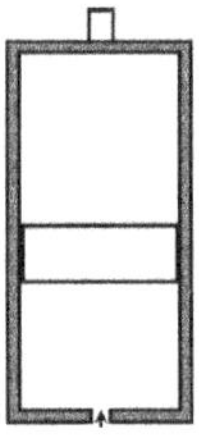	Bladder 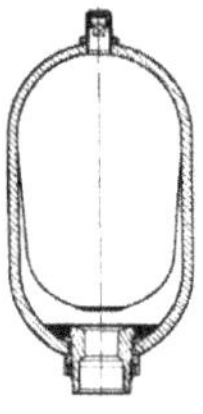	Membrane 

TABLE 6.1 Basic Types of Hydraulic Accumulators

Weight-loaded and spring-type accumulators are not widely used, regardless of the simplicity of their construction and their possible fabrication using standard hydraulic cylinder barrels. This is due to their low response, large sizes, and working constraints.

The most widely used accumulators are the gas-charged types, where the oil is stored under the pressure of a gas, usually nitrogen. The air can be used as a charging gas in the case of fire-resistant oils. Gas-charged accumulators are classified into four types according to the oil–gas separation: piston type, bladder type, diaphragm type, and accumulators without oil–gas separation. The last type operates under restricted conditions where the oil should not be allowed to be completely discharged.

Gas-charged accumulators with separating elements consist of a steel body containing two chambers for the oil and the compressed nitrogen. The gas chamber is pre-charged with compressed nitrogen through a charging check valve. The charging process is carried out while the accumulator is completely empty of oil. The gas and oil chambers are completely separated. During operation, the oil is pumped into the oil chamber. When the oil pressure exceeds the gas-charging pressure, the oil flows into the accumulator, decreasing the gas volume and increasing its pressure. The steady-state equilibrium is reached when the oil pressure is equal to the gas pressure. The oil is stored at high pressure under the action of the compressed gas. The following nomenclatures are used in this chapter:

P_o = Accumulator charging pressure, gas pressure, Pa (abs)
P_1 = Minimum system pressure, Pa (abs)
P_2 = Maximum system pressure, Pa (abs)
V_o = Accumulator size, volume of charging gas at pressure P_o, m^3
V_1 = Volume of gas at pressure P_1, m^3
V_2 = Volume of gas at pressure P_2, m^3

The following equation describes the gas compression process.

$$P_o V_o^n = P_1 V_1^n = P_2 V_2^n = \text{const.} \tag{6.2}$$

According to the type of compression process, the value of the exponent n varies in the range from 1 to 1.4. For an isothermal process, $n = 1$; for a polytropic process, $1 < n < 1.4$, and for an adiabatic process, $n = \gamma = 1.4$. The pressure is absolute, whenever a gas process is considered. If the compression process is so slow that the gas temperature is kept constant, the process is isothermal and the gas pressure and volume are related by the following relation.

$$P_o V_o = P_1 V_1 = P_2 V_2 = \text{Const.} \tag{6.3}$$

6.2.2 The Volumetric Capacity of Accumulators

The accumulator operates usually between two pressure levels: minimum pressure, P_1, and maximum pressure, P_2 (see Fig. 6.1).

The volumetric capacity of the accumulator, V_a, is defined as the volume of oil delivered to/from the accumulator for operating pressure changing from P_1 to P_2 or from P_2 to P_1.

$$V_a = V_1 - V_2 = V_o \left\{ \left(\frac{P_o}{P_1} \right)^{\frac{1}{n}} - \left(\frac{P_o}{P_2} \right)^{\frac{1}{n}} \right\} \quad \text{For polytropic process} \tag{6.4}$$

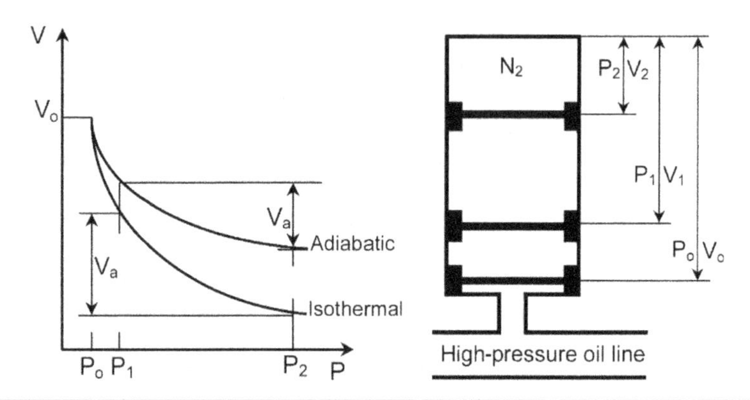

Figure 6.1 Variation of gas volume with operating pressure in a hydraulic accumulator.

$$V_a = V_1 - V_2 = V_o\left\{\left(\frac{P_o}{P_1}\right) - \left(\frac{P_o}{P_2}\right)\right\} \qquad \text{For isothermal process} \tag{6.5}$$

The charging pressure, P_o, should be lower than the minimum operating pressure, P_1, to insure the correct exploitation of the accumulator under all of the operating conditions. If this condition is not satisfied, then whenever the operating pressure becomes smaller than P_o, the compressed gas expands and fills the inner space of the accumulator. Then, the accumulator stops operating. Therefore, the charging pressure is usually selected in the range

$$P_o = (0.7 \text{ to } 0.9)\, P_1 \tag{6.6}$$

If for any reason the charging pressure becomes greater than or equal to the minimum system pressure, the expression for the volumetric capacity becomes

$$V_a = V_1 - V_2 = V_o\left\{1 - \left(\frac{P_o}{P_2}\right)^{\frac{1}{n}}\right\} \tag{6.7}$$

6.2.3 The Construction and Operation of Accumulators

Piston-Type Accumulators

The piston-type accumulator (as shown in Fig. 6.2) consists of a cylindrical body separated internally into two chambers by means of a piston with perfect sealing rings. This type can operate at a very high compression ratio: P_2/P_o. In addition, the oil can be completely discharged during operation without fear of damaging the oil–gas separating element.

The piston type accumulator has the following disadvantages:

- The piston mass and seals slow down the accumulator response.
- The piston seals are subjected to wear and leakage of the compressed gas.

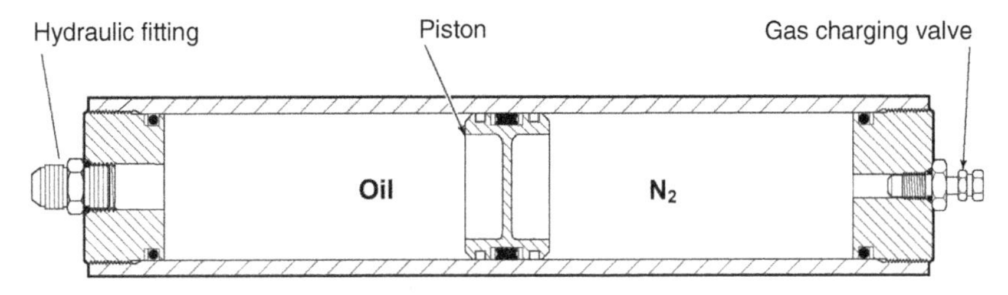

FIGURE 6.2 Typical construction of a piston-type accumulator.

Therefore, piston-type accumulators require more frequent checking of the gas pre-charge pressure.

Bladder-Type Accumulators

In this class of accumulators, a bladder is used as the elastic separation of the oil and compressed gas. The bladder is fastened inside a steel body by means of the vulcanized gas-charging valve assembly. It can be removed and replaced through an opening in the steel body at the oil-valve assembly. Initially, the bladder is charged with compressed gas while the oil port is drained. The bladder is stretched until it comes in contact with the vessel walls.

The bladder material withstands high-compression stresses, but its resistance to shear and tensile stresses is very low. Therefore, the bladder is protected against extruding through the oil connection port in one of two ways:

- By closing the oil port by a hemispherical steel plate with a great number of small diameter holes that allow for free oil flow. The holes' diameters are small enough that the resulting shear stress acting on the bladder walls is less than the allowable value (see Fig. 6.3).
- By using a mushroom-shaped protection valve that seats when the bladder is pre-charged (see Figs. 6.4 and 6.5).

When pumping the oil to the accumulator at pressures higher than the pre-charge pressure, the oil enters the accumulator, compressing the gas and reducing the bladder volume (see Fig. 6.5). The internal tightness of the bladder accumulator is perfect as long as the bladder is not damaged.

The bladder damage can be avoided by taking into consideration the following precautions:

- The minimum operating pressure P_1 should be greater than the pre-charge pressure P_o. In this way, the bladder does not contact the extrusion protection element during normal operation.
- Excessive deformation of the bladder should be avoided (see Fig. 6.5). Therefore, the maximum compression ratio of gas should be limited: $P_2/P_o < 4$. The over-flexing of the bladder can be avoided if the maximum operating pressure P_2 does not exceed three times the minimum pressure P_1; $(P_2 \leq 3P_1)$.

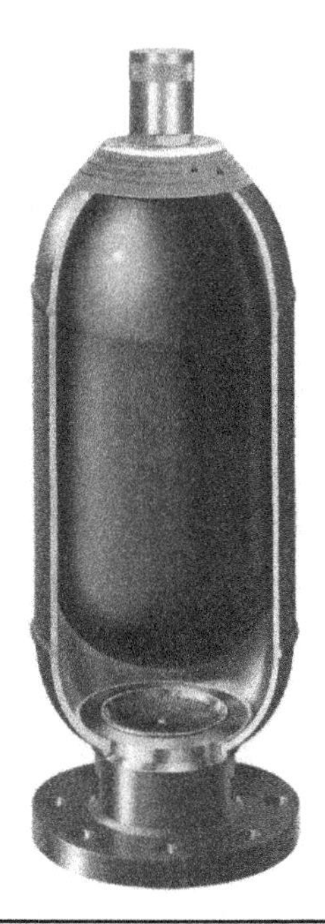

Figure 6.3 A bladder-type accumulator with perforated disc bladder protection. (*Courtesy of Olaer Industries, France.*)

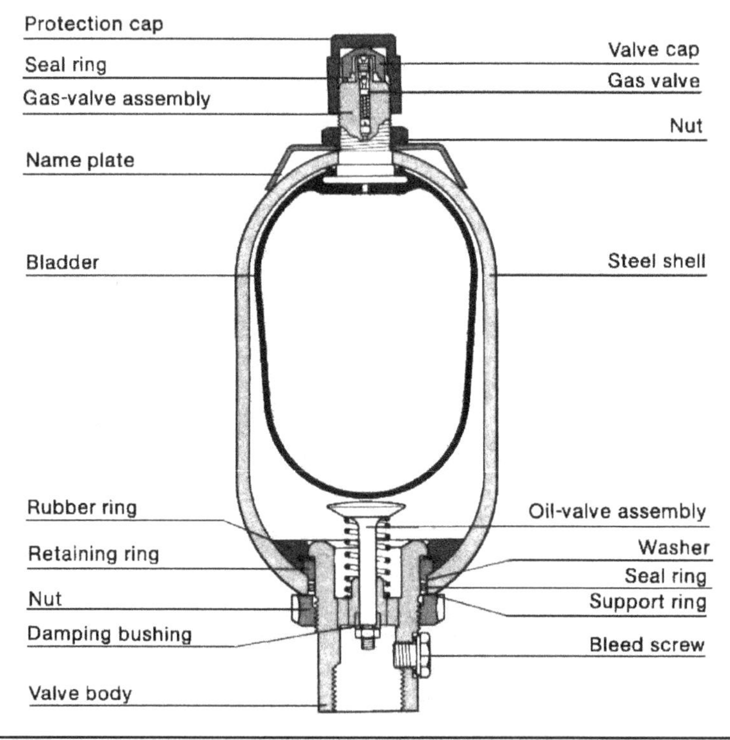

Figure 6.4 Construction of a bladder-type accumulator with a mushroom-shaped bladder protection valve. (*Courtesy of Bosch Rexroth AG.*)

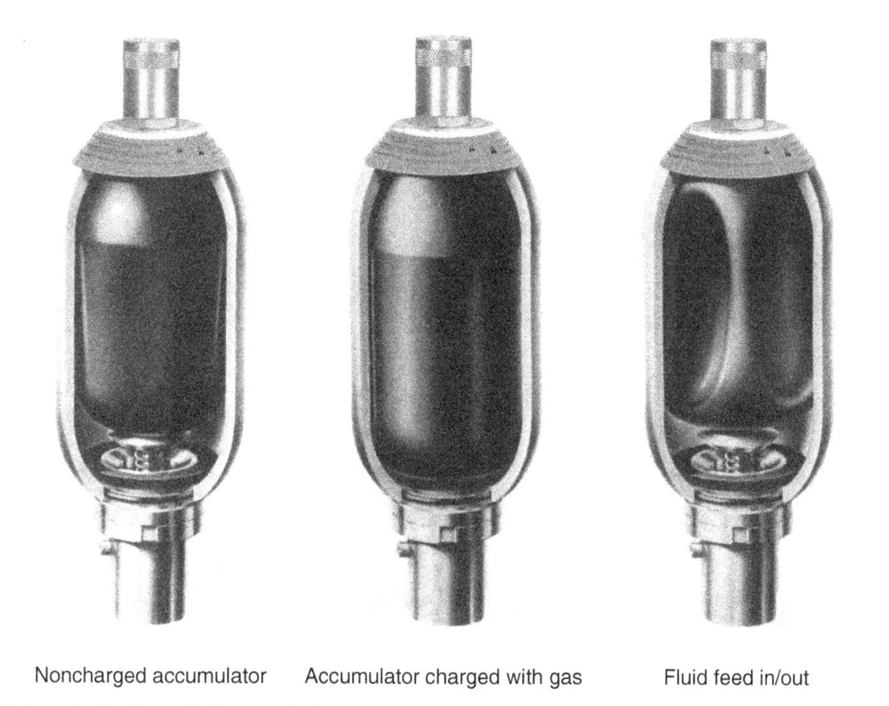

Figure 6.5 Operation of the bladder-type accumulator with a mushroom-shaped protection valve. (*Courtesy of Olaer Industries, France.*)

The size of the accumulator is the volume of the gas-charging accumulator at the charging pressure. This volume is determined by the gas volume rather than the oil flow rate. The flow rate of an accumulator is determined solely by the pressure conditions and the system resistance. For low system resistance and high accumulator pressure, the flow rate could be quite high. According to the accumulator size, the maximum flow rate is limited in order to increase the service life of the bladder. A one-liter-size accumulator is allowed to receive and discharge a flow of up to 240 L/min, while an accumulator of 50 liters in size has a maximum flow rate of 900 L/min.

Diaphragm-Type Accumulators

A diaphragm is clamped between the walls of the pressure vessel and serves as an elastic separator between the hydraulic fluid and gas (see Fig. 6.6). The membrane is fixed to the pressure vessel either through welding (nonreplaceable), or by screwing (replaceable membrane). A shutoff button (plate) is fixed to the base of the diaphragm. This button serves to block the inlet opening at the connection to the piping when the diaphragm is fully expanded. In this way, the diaphragm is protected from being extruded into the opening when in the pre-charged state: $P < P_o$. The operation of this type is explained in Fig. 6.7. The accumulator flow rate should not exceed 40 L/min in order to insure a long service life of the membrane.

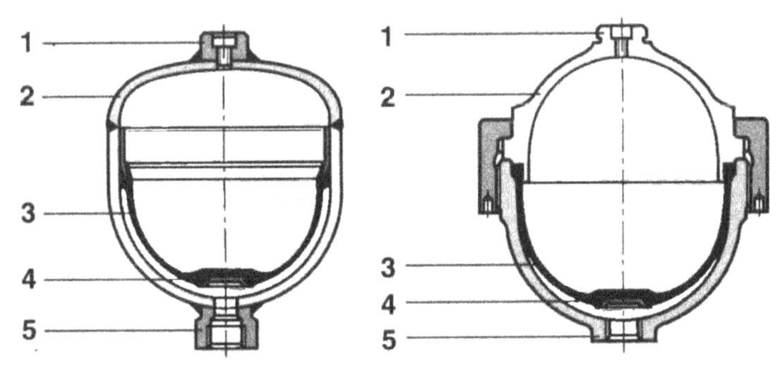

FIGURE 6.6 Diaphragm-type accumulator. (*Courtesy of Bosch Rexroth AG.*)

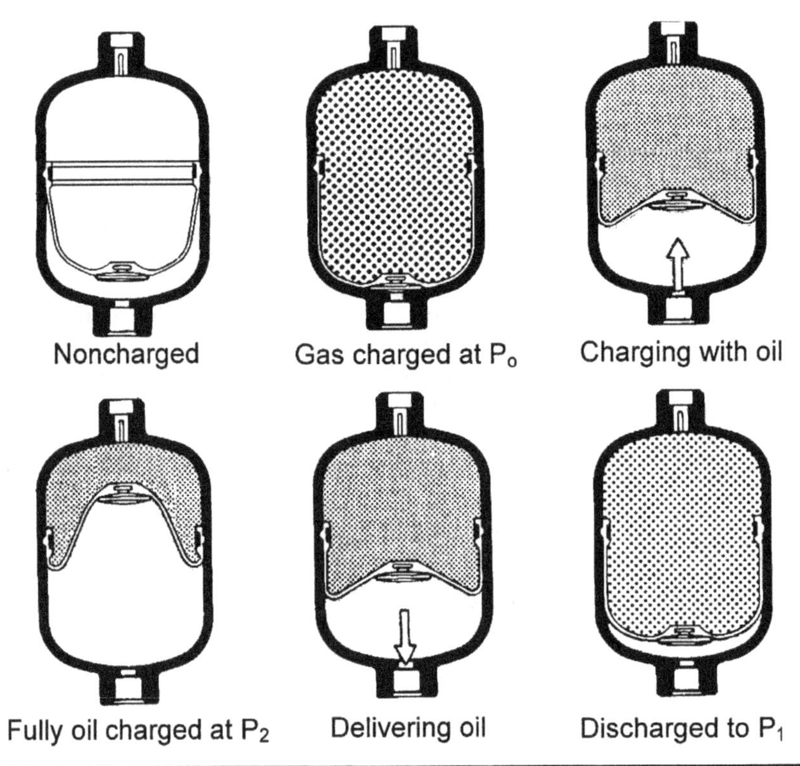

FIGURE 6.7 Operation of the membrane (diaphragm)-type accumulator. (*Courtesy of Bosch Rexroth AG.*)

6.2.4 Applications of Hydraulic Accumulators

Hydraulic accumulators are installed in the hydraulic systems to fulfill several functions. The following are the main applications of hydraulic accumulators:

1. Energy storage
   a) Reserve source of energy
   b) Compensation of the short duration large flow demands, reducing the required pump size and the driving power
   c) Pump unloading
   d) Reducing the response time of actuators placed at long distances from the pump
2. Maintaining constant pressure; compensation for leakage losses
3. Thermal compensation
4. Smoothing of the pressure and flow pulsation
5. Load suspension on load-transporting vehicles
6. Absorption of hydraulic shocks
7. Hydraulic spring in car suspension

Energy Storage

Theoretical Background

Hydraulic accumulators can be used to reduce the required pump size and driving power in the case of systems having intermittent flow demands or systems requiring short-duration, great flow rates during their operating cycle. The accumulator is charged during periods of low flow rate demands and discharges during periods of higher flow rate demands. An accumulator charging valve can be installed to bypass the pump when reaching maximum pressure (see Sec. 5.2.5). The charging valve reconnects the pump to the system when the pressure level declines to its minimum value. Then, the accumulator is recharged. In this case, the accumulator acts as the main source of hydraulic energy in the system.

When using the accumulator as an energy storage element, it is important to deduce the mathematical expressions for the total stored energy and the useful energy delivered by the accumulator. The total energy stored in the hydraulic accumulator is the increase in the compressed air energy when compressed from pressure P_o to pressure P_2. An expression for this energy could be deduced as follows.

$$dE = -PdV \tag{6.8}$$

The negative sign indicates that the stored energy increases with the decrease of gas volume. Considering a polytropic compression process, then

$$PV^n = P_o V_o^n = P_1 V_1^n = P_2 V_2^n \tag{6.9}$$

$$V^n dP + nPV^{n-1} dV = 0 \quad \text{or} \quad dV = -\frac{V}{nP} dP \tag{6.10}$$

$$V = \left(\frac{P_o}{P}\right)^{\frac{1}{n}} V_o \quad \text{Then} \quad dV = -\frac{V_o P_o^{1/n}}{nP^{(n+1)/n}} dP \tag{6.11}$$

Then,
$$E = \frac{V_o P_o^{1/n}}{n} \int_{P_o}^{P_2} P^{-1/n}\, dP \tag{6.12}$$

or
$$E = \frac{V_o P_o^{1/n}}{n-1} [P_2^{(n-1)/n} - P_o^{(n-1)/n}] \tag{6.13}$$

This expression shows that the stored energy is highly affected by the charging pressure P_o. It becomes zero for $P_o = 0$ or $P_o = P_2$. The value of the charging pressure for the maximum value of stored energy is found as follows:

For maximum stored energy, $dE/dP_o = 0$,

or
$$\frac{P_o}{P_2} = n^{-n/(n-1)} \tag{6.14}$$

For an adiabatic process, $n = 1.4$, the stored energy is maximum for $P_o = 0.308P_2$. An expression for the maximum energy could be deduced by substituting Eq. (6.14) into Eq. (6.13).

$$E_{\max} = \frac{V_o P_2}{n^{n/(n-1)}} \tag{6.15}$$

By defining $\bar{E} = E / E_{\max}$ and $\bar{P}_o = P_o / P_2$, the following expression is deduced for the nondimensional energy, assuming a polytropic gas compression process:

$$\bar{E} = \frac{n^{n/(n-1)}}{n-1} \bar{P}_o^{1/n} [1 - \bar{P}_o^{(n-1)/n}] \tag{6.16}$$

In the case of an isothermal gas process, $n = 1$, an expression for the total energy stored in the accumulator could be deduced as follows:

$$PV = P_o V_o = P_2 V_2 \tag{6.17}$$

$$PdV + Vdp = 0 \tag{6.18}$$

$$dV = -\frac{P_o V_o}{P^2} dP \tag{6.19}$$

$$E = -\int_{P_o}^{P_2} P\, dV = P_o V_o \ln(P_2/P_o) = -V_o P_2 \frac{P_o}{P_2} \ln\left(\frac{P_o}{P_2}\right) = -V_o P_2 \bar{P}_o \ln(\bar{P}_o) \tag{6.20}$$

For maximum energy,

$$\frac{dE}{d\bar{P}_o} = 0 \quad \text{or} \quad \ln(\bar{P}_o) = -1 \quad \text{and} \quad \bar{P}_o = 1/e \tag{6.21}$$

Then, by substitution in Eq. (6.20), the expression for the maximum energy is obtained.

$$E_{\max} = P_2 V_o / e \quad \text{and} \quad \bar{E} = E/E_{\max} = -e \bar{P}_o \ln(\bar{P}_o) \tag{6.22}$$

Figure 6.8 shows the variation of the total energy stored in the accumulator with the charging pressure, for polytropic and isothermal gas process. The total energy is found to be maximal for $\bar{P}_o = 0.308$ *to* 0.37, depending on the type of gas process.

When using the accumulator as a reserve source or as an emergency source of energy, it is important to select the operating parameters to deliver the maximum energy.

The minimum operating pressure should be sufficient to drive the hydraulic motors and cylinders. Then, the effective energy delivered by the accumulator can be calculated by the following expression:

$$E_e = P_1(V_1 - V_2) \tag{6.23}$$

For a polytropic gas process, the following expressions for effective accumulator energy can be deduced:

$$E_e = P_1 V_1 \left[1 - \left(\frac{P_1}{P_2} \right)^{1/n} \right] \tag{6.24}$$

The effective energy is maximum if $dE_e/dP_1 = 0$, or

$$P_1 = \left(\frac{n}{n+1} \right)^n P_2 \quad \text{or} \quad P_1 = 0.47\, P_2 \quad \text{for } n = \gamma = 1.4 \tag{6.25}$$

Then,

$$E_{e\max} = P_2 V_1 \frac{n^n}{(n+1)^{n+1}} \tag{6.26}$$

and

$$\bar{E}_e = \frac{(n+1)^{n+1}}{n^n} \bar{P}_1 \left(1 - \bar{P}_1^{\,1/n} \right) \tag{6.27}$$

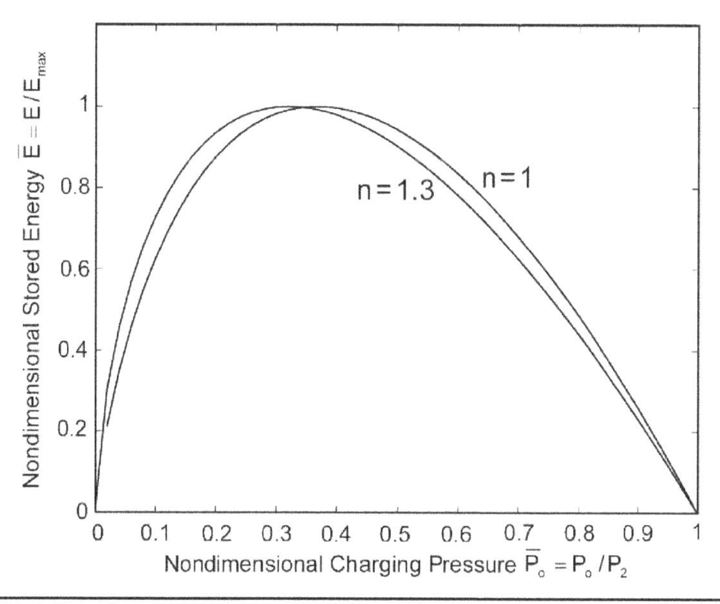

Figure 6.8 Variation of the total energy stored in a hydraulic accumulator with the charging pressure, calculated.

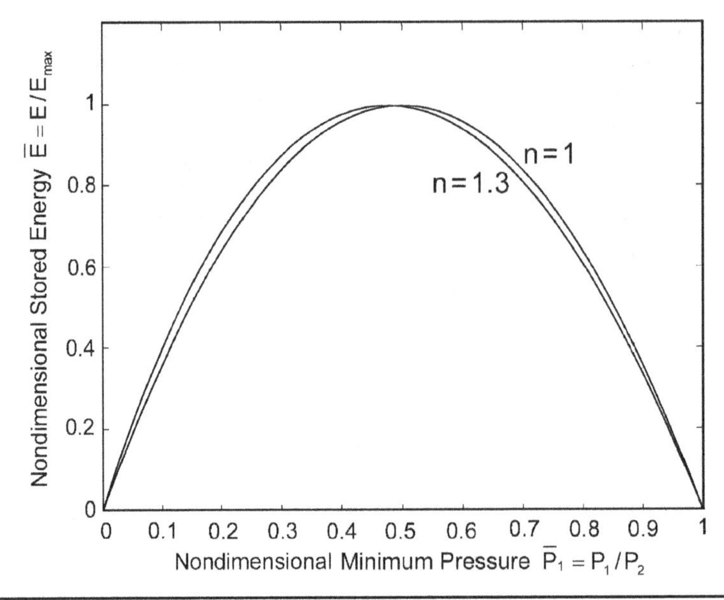

FIGURE 6.9 Variation of effective accumulator energy with the ratio of the minimum operating pressure to the maximum pressure, P_1/P_2, calculated.

For an isothermal process, $n = 1$. Then, by substituting in Eqs. (6.25) through (6.27), the effective energy is maximal for $P_1 = 0.5P_2$.

$$\bar{E}_e = P_1V_1\left(1-\frac{P_1}{P_2}\right) \tag{6.28}$$

Then,

$$E_{e\max} = 0.5\,P_1V_1 \tag{6.29}$$

and

$$\bar{E}_e = 4\bar{P}_1\left(1-\bar{P}_1\right) \tag{6.30}$$

where

$$\bar{E}_e = E_e/E_{e\max} \quad \text{and} \quad \bar{P}_1 = P_1/P_2 \tag{6.31}$$

Figure 6.9 shows the variation of the effective accumulator energy with the minimum operating pressure for a polytropic and isothermal gas process. The effective energy is found to be maximal for $\bar{P}_1 = 0.47$ *to* 0.5, depending on the type of gas process.

Emergency Sources of Energy

In the case of some critical applications, the operating cycle should be completed. It might be necessary to install a redundant source of hydraulic energy to insure the required reliability. The installation of a hydraulic accumulator of convenient size is one of the possible solutions. The accumulator is charged during the normal operation of the hydraulic power system. It is allowed to discharge only under emergency conditions (see Fig. 6.10).

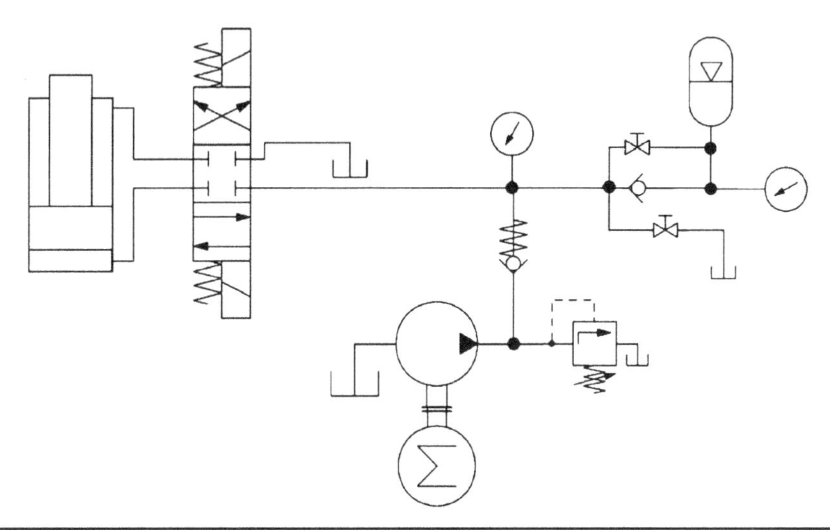

FIGURE 6.10 Application of a hydraulic accumulator as a reserve source of energy.

Compensation for Large Flow Demands

Sometimes, in the case of systems with intermittent operations requiring short duration high flow rates, a large pump is recommended. The prime mover required to drive the pump should have enormous power. The required pump's geometric volume and driving power are given by the following relations:

$$V_g = Q_{\max}/n\eta_v \tag{6.32}$$

$$N = Q_{\max}P/\eta_T \tag{6.33}$$

where n = Pump speed, rev/s
N = Pump driving power, W
P = Pump exit pressure, Pa
$Q_{\max}$ = Maximum required flow rate, m^3/s
V_g = Pump geometric volume, m^3/rev
η_v = Pump volumetric efficiency
η_T = Pump total efficiency

The required pump size and driving power can be reduced by installing a hydraulic accumulator of convenient size. The accumulator is charged by the pump during periods of low flow demands. Both the pump and accumulator deliver the flow rate required for system operation during the large flow demand period. A typical hydraulic circuit for this application is given in Fig. 6.11.

Example 6.1 A hydraulic system operates in a regular operating cycle of 60 s duration. The flow demand during the operating cycle is shown in Fig. 6.12. The maximum pump delivery pressure

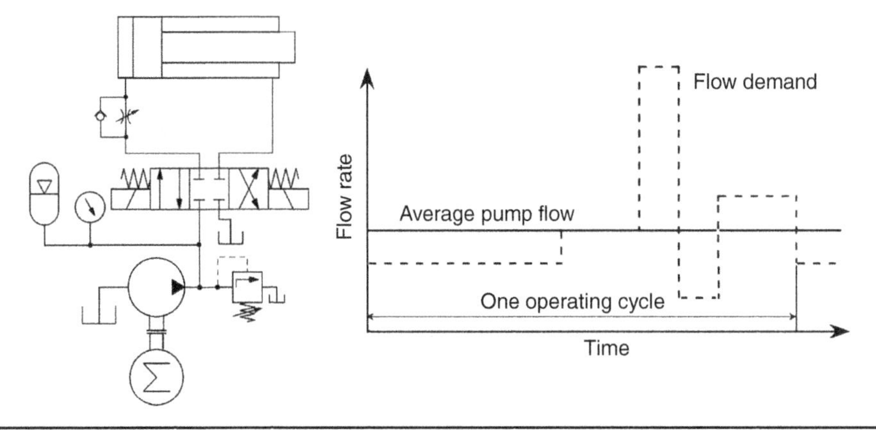

Figure 6.11 A hydraulic circuit using an accumulator to compensate for large flow demands.

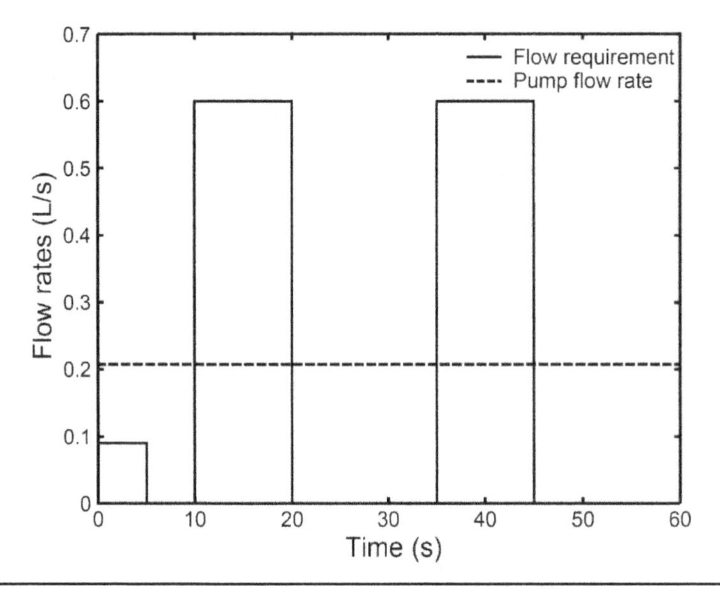

Figure 6.12 System flow demand and average recommended pump flow rate.

during the operating cycle is 140 bar. The flow rates are set by flow control arrangements. The system has a fixed displacement pump. Determine the required pump flow rate in the following cases:

a) When using the pump only for the hydraulic power supply.

b) If a hydraulic accumulator is used to compensate for the short duration flow demands, calculate the suitable size of the accumulator if the maximum allowable pressure is 210 bar.

Solution

a) The pump should supply a constant flow rate of 0.6 L/s, which covers the maximum required flow rate. The pump loading pressure is 140 bar. The required hydraulic power is $N = 0.6 \times 10^{-3} \times 140 \times 10^{5} = 8400$ W.

b) When using the accumulator, the pump flow rate will be calculated as follows:

Total oil volume required during one cycle = 0.09 × 5 + 2 × 0.6 × 10 = 12.45 L

Cycle duration = 60 s

Recommended pump flow rate = 12.45/60 = 0.2075 L/s

When installing an accumulator, it will cover, together with the pump, the flow demand at all instances. The variation of oil volume in the accumulator (ΔV_o) will be calculated as explained by Table 6.2 and illustrated by Fig. 6.13.

During one complete operating cycle, the net variation of oil volume in the accumulator is zero, as the pump displaces the oil volume required for one complete operating cycle. The maximum variation of oil volume in the accumulator is

$$V_a = 1.625 - (-3.1125) = 4.7375 \text{ L}$$

This is the volumetric capacity of the used accumulator. It is the volume delivered by the accumulator between the maximum pressure (210 bar) and the minimum pressure (140 bar). The accumulator charging pressure = 0.9 × 140 = 126 bar.

The accumulator size is then calculated as follows:

$$V_o = V_a/[(P_o/P_1)^{1/n} - (P_o/P_2)^{1/n}]$$

$$V_o = \frac{4.375}{\left\{\left(\frac{127}{141}\right)^{1/1.3} - \left(\frac{127}{211}\right)^{1/1.3}\right\}} = 17.78 \text{ L}$$

Thus, a 20-liter-size accumulator would be convenient for this application.

Period (s)	Volume Delivery	$\Sigma\Delta V_o$
0–5	The pump flow is greater than the flow demand. The excess pump flow (0.2075 – 0.09 = 0.1175 L/s) is supplied to the accumulator. By the end of this period, the accumulator gains (0.1175 × 5 = 0.5875 L).	0.5875 L
5–10	The whole pump flow is delivered to the accumulator. By the end of this period, the accumulator gains (0.2075 × 5 = 1.0375 L).	1.625 L
10–20	The pump flow is less than the flow demand. The accumulator compensates for this difference. It delivers (0.6 – 0.2075 = 0.3925 L/s). By the end of this period, the accumulator loses (0.3925 × 10 = 3.925 L).	–2.3 L
20–35	The whole pump flow is delivered to the accumulator. By the end of this period, the accumulator gains (0.2075 × 15 = 3.1125 L).	0.8125 L
35–45	The pump flow is less than the flow demand. The accumulator compensates for this difference. It delivers (0.6–0.2075 = 0.3925 L/s). By the end of this period, the accumulator loses (0.3925 × 10 = 3.925 L).	–3.1125 L
45–60	The whole pump flow is delivered to the accumulator. By the end of this period, the accumulator gains (0.2075 × 15 = 3.1125 L).	0 L

TABLE 6.2 Variation of Oil Volume in the Accumulator over a Single Operating Cycle

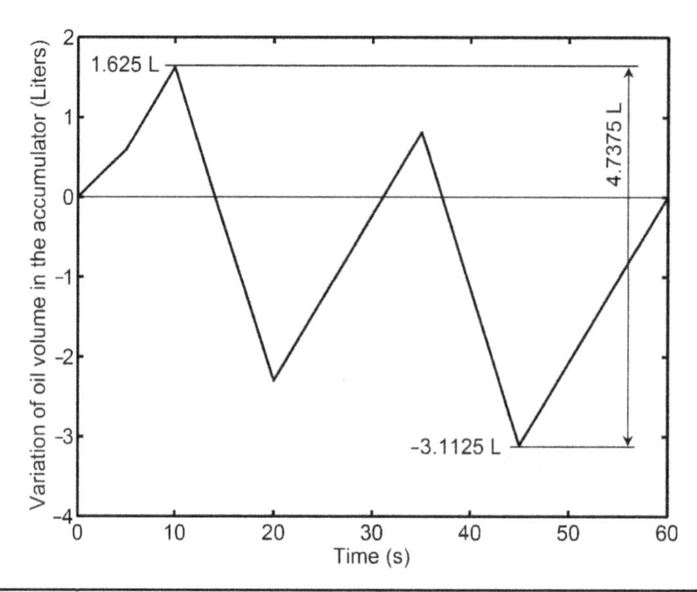

FIGURE 6.13 Variation of volume of oil in the hydraulic accumulator during one operational cycle.

Pump Unloading

In this application, the pump serves mainly to charge the accumulator. The system is equipped with an accumulator charging valve which bypasses the pump to the tank when the pressure in the accumulator reaches a certain maximum value: P_2. The pump is then unloaded, and the accumulator operates as a source of the hydraulic energy needed for the system operation. When operating any of the system organs, the volume of oil in the accumulator decreases and its pressure decreases. When the system pressure decreases to the minimum value, P_1, the accumulator-charging valve cuts the pump bypass and reconnects the pump to recharge the accumulator. A check valve is installed to force the accumulator to discharge only in the direction of the system. In this way, the pump runs idling most of the time. The typical construction of an accumulator-charging valve is given in Fig. 5.21.

Reducing the Actuator's Response Time

Some hydraulic systems include hydraulic cylinders placed at long distances from the hydraulic generator. The time of response of these cylinders is relatively long due to the effect of the long transmission lines. The time of response may be reduced by placing an accumulator near the hydraulic cylinders (see Fig. 6.14). The accumulator is recharged during the time interval between two successive operations. It discharges in the direction of the hydraulic cylinder, reducing its time of response.

Maintaining Constant Pressure

In some applications, it is necessary to maintain constant pressure in a particular branch of a circuit over a certain period. The installation of an accumulator permits the pump to be used for the operation of other elements, bypassing the pump or even switching its driving motor off while satisfying the constant pressure requirement.

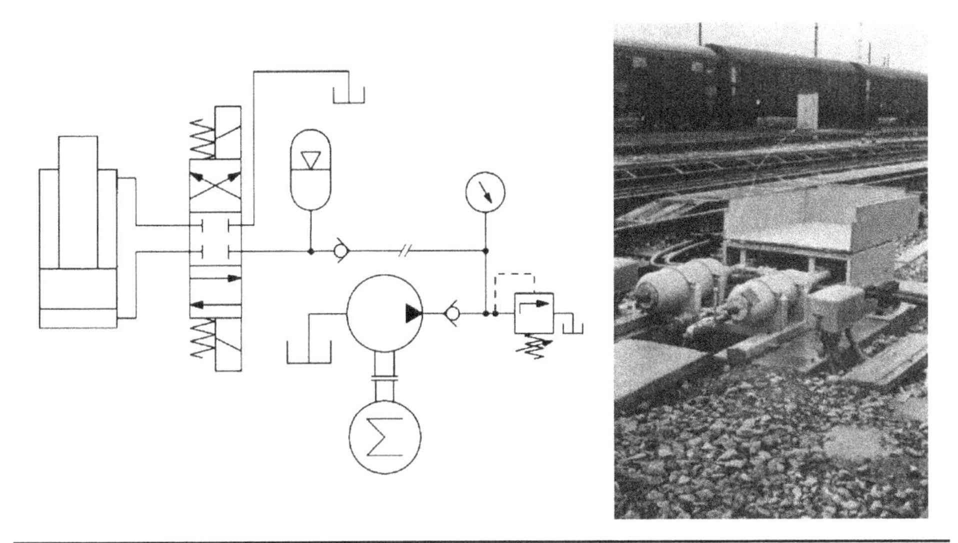

FIGURE 6.14 Accumulators used to reduce the actuator's response time. (*Courtesy of Bosch Rexroth AG.*)

A typical example for using the accumulators to maintain a constant pressure is given in Fig. 6.15. The compressed gas in the accumulator maintains the pressure in the hydraulic cylinder and compensates for leakage. The accumulator size is determined according to the permissible pressure drop, the duration of the application of pressure, and the resistance to internal leakage.

Thermal Compensation

During the normal operation of the hydraulic systems, the position of some hydraulic actuators is locked hydraulically and/or mechanically. When locking the position hydraulically, a volume of oil is trapped in the hydraulic cylinders and pipes. If this oil is subjected to a considerable increase in temperature, its pressure rises due to the effect of volumetric thermal expansion and oil compressibility (see Sec. 2.2.4). The resulting pressure increment is

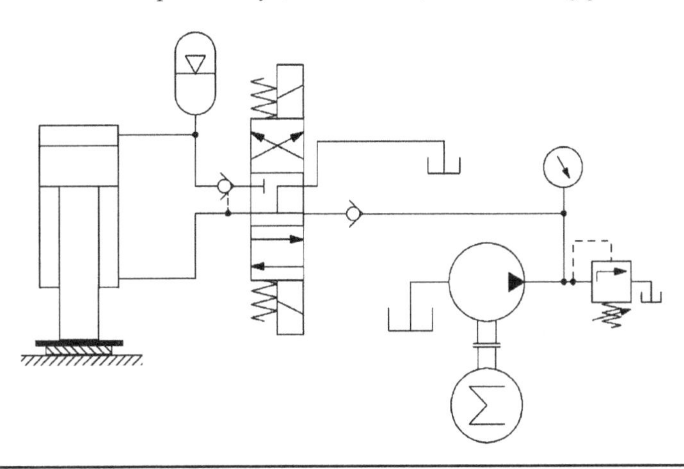

FIGURE 6.15 Illustration of the application of hydraulic accumulators for maintaining constant pressure, leakage compensation, and thermal expansion compensation.

given by the following expression, neglecting the volumetric expansion of the cylinder and pipe material:

$$\Delta P = \alpha B \Delta T \tag{6.34}$$

where α = Oil thermal expansion coefficient, K^{-1}
B = Bulk modulus of oil, Pa
ΔT = Temperature increment, K
ΔP = Resulting pressure increment, Pa

Example 6.2 The extension position of a hydraulic cylinder is locked hydraulically. The cylinder is subjected to a temperature increment of 50 K. The oil thermal expansion coefficient is 6×10^{-4} K^{-1}, and the bulk modulus of oil is 1.4 GPa. The resulting pressure increment can be predicted as follows, assuming rigid cylinder and pipe walls:

$$\Delta P = \alpha B \Delta T = 6 \times 10^{-4} \times 1.4 \times 10^{9} \times 50 = 42 \times 10^{6} \text{ Pa} = 420 \text{ bar}$$

The system can be protected against this great pressure rise by using a relief valve or a hydraulic accumulator. The hydraulic accumulator installation compensates the volumetric variations of the hydraulic oil resulting from thermal expansion (see Fig. 6.15).

Consider a line of length L and cross-sectional area A, subjected to a temperature increment ΔT. The resulting volumetric expansion due to the temperature rise is ($V_a = AL\alpha\Delta T$). This volume is the volumetric capacity of the accumulator needed to compensate the volumetric variations due to thermal expansion. Referring to Eq. (6.4), the size of the needed accumulator is then given by the following expression, assuming a polytropic gas process.

$$V_o = \frac{AL\alpha\Delta T}{(P_o/P_1)^{1/n} - (P_o/P_2)^{1/n}} \tag{6.35}$$

When taking into consideration the thermal expansion of pipe material, this expression becomes

$$V_o = \frac{AL(\alpha - \alpha_P)\Delta T}{(P_o/P_1)^{1/n} - (P_o/P_2)^{1/n}} \tag{6.36}$$

where α_P= Pipe material volumetric thermal expansion coefficient, K^{-1}

Smoothing of the Pressure Pulsations

The displacement pumps, used in the hydraulic power systems, deliver pulsating flow. The flow pulsations lead to considerable pressure pulsations as well the pulsation of the speed of the hydraulic cylinders and motors (see Sec. 4.5). Some applications necessitate a nonpulsating hydraulic supply. In this case, the hydraulic accumulator can be used. It acts as a capacitive element. Together with the transmission line resistance, they act as a low pass filter.

Capacitance of Hydraulic Accumulators

The hydraulic accumulator acts as a capacitive element. An expression for its capacitance can be deduced as follows:

For the polytropic compression of gas, $PV_g^{\,n} = P_o V_o^{\,n} = \text{const.}$

$$V_g^{\,n} dp + npV_g^{\,n-1} dV_g = 0 \tag{6.37}$$

$$dV_g = -(V_g/np)dp \tag{6.38}$$

$$V_g + V_L = V_o = \text{const.} \tag{6.39}$$

$$dV_g/dt = -dV_L/dt = -q \tag{6.40}$$

$$q = \frac{V_g}{np}\frac{dp}{dt} \tag{6.41}$$

or

$$C_A = \frac{V_g}{nP} = \frac{V_o P_o^{1/n}}{nP^{(n+1)/n}} \tag{6.42}$$

where V_g = Volume of gas, m^3
C_A = Hydraulic capacitance of the accumulator, m^3/Pa

Smoothing of Pressure and Flow Pulsations

Figure 6.16 shows a hydraulic accumulator installed at the pump exit to smooth the pressure pulsation resulting from the pump pulsating delivery. The following are the equations describing this system, assuming zero return pressure and neglecting the effect of the inertia and capacitance of the transmission line:

$$q_T = C_d A_T \sqrt{2p/\rho} \tag{6.43}$$

Equation (6.43) can be linearized, assuming zero initial conditions,

$$q_T = \frac{dq_T}{dp}\Delta p = p/R_T \tag{6.44}$$

where R_T = Throttle element resistance, Pa s/m^3
q_T = Flow rate through the throttle valve, m^3/s
A_T = Throttle area, m^2
p = Pressure, Pa
ρ = Oil density, kg/m^3

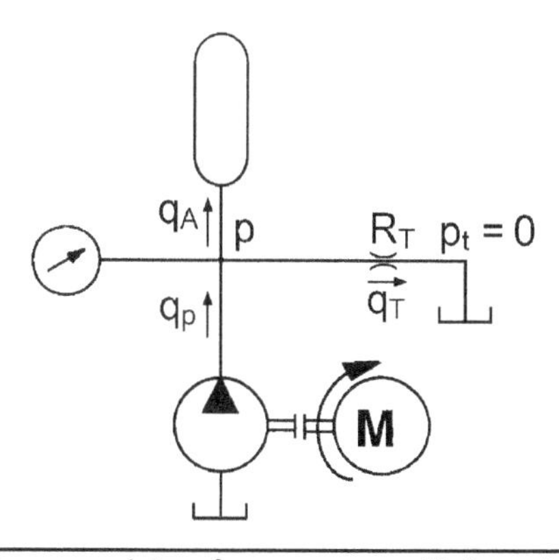

Figure 6.16 Using hydraulic accumulators for pressure oscillations damping.

$$q_p = q_T + q_A \tag{6.45}$$

$$q_A = C_A \frac{dp}{dt} \tag{6.46}$$

$$p = q_T R_T \tag{6.47}$$

Considering the effect of the accumulator and applying the Laplace transform to Eqs. (6.45) through (6.47), the following transfer function is deduced:

$$\frac{P(s)}{Q_p(s)} = \frac{R_T}{R_T C_A s + 1} = R_T\, G(s) \tag{6.48}$$

where
$$G(s) = \frac{1}{Ts+1} \quad \text{and} \quad T = R_T C_A \tag{6.49}$$

According to the given assumptions, the transfer function $G(s)$ relating the delivery pressure to the pump pulsating exit flow is a first-order transfer function. Its frequency response is shown in Fig. 6.17. This figure shows that the high-frequency components of the input signal are highly damped, while the low-frequency components are slightly damped or allowed to pass without considerable change in magnitude. In this arrangement, the accumulator acts as a low pass filter.

When introducing an orifice at the accumulator inlet (see Fig. 6.18) for a very short pump exit line, the line inertia is negligible. Thus, the system can be described by the following relations, neglecting the accumulator inlet line inertia and capacitance:

The continuity equation applied to the pump exit line is

$$q_p - q_A - q_T = \frac{V_p}{B}\frac{dP}{dt} = C_p \frac{dp}{dt} \tag{6.50}$$

where C_p = Hydraulic capacitance of the pump exit line, m^5/N.

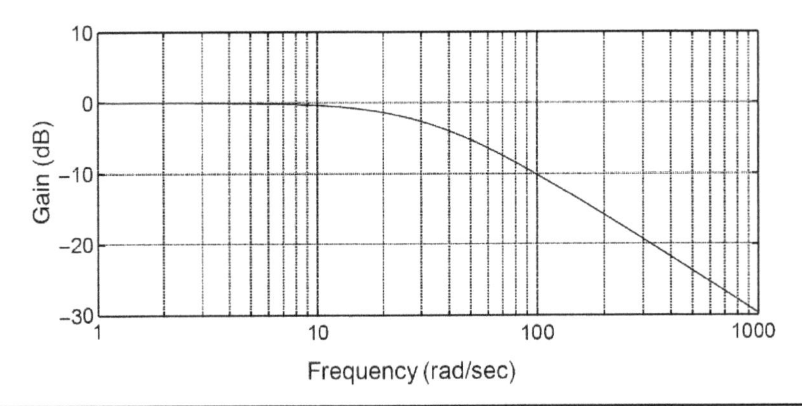

FIGURE 6.17 Gain plot of the frequency response of the accumulator when used as a damper for the pressure oscillations.

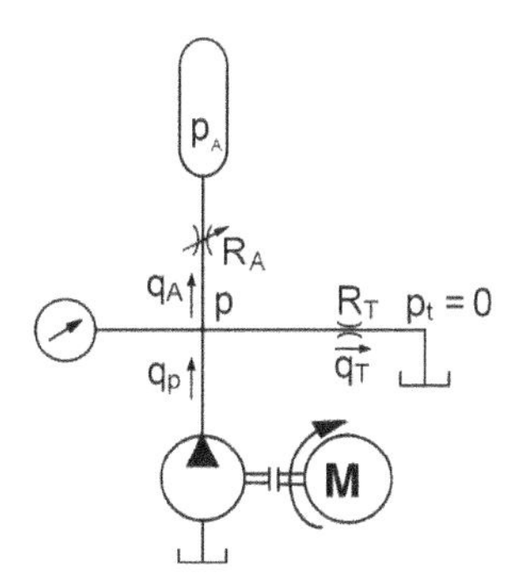

FIGURE 6.18 Damping of the pressure pulsation by using a hydraulic accumulator with an inlet orifice.

The flow rate through the accumulator inlet throttle is

$$q_A = \frac{p - p_A}{R_A} \tag{6.51}$$

where R_A = Resistance of the accumulator inlet orifice, Pa

Also,
$$q_A = C_A \frac{dp_A}{dt} \tag{6.52}$$

The flow rate through the return orifice is

$$q_T = \frac{p}{R_T} \tag{6.53}$$

Applying the Laplace transform to Eqs. (6.50) through (6.53), the following transfer function can be deduced:

$$\frac{P(s)}{Q_p(s)} = \frac{R_T\left(R_A C_A s + 1\right)}{R_A R_T C_A C_p s^2 + \left(R_T C_A + R_A C_A + R_T C_p\right)s + 1} \tag{6.54}$$

The hydraulic capacitance of the pump exit line C_p is too small compared with that of the accumulator C_A. For a pump exit zone of $V_p = 0.5$ liters, a hydraulic accumulator of $V_g = 1$ liters, oil bulk modulus $B = 1.3$ GPa, operating pressure of $p = 10$ MPa, and assuming a polytropic process, $n = 1.3$, the ratio of these two capacitances is

$$\frac{C_A}{C_p} = \frac{V_g B}{npV_p} = 200 \tag{6.55}$$

Thus, the contribution of the pump exit line capacitance is negligible. Therefore, considering this result by neglecting the term C_p, Eq. (6.54) becomes:

$$\frac{P(s)}{Q_p(s)} = \frac{R_T\left(R_A C_A s + 1\right)}{\left(R_T + R_A\right)C_A\, s + 1} \tag{6.56}$$

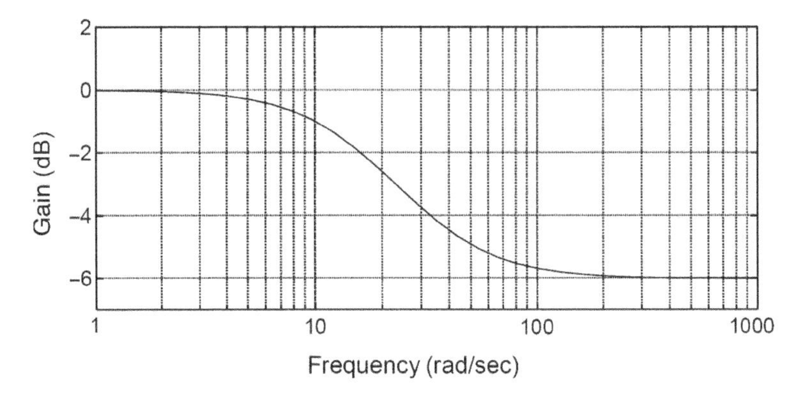

FIGURE 6.19 Gain plot of the frequency response of the accumulator with inlet restriction when used as a pressure pulsation damper.

or

$$\frac{P(s)}{Q_p(s)} = \frac{R_T(T_2 s+1)}{T_1 s+1} = R_T\, G(s) \tag{6.57}$$

$$G(s) = \frac{T_2 s+1}{T_1 s+1} \tag{6.58}$$

The time constants T_1 and T_2 are

$$T_1 = (R_T + R_A)C_A \quad \text{and} \quad T_2 = R_A C_A \tag{6.59}$$

The frequency response of the system described by the transfer function in Eq. (6.58) is shown in Fig. 6.19. The study of this figure shows that

1. The accumulator acts as a low pass filter.
2. There are two corner frequencies: $\omega_1 = 1/T_1$ and $\omega_2 = 1/T_2$. By increasing the accumulator inlet resistance, R_A, these two corner frequencies become nearer, which decreases the maximum attenuation of the pressure oscillation. Therefore, the value of this resistance should be calculated to produce the recommended damping.
3. The accumulator dampens the pressure pulsation of high frequencies. However, the reduction in the magnitude ratio is limited due to the presence of a phase lead element, imposed by the accumulator input resistance.

Load Suspension on Load Transporting Vehicles

In the case of load transporting vehicles (see Fig. 6.20), the suspension elements are subjected to great inertia loads and pressure oscillations in the lifting cylinders, due to the irregularity of the road. These elements are subjected to severe fatigue loads. The installation of a hydraulic accumulator results in smoother pressure variations and a considerable reduction in fatigue loads, which allows faster movement with fragile loads.

Absorption of Hydraulic Shocks

The rapid closure of control valves results in a rapid deceleration of the moving fluid columns in hydraulic transmission lines. When the fluid lines are long enough, the

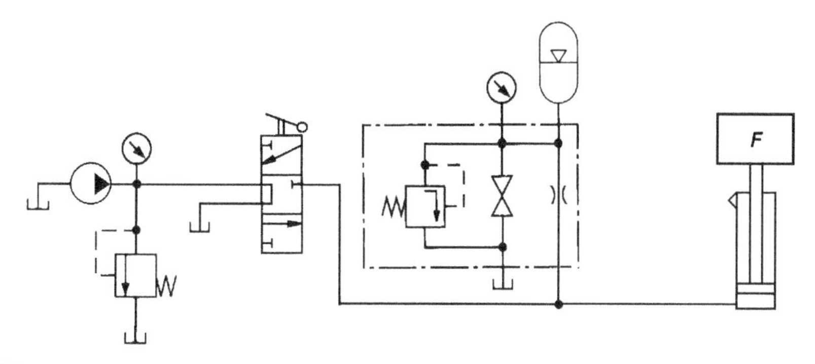

Figure 6.20 Load suspension using hydraulic accumulators.

resulting pressure shock can be dangerous in effect. Consider that the oil is moving with a mean speed v in a line of cross-sectional area A and length L. The end of the line is closed rapidly during time period t_c. The valve closure creates a pressure wave, which propagates through the line. The pressure wave travels to the line end and reflects back at sonic speed c during a time interval t_p.

$$t_p = 2L/c \tag{6.60}$$

If $t_c \le t_p$, the pressure at the valve inlet increases and the line is subjected to a hydraulic shock. The rapid closure of the valve produces a pressure wave that travels up the pipe with velocity c. During a short time interval (dt), an element of liquid of length L is brought to rest. An expression for the decelerating force acting on the oil and the pressure rise at the valve inlet can be deduced as follows:

$$F = m\frac{dv}{dt} \tag{6.61}$$

$$PA - (P + dP)A = \rho Acdt\frac{dv}{dt}, \text{ where } L = c\,dt \tag{6.62}$$

$$dP = -\rho c dv \tag{6.63}$$

or

$$\Delta P = -\rho c \Delta v \tag{6.64}$$

When the liquid is fully stopped, $\Delta v = -v$, then

$$\Delta P = \rho v c \tag{6.65}$$

Example 6.3 The liquid flows in a hydraulic transmission line with $v = 10$ m/s, $\rho = 900$ kg/m^3, and $c = 1300$ m/s, then the sudden closure of the line end valve results in a pressure rise $\Delta P = 11.7$ MPa.

If the closure time t_c is greater than the pressure wave propagation time t_p, ($t_c > t_p$), the peak pressure can be predicted as follows.

$$\Delta P = \frac{t_p}{t_c}\rho v c = \frac{2L}{t_c}\rho v \tag{6.66}$$

or $$\Delta P = \rho v c \quad \text{for} \quad t_c \le t_p \tag{6.67}$$

and $$\Delta P = \frac{2L}{t_c}\rho v \quad \text{for} \quad t_c > t_p \tag{6.68}$$

These expressions show that the pressure rise due to the sudden closure of line is independent of the steady-state pressure level. Therefore, when it is not feasible to close the valve slowly, the hydraulic accumulators are used to absorb most of the transient pressure rise. When using the hydraulic accumulator to absorb the resulting hydraulic shock, it should be fitted as closely as possible to the source of the shock. The size of the accumulator should be calculated such that it can effectively absorb the resulting pressure rise. Figure 6.21 shows a hydraulic accumulator installed near the control valve to protect the transmission line against hydraulic shocks.

Initially, before the valve closure, the steady-state pressure at the accumulator inlet (just before the valve) is P_1. The valve closure increases the pressure to P. This pressure increase starts to decelerate the moving liquid column. The application of Newton's second law to the moving mass of liquid yields

$$(P - P_1)A = -\rho A L \frac{dv}{dt} \tag{6.69}$$

The oil flows into the accumulator due to the pressure increase, resulting from the rapid valve closure. Consider the severest possible operating conditions, where the valve is closed during a nearly zero time interval, and the oil flow rate into the accumulator is Av, then

$$\frac{dV_L}{dt} = Av \tag{6.70}$$

$$V + V_L = V_o = Const. \tag{6.71}$$

where V = Volume of gas in the accumulator, m^3
V_L = Volume of oil in the accumulator, m^3
V_o = Total volume, size, of accumulator, m^3

$$\frac{dV}{dt} = -\frac{dV_L}{dt} = -Av \tag{6.72}$$

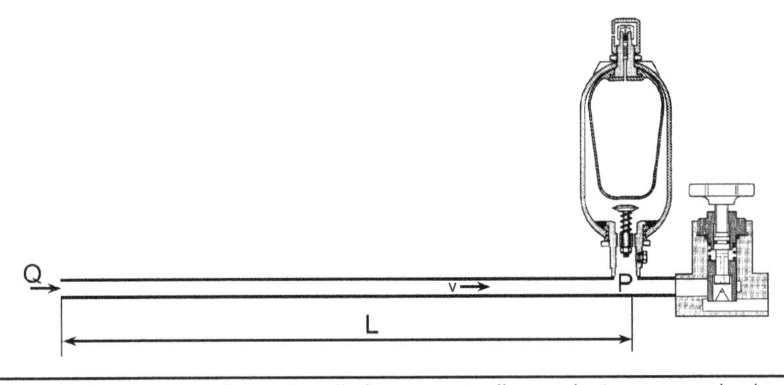

FIGURE 6.21 Hydraulic accumulators installed to protect a line against pressure shocks.

The treatment of Equations (6.69) and (6.72) yields

$$(P - P_1)dV = \rho ALvdv \tag{6.73}$$

The accumulator is installed to limit the pressure to a maximum value of P_2. The size of the needed accumulator is found as follows.

For an isothermal process . . .

$$PV = P_o V_o = \text{const.} \tag{6.74}$$

Then,

$$dV = -\frac{P_o V_o}{P^2} dP \tag{6.75}$$

$$(P - P_1)\left(-\frac{P_o V_o}{P^2}\right) dP = \rho ALvdv \tag{6.76}$$

Thus,

$$\int_{P_1}^{P_2} \left(\frac{1}{P} - \frac{P_1}{P^2}\right) dP = -\frac{\rho AL}{P_o V_o} \int_v^0 v\,dv \tag{6.77}$$

$$V_o = \frac{\rho A L v^2}{2P_o \left\{ \ln\left(\frac{P_2}{P_1}\right) + \frac{P_1}{P_2} - 1 \right\}} \tag{6.78}$$

where P_1 and P_2 are the initial and maximum pressures, respectively, in Pa (abs).

For a polytropic process . . .

$$PV^n = P_o V_o^{\,n} = \text{const.} \quad \text{or} \quad dV = -\frac{P_o^{1/n} V_o}{n P^{\frac{n+1}{n}}} dP \tag{6.79}$$

Then

$$(P - P_1)\left(-\frac{P_o^{1/n} V_o}{nP^{\frac{n+1}{n}}}\right) dP = \rho ALvdv \tag{6.80}$$

$$\int_{P_1}^{P_2} (P^{-\frac{1}{n}} - P_1 P^{-\frac{n+1}{n}}) dP = -\frac{n\rho AL}{P_o^{1/n} V_o} \int_v^0 v\,dv \tag{6.81}$$

The following expression for the recommended accumulator size can be deduced:

$$V_o = \frac{n\rho A L v^2}{2P_o^{1/n}\left\{\frac{n}{n-1}\left(P_2^{(n-1)/n} - P_1^{(n-1)/n}\right) + n\left(P_1 P_2^{-1/n} - P_1^{(n-1)/n}\right)\right\}} \tag{6.82}$$

These expressions for the accumulator size were derived neglecting the effect of the fluid compressibility and the pipe wall elasticity. The friction losses in the line and accumulator inlet local losses were also neglected. The friction and local pressure losses assist the fluid deceleration, while the walls' elasticity and oil compressibility would accept some oil during the transient periods. Therefore, when neglecting these parameters, the deduced formulas result in an accumulator size greater than that actually required. Nevertheless, it is safer and counts for the possible approximations and calculation inaccuracy.

Example 6.4 A hydraulic transmission line has the following parameters. Calculate the suitable accumulator size for protection against hydraulic shocks if the maximum allowable pressure increment is 5 bar: $v = 2$ m/s, $A = 4$ cm^2, $L = 100$ m, $\rho = 800$ kg/m^3, $P_1 = 5$ bar (abs), and $P_o = 4$ bar (abs).

$V_o = 0.828$ liters calculated assuming an isothermal process, $n = 1$
$V_o = 0.857$ liters calculated assuming a polytropic process, $n = 1.1$
$V_o = 0.921$ liters calculated assuming a polytropic process, $n = 1.3$

Hydraulic Springs

The hydraulic accumulator is being used as a suspension element in the automotive sector, replacing the mechanical springs. Figure 6.22 shows a hydro-pneumatic wheel suspension system with leveling control. Figure 6.23 illustrates a typical connection of the accumulator when used as a hydraulic spring. An expression for the stiffness of this spring is deduced in the following:

$$V_L = Ax \tag{6.83}$$

$$V_g = V_o - V_L = V_o - Ax \tag{6.84}$$

where V_g = Gas volume at pressure P, m^3

Then,

$$P(V_o - Ax)^n = P_o V_o^n \tag{6.85}$$

$$F = PA = \frac{A P_o V_o^n}{(V_o - Ax)^n} \tag{6.86}$$

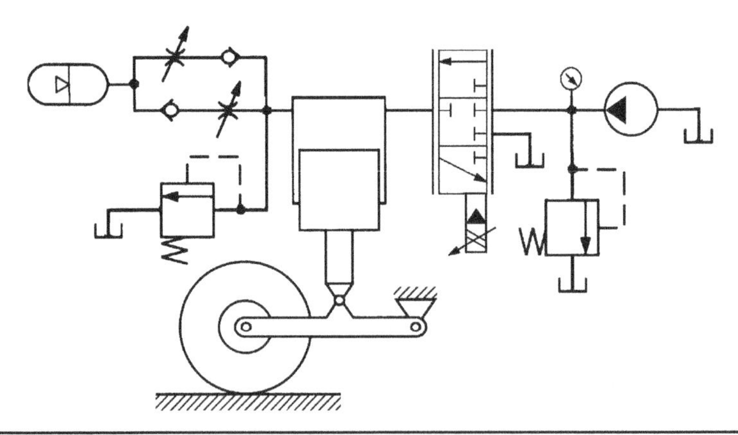

Figure 6.22 Using the accumulator as a hydraulic spring in car suspension.

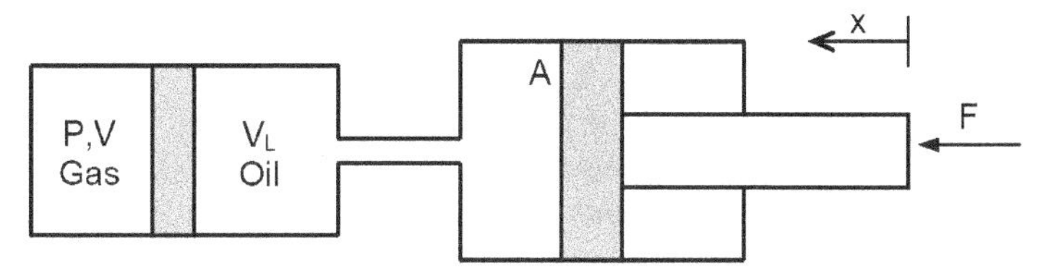

FIGURE 6.23 A hydraulic cylinder and accumulator operating as a hydraulic spring.

The equivalent spring stiffness, k, is

$$k = \frac{dF}{dx} = \frac{nV_o^{\,n} A^2}{(V_o - Ax)^{n+1}} P_o \tag{6.87}$$

where k = Equivalent stiffness, N/m
V_L = Volume of liquid in the accumulator, m³
A = Piston area, m²
x = Piston displacement, m
V_o = Initial gas volume, accumulator size, m³
P_o = Gas-charging pressure, Pa (abs)
P = Actual pressure, Pa (abs)
n = Polytropic exponent
F = Spring force, N

The stiffness of a hydraulic spring was calculated for different sizes of hydraulic accumulator, considering a 10 cm piston diameter, a 2 MPa charging pressure, and a polytropic exponent of 1.3. The calculation results are plotted in Fig. 6.24. These results show that the spring stiffness increases with the piston displacement. The smaller-size accumulator presents greater stiffness due to the rapid increase of oil volume in the accumulator.

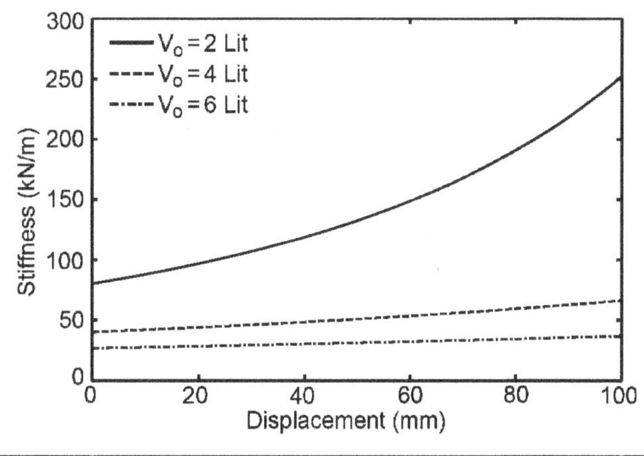

FIGURE 6.24 Effect of the piston displacement on the stiffness of the hydraulic spring for different sizes of accumulators.

6.3 Hydraulic Filters

Hydraulic filters are used to limit the contamination of hydraulic oil. They are installed on the pump suction line, delivery pressure line, or system return line. Pump inlet with suction line filter should be carefully designed and maintained to avoid cavitation. The air breathers of nonpressurized hydraulic tanks are also equipped with hydraulic filters. The filters serve mainly to control the size distribution of impurities in hydraulic oils, to minimize wear and prevent the clogging of fine orifices by contaminants. A sample of return line and pressure line filters is shown in Figs. 6.25 and 6.26.

The return filter is designed for mounting either on the return pipe line or directly on the reservoir (see Fig. 6.25). It is comprised of a housing (1), a cover (2) with a connection for the clogging indicator (3), filter element (4), contamination retaining basket (5), and clogging indicator (6). Additionally, a bypass valve (7) is integrated. The fluid passes via port A to the filter element (4). The contamination particles, which have been filtered out, are kept in the retaining basket (5) and filter element (4). The filtered fluid flows to the reservoir via port B. The retaining basket (5) should be cleaned when removing the filter element (4), so the settled contamination does not pass into the reservoir.

The pressure line filters are suitable for direct mounting on the pressure lines (see Fig. 6.26). They consist of the filter head (1), a filter housing (2), a filter element (3), a clogging indicator (4), and a bypass valve (5) for filters with a low-pressure differential filter element. The fluid passes from port A to the filter element. The contaminating particles are separated. The retained contamination particles settle into the filter housing (2) and the filter element (3). The filtered fluid returns to the circuit through port B.

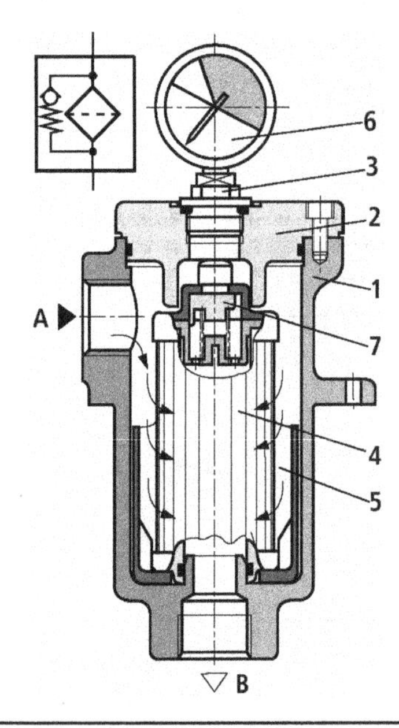

FIGURE 6.25 A return line filter. (*Courtesy of Bosch Rexroth AG.*)

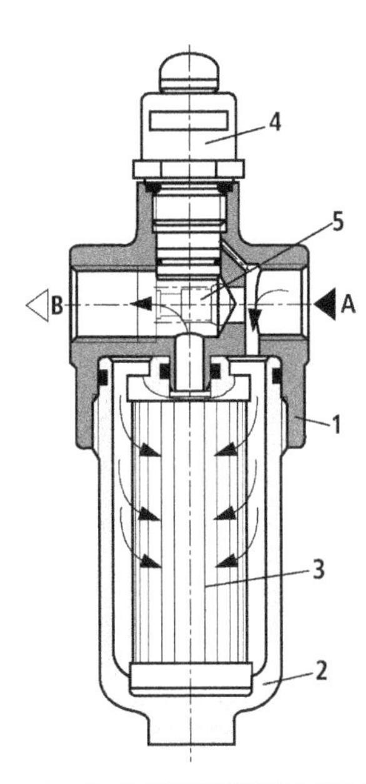

FIGURE 6.26 A high-pressure filter. (*Courtesy of Bosch Rexroth AG.*)

6.4 Hydraulic Pressure Switches

6.4.1 Piston-Type Pressure Switches

This class of pressure switch is actuated by a pressure-loaded piston. They may have normally open or normally closed contacts. They activate an electrical contact at an adjustable pressure setting. The pressure switch (as shown in Fig. 6.27a) consists of a housing (1), a micro switch (2), an adjustment mechanism (3), a plunger (4), a piston (5), and a spring (6). The pressure acts against the piston (5), which extends the plunger (4) against the spring force. As the pressure exceeds the spring force, the plunger displaces and activates the micro switch. The mechanical stop (7) protects the micro switch from over-travel.

Another example of piston-type hydroelectric pressure switches is shown in Fig. 6.27b. It consists of a housing (1), a piston (2), a spring (3), an adjustment element (4), and a micro switch (5). The micro switch is initially contacted for low pressures. The pressure is applied to the piston via the orifice (7). The piston acts against the spring force. The plate (6) transfers the piston movement and releases to the micro switch upon reaching the set pressure. The electric circuit is switched on or off according to the field wiring.

6.4.2 Bourdon Tube Pressure Switches

Bourdon tube actuated pressure switches are suitable for sustained pressure, contamination resistance, and high accuracy. They consist of a housing (1), a bourdon tube (2), a striker plate (3), and a micro switch (4) as shown in Fig. 6.28. The pressure signal acts

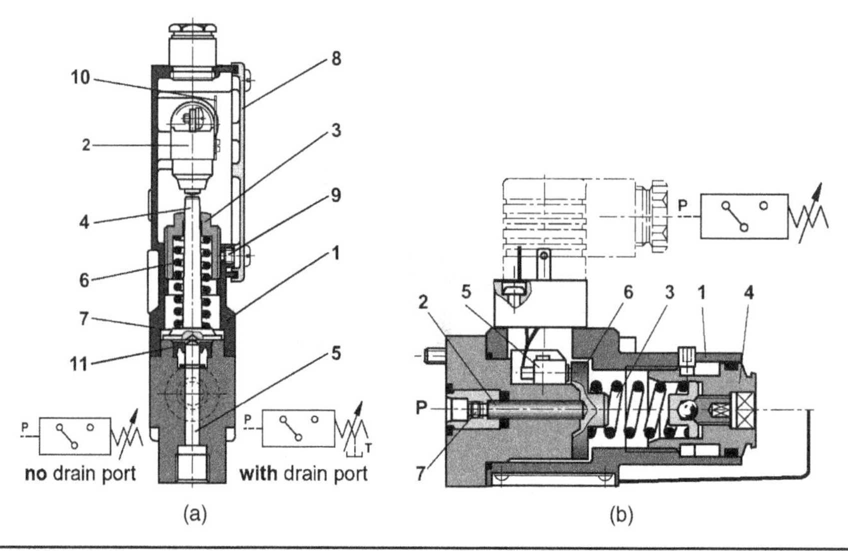

FIGURE 6.27 A piston-type electrohydraulic pressure switch. (*Courtesy of Bosch Rexroth AG.*)

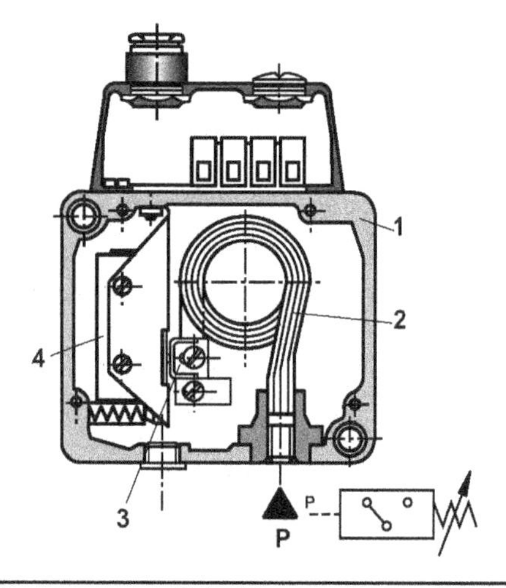

FIGURE 6.28 A Bourdon tube pressure switch. (*Courtesy of Bosch Rexroth AG.*)

upon the bourdon tube. As the pressure increases, the bourdon tube expands. The attached lever converts the expansion of the bourdon tube to a linear movement to activate the micro switch.

6.4.3 Pressure Gauge Isolators

The pressure gauge isolator valve is a three-way two-position directional control valve used to isolate a pressure gauge from the system. The pressure can be monitored by depressing the push button. These valves consist of a housing (1), a spool (2), a spring (3), a push button (4) and a pressure gauge connection (5) as shown in Fig. 6.29. In the

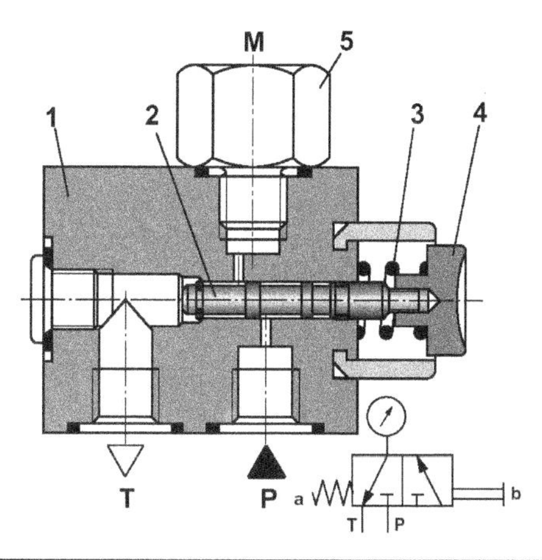

FIGURE 6.29 A pressure gauge isolation valve. (*Courtesy of Bosch Rexroth AG.*)

un-actuated position, the port (P) is blocked and the pressure gauge is connected to the tank port (T). By depressing the push button, the spool shifts, transmitting the pressure signal to the gauge port. After release, the spool returns to its neutral position under the action of the spring force and the port (M) is connected to the tank line (T).

6.5 Exercises

1. Explain the principals of operation and the possible applications of the hydraulic accumulators.

2. Define and derive an expression for the volumetric capacity of an oleopneumatic accumulator.

3. Explain the construction and operation of the piston type accumulators. (See Fig. 6.2.)

4. Explain the construction and operation of bladder-type accumulators. (See Figs. 6.3 through 6.5.)

5. Explain the construction and operation of the diaphragm type accumulators (as shown in Figs. 6.6 and 6.7.)

6. Discuss in detail the applications of the hydraulic accumulators as energy storage elements. Draw a hydraulic circuit for this application.

7. Derive an expression for the total energy stored in a hydraulic accumulator, assuming isothermal gas compression. Find the condition for the maximum energy stored.

8. Derive an expression for the total energy stored in a hydraulic accumulator, assuming polytropic compression of gas. Find the condition for the maximum energy stored.

9. Derive an expression for the useful energy stored in a hydraulic accumulator, assuming the isothermal compression of gas. Find the condition for the maximum energy stored.

10. Derive an expression for the useful energy stored in a hydraulic accumulator, assuming polytropic compression of gas. Find the condition for the maximum energy stored.

11. Discuss in detail the application of a hydraulic accumulator for damping pressure oscillation at the delivery line of displacement pumps.

12. Discuss in detail the application of a hydraulic accumulator for protection against hydraulic shocks.

13. Discuss in detail the application of hydraulic accumulators in protecting against thermal expansion.

14. Discuss in detail the application of a hydraulic accumulator for internal leakage compensation and the application of constant pressure.

15. Discuss in detail the application of a hydraulic accumulator as a hydraulic spring.

16. A hydraulic accumulator is installed to protect a hydraulic line against an excessive rise in pressure due to thermal expansion. Derive an expression for the proper size of this accumulator if the permissible pressure increment is ΔP for a temperature increase of ΔT.

17. Calculate the size of a hydraulic accumulator necessary to deliver five liters of oil between pressures of 200 and 100 bar if the charging pressure is 90 bar (gauge pressure), assuming an adiabatic compression process.

18. A hydraulic system operates in a regular operating cycle of 50 s duration. The flow demand during the operating cycle is shown in the figure that follows. The maximum pump delivery pressure during the operating cycle is 160 bar, and the flow rates are set by a flow control arrangement. The system has a fixed displacement pump. Determine the required pump flow rate in the following cases:

a) When using the pump only for hydraulic power supply.

b) If a hydraulic accumulator is used to compensate for the short duration flow demands. Calculate the suitable size of the accumulator if the maximum allowable pressure is 240 bar.

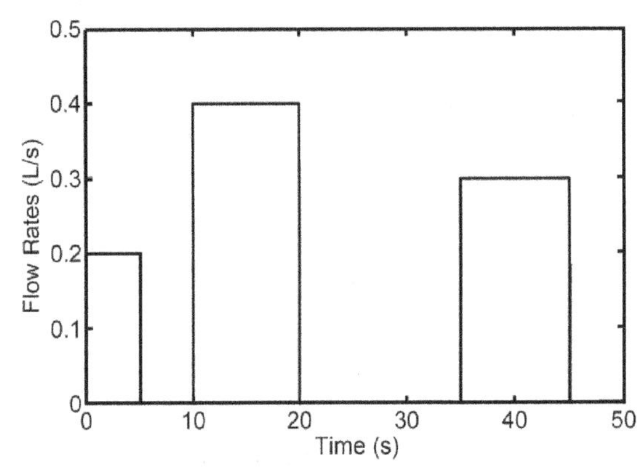

19. Discuss briefly the function of hydraulic filters.

20. Explain the construction and operation of the pressure switches illustrated in Fig. 6.27.

21. A hydraulic power generator has the following operating parameters:

Pump flow rate $Q_P = 0.4$ lit/s

Maximum operating pressure $P_{max} = 70$ bar

Minimum operating pressure $P_{min} = 60$ bar

Pump total efficiency $\eta_T = 0.9$

There is a demand for 0.8 liter of oil over a period of $\Delta T_1 = 0.1$ s, at intermittent intervals. The minimum time interval between two successive demands is $\Delta T_2 = 30$ seconds. Calculate the size of the suitable accumulator and the maximum required power when operating with and without the accumulator.

6.6 Nomenclature

A = Piston area, m^2
A_T = Throttle area, m^2
F = Spring force, N
k = Equivalent stiffness, N/m
n = Pump speed, rev/s
n = Polytropic exponent
N = Pump driving power, W
P = Pressure, Pa
P_1 = Minimum system pressure, Pa (abs)
P_2 = Maximum system pressure, Pa (abs)
P_o = Accumulator charging pressure, gas pressure, Pa (abs)
Q_{max} = Maximum required flow rate, m^3/s
q_T = Flow rate through the throttle valve, m^3/s
R_T = Throttle element resistance, Pa s/m^3
V = Oil volume in the accumulator, m^3
V_1 = Volume of gas at pressure P_1, m^3
V_2 = Volume of gas at pressure P_2, m^3
V_g = Pump geometric volume, m^3/rev
V_g = Gas volume, m^3
V_L = Volume of liquid in the accumulator, m^3
V_o = Accumulator size, volume of charging gas at pressure P_o, m^3
x = Piston displacement, m
x_o = Spring pre-compression, m
ΔT = Temperature increment, K
ρ = Oil density, kg/m^3
η_v = Pump volumetric efficiency
η_T = Pump total efficiency
α = Oil thermal expansion coefficient, K^{-1}

Appendix 6A Smoothing Pressure Pulsations by Accumulators

Figure 6A.1 shows a hydraulic accumulator installed at the pump exit. It serves to smooth the pressure pulsation resulting from the pump pulsating delivery. This appendix illustrates the effect of hydraulic accumulators on the pressure oscillation damping.

Consider a z-piston axial piston pump, where the pistons perform simple harmonic motion. The displacement of each piston is described by the following equation:

$$x_i = h \sin\left\{\omega t + \frac{2\pi(i-1)}{z}\right\};\ i = 1 \text{ to } z \quad (6A.1)$$

$$\omega = 2\pi n_p \quad (6A.2)$$

The flow rate delivered by each piston is Q_i, where $Q_i \geq 0$.

$$Q_i = A_p \frac{dx_i}{dt} = \omega h A_p \cos\left\{\omega t + \frac{2\pi(i-1)}{z}\right\} \quad (6A.3)$$

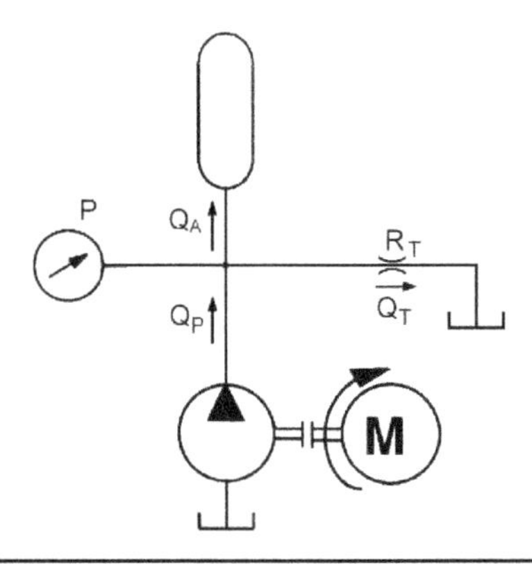

FIGURE 6A.1 Using hydraulic accumulators to damp pressure oscillations.

Neglecting the internal leakage, the pump flow rate is

$$Q_p = \sum_{i=1}^{i=z} Q_i = \sum_{i=1}^{i=z} \omega h A_p \cos\left\{\omega t + \frac{2\pi(i-1)}{z}\right\} \tag{6A.4}$$

where A_p = Piston area, m^2
h = Half of the piston stroke, m
n_p = Pump speed, rps
Q_i = Flow rate delivered by a single piston, m^3/s
Q_p = Pump flow rate, m^3/s
x_i = Piston displacement, m
z = Number of pistons
ω = Pump driving shaft speed, rad/s

The flow rate, Q_T, through the loading orifice is

$$Q_T = C_d A_T \sqrt{2P/\rho} \tag{6A.5}$$

The increase in oil volume in the accumulator is

$$\Delta V_L = \int (Q_p - Q_T)\,dt \tag{6A.6}$$

The accumulator pressure is given by the following equation:

$$P = P_o \left(\frac{V_o}{V_L}\right)^n \tag{6A.7}$$

where A_T = Throttle valve area, m^2
n = Polytropic exponent
P = Pump exit pressure, Pa (abs)
P_o = Accumulator charging pressure, Pa (abs)
Q_T = Flow rate through the throttle element, m^3/s
V_L = Oil volume in the accumulator, m^3
V_o = Accumulator size, m^3
ρ = Oil density, kg/m^3

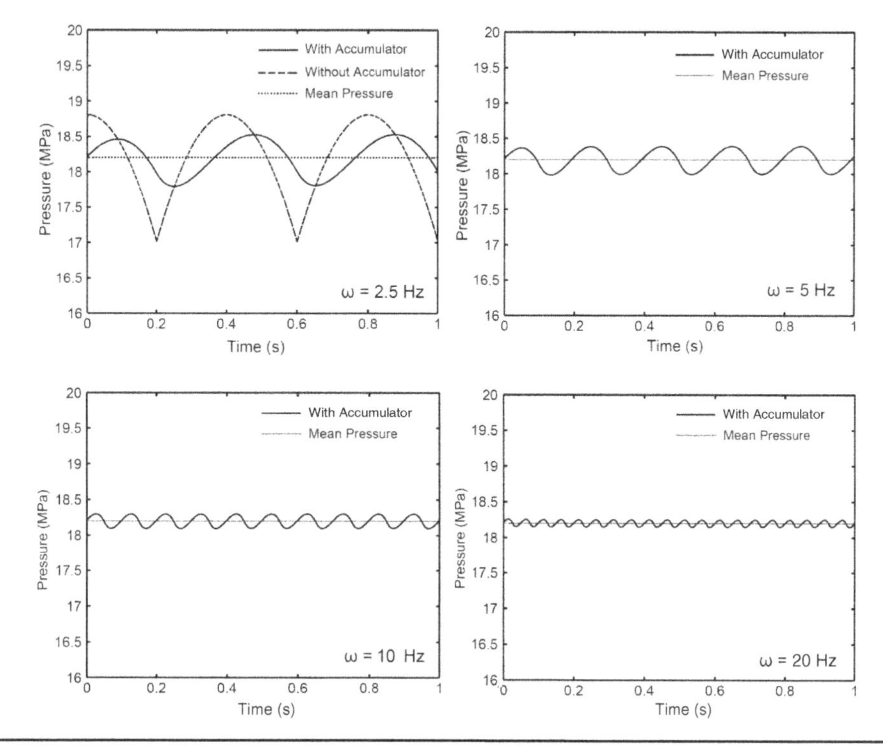

Figure 6A.2 Simulation results of the damping effect of a hydraulic accumulator on the pressure oscillations of the oscillating flow of different frequencies.

The system shown by Fig. 6A.1 is described mathematically by Eqs. (6A.1) through (6A.7). These equations were used to develop a computer simulation program for the studied system using the SMULINK program. The frequency of the pressure oscillations was adjusted by changing the pump speed. The average flow rate of the pump was kept unchanged by resetting the pistons stroke, for comparison purposes. The simulation results are shown in Fig. 6A.2. These results show that the low-frequency signals are slightly attenuated. Meanwhile, the damping effect increases with the increase of the pressure oscillation frequency. Then, in this case, the accumulator acts as a low pass filter, as explained also by Fig. 6.17.

Appendix 6B Absorption of Hydraulic Shocks by Accumulators

Herein, the effectiveness of the application of oleo-pneumatic accumulators for the protection against hydraulic shocks is investigated. The proper accumulator size can be calculated using Eq. (6.82). The studied hydraulic transmission line (see Fig. 6B.1) has the following parameters:

Accumulator charging pressure	= 5.1 MPa
Allowable end pressure increment	= 2.2 MPa
Bulk modulus of oil	= 1.6 GPa
Pipe line diameter	= 1 cm
Initial line-end pressure	= 6.2 MPa

Initial oil velocity	= 6 m/s
Inlet constant pressure	= 8 MPa
Kinematic Viscosity	= 56 cSt
Pipe line length	= 18 m
Line capacitance: C	= 8.84×10^{-13} m³/Pa
Line inertia: I	= 1.99×10^{8} kg/m⁴
Line resistance: R	= 3.56×10^{9} Ns/m⁵
Oil density	= 868 kg/m³
Polytropic exponent	= 1.3

The equations describing the line are given in App. 3B. When the accumulator is installed, the equation describing the last capacitor [Eq. (3B.13) for the two-lump model] will be replaced by the equations describing the accumulator as follows:

$$V = V_o - \int (Q_{1L} - Q_L)\,dt \tag{6B.1}$$

$$P_L = P_o (V_o/V)^n \tag{6B.2}$$

The studied line is supplied by pressurized oil at a constant pressure. The line-end directional control valve DCV is open and the throttle valve was partially opened to control the oil speed at 6 m/s. The DCV is suddenly closed at $t = 0.2$s. Figure 6B.2 shows the transient response of the line-end pressure in the two cases: with and without an accumulator.

In the case of a line without an accumulator, the transient response shows an overshoot of 63.9 bar and a settling time of 353 ms. The installation of the hydraulic accumulator reduced the pressure overshoot to 8 bar and the settling time to 169 ms. In addition to this improvement, the transient response oscillations are substantially reduced, which provides an important increase in the fatigue life of the pipe.

The effect of accumulator size is investigated by calculating the integral error squared (IES), which is defined as

$$\text{IES} = \int_0^T (P_L - P_{Lss})^2\,dt \tag{6B.3}$$

The calculation results are shown in Fig. 6B.3. Generally, the installation of the accumulator very close to the valve reduces the duration and amplitude of the transient pressure oscillations. The minimum value of IES is obtained for the accumulator size

FIGURE 6B.1 Schematic of the experimentation setup.

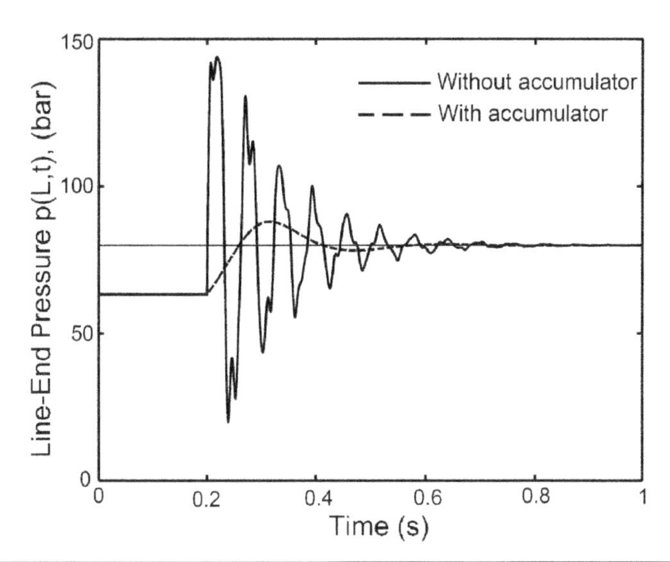

Figure 6B.2 Transient response of a transmission line to the sudden closure of the exit throttle valve.

estimated by Eq. (6.82) (0.152 liters in this case study, while the estimated accumulator size is 0.126 L).

The effect of accumulator size on the IES can be explained as follows: The increase of the accumulator size, V_o, over the minimum value increases the accumulator capacitance [see Eq. (6.42)] and decreases the system stiffness and natural frequency. Meanwhile, the system resistance is not affected by the accumulator size, which results in an increased damping coefficient. Thus, the increase in accumulator size results in a more damped response, a longer settling time, and increased IES.

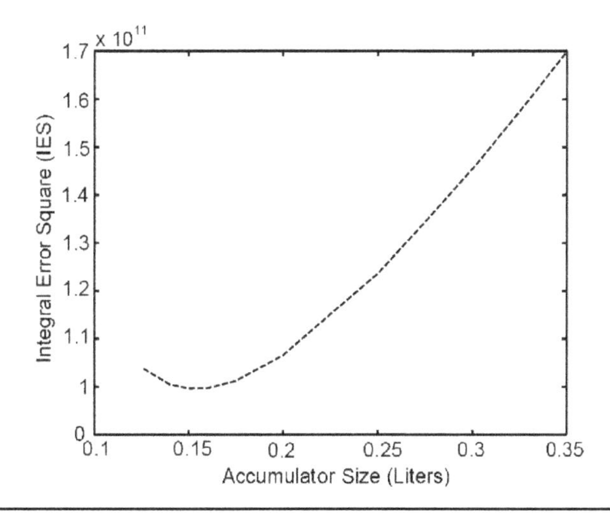

Figure 6B.3 Effect of the accumulator size on the integral square error of the transient response.

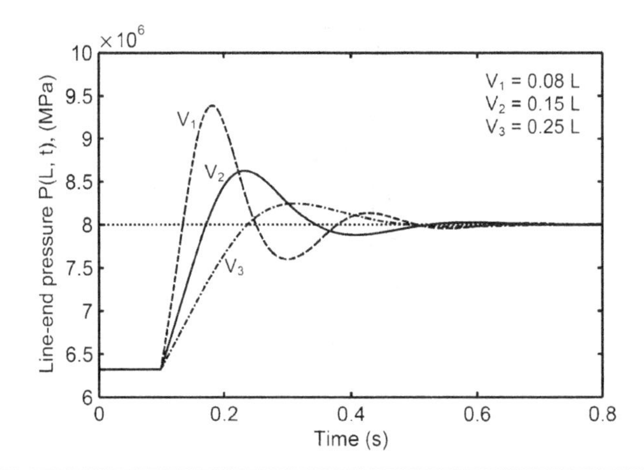

FIGURE 6B.4 The transient response of a transmission line, equipped with a hydraulic accumulator of different sizes, to step closure of a throttle valve at $t = 0.1$ s.

The reduction of the accumulator size, below the calculated minimum size, reduces the damping coefficient and leads to a greater maximum percentage overshoot and increased IES (see Fig. 6B.4). Therefore, both the minimum IES and maximum percentage overshoot should be considered when selecting the proper accumulator size.

Nomenclature and Abbreviations

n = Polytropic exponent
P = Pressure, Pa
P_L = Line-end pressure, Pa
P_{Lss} = Steady-state pressure at the closed end, Pa
P_o = Accumulator charging pressure, Pa (abs)
Q = Flow rate, m^3/s
T = Time duration of response, s
V_o = Accumulator size, m^3

CHAPTER 7

Hydraulic Actuators

7.1 Introduction

Hydraulic actuators are installed to drive loads by converting the hydraulic power into mechanical power. The mechanical power delivered to the load is managed by controlling the fluid pressure and flow rate, by using various hydraulic control valves. The hydraulic actuators are classified into three main groups according to motion type:

- Hydraulic cylinders, performing linear motion
- Hydraulic motors, performing continuous rotary motion
- Hydraulic rotary actuators, performing limited angular displacement

7.2 Hydraulic Cylinders

The hydraulic cylinders convert the hydraulic power into mechanical power, performing rectilinear motion. The pressure of input oil is converted into the force acting on the piston (see Fig. 7.1). The following relations describe the steady-state motion of a frictionless leakage-free hydraulic cylinder:

$$F = P_1 A_p - P_2 A_r \quad \text{and} \quad v = \frac{Q_1}{A_p} = \frac{Q_2}{A_r} \tag{7.1}$$

In the case of steady-state operation of real cylinders, the internal leakage, Q_L, and friction forces, F_f, should be taken into consideration. Therefore, the mechanical power delivered to the load (vF) is less than the hydraulic power supplied to the cylinder ($P_1Q_1 - P_2Q_2$). The cylinder is described by the following relations:

$$F = P_1 A_p - P_2 A_r - F_f \quad \text{and} \quad v = \frac{Q_1 - Q_L}{A_p} = \frac{Q_2 - Q_L}{A_r} \tag{7.2}$$

where A_p = Piston area, m^2
A_r = Rod-side area, m^2
F = Piston driving force, N
F_f = Friction force, N
P = Pressure, Pa

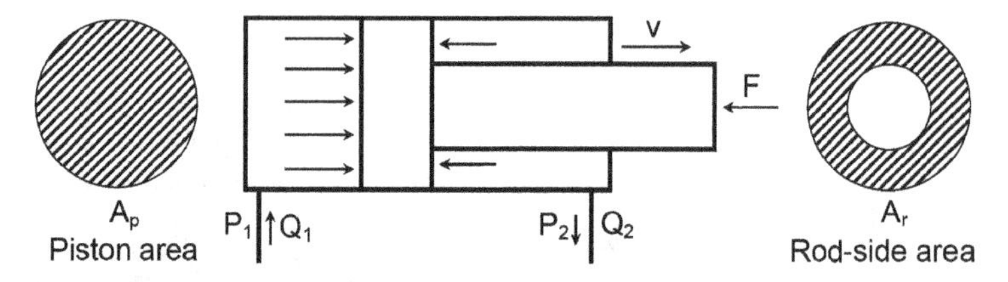

FIGURE 7.1 Functional scheme of a hydraulic cylinder.

Q = Flow rate, m³/s
Q_L = Internal leakage flow rate, m³/s
v = Piston speed, m/s

7.2.1 The Construction of Hydraulic Cylinders

Figure 7.2 illustrates the construction of a typical hydraulic cylinder. It consists mainly of the piston, a piston rod, cylinder barrel, cylinder head, and cylinder cap. The piston rod is extruded through the cylinder head. The piston carries a convenient sealing assembly to insure the required internal tightness, while the cylinder head is equipped with convenient seals to resist external leakage. The two basic types of hydraulic cylinders are tie-rod and mill-type. This classification depends on the method of assembling cylinder parts. In the tie-rod cylinders, the cylinder head and cylinder cap are connected together by tie rods; the sealing of the cylinder barrel-head-cap contact surfaces is insured by static seals (see Fig. 7.3). On the other side, the mill-type cylinders are assembled by various methods, including:

- The cylinder cap is welded to the cylinder barrel, and the cylinder head is bolted to a bush screwed to the barrel (see Fig. 7.4).
- The cylinder cap is welded to the cylinder barrel, and the cylinder head is screwed to the barrel (see Fig. 7.5).

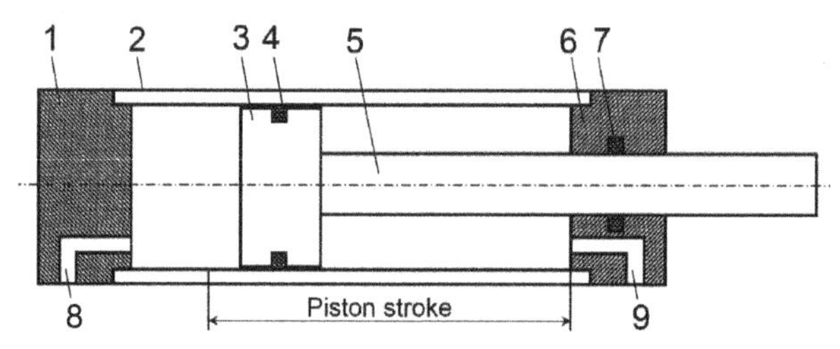

1. Cylinder cap, 2. Barrel, 3. Piston, 4. Piston seal, 5. Piston rod, 6. Cylinder head, 7. Rod seal, 8. and 9. Cylinder ports

FIGURE 7.2 Construction of a typical hydraulic cylinder.

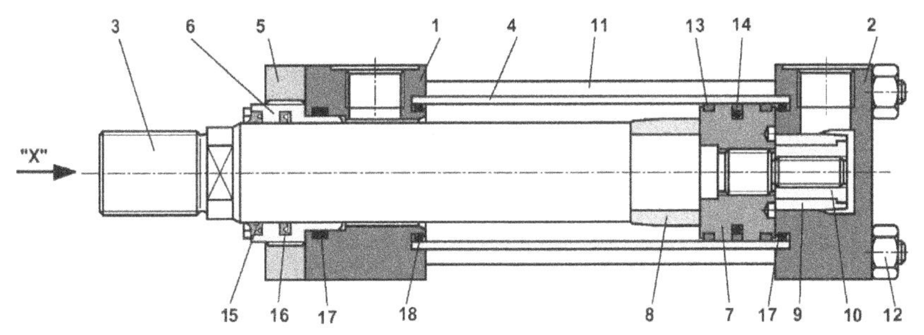

1. Head, 2. Cap, 3. Piston rod, 4. Cylinder tube, 5. Flange, 6. Guide bush, 7. Piston, 8. and 9. Cushioning bush; spear, 10. Threaded ring, 11. Tie rod, 12. Nut, 13. Guide ring, 14. Piston seal, 15. Wiper, 16. Piston rod seal, 17 and 18. O ring; static seal

FIGURE 7.3 Construction of a tie-rod cylinder. (*Courtesy of Bosch Rexroth AG.*)

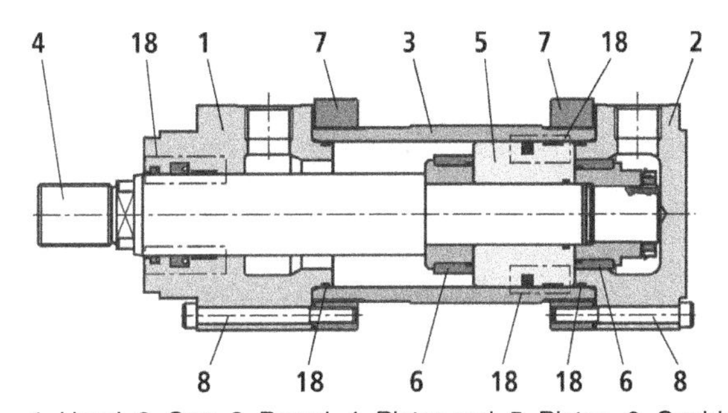

1. Head, 2. Cap, 3. Barrel, 4. Piston rod, 5. Piston, 6. Cushioning spear, 7. Flange, 8. Bolts, 18. Seal kit

FIGURE 7.4 Mill-type hydraulic cylinder with bolted head. (*Courtesy of Bosch Rexroth AG.*)

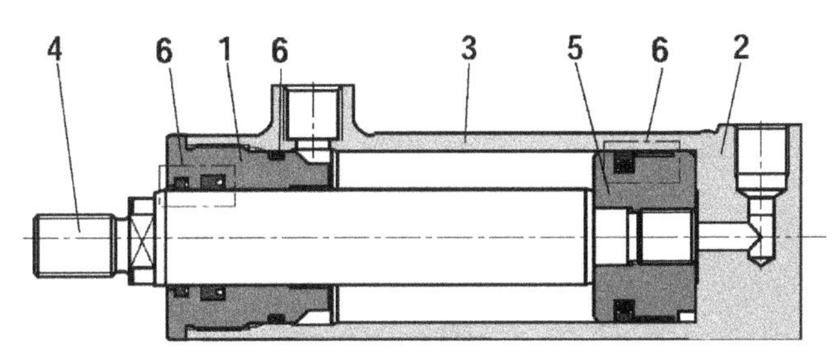

1. Head, 2. Cap, 3. Barrel, 4. Piston rod, 5. Piston, 6. Seal kit: Wiper, Rod seal, Piston seal, O-ring and Guide bush

FIGURE 7.5 Mill-type hydraulic cylinder with screwed head. (*Courtesy of Bosch Rexroth AG.*)

7.2.2 Cylinder Cushioning

The extension and retraction speeds of hydraulic cylinders are managed by controlling the inlet or exit-oil flow rates. When reaching its end position, the piston is suddenly stopped. In the case of high speed and/or great inertia, the sudden stopping of the piston results in a severe impact force. This force is proportional to the mass and the square of the velocity of the moving parts. It affects both the cylinder and the driven mechanism. Therefore, a cushioning arrangement might be necessary to reduce the piston speed to a limiting value before reaching its end position. The cushion dissipates the kinetic energy of the moving parts.

Figure 7.6 shows a hydraulic cylinder with an adjustable cushioning element. The piston (1) is fitted with a conical end-position cushioning bush (spear) (2). During the cylinder retraction stroke, the returned oil flows freely to the return line. When approaching the end position, the spear enters a cylindrical cave in the cylinder cap (4). The oil returned from the piston chamber (5) is forced to flow through the passage (6) to the throttle valve (7), and through the radial clearance between the spear and cylindrical cave. The pressure in the piston chamber increases and acts to decelerate the moving piston.

The check valve (3) permits by-pass of the throttle valve at the start of the motion in the opposite direction. When extending the cylinder, the oil flows to the piston chamber through the check valve. In this way, the whole piston area is acted on by the pressurized oil. The piston may be equipped with a cushioning element at one side or at both sides. The throttle valve is either of fixed or of adjustable area. The illustrated

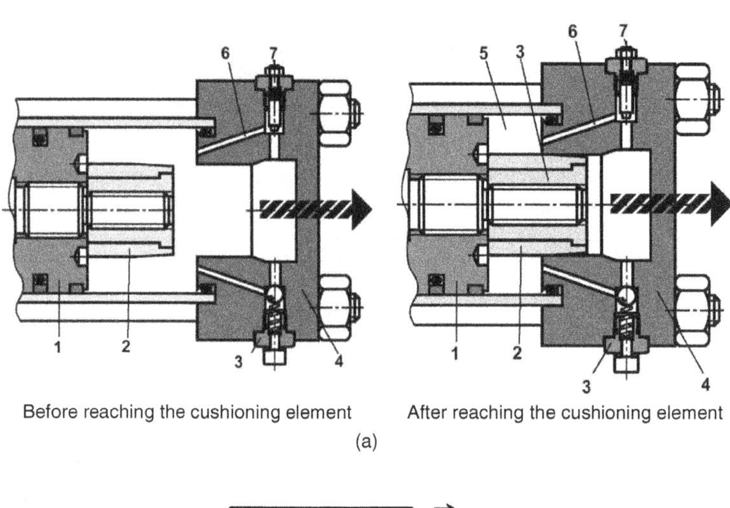

(a)

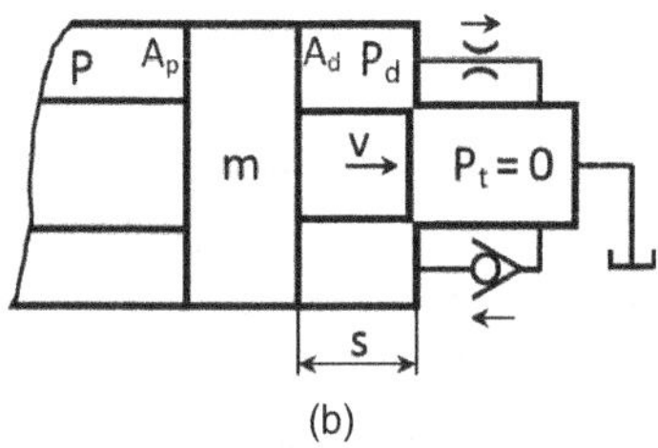

(b)

Figure 7.6 (a) Cushioning of hydraulic cylinders. (*Courtesy of Bosch Rexroth AG.*) (b) Functional scheme of the cushioning element.

construction has a throttle valve area adjustable by a screw (see Fig. 7.6a). The kinetic energy of the moving parts should not exceed the capacity of the cushioning element, defined as the work done during cushioning period.

The cushioning should produce a controlled deceleration of the cylinder, near one or both end positions. This is done by creating a decelerating pressure force. In doing this, the pressure must not exceed its limiting value. The kinetic energy of the moving parts is converted into heat by throttling the out-going fluid in the cushioning zone. For an ideal operation, the piston will be fully stopped at the end of the cushioning stroke, s. If the piston will be fully stopped at the end of the cushioning stroke, the required deceleration, a, is

$$a = \frac{v^2}{2s} \tag{7.3}$$

When the cylinder is installed horizontally, the decelerating (cushioning) force can be calculated, considering the equilibrium of forces acting on the piston (see Fig. 7.6b), as follows:

$$P_d A_d = ma + PA_P \tag{7.4}$$

Normally, the damping pressure may not exceed the nominal pressure of the cylinder. An average value of the damping pressure, P_d, is given as follows:

$$P_d = \frac{1}{A_d}\left(\frac{mv^2}{2s} + PA_P\right) \tag{7.5}$$

where a = Deceleration, m/s^2
A_d = Piston area subjected to pressure P_d, m^2
A_P = Piston area subjected to pressure P, m^2
m = Moving mass, kg
P = Driving pressure, Pa
P_d = Mean pressure in the cushioning volume, Pa
s = Damping length, m
v = Piston velocity at the beginning of the cushioning stroke, m/s

For vertical mounting of the cylinder, the pressure generated by the weight of the moving parts must be added or subtracted depending on the direction of motion. The cylinder friction is ignored in these calculations. If the calculations give an unacceptably high mean damping pressure, the damping length must be increased or the speed must be reduced.

7.2.3 Stop Tube

In the case of flange and foot mounting or where the cylinder is rigidly fixed, there is a possibility of excessive side loading forces. The side loading force, perpendicular to the cylinder axis, results in high loads on the rod bearing, especially in the full extension position. This increased load has a harmful effect on the rod bearing. This undesirable effect can be reduced by using a stop tube, fitted inside the cylinder body. It stops the piston before reaching the cylinder head, which increases the minimum distance between the rod

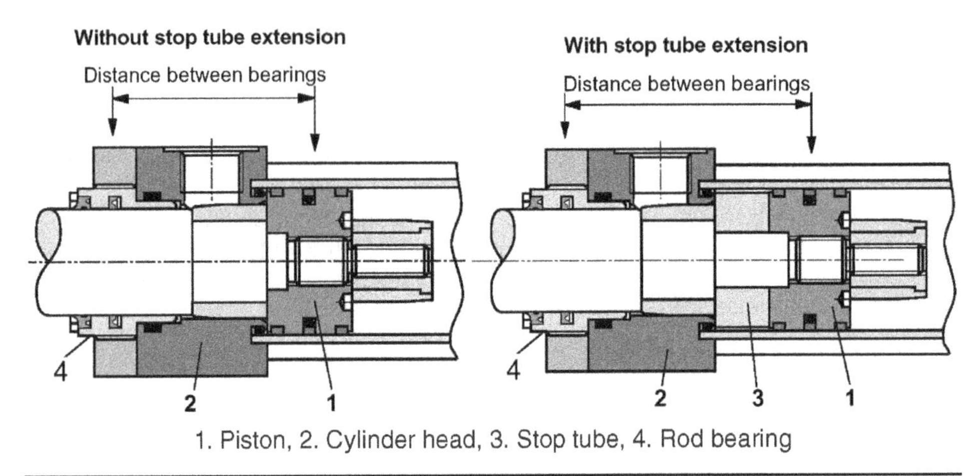

FIGURE 7.7 Illustration of the function of the stop tube. (*Courtesy of Bosch Rexroth AG.*)

bearing and the piston. Figure 7.7 illustrates the function of the stop tube. A stop tube (3) is inserted between the piston (1) and the cylinder head (2). The stop tube extends the lever arm and thus reduces the bearing loads. The increase in stop tube length decreases the reaction force on the piston and rod bearing. Meanwhile, the available stroke is reduced.

7.2.4 Cylinder Buckling

The maximum axial load acting on the hydraulic cylinder must not exceed the limit at which buckling takes place. This limiting force should be calculated and considered, otherwise catastrophic damage may take place. The limiting load, for buckling, is calculated as follows:

$$F = \frac{\pi^2 EJ}{nL_K^2} \quad \text{for} \quad \lambda \geq \lambda_g \quad \text{according to Euler} \tag{7.6}$$

$$F = \frac{\pi d^2 (335 - 0.62\lambda)}{4n} \quad \text{for} \quad \lambda \leq \lambda_g \quad \text{according to Tetmajer} \tag{7.7}$$

$$J = \frac{\pi d^4}{64} = 0.0491\, d^4 \quad \text{for a circular cross-sectional area} \tag{7.8}$$

$$\lambda = 4\frac{L_K}{d} \tag{7.9}$$

$$\lambda_g = \pi\sqrt{1.25\, E/R} \tag{7.10}$$

The free buckling length, L_K, is found from the Euler loading table shown in Fig. 7.8. The reinforcement by the cylinder tube is not taken into consideration in calculation. This gives an allowance for superimposed bending stress due to the cylinder installation.

where n = Safety factor = 3.5
d = Piston rod diameter, m

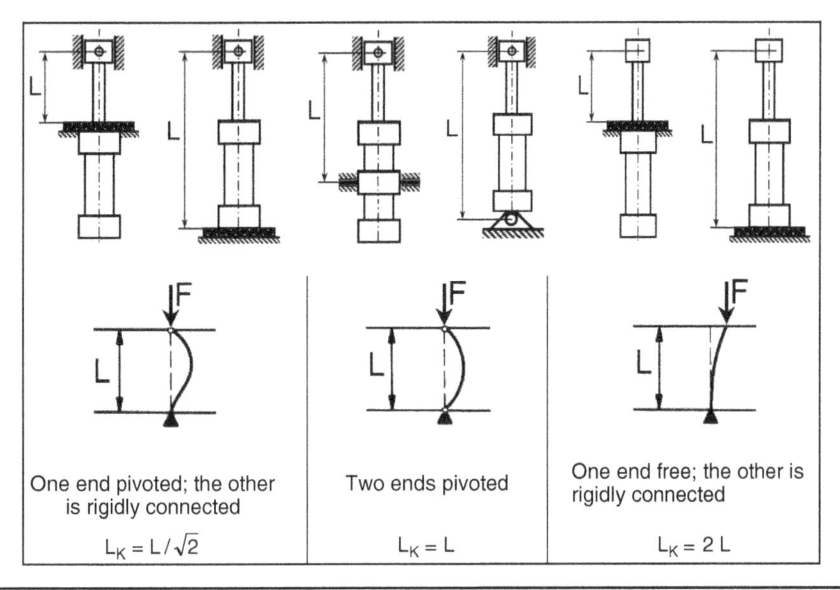

FIGURE 7.8 Influence of the mounting type on the free buckling length.

E = Modulus of elasticity, N/m^2 ($E = 2.1 \times 10^{11}$, for steel)
J = Area moment of inertia or second moment of area, m^4
L_K = Free buckling length, m
R = Yield strength of the piston rod material, N/m^2
λ = Slenderness ratio

7.2.5 Hydraulic Cylinder Stroke Calculations

The double-acting hydraulic cylinder shown in Fig. 7.9 illustrates the various "dead" or wasted lengths. The minimum length of this hydraulic cylinder, L, is composed of the following terms:

$$L = L_4 + \text{adjustment allowance} + \text{dead lengths} \tag{7.11}$$

$$L = 2L_1 + L_2 + L_3 + L_4 + L_5 + L_6 + L_7 + L_8 \tag{7.12}$$

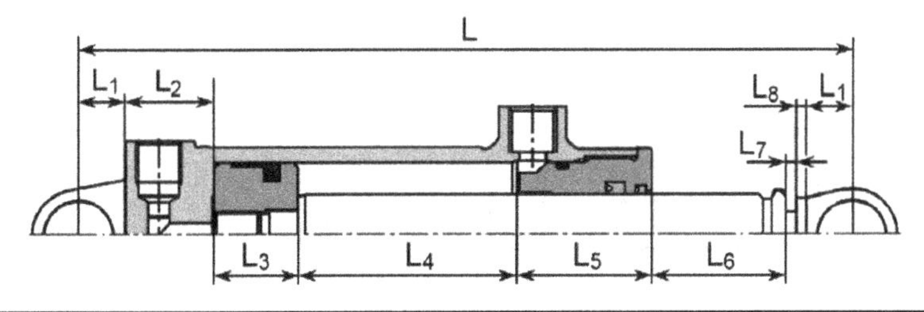

FIGURE 7.9 Basic dimensions for the dead length and stroke calculations.

where L_1 = Radius of attachment lugs + a clearance of 2.5 mm, m
L_2 = Cylinder cap length, m
L_3 = Piston length, m
L_4 = Stroke, m
L_5 = Cylinder head length, m
L_6 = Length of extruded part of piston rod + thickness of lock nut, m
L_7 = Allowance for length adjustment, m
L_8 = Thickness of end pieces, m

The length between the pin centers may be reduced if the cylinder attachment becomes a trunnion (see Fig. 7.21).

7.2.6 Classifications of Hydraulic Cylinders

The hydraulic cylinders are classified into the following types: single acting, double acting, tandem, three position, and telescopic. They may be equipped with a mechanical position locking element.

Single-Acting Cylinders

Figures 7.10 and 7.11 illustrate the construction of single-acting hydraulic cylinders. The piston, or plunger, is driven hydraulically in one direction. In the other direction, the piston moves under the action of an external force or a built-in spring.

Double-Acting Hydraulic Cylinders

The piston of a double-acting hydraulic cylinder is driven hydraulically in both directions of motion. This cylinder may be single rod (Fig. 7.12), twin-rod symmetrical (Fig. 7.13), or twin-rod nonsymmetrical (Fig. 7.14). The twin-rod cylinder is said to be symmetrical if the diameters of the piston rods are equal. It is usually used in hydraulic servo systems.

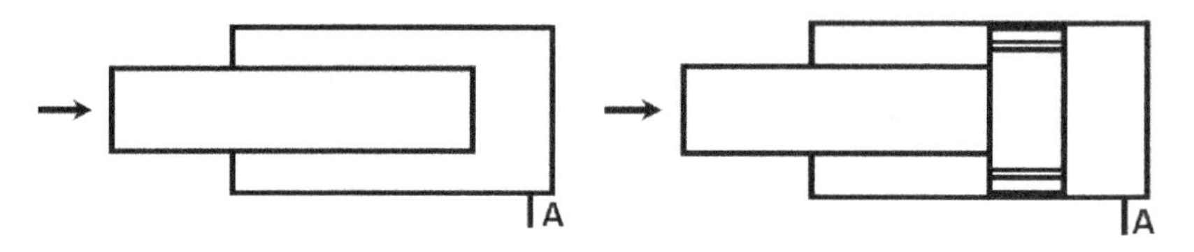

Figure 7.10 Single-acting plunger-type cylinder, returned by external force.

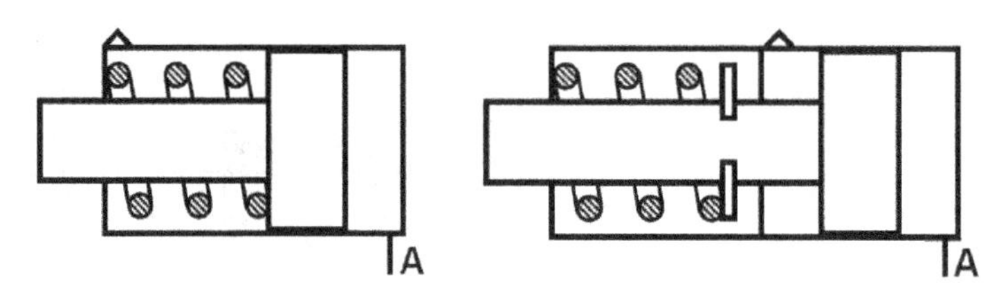

Figure 7.11 Single-acting piston-type cylinder, spring returned.

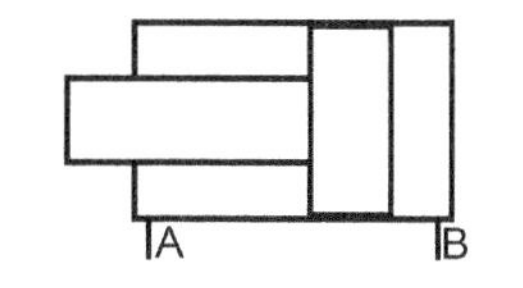

FIGURE 7.12 Single rod cylinder.

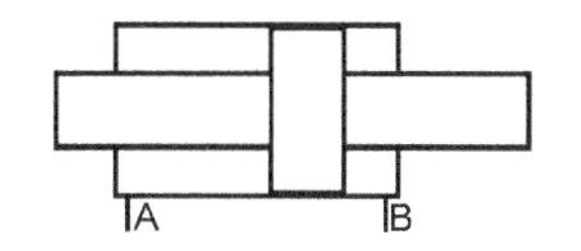

FIGURE 7.13 Twin-rod symmetrical cylinder.

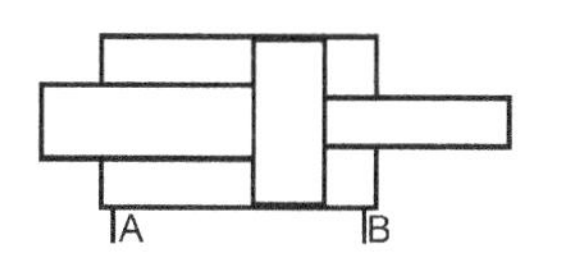

FIGURE 7.14 Twin-rod nonsymmetrical cylinder.

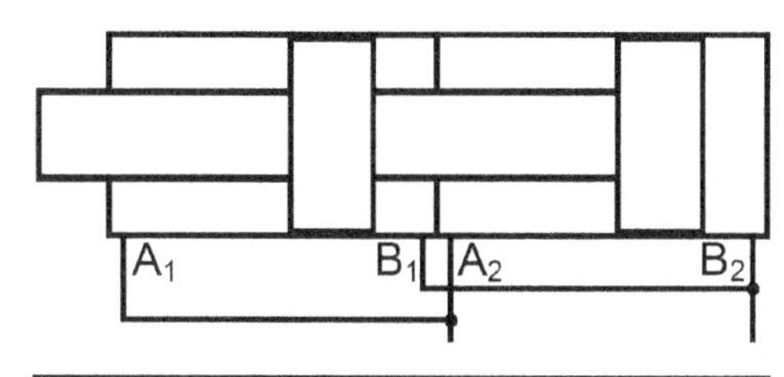

FIGURE 7.15 A differential tandem cylinder.

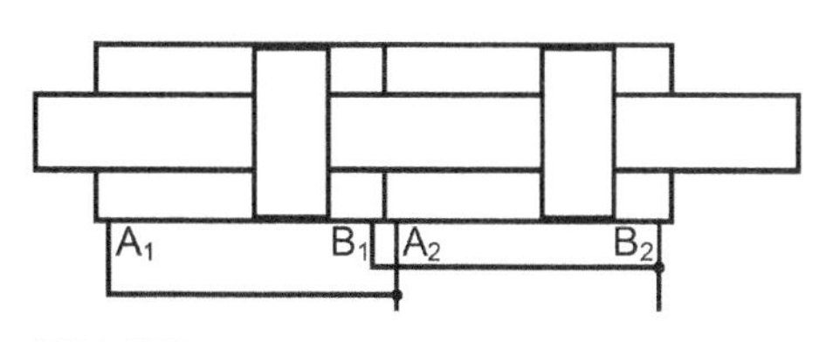

FIGURE 7.16 A symmetrical tandem cylinder.

Tandem Cylinders

The tandem cylinder duplicates the pressure force, for the same barrel diameter (see Figs. 7.15 and 7.16).

Three-Position Hydraulic Cylinders

Some of the operating organs may have three operational positions. In this case, the ordinary double-acting hydraulic cylinder does not give the required controllability. Figure 7.17 shows the typical construction of a three-position cylinder. The cylinder has two separate pistons and piston rods. The three positions are obtained by pressurizing the cylinder chambers, as shown in the table in Fig. 7.17.

Cylinders with Mechanical Locking Elements

The position locking of hydraulic cylinders can be realized hydraulically or mechanically. For hydraulic position locking, single- or twin-pilot operated check valves are used.

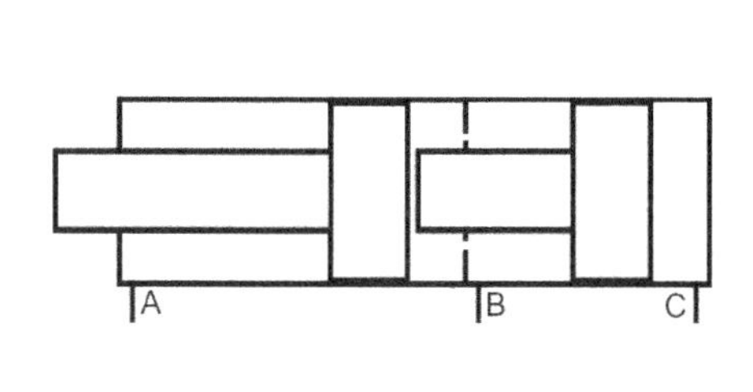

Position	Port pressurization		
	A	B	C
Retracted	P	T	T
Mid-position	P	T	P
Extended	T	P	T
Extended	T	P	P

FIGURE 7.17 Operation of a three-position cylinder. (Note: P and T are the pressure and tank lines.)

Mechanical locking elements keep the cylinder piston in the required position regardless of the variation of the loading force. Sometimes, both the hydraulic and mechanical locking are used. Mechanical locks are installed at either or both sides of the cylinder.

Telescopic Cylinders

Telescopic cylinders are used in industrial and mobile equipment hydraulic systems. This class of cylinders provides long cylinder strokes with relatively small installation space. The telescopic cylinder may be either single acting (Fig. 7.18) or double acting (Fig. 7.19). If the pressure affects the pistons via port A, they travel outward one after another. The double-acting cylinder retracts by pressurizing the port B.

7.2.7 Cylinder Mounting

Eye or Clevis Cylinder Mounting

Figure 7.20 illustrates the possible combinations of mounting a plane bearing and a spherical bearing at the cylinder cap and rod eye.

Trunnion Mounting

The trunnion mounting allows angular movement of the cylinder and a shorter retraction length (see Fig. 7.21).

Flange Mounting

The flange mounting is preferred for vertical cylinder mounting. When the cylinder is loaded mainly by thrust force (tension or compression), the mounting bolts at the flange should be unloaded. The mounting positions shown by Fig. 7.22 are recommended for this purpose.

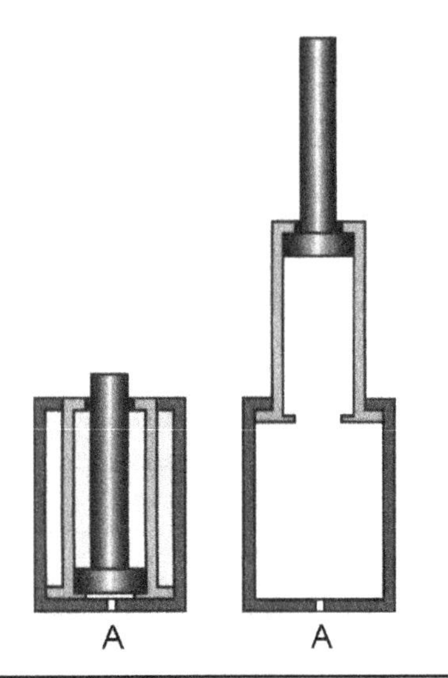

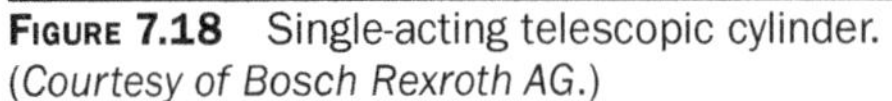

Figure 7.18 Single-acting telescopic cylinder. *(Courtesy of Bosch Rexroth AG.)*

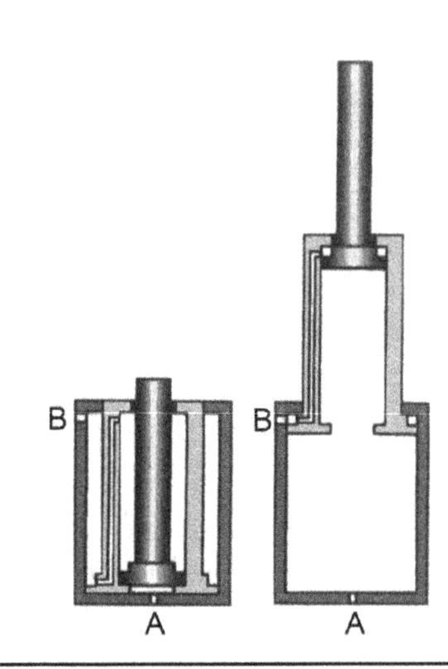

Figure 7.19 Double-acting telescopic cylinder. *(Courtesy of Bosch Rexroth AG.)*

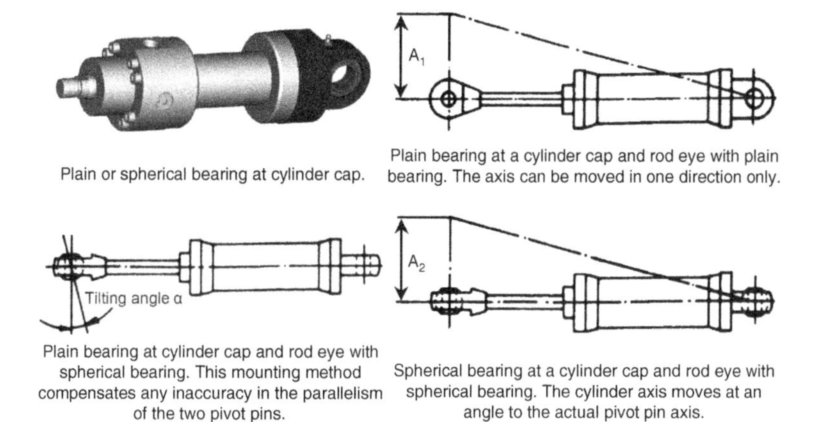

Figure 7.20 Eye or clevis cylinder mounting. (*Courtesy of Bosch Rexroth AG.*)

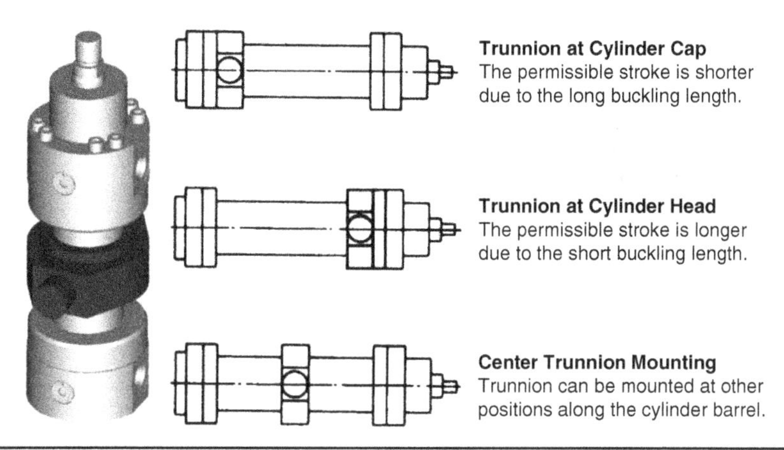

Figure 7.21 Trunnion mounting. (*Courtesy of Bosch Rexroth AG.*)

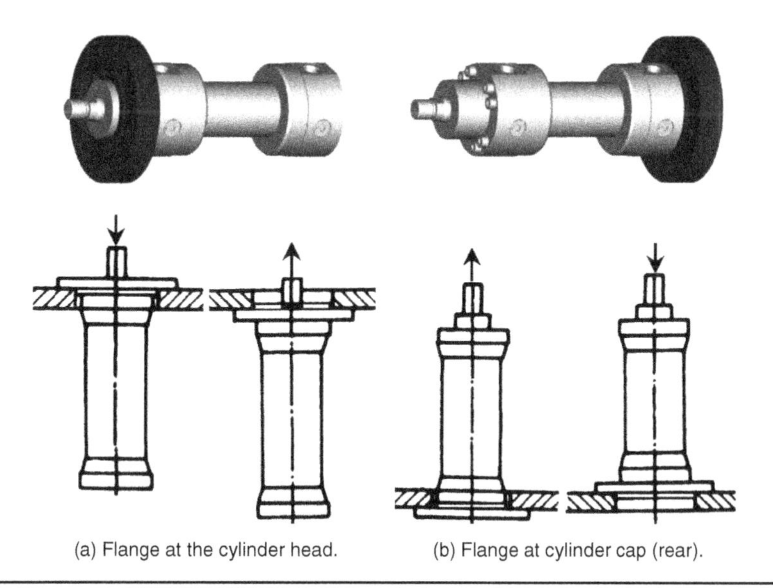

Figure 7.22 Flange mounting. (*Courtesy of Bosch Rexroth AG.*)

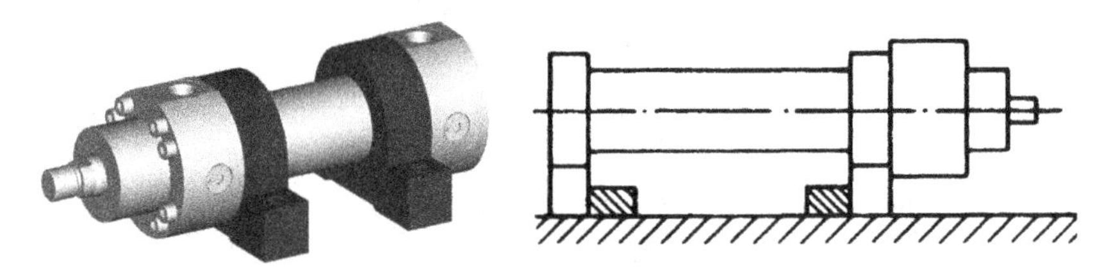

Figure 7.23 Foot mounting. (*Courtesy of Bosch Rexroth AG.*)

Foot Mounting

The foot mounting of hydraulic cylinders is illustrated in Fig. 7.23. The mounting bolts should be protected against shear stress. Therefore, thrust keys are provided to absorb the cylinder forces as shown.

7.2.8 Cylinder Calibers

Manufacturers produce a wide range of dimensions for hydraulic cylinders. The commonly produced piston and piston rod dimensions are shown in Table 7.1. Meanwhile, the stroke is determined by the user.

Piston ΦD (mm)	Rod Φd (mm)	Piston ΦD (mm)	Rod Φd (mm)	Piston ΦD (mm)	Rod Φd (mm)	Piston ΦD (mm)	Rod Φd (mm)
25	12	40	16	63	25	100	45
	14		18		28		50
	16		22		36		56
			25		45		70
32	18	50	22	80	36	125	50
	22		25		45		56
	25		28		56		63
			36				70
							90
150	63	200	90	280	110	400	160
	70		100		125		180
	80		110		140		200
	100		140		180		
					200		
160	90	220	90	320	125	450	180
			100		140		200
			110		160		220
			140		200		
			160		220		
180	80	250	100	360	140	500	200
	90		110		160		220
	125		125		180		250
			160				
			180				

Table 7.1 Commonly Produced Cylinder Diameters

7.3 Hydraulic Rotary Actuators

Hydraulic actuators are elements converting the hydraulic power into mechanical power with a rotary motion of limited rotation angle.

7.3.1 Rotary Actuator with Rack and Pinion Drive

In this design, the central part of the piston is formed into a rack. The rectilinear motion of the piston is converted into the rotary motion of a pinion. Swivel angles up to 360° and more are possible, depending on the piston stroke and gear ratio (see Fig. 7.24).

7.3.2 Parallel Piston Rotary Actuator

In this type, two pistons move parallel to each other (see Fig. 7.25). They are alternatively pressurized hydraulically. The pressure force is transmitted through the piston rods, and then transformed into the output torque. This class of rotary actuators rotates within 100°.

7.3.3 Vane-Type Rotary Actuators

The vane-type actuator consists of a single- or double-vane rotor connected to an output shaft (see Fig. 7.26). The rotation angle of a single-vane unit is limited to about 320°, while that of a double vane is limited to 150°.

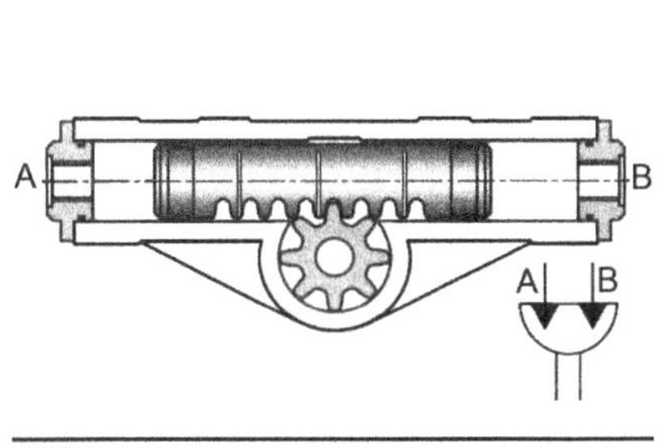

Figure 7.24 Rotary actuator with rack and pinion drive. (*Courtesy of Bosch Rexroth AG.*)

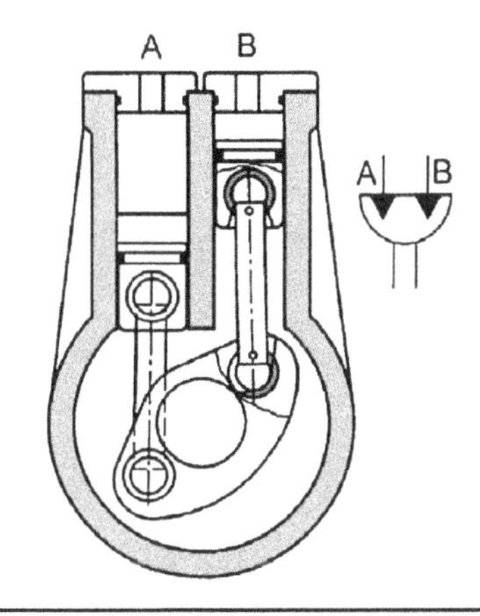

Figure 7.25 Piston-type rotary actuator.

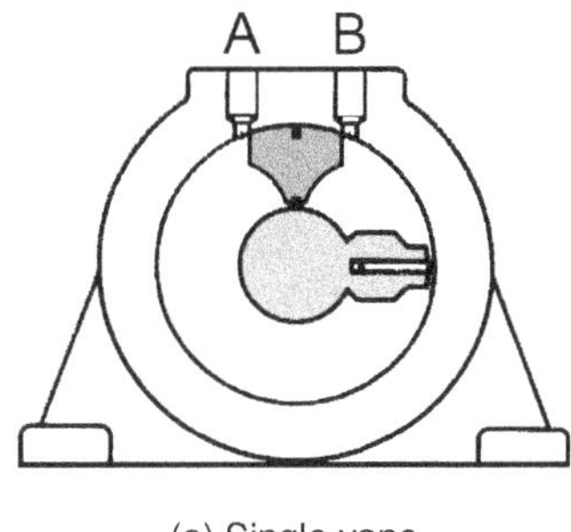

(a) Single vane

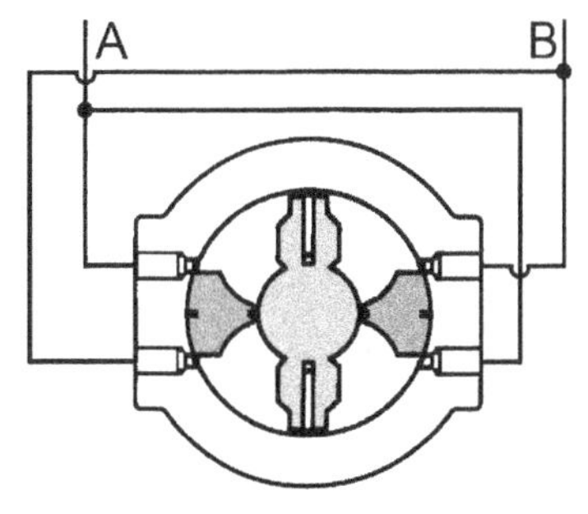

(b) Double vane

Figure 7.26 Vane rotary actuators. (*Courtesy of Bosch Rexroth AG.*)

7.4 Hydraulic Motors

7.4.1 Introduction

The function of hydraulic motors is the reverse of that of the pump. Hydraulic motors are displacement machines converting the supplied hydraulic power into mechanical power. They perform continuous rotary motion. The displacement (or geometric volume) of a hydraulic motor is the volume of oil needed to rotate the motor shaft by one complete revolution. The motor speed depends on the flow rate, while the supply pressure depends mainly on the motor loading torque. In the case of an ideal motor with no leakage and no friction, the following relations are used:

$$n_m = Q_t / V_m \tag{7.13}$$

$$\Delta P = \frac{2\pi}{V_m} T \tag{7.14}$$

where n_m = Motor speed, rev/s
ΔP = Applied pressure difference, Pa
V_m = Geometric volume of motor, m^3/rev
T = Loading torque, Nm
Q_t = Theoretical flow rate, m^3/s

The theoretical motor flow rate is less than the real flow due to the internal leakage. The volumetric efficiency of the motor is defined as follows:

$$\eta_v = \frac{Q_t}{Q} \tag{7.15}$$

or

$$n_m = \frac{Q\eta_v}{V_m} \tag{7.16}$$

The motor output mechanical power is less than the input hydraulic power due to the volumetric, mechanical, and hydraulic losses. The power losses are evaluated by the total efficiency η_T:

$$Q\Delta P\eta_T = 2\pi n_m T \tag{7.17}$$

Then

$$\Delta P = \frac{2\pi}{V_m \eta_m \eta_h} T \tag{7.18}$$

where Q = Real motor flow rate, m^3/s
η_T = Total motor efficiency
η_m = Motor mechanical efficiency
η_v = Motor volumetric efficiency
η_h = Motor hydraulic efficiency

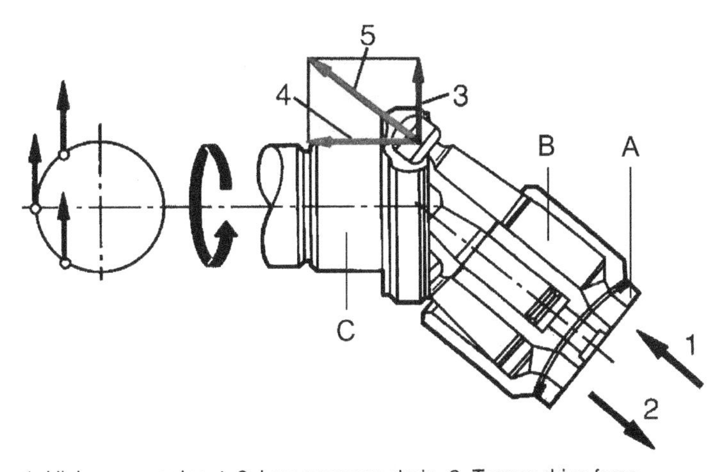

1. High-pressure input, 2. Low-pressure drain, 3. Torque drive force,
4. Bearing supported force component, 5. Piston force; (A) Port plate,
(B) Cylinder bores, (C) Motor shaft

FIGURE 7.27 Bent-axis axial piston motor. (*Courtesy of Bosch Rexroth AG.*)

7.4.2 Bent-Axis Axial Piston Motors

The hydraulic motors convert the input pressure to an equivalent torque while the inlet flow rate determines the motor speed. Figure 7.27 shows a bent-axis axial piston motor. The pressurized oil flows through one of the kidney-shaped holes on the port plate (A) to the cylinder bores (B). The piston chambers connected with the inlet port are pressurized. The pressure forces acting on the pistons are resolved at the drive flange, which is connected to the drive-shaft (C). These forces are converted into a torque, which acts on the motor shaft.

7.4.3 Swash Plate Axial Piston Motors

The construction of the swash plate class of motors is illustrated by Fig. 7.28. The fluid is fed from the hydraulic system to the hydraulic motor. The pressure and return lines are

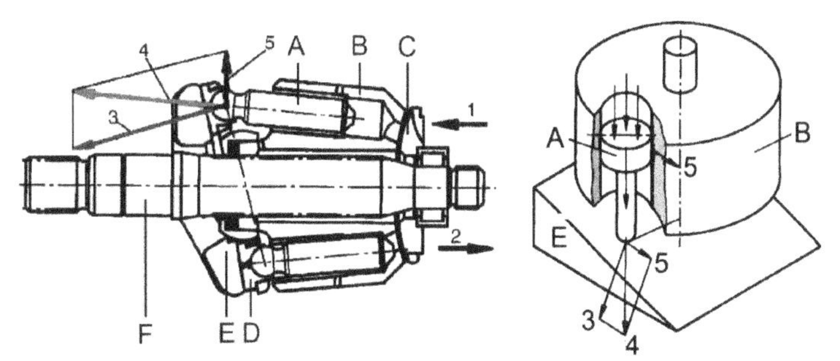

A. Pistons, B. Cylinder block, C. Fixed port plate, D. Slipper pads,
E. Swash plate, F. Drive shaft

FIGURE 7.28 Swash plate axial piston motors. (*Courtesy of Bosch Rexroth AG.*)

connected to the two kidney-shaped ports on the fixed port plate (C). In the case of a nine-piston motor, four or five cylinder bores are connected with the kidney-shaped control opening on the pressure side. The rest of the cylinder block bores are connected to the return line through the other opening. The swash plate (E) does not rotate. By pressurizing the pistons (A), they slide down the swash plate and rotate the cylinder block (B). The cylinder block and the pistons rotate with the drive shaft (F). The pressure forces create the torque at the cylinder block and, hence, at the motor shaft. The flow rate fed to the motor determines the output shaft speed.

The resolution of the forces takes place at the swash plate in the slipper pads (D) and cylinder block. The piston slipper pads have hydrostatic bearings that reduce the friction and wear and increase the motor's service life.

The piston is fed with fluid from the pump and hence pushed against the sloping surface. Resolving the forces at the point of contact (friction bearing) with the sloping surface, a bearing force and a torque force component (3 and 5) are obtained. The piston slides down the sloping surface, and thus drives the cylinder block and drive shaft along with it.

7.4.4 Vane Motors

The construction of the vane motors is, in general, similar to that of the vane pumps. The typical construction of a vane motor is shown in Fig. 7.29. The motor torque results from the action of the high pressure of the inlet oil on the vanes. The rotor is thus driven by this torque against the external load. The vanes are pushed radially outward by the springs so they are in contact with the cam ring at the start of the operation. During the motor operation, an additional pressure force assists the spring force to reach the required tightness.

7.4.5 Gear Motors

Gear motors are very similar in design to the gear pumps, but motors are usually designed to have a case drain port and a reversible direction of rotation. Figure 7.30 shows an external gear motor.

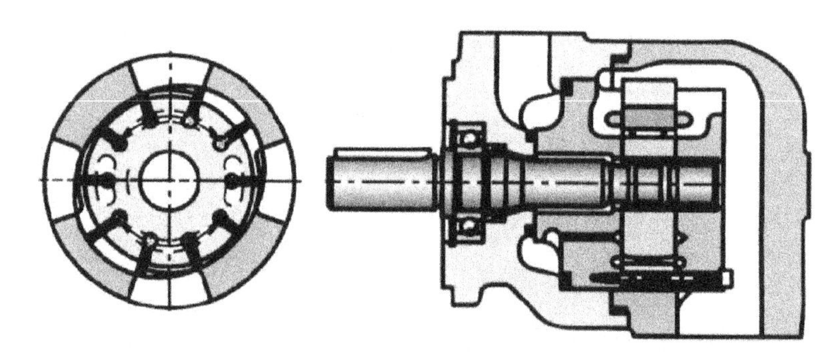

Figure 7.29 Vane motors.

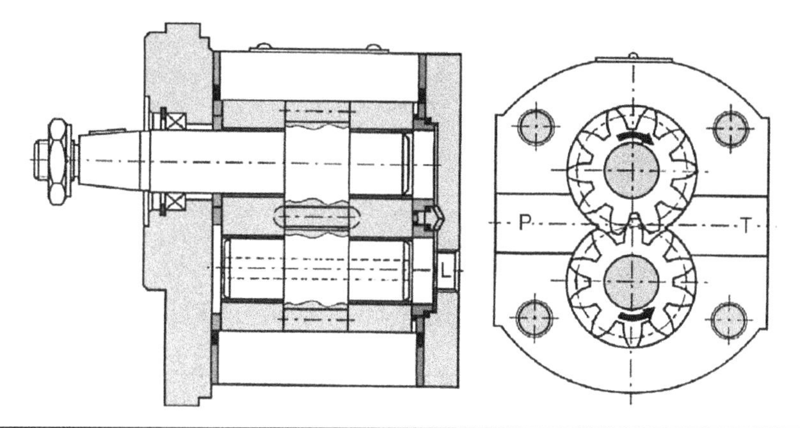

Figure 7.30 External gear motor. (*Courtesy of Bosch Rexroth AG.*)

7.5 Exercises

1. Discuss briefly the function, construction, and operation of hydraulic cylinders.
2. Deal with the cushioning in hydraulic cylinders, giving the necessary schemes.
3. Explain the buckling calculations in hydraulic cylinders.
4. Discuss the different constructions of hydraulic cylinders, giving the necessary schemes.
5. Deal with the calculation of the cylinder stroke. (See Fig. 7.9.)
6. Discuss the different methods of mounting hydraulic cylinders.
7. Explain briefly the construction and operation of rotary actuators.
8. Discuss briefly the function of the hydraulic motors, giving mathematical expressions describing ideal and real motors.
9. Explain the construction and operation of the bent-axis hydraulic motor. (See Fig. 7.27.)
10. Explain the construction and operation of the swash plate hydraulic motor. (See Fig. 7.28.)
11. Explain the construction and operation of vane motors. (See Fig. 7.29.)

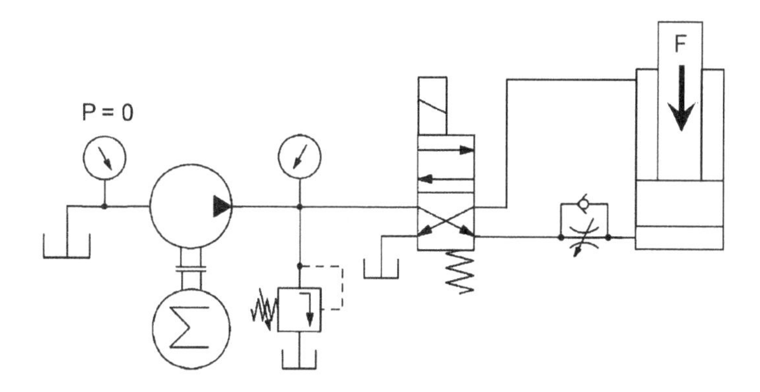

12. Shown is the load-lifting mode of a hydraulic system having the following parameters:

Pump: a swash plate axial piston pump, with piston diameter = 8 mm, pitch circuit diameter = 3 cm, swash plate inclination angle = 20°, mechanical efficiency = 0.9, total efficiency = 0.81, number of pistons = 7, pump speed = 3000 rpm.

Relief valve: preset at a relief pressure of 20 MP, with zero override pressure.

Hydraulic cylinder: an ideal cylinder, loaded by a constant load of 60 kN; the piston and piston rod diameters are 10 cm and 7 cm, respectively.

Check valve: of zero cracking pressure.

Throttle valve: sharp-edged with a 3 mm² cross-sectional area.

Hydraulic oil: a 850 kg/m³ density.

(a) Explain the function of the system.

(b) Calculate the piston speed and pump driving power at each of the two positions of the DCV if the pressure in the pump delivery line does not reach the preset relief pressure. Neglect the losses in lines and DCV.

13. For the following system, calculate the pump exit pressure, and the over-ride pressure of the relief valve, given:

Pump: pump speed = 1000 rpm, volumetric efficiency = 0.95, and pump displacement = 8 cm³

Relief valve: P_r = 22.5 MPa, and $Q_r = K$(Override pressure)$\sqrt{P_P}$; $K = 10^{-13}$

where P_p is the pump exit pressure and $(P_p - P_r)$ is the override pressure

Throttle valve: throttle area a = 1 mm² and $Q_m = 0.029a\sqrt{P_p - P_1}$

Hydraulic motor: motor displacement V_m = 80 cm³/rev, total efficiency = 0.68, volumetric efficiency = 0.93, and loading torque = 200 Nm

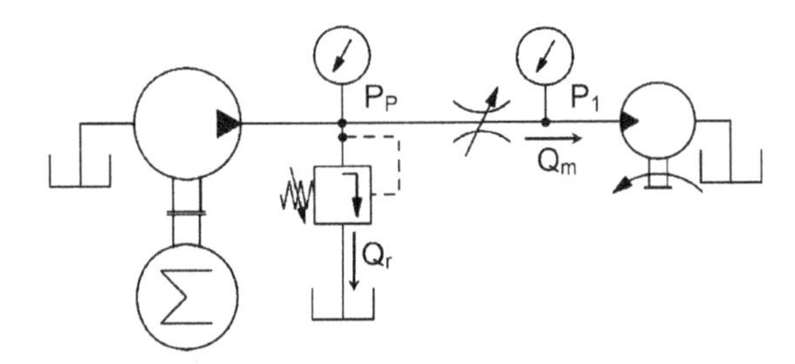

14. A 50-kN hydraulic press performs pressing and clamping actions. The clamping cylinder force is 4 kN. The pressing cylinder stroke is 30 cm and its extension speed is 8 cm/s. Design the hydraulic circuits, perform the preliminary calculations, and select the needed hydraulic elements. Then, calculate the different operating modes of the system.

Assume, always, reasonable values for any missing data.

7.6 Nomenclature

a = Deceleration, m/s²

A_d = Piston area subjected to pressure P_d, m²

A_p = Piston area m^2
A_p = Piston area subjected to pressure P, m^2
d = Piston rod diameter, m
E = Modulus of elasticity, N/m^2
F = Piston driving force, N
F_f = Friction force, N
J = Second moment of area, m^4
L_1 = Radius of attachment lugs + a clearance of 2.5 mm, m
L_2 = Cylinder cap length, m
L_3 = Piston length, m
L_4 = Stroke, m
L_5 = Cylinder head length, m
L_6 = Length of extruded part of piston rod + thickness of lock nut, m
L_7 = Allowance for length adjustment, m
L_8 = Thickness of end pieces, m
L_K = Free buckling length, m
m = Mass, kg
n = Safety factor = 3.5
n_m = Motor speed, rev/s
P = Supply pressure, driving pressure, Pa
P_d = Mean pressure in the cushioning volume, Pa
Q = Input flow rate, real motor flow rate, m^3/s
Q_L = Internal leakage flow rate, m^3/s
Q_t = Theoretical flow rate, m^3/s
R = Yield strength of the piston rod material, N/m^2
s = Damping length, m
T = Loading torque, Nm
v = Piston speed, m/s
v = Piston velocity at the beginning of the cushioning stroke, m/s
V_m = Geometric volume of motor, m^3/rev
ΔP = Applied pressure difference, Pa
λ = Slenderness ratio
η_T = Total motor efficiency
η_m = Motor mechanical efficiency
η_v = Motor volumetric efficiency
η_h = Motor hydraulic efficiency

Appendix 7A Case Studies: Hydraulic Circuits

This appendix presents the hydraulic circuits of typical systems for discussion regarding their construction, operation, and possible faults, as well as the ins and outs of fault diagnostics. These circuits were developed using the Automation Studio™ software.

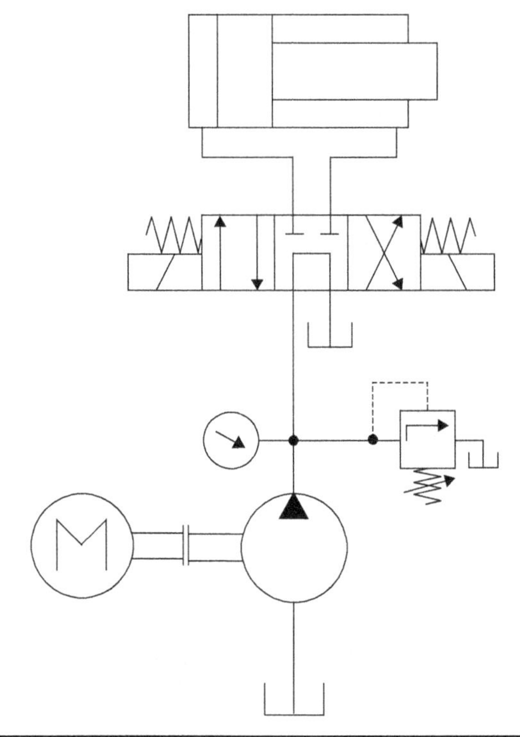

FIGURE 7A.1 A typical circuit with a pump by-pass.

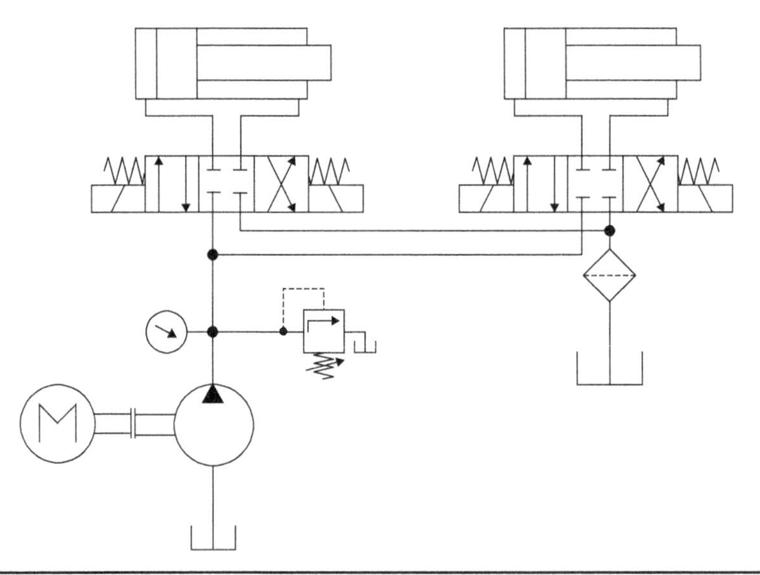

FIGURE 7A.2 A typical circuit including the parallel connection of DCVs.

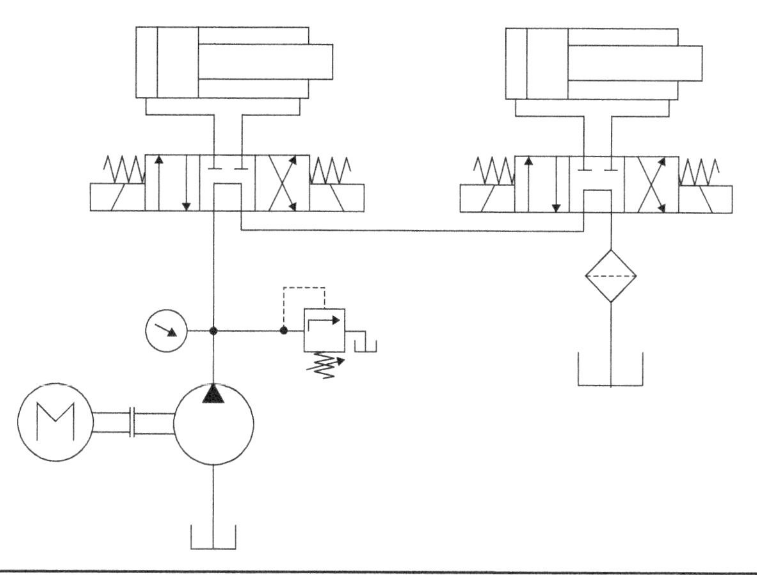

Figure 7A.3 A typical circuit including a series connection of DCVs with a pump by-pass.

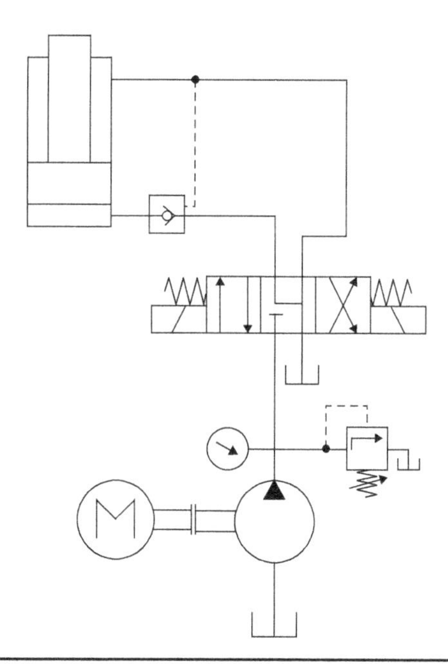

Figure 7A.4 A typical circuit including position holding by a pilot-operated check valve.

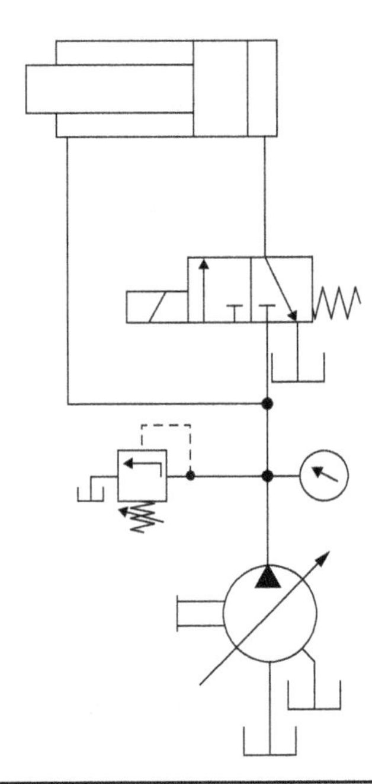

Figure 7A.5 A typical circuit of a system with a regenerative connection (semi-open circuit).

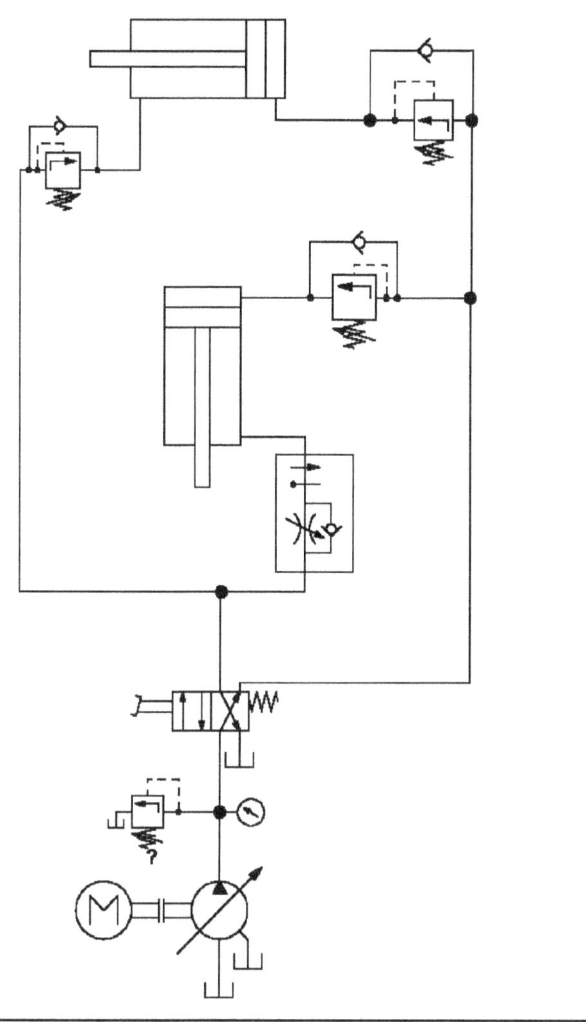

Figure 7A.6 A typical circuit of a system with a sequence of operation.

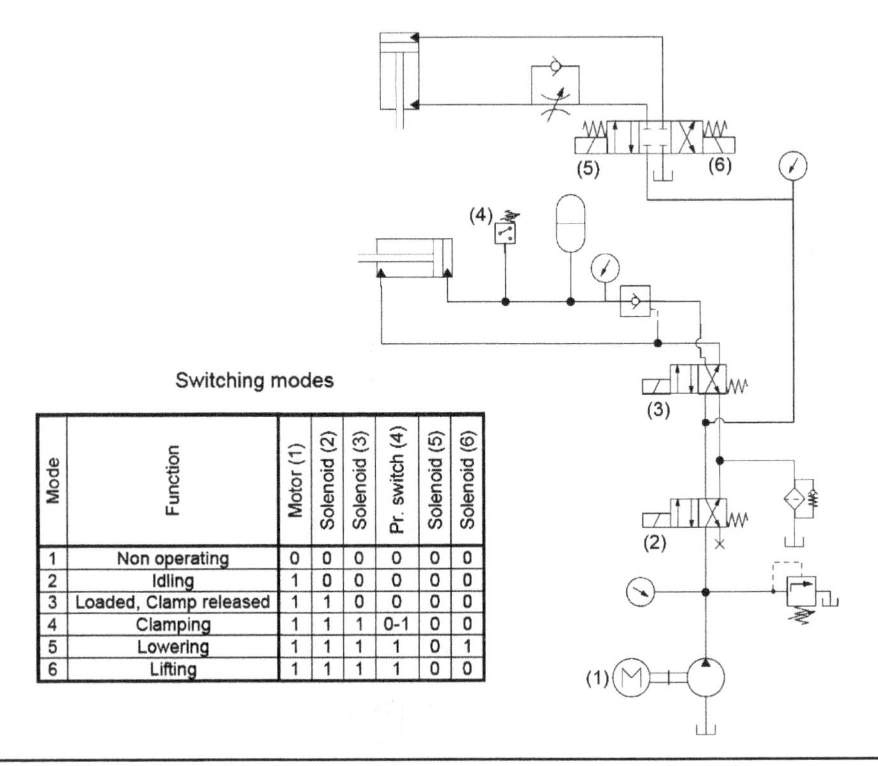

Switching modes

Mode	Function	Motor (1)	Solenoid (2)	Solenoid (3)	Pr. switch (4)	Solenoid (5)	Solenoid (6)
1	Non operating	0	0	0	0	0	0
2	Idling	1	0	0	0	0	0
3	Loaded, Clamp released	1	1	0	0	0	0
4	Clamping	1	1	1	0-1	0	0
5	Lowering	1	1	1	1	0	1
6	Lifting	1	1	1	1	0	0

Figure 7A.7 The hydraulic circuit of a simple hydraulic press with a clamping device.

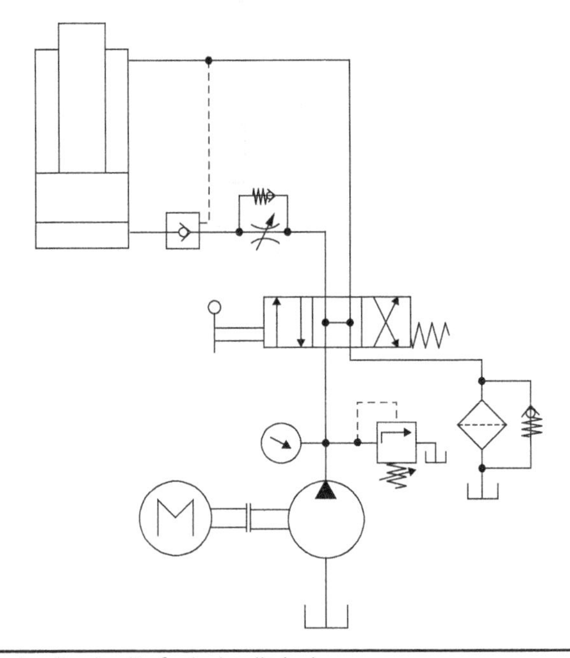

Figure 7A.8 The hydraulic circuit of a hydraulic jack.

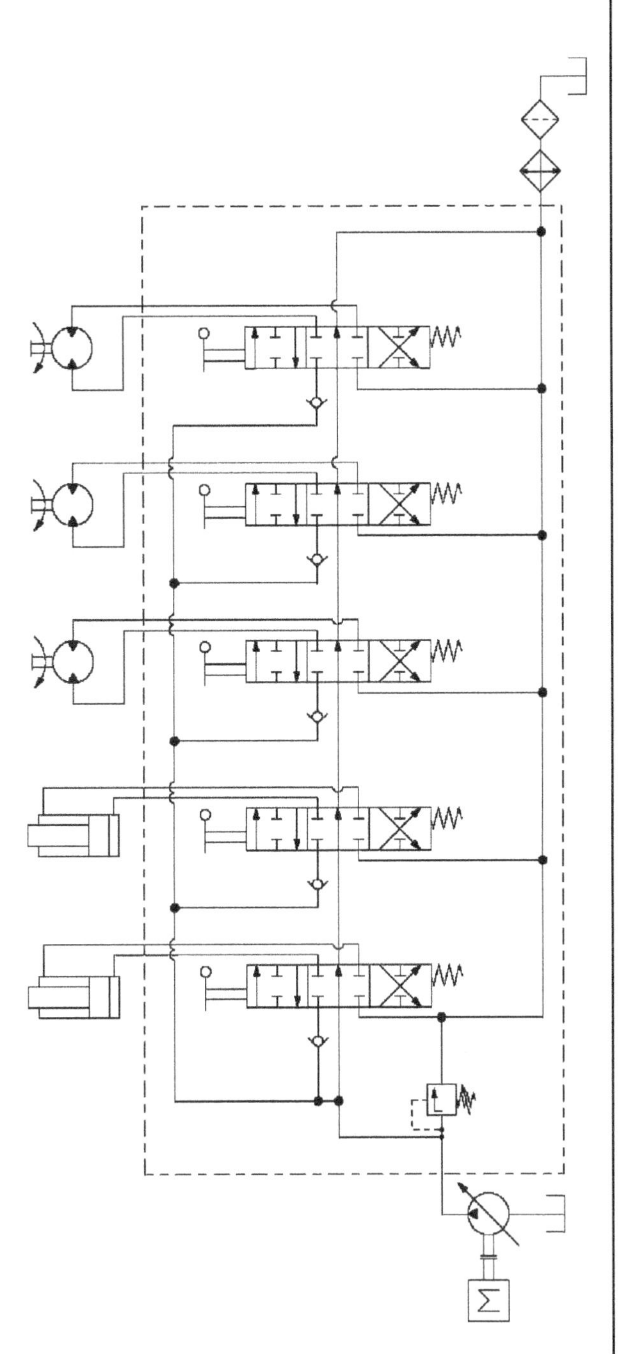

FIGURE 7A.9 The typical circuit of a mobile system with a parallel connection.

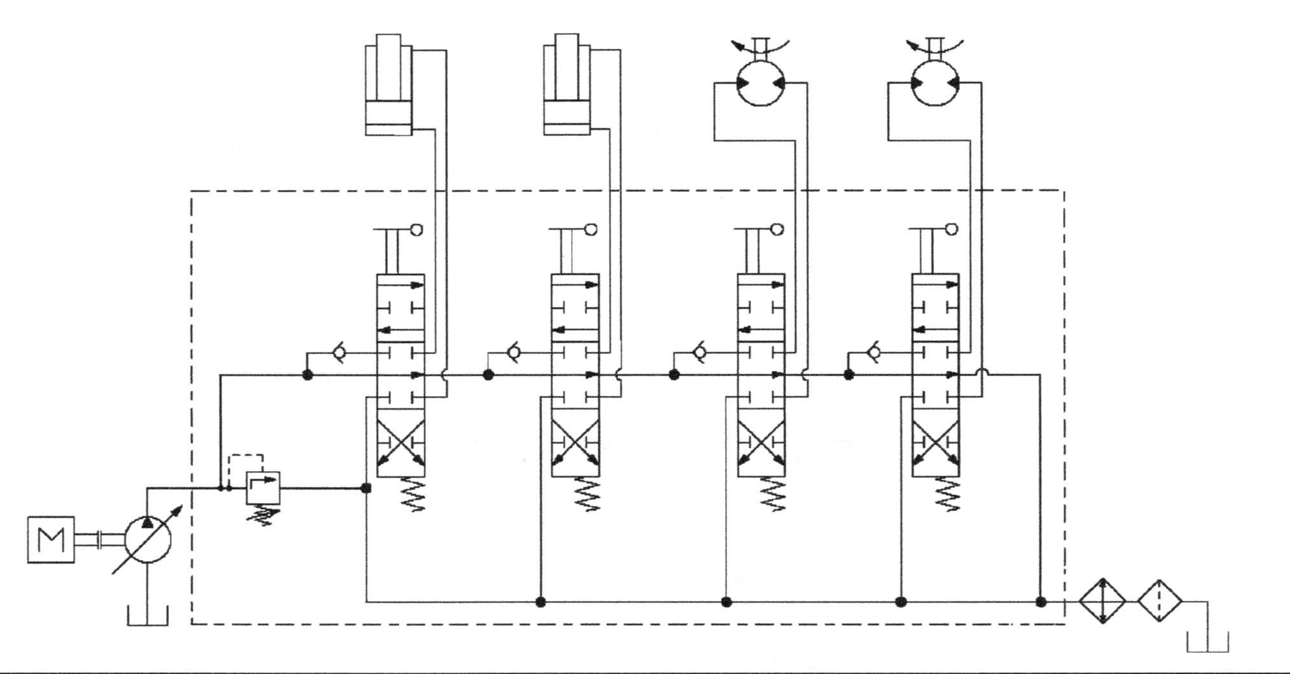

Figure 7A.10 A typical circuit of a mobile system with or connection; only one actuator runs at a time with priority to the upstream valve.

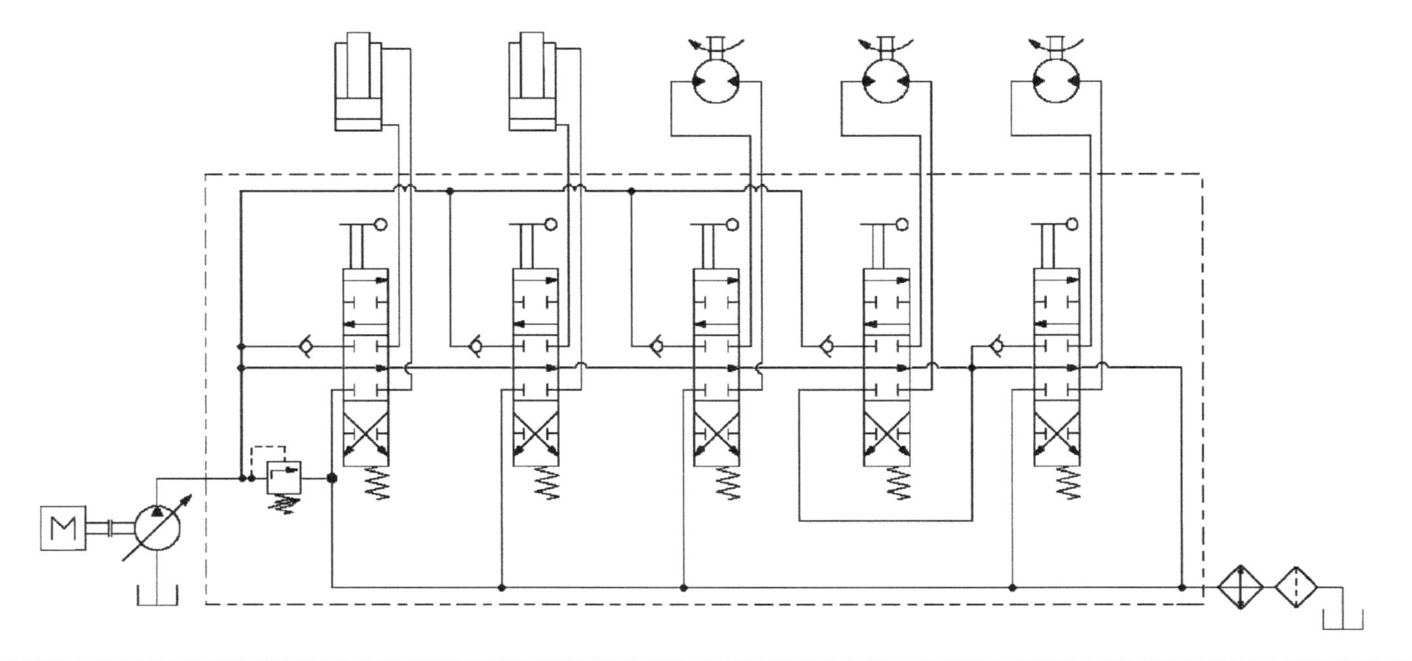

Figure 7A.11 A typical circuit of a mobile system with parallel connections and a series of connected traction motors.

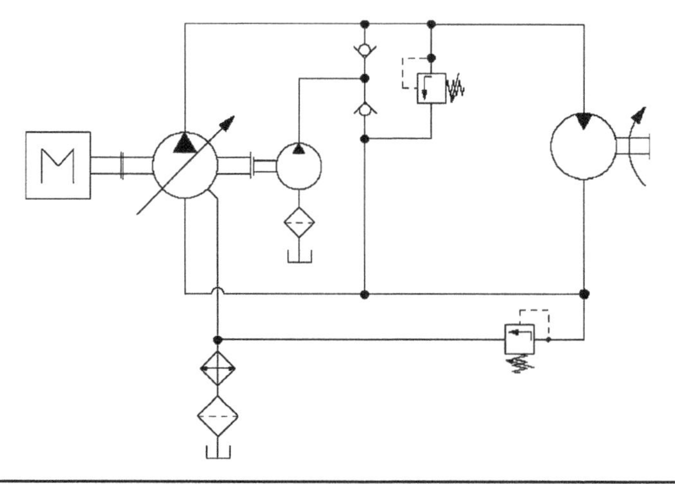

Figure 7A.12 A typical closed circuit with a unidirectional motor.

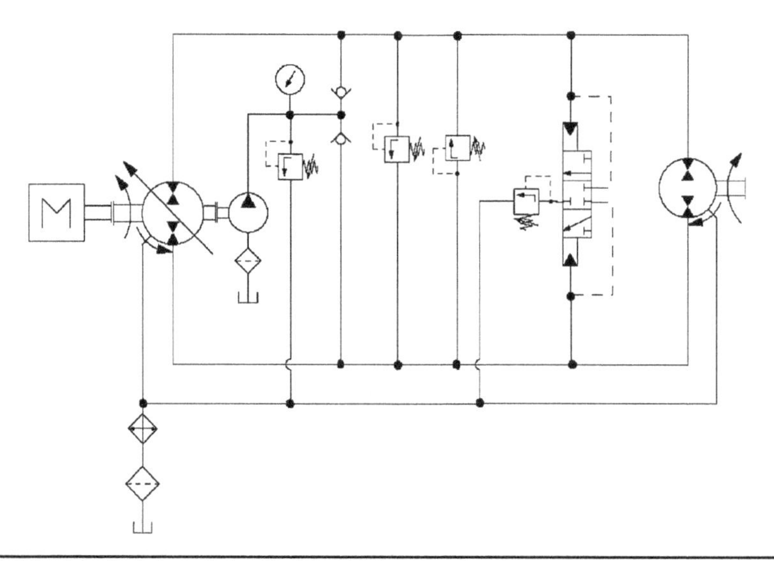

Figure 7A.13 A typical closed circuit with a bidirectional motor.

CHAPTER 8

Hydraulic Servo Actuators

8.1 Construction and Operation

Hydraulic servo actuators (HSAs) are used to precisely control displacement in a wide range of equipment. Generally, a hydraulic servo actuator consists of a hydraulic actuator controlled by a directional control valve of an infinite number of positions and equipped with a feedback arrangement. Figures 8.1 through 8.3 show typical constructions and symbol of hydraulic servo actuators.

When displacing the spool (2) to the right by a distance z (see Fig. 8.1), the spool valve connects the high-pressure line (P) with the left piston chamber (B). The oil flows from the high-pressure line to this chamber, increasing the pressure, P_B. The right piston chamber (A) is connected simultaneously with the return line (T). The pressure, P_A, decreases and the pressure difference $(P_B - P_A)$ acts to drive the piston (6) to the right. The body of the directional control valve (4) is rigidly attached to the piston rod and they move as one body. This displacement causes a gradual decrease in the spool valve opening distance, throttling area, inlet flow rate, and piston speed. Finally, when the total piston displacement equals that of the spool, the spool valve ports are almost closed. The pressure difference in the piston chambers produces a force equal to the loading force. The fluid flow to the cylinder chambers is cut and the piston is stopped.

The HSA is operated by displacing the spool relative to the sleeve (or relative to the valve body). In the steady state, for the piston to stop moving, the spool should be brought to its neutral position. Therefore, the feedback acts to bring the spool to the neutral position when the piston displaces to the required position. This feedback action can be realized in any of the following ways:

- By displacing the sleeve of the DCV (depending on the piston rod displacement) in the same direction of motion as the spool (see Fig. 8.1).
- By displacing the spool back to its neutral position (see Figs. 8.4 and 8.6).
- By the simultaneous displacement of both the sleeve and spool until they reach the neutral position (see Figs. 8.3 and 8.5).

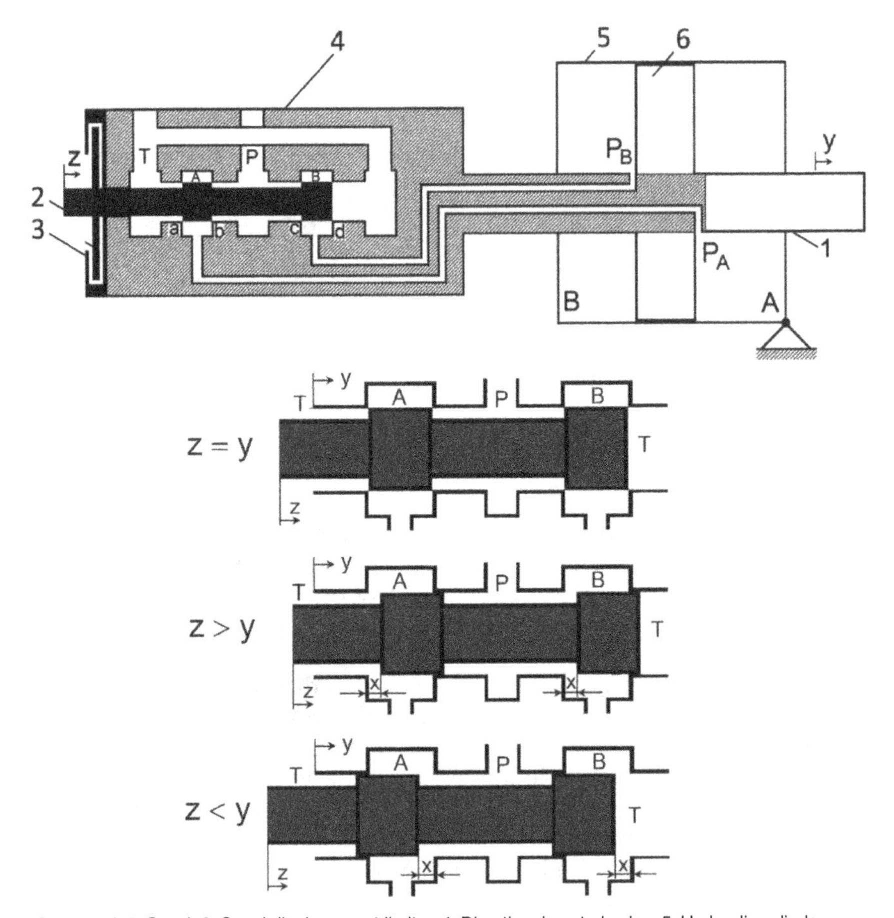

1. Piston rod, 2. Spool, 3. Spool displacement limiter, 4. Directional control valve, 5. Hydraulic cylinder, 6. Piston, z = Spool displacement, y = Piston displacement, x = Valve opening distance.

Figure 8.1 Functional schematic of a hydraulic servo actuator.

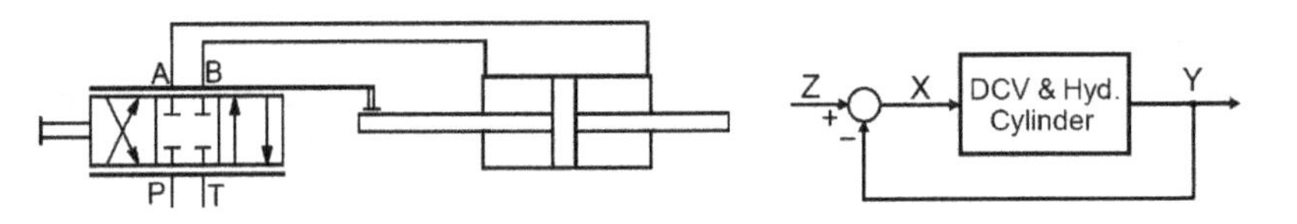

Figure 8.2 A symbol and functional block diagram of the HSA.

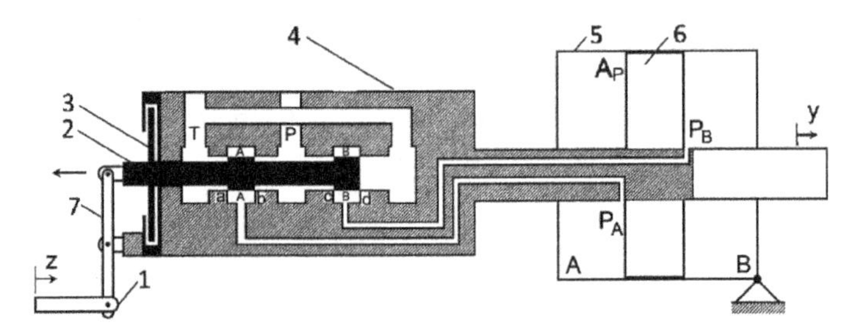

1. Control rod (input displacement), 2. Spool, 3. Spool displacement limiter, 4. Directional control valve, 5. Hydraulic cylinder, 6. Piston, 7. Feedback rod

Figure 8.3 Typical design of an HSA with mechanical feedback linkage.

In the actual operating conditions, the force needed to displace the spool is negligible, while the pressure force applied by the piston is high enough to drive the piston against the load.

8.2 Applications of Hydraulic Servo Actuators

Hydraulic servo actuators have a wide range of applications in different fields, such as:

- The steering systems of mobile equipment
- Machine tools, such as the copying machines
- Variable-displacement pump control

In aerospace and marine applications, the HSA is used to control the rotating blades' pitch angles, thrust deflectors, and the displacement of different control surfaces, such as rudders, ailerons, and elevators.

8.2.1 The Steering Systems of Mobile Equipment

Figure 8.4 shows an application of the hydraulic servo actuator in the steering system of mobile equipment. It consists of a hydraulic generator, a 4/3 directional control valve, a rotary actuator and a mechanical feedback mechanism. The figure shows the system operation during a right turn. The steering wheel is turned in the clockwise direction by the operator. The spool end, engaged with the worm, is forced out of the nut. The valve spool is then displaced to the left, which directs the high pressure oil to port (B) of the steering cylinder.

The other side of the steering cylinder (port A) is connected to the reservoir through the spool valve. The pistons and the rack move to the right and rotate the pinion to turn the front axle, or wheels, to the right. The wheels continue to turn as long as the steering wheel is rotating. When the steering wheel is stopped, the pressurized oil will continue to flow to the steering cylinder, which moves the pistons and rack and turns the wheels

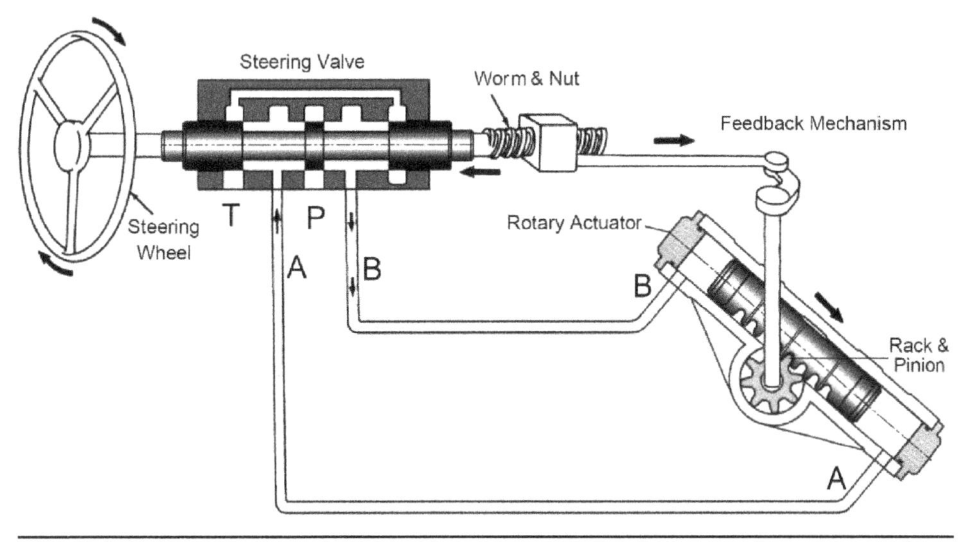

Figure 8.4 The application of hydraulic servo actuators in the steering systems of mobile equipment.

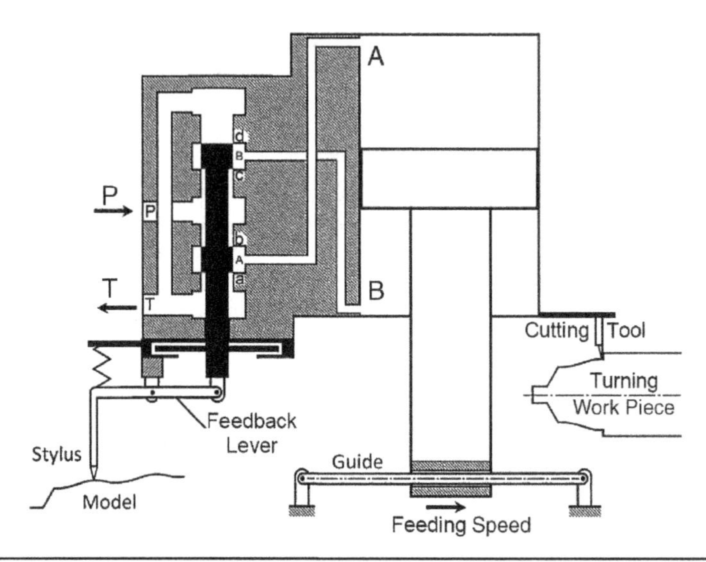

FIGURE 8.5 Application of the HSA in copying machines.

slightly to the right. This motion of the rack causes further pinion rotation, forward motion of the feedback linkage, and pull of the nut and the spool back until the spool attains its neutral position.

8.2.2 Applications in Machine Tools

Figure 8.5 illustrates the application of hydraulic servo actuators in the copying machines of turning machining process. The shape of the model is followed up by the stylus. The cylinder body, which carries the tool, moves vertically following the stylus displacement. The system settles only when the feedback lever adopts a horizontal position and the cylinder attains the same displacement of the stylus: magnitude and direction. The feeding speed should not be very fast, considering the settling time of the hydraulic servo actuator.

8.2.3 Applications in Displacement Pump Controls

In the case of the closed hydraulic circuits, Figs. 7A.12 and 7A.13, the speed of the hydraulic motors is set by controlling the displacement of the hydraulic pump and/or hydraulic motor. The pump displacement can be managed by a wide range of controllers, among them hydraulic servo actuators. Figure 8.6 illustrates the application of the hydraulic servo actuator (HSA) in the control of the swash plate angle of an axial piston pump. The system consists of a directional control valve (1), two hydraulic cylinders (A and B), and a mechanical feedback system. The system is supplied by hydraulic power from a charge pump. The control is carried out by means of a control lever (3). When this lever is put in the vertical position, the feedback rod (4) becomes vertical and the spool has assumed a neutral position. This initial position corresponds to the zero-deflection angle of the swash plate. When the control lever is deflected to the right, the lever-mechanism pulls the spool (5) to the left. The cylinders chambers (A and B) connect to the pressure (P) and return line (T), respectively.

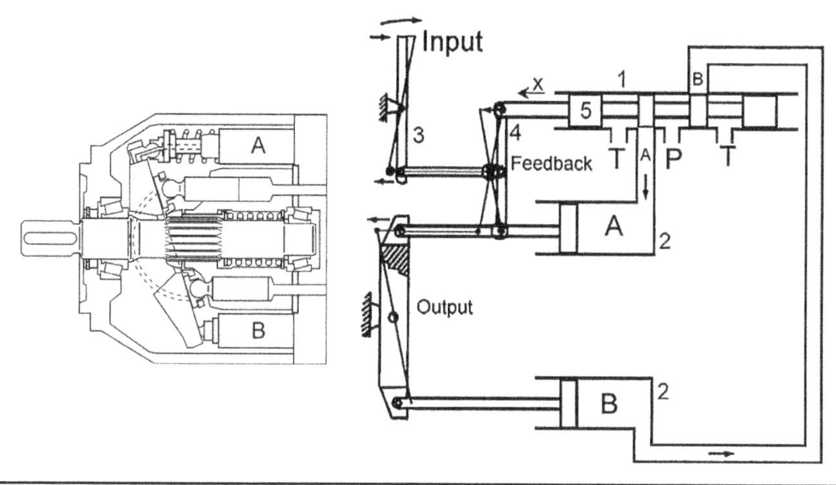

FIGURE 8.6 Control of a swash plate angle by using a hydraulic servo actuator.

The pistons start to move, rotating the swash plate in a counterclockwise direction. Simultaneously, the feedback rod (4) rotates clockwise and displaces the spool to the right. The system reaches a new steady-state position when the spool (5) regains its neutral position.

8.3 The Mathematical Model of HSA

Figure 8.7 shows the functional schematic of an HSA with mechanical feedback. A mathematical model describing the dynamic behavior of this HSA is deduced as follows.

Flow Rate Through the DCV Restriction Areas

The HSA is equipped with a zero-lap spool-type directional control valve. Neglecting the effect of the inner hydraulic transmission lines, the flow rates through the valve restrictions (a, b, c, and d) are given by the following equations:

$$Q_a = C_d A_a(x)\sqrt{2(P_A - P_t)/\rho} \tag{8.1}$$

$$Q_b = C_d A_b(x)\sqrt{2(P_s - P_A)/\rho} \tag{8.2}$$

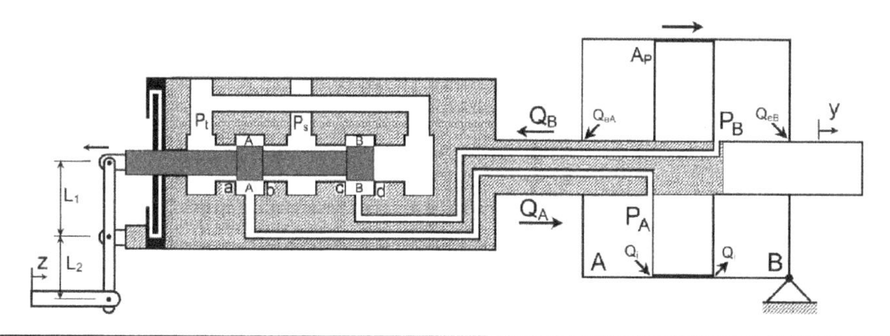

FIGURE 8.7 Functional schematic of an HSA with mechanical feedback.

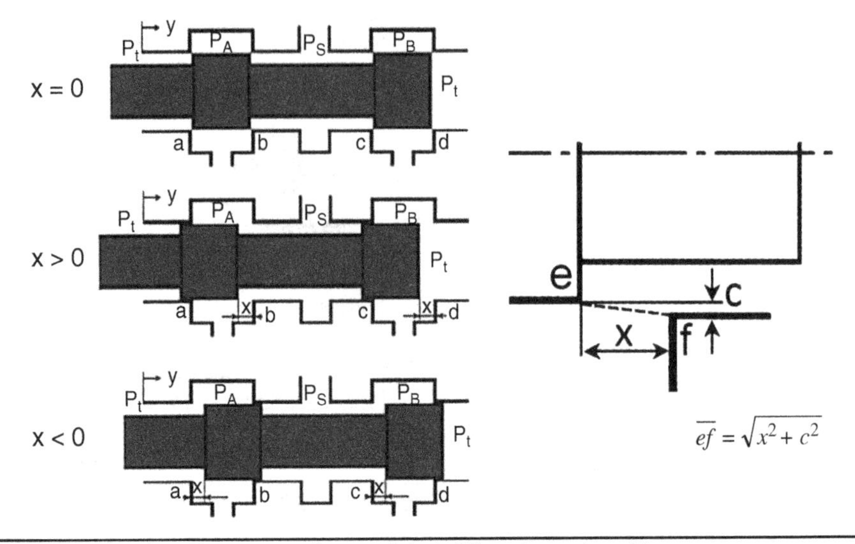

FIGURE 8.8 Valve throttling areas a, b, c, and d at different spool positions.

$$Q_c = C_d A_c(x)\sqrt{2(P_s - P_B)/\rho} \tag{8.3}$$

$$Q_d = C_d A_d(x)\sqrt{2(P_B - P_t)/\rho} \tag{8.4}$$

where Q = Flow rate, m^3/s
P_s = Supply pressure, Pa
P_t = Return pressure, Pa
A = Restriction areas, m^2
ρ = Oil density, kg/m^3
C_d = Discharge coefficient
x = Spool valve opening distance, m

Usually, the directional control valve of the HSA is a zero-lapping matched symmetrical type. The matched valve has $A_a(x) = A_c(x)$ and $A_b(x) = A_d(x)$, while the symmetrical valve has $A_a(-x) = A_b(x)$ and $A_c(-x) = A_d(x)$. Then, the valve restriction areas are given by the following equations (see Fig. 8.8).

$$\left.\begin{aligned} A_a &= A_c = A_r \\ A_b &= A_d = \omega\sqrt{(x^2 + c^2)} \end{aligned}\right\} \text{For } x \ge 0 \tag{8.5}$$

$$\left.\begin{aligned} A_a &= A_c = \omega\sqrt{(x^2 + c^2)} \\ A_b &= A_d = A_r \end{aligned}\right\} \text{For } x \le 0 \tag{8.6}$$

where ω = Width of the port, m
c = Spool radial clearance, m
x = Valve opening distance, m
A_r = Radial clearance area, m^2

The Continuity Equation Applied to the Cylinder Chambers

Actually, the deformation of cylinder wall material is negligible, compared with the oil volumetric variation due to the oil compressibility. Then, neglecting the effect of the inner conduits and assuming that the piston is initially at its mid position, the application of the continuity equation to the cylinder chambers yields the following equations:

$$Q_b - Q_a - A_p \frac{dy}{dt} - Q_i - Q_{eA} = \frac{V_o + A_p y}{B} \frac{dP_A}{dt} \tag{8.7}$$

$$A_P \frac{dy}{dt} + Q_i + Q_c - Q_d - Q_{eB} = \frac{V_o - A_p y}{B} \frac{dP_B}{dt} \tag{8.8}$$

Assuming that the leakage flow rate is linearly proportional to the pressure difference, the leakage flow rates are given by the following relations:

$$Q_i = (P_A - P_B)/R_i \tag{8.9}$$

$$Q_{eA} = P_A / R_e \tag{8.10}$$

$$Q_{eB} = P_B / R_e \tag{8.11}$$

where A_p = Piston area, m^2
Q_e = External leakage flow rate, m^3/s
Q_i = Internal leakage flow rate, m^3/s
R_e = Resistance to external leakage, Pa s/m^3
R_i = Resistance to internal leakage, Pa s/m^3
V_o = Half of volume of oil filling the cylinder, m^3

The Equation of the Feedback Mechanism

Figure 8.9, explains the kinematics of the feedback mechanism. The control rod displacement (z) rotates the feedback lever counter clockwise around point (2) which causes the spool attached to point (1) to move by distance (d_1) to the left. The valve body displaces with the piston to the right by distance (y). This displacement rotates the feedback lever

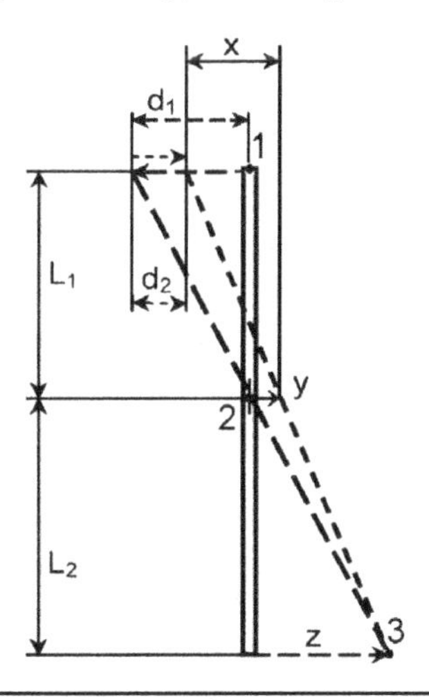

FIGURE 8.9 Kinematics of the feedback mechanism.

clockwise around point (3) which results in a spool displacement to the right by distance (d_2). Assuming an ideal zero-lapping valve, the spool valve opening (x) equals the net spool displacement relative to the valve housing.

$$d_1 = z\frac{L_1}{L_2} \quad \text{and} \quad d_2 = y\frac{L_1 + L_2}{L_2} \tag{8.12}$$

$$x = d_1 - d_2 + y \tag{8.13}$$

or

$$x = z\frac{L_1}{L_2} - y\frac{L_1 + L_2}{L_2} + y \tag{8.14}$$

Then,

$$x = (z - y)\frac{L_1}{L_2} \tag{8.15}$$

or

$$x = k_f(z - y), \quad \text{where } k_f = L_1/L_2 \tag{8.16}$$

for

$$L_1 = L_2, \qquad x = z - y \tag{8.17}$$

Equation of Motion of the Piston

$$A_p(P_A - P_B) = m\frac{d^2y}{dt^2} + f_v\frac{dy}{dt} + F_L \tag{8.18}$$

where m = Reduced mass of the moving parts, kg
f_v = Friction coefficient, Ns/m
F_L = External loading force, N
y = Piston displacement, m

8.4 The Transfer Function of HSA

The transfer function of the hydraulic servo actuator can be deduced in two ways:

- By identifying the hydraulic servo actuator based on its measured or calculated step response
- By the mathematical treatment of the mathematical model

8.4.1 Deduction of the HSA Transfer Function, Based on the Step Response

The dynamic behavior of the HSA is described mathematically by Eqs. (8.1) to (8.18). A simulation program could be developed, based on the deduced mathematical model. The available simulation programs, such as the SIMULINK, enable the simulation of highly nonlinear systems. The simulation program can be used to study both the static and dynamic behaviors of the HSA. In addition, the transfer function of the HSA can be deduced using the simulation results. This is done by calculating the step response, and then identifying the obtained response by a simple transfer function. This process is explained in App. 8A.

8.4.2 Deducing the HSA Transfer Function Analytically

The studied hydraulic servo actuator (see Figs. 8.7 and 8.8) is described mathematically by a set of nonlinear implicit differential and algebraic relations [see Eqs. (8.1) through (8.18)]. The HSA can be described by a transfer function only if its mathematical model is linear, with zero initial conditions. Therefore, it is necessary to carry out a linearization process assuming the following simplifying assumptions and considering that the HSA is excited by small perturbations:

1. The DCV is matched symmetrically.
2. The spool of the DCV is ideal, with zero lapping.
3. The spool valve radial-clearance leakage is negligible and the inlet and return flow rates to the cylinder are equal…

   for $x > 0$ $\quad Q = Q_b = Q_d$ and $Q_a = Q_c = 0$
   for $x < 0$ $\quad Q = Q_a = Q_c$ and $Q_b = Q_d = 0$
4. The return pressure is null: $P_t = 0$.
5. The DCV restriction areas are linearly proportional to the spool valve opening.

$$A(x) = \omega x \tag{8.19}$$

6. The piston is initially at the middle of the cylinder.
7. The loading force, F_L, is linearly proportional to the piston displacement, and the load pressure, P_L, is defined as follows:

$$F_L = A_p\,(P_A - P_B) \tag{8.20}$$

$$P_L = P_A - P_B = F_L/A_P \tag{8.21}$$

The transfer function of the HSA is deduced mathematically as follows.

Flow Rate Equations

$$Q_b = C_d \omega x \sqrt{2(P_s - P_A)/\rho} \tag{8.22}$$

$$Q_d = C_d \omega x \sqrt{2P_B/\rho} \tag{8.23}$$

$$Q = Q_b = Q_d \tag{8.24}$$

Therefore, the following relations could be reached, by substituting for Q_b and Q_d, from Eqs. (8.22) and (8.23) into Eq. (8.24).

$$P_s = P_A + P_B \quad \text{and} \quad P_L = P_A - P_B \tag{8.25}$$

or

$$P_A = (P_s + P_L)/2 \quad \text{and} \quad P_B = (P_s - P_L)/2 \tag{8.26}$$

Substituting for P_A and P_B from Eq. (8.26) into Eq. (8.22) or (8.23), the following equation is obtained:

$$Q = C_d \omega x \sqrt{(P_s - P_L)/\rho} \tag{8.27}$$

By linearizing this nonlinear equation in the vicinity of a steady-state operating point (x_i, Q_i, P_{Li}), the following relation results:

$$\Delta Q = \frac{\partial Q}{\partial x}\Delta x + \frac{\partial Q}{\partial P_L}\Delta P_L = k_x \Delta x - k_p \Delta P_L \tag{8.28}$$

where

$$k_x = C_d \omega \sqrt{(P_s - P_L)/\rho} \tag{8.29}$$

$$k_p = \frac{0.5 C_d\, \omega\, x}{\sqrt{\rho (P_s - P_L)}} \tag{8.30}$$

Continuity Equations

The following are the continuity equations applied to both chambers of the hydraulic cylinder:

$$Q - A_p \frac{dy}{dt} - \frac{P_A - P_B}{R_i} - \frac{P_A}{R_e} = \frac{V_o + A_p y}{B}\frac{dP_A}{dt} \tag{8.31}$$

$$-Q + A_p \frac{dy}{dt} + \frac{P_A - P_B}{R_i} - \frac{P_B}{R_e} = \frac{V_o - A_p y}{B}\frac{dP_B}{dt} \tag{8.32}$$

For small displacements, the swept volume, $A_p y$, is negligible with respect to the volume of oil in the cylinder chambers ($V_o >> A_p y$). Then, by subtracting Eq. (8.31) from Eq. (8.32), the following equation is obtained:

$$Q - A_p \frac{dy}{dt} - \frac{P_L}{R} - \frac{V}{4B}\frac{dP_L}{dt} = 0 \tag{8.33}$$

where

$$R = \frac{2 R_i R_e}{R_i + 2R_e} \tag{8.34}$$

$V = 2V_o$ = total volume of oil in the cylinder, m^3

The Equation of Motion of a Piston

The piston is driven by the pressure difference in the hydraulic cylinder chambers. Assuming that the loading force is linearly proportional to the piston displacement, $F_L = k_L y$, the equation of motion of the piston becomes as follows:

$$P_L A_p = m\frac{d^2 y}{dt^2} + f_v \frac{dy}{dt} + k_L y \tag{8.35}$$

In the case of the linearized mathematical model, the system variables are looked at as the deviation of these variables from their steady-state values. Then, ΔQ, Δx, and ΔP are replaced by Q, x, P, and Eq. (8.28) becomes

$$Q = k_x x - k_p P_L \tag{8.36}$$

The Feedback Equation

$$x = k_f (z - y) \tag{8.37}$$

Applying Laplace's transform to Eq. (8.33) and Eqs. (8.35) through (8.37), then, after rearrangement, the following transfer function is obtained:

$$\frac{Y}{Z}(s) = \frac{k_f k_x}{a_3 s^3 + a_2 s^2 + a_1 s + a_o} \tag{8.38}$$

where

$$a_3 = \frac{Vm}{4BA_p} \tag{8.39}$$

$$a_2 = \frac{m(1/R + k_p)}{A_p} + \frac{Vf_v}{4BA_p} \tag{8.40}$$

$$a_1 = A_p + \frac{f_v(1/R + k_p)}{A_p} + \frac{k_L V}{4BA_p} \tag{8.41}$$

$$a_o = k_f k_x + \frac{k_L(1/R + k_p)}{A_p} \tag{8.42}$$

The dynamic behavior of the HSA can be evaluated on the basis of the deduced linearized model, considering only small disturbances around the considered steady-state operating point. The coefficients of the transfer function should be recalculated if the initial conditions are changed.

8.5 Valve-Controlled Actuators

8.5.1 Flow Characteristics

In the case of the HSA, the maximum displacement of the spool of the DCV, relative to the valve body, is limited by mechanical position limiters, usually within ±1 mm. When greater displacement is required, the control rod should be continuously displaced by applying the needed force, until the piston reaches the required position. During this period, the spool valve is fully open, the spool is continuously displaced, and the piston rod follows this displacement. Therefore, it is necessary to study the behavior of the system at these conditions. During this operating mode, the HSA acts as a simple valve-controlled actuator.

Figure 8.10 shows a symmetrical hydraulic cylinder controlled by an ideal (zero-lapping) 4/3 directional control valve. The flow rates through the valve restrictions are calculated by the following equations:

$$Q_a = C_d A_a(x)\sqrt{2(P_A - P_t)/\rho} \tag{8.43}$$

$$Q_b = C_d A_b(x)\sqrt{2(P_s - P_A)/\rho} \tag{8.44}$$

$$Q_c = C_d A_c(x)\sqrt{2(P_s - P_B)/\rho} \tag{8.45}$$

$$Q_d = C_d A_d(x)\sqrt{2(P_B - P_t)/\rho} \tag{8.46}$$

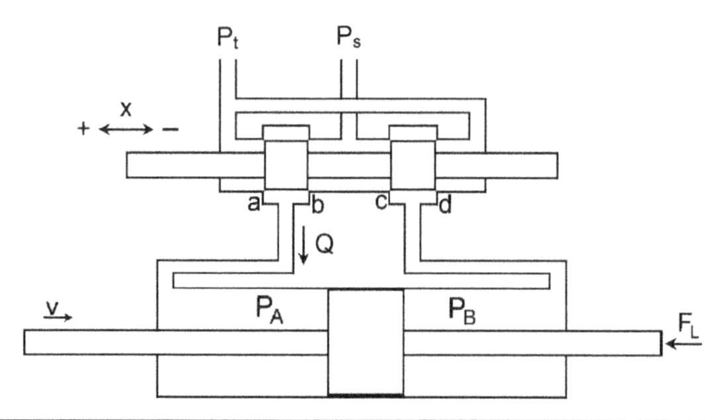

FIGURE 8.10 A valve-controlled actuator.

The flow characteristics of this system are investigated in the steady-state operation, taking into consideration the following assumptions:

1. The valve is matched symmetrically; $A_a(x) = A_c(x)$, $A_b(x) = A_d(x)$, and $A_a(-x) = A_b(x)$, $A_c(-x) = A_d(x)$.
2. The hydraulic cylinder is ideal; no friction and no leakage.
3. The throttling areas of the valve ports are linearly proportional to the spool displacement; $(A = \omega x)$.
4. The spool valve is of the zero lapping type, with no radial-clearance leakage.
5. The return line pressure is null: $P_t = 0$.

In the steady state, for positive spool displacement, the flow rates Q_b and Q_d are equal due to the symmetry of the cylinder, and the areas A_b and A_d are also equal for matched valves.

$$C_d A_b(x)\sqrt{2(P_s - P_A)/\rho} = C_d A_d(x)\sqrt{2(P_B - P_t)/\rho} \tag{8.47}$$

then,

$$P_s = P_A + P_B \tag{8.48}$$

Defining the load pressure, $P_L = P_A - P_B$, then

$$P_A = (P_s + P_L)/2 \quad \text{and} \quad P_B = (P_s - P_L)/2 \tag{8.49}$$

The load flow is defined by: $Q = Q_b - Q_a = Q_c - Q_d$. Then, by substituting for P_A and P_B, the load flow is given by the following equation:

$$Q = C_d A_b(x)\sqrt{(P_s - P_L)/\rho} - C_d A_a(x)\sqrt{(P_s + P_L)/\rho} \tag{8.50}$$

In the case of an ideal valve, for positive spool displacement,

$$A_c = A_a = 0, \qquad A_b = A_d = \omega x \quad \text{and} \quad Q = Q_b = Q_d \tag{8.51}$$

then,
$$Q = C_d \omega x \sqrt{(P_s - P_L)/\rho} \tag{8.52}$$

The maximum flow rate Q_{max} is obtained at $x = x_{max}$ and $P_L = 0$

or
$$Q_{max} = C_d \omega x_{max} \sqrt{P_s/\rho} \tag{8.53}$$

The flow equation is written in nondimensional form by defining the following nondimensional parameters:

$$\bar{Q} = Q/Q_{max}, \quad \bar{x} = x/x_{max} \quad \text{and} \quad \bar{P} = P_L/P_s \tag{8.54}$$

then,
$$\bar{Q} = \bar{x}\sqrt{1-\bar{P}} \tag{8.55}$$

The nondimensional piston speed is given by $\bar{v} = v/v_{max}$, where $v = Q/A_p$ and $v_{max} = Q_{max}/A_p$

then,
$$\bar{v} = v/v_{max} = \bar{Q} = \bar{x}\sqrt{1-\bar{P}} \tag{8.56}$$

For negative spool displacement (see Fig. 8.10), the piston moves in the same direction of the loading force. In this case, the nondimensional valve flow rate is given by the following relation:

$$\bar{Q} = \bar{x}\sqrt{1+\bar{P}} \tag{8.57}$$

The steady-state flow characteristics of the valve-controlled actuator are plotted in Fig. 8.11.

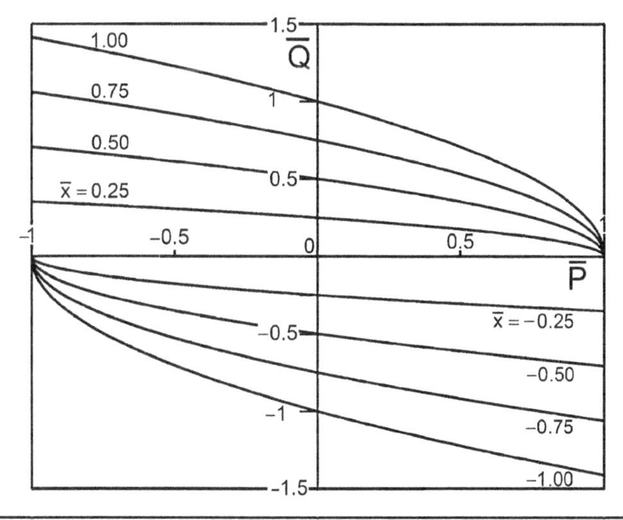

Figure 8.11 Steady-state flow characteristics of a valve-controlled actuator.

8.5.2 Power Characteristics

The power characteristics of the valve-controlled actuators describe the relation between the output power, the spool displacement, and the load pressure. The output power is given by the following equation:

$$N = F_L v = P_L A_P \frac{Q}{A_p} = P_L Q \tag{8.58}$$

or

$$N = C_d \omega x P_L \sqrt{(P_s - P_L)/\rho} \tag{8.59}$$

The output power is null if either x or P_L becomes zero or $P_L = P_s$. The maximum power is obtained at $x = x_{max}$ and the load pressure P_L in the range $0 < P_L < P_s$. The maximum power is obtained as follows:

$$\frac{\partial N}{\partial P_L} = 0 \tag{8.60}$$

$$C_d \omega x \sqrt{(P_s - P_L)/\rho} - \frac{1}{2\rho} C_d \omega x P_L \left\{ \frac{1}{\rho}(P_s - P_L) \right\}^{-1/2} = 0 \tag{8.61}$$

then,

$$P_L = \frac{2}{3} P_s \quad \text{and} \quad \bar{P} = \frac{2}{3} \tag{8.62}$$

$$N_{max} = \frac{2}{3\sqrt{3}} C_d \omega x_{max} P_s \sqrt{\frac{1}{\rho} P_s} \tag{8.63}$$

The nondimensional power $\bar{N}$ is defined as $\bar{N} = N/N_{max}$

then,

$$\bar{N} = \frac{3\sqrt{3}}{2} \bar{x} \bar{P} \sqrt{1 - \bar{P}} \tag{8.64}$$

The power characteristics of a valve-controlled actuator are plotted in Fig. 8.12.

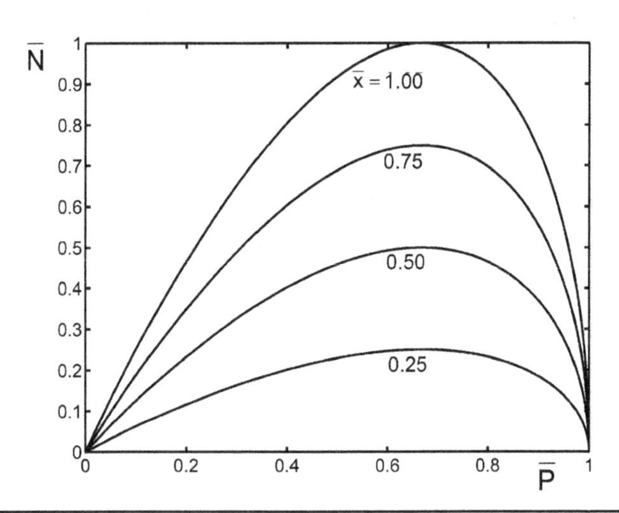

FIGURE 8.12 The power characteristics of a valve-controlled actuator.

8.6 Exercises

1. Draw the functional scheme of a hydraulic servo actuator and explain its function.

2. Draw a scheme of an HSA with mechanical feedback, explain its function, and derive a mathematical model describing its dynamic behavior.

3. Deduce a mathematical model describing the hydraulic servo actuator illustrated by Fig. 8.1, develop a simulation program for this HSA, then,

a. Discuss the transient performance of the HSA.

b. Discuss the effect of internal leakage on the HSA's behavior.

c. Discuss the effect of the load coefficient on the HSA's behavior, given the following:

$\omega = 1$ mm $\quad C_d = 0.611 \quad \rho = 867$ kg/m$^3 \quad P_s = 15$ MPa

$A_p = 20$ cm$^2 \quad m = 6$ kg $\quad V_o = 120$ cm$^3 \quad P_t = 0$

$f_v = 2000$ Ns/m $\quad B = 1.5 \times 10^9$ Pa $\quad R_e = 10^{18}$ Pa s/m$^3 \quad R_i = 10^{12}$ Pa s/m^5

$k_L = 1.5 \times 10^5$ N/m $\quad$ Assume convenient values for any missing data.

4. Derive the transfer function of the HSA (see Fig. 8.1) and state clearly the simplifying assumptions.

5. Discuss in detail the flow characteristics of the valve-controlled actuators, derive the necessary relations, and draw the needed schemes.

6. Discuss in detail the power characteristics of the valve-controlled actuators, derive the necessary relations, and draw the needed schemes.

8.7 Nomenclature

A = Restriction area, m^2
A_p = Piston area, m^2
A_r = Radial clearance area, m^2
B = Bulk's modulus of oil, Pa
C_d = Discharge coefficient
d_s = Spool diameter, m
f_v = Friction coefficient, Ns/m
F_L = Loading force, N
k_L = Piston loading coefficient, N/m
k_p = Pressure gain, m^5/Ns
k_x = Displacement gain, m^2/s
k_f = Feedback gain
L = Length, m
m = Reduced mass of the moving parts, kg
N = Power, W
P_L = Load pressure, Pa
P_s = Supply pressure, Pa
P_t = Return pressure, Pa
Q = Flow rates, m^3/s
Q_e = External leakage flow rate, m^3/s

Q_i = Internal leakage flow rate, m^3/s
R = Equivalent leakage resistance, Pa s/m^3
R_e = Resistance to external leakage, Pa s/m^3
R_i = Resistance to internal leakage, Pa s/m^3
v = Piston speed, m/s
V = Total volume of oil in the cylinder, m^3
V_o = Half of volume of oil filling the cylinder, m^3
x = Spool valve opening distance, m
y = Piston displacement, m
ω = Throttling area width, m
ρ = Oil density, kg/m^3

Appendix 8A Modeling and Simulation of a Hydraulic Servo Actuator

This appendix presents a case study for the mathematical modeling and simulation of a hydraulic servo actuator with solid negative feedback. The construction and operation of the studied HSA are explained by the functional schemes presented by Figs. 8A.1 and 8A.2. The studied HSA has the following parameters:

Bulk modulus of oil	$B = 1.5 \times 10^9$	N/m^2
External load coefficient	$k_L = 400$	kN/m
Flow coefficient $[K_o = C_d\sqrt{2/\rho}]$	$K_o = 0.0293$	$m^{3/2}kg^{-1/2}$
Half of the cylinder's inner volume	$V_o = 100$	cm^3
Limit of spool displacement	$x_{lim} = 1$	mm
Piston area	$A_p = 30$	cm^2
Piston friction coefficient	$f_v = 2$	kNs/m
Radial clearance	$c = 4$	μm
Reduced mass of piston and moving parts	$m = 10$	kg
Resistance to external leakage	$Re = 1 \times 10^{14}$	Ns/m^5
Resistance to internal leakage	$R_i = 1 \times 10^{14}$	Ns/m^5
Return pressure	$P_t = 0$	Pa
Supply pressure	$P_s = 20$	MPa
Width of valve port	$\omega = 1$	mm

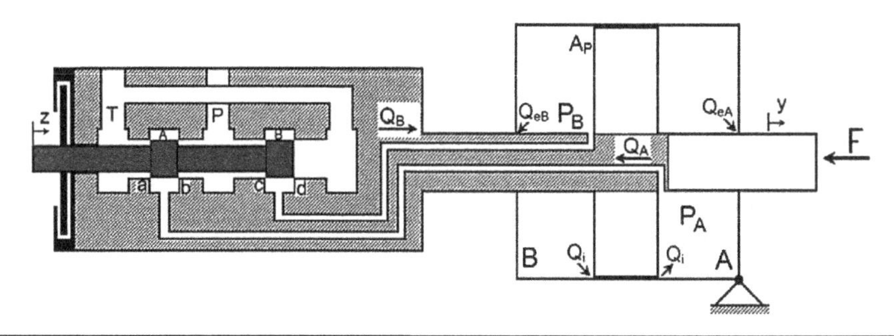

FIGURE 8A.1 Schematic of a hydraulic servo actuator.

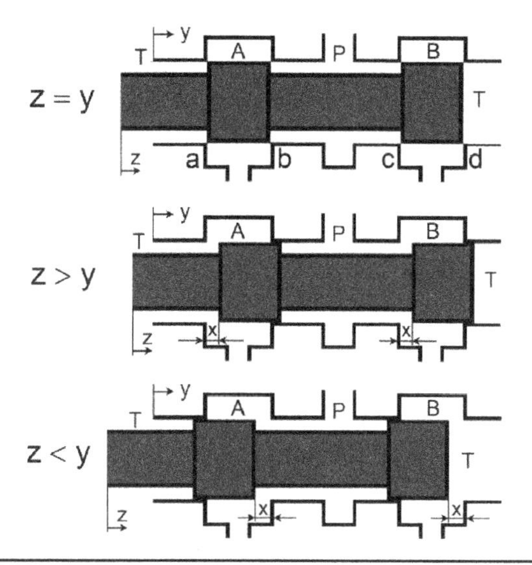

FIGURE 8A.2 Operating positions of the ports of HSA spool valve.

A Mathematical Model of the HSA

A mathematical model describing the dynamic behavior of the studied HSA can be deduced as follows.

The Flow Rate Through DCV Restriction Areas

The HSA is equipped with a zero-lap spool type directional control valve. Neglecting the effect of the inner hydraulic transmission lines, the flow rates through the valve restriction areas are given by

$$Q_a = C_d A_a(x)\sqrt{2(P_A - P_t)/\rho} \tag{8A.1}$$

$$Q_b = C_d A_b(x)\sqrt{2(P_s - P_A)/\rho} \tag{8A.2}$$

$$Q_c = C_d A_c(x)\sqrt{2(P_s - P_B)/\rho} \tag{8A.3}$$

$$Q_d = C_d A_d(x)\sqrt{2(P_B - P_t)/\rho} \tag{8A.4}$$

The directional control valve of the HSA is of the zero-lapping matched symmetrical type. The matched valve is $A_a(x) = A_c(x)$ and $A_b(x) = A_d(x)$, while the symmetrical valve is $A_a(x) = A_b(-x)$ and $A_c(x) = A_d(-x)$.

$$\left.\begin{array}{l} A_a = A_c = \omega\sqrt{(x^2 + c^2)} \\ A_b = A_d = A_r \end{array}\right\} \quad \text{For } x \geq 0;\, z \geq y \tag{8A.5}$$

$$\left.\begin{array}{l} A_a = A_c = A_r \\ A_b = A_d = \omega\sqrt{(x^2 + c^2)} \end{array}\right\} \quad \text{For } x \leq 0;\, z \leq y \tag{8A.6}$$

Continuity Equations Applied to the Cylinder Chambers

Actually, the deformation of cylinder wall material is negligible compared with the oil volumetric variation due to oil compressibility. The application of the continuity equation to the cylinder chambers yields the following equations:

$$Q_B - A_P\frac{dy}{dt} - Q_i - Q_{eB} = \frac{V_o + A_p y}{B}\frac{dP_B}{dt} \tag{8A.7}$$

$$A_P\frac{dy}{dt} + Q_i - Q_A - Q_{eA} = \frac{V_o - A_p y}{B}\frac{dP_A}{dt} \tag{8A.8}$$

The flow rates Q_A and Q_B are given by

$$Q_B = Q_c - Q_d \tag{8A.9}$$

$$Q_A = Q_a - Q_b \tag{8A.10}$$

Assuming that the leakage flow rate is linearly proportional to the pressure difference, the leakage flow rates could be given by the following relations:

$$Q_i = (P_B - P_A)/R_i \tag{8A.11}$$

$$Q_{eB} = P_B/R_e \tag{8A.12}$$

$$Q_{eA} = P_A/R_e \tag{8A.13}$$

The Equation of Motion of a Piston

$$A_p(P_B - P_A) = m\frac{d^2y}{dt^2} + f_v\frac{dy}{dt} + F_L \tag{8A.14}$$

The Feedback Equation

$$x = z - y, \quad \text{where} \quad |x| \le x_{\text{Lim}} \tag{8A.15}$$

Simulation of the HSA

The dynamic behavior of the studied HSA is described by Eqs. (8A.1) through (8A.15). These equations were applied to develop a computer simulation program using the SIMULINK program. The transient response of the HSA to a step input displacement was calculated using the simulation program. The response was calculated for input steps of different magnitudes. The calculation results are plotted in Figs. 8A.3 through 8A.5.

Figure 8A.3 shows that the HSA presents an over-damped response, apparently of the first order of about a 32-ms time constant. The response to step inputs of greater magnitudes are plotted in Figs. 8A.4 and 8A.5. These two figures show the operation at the saturation conditions imposed by the spool displacement limiter.

The HSA can be identified, based on the calculated response, by a first-order transfer function. The system includes a displacement limiter, which limits the applications of the transfer function to very small displacements. Therefore, a representative model

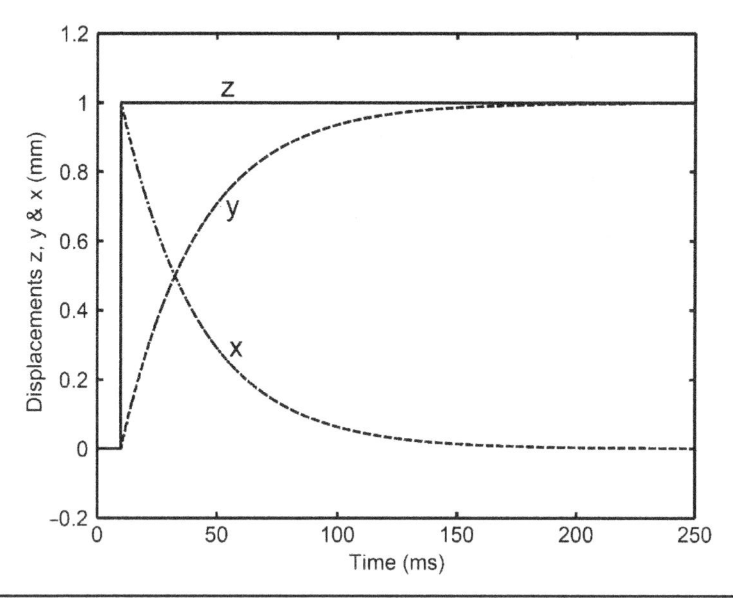

FIGURE 8A.3 A step response of the HSA for a step input of 1 mm magnitude, applied at time t = 10 ms.

(of the style presented by Fig. 8A.6) is more practical. This model presented a transient response, almost, coinciding with that of the detailed model, over the whole range of operation. When operating within the limiting spool displacement—$|x| \le x_{\text{Lim}}$—the studied HSA can be precisely described by the following transfer function:

$$G(s) = \frac{Y}{Z} = \frac{1}{0.033s + 1}$$

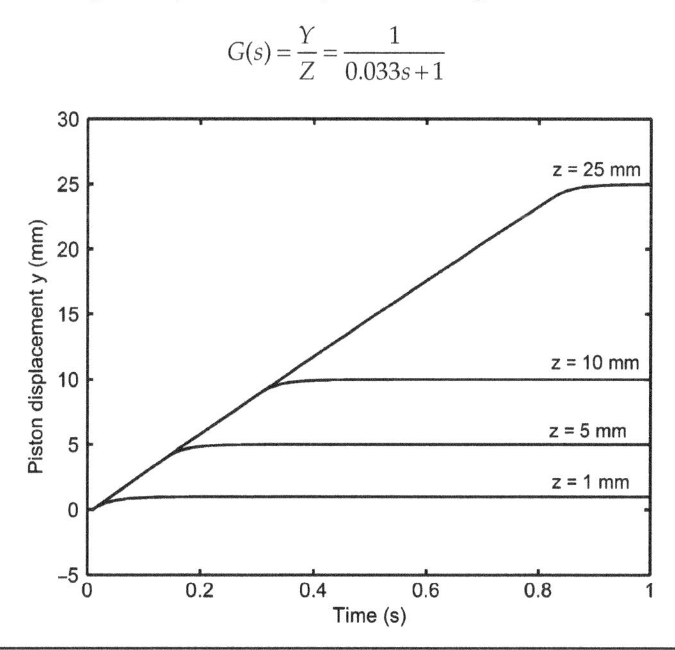

FIGURE 8A.4 The step response of the HSA for step inputs of different magnitudes, applied at time t = 10 ms.

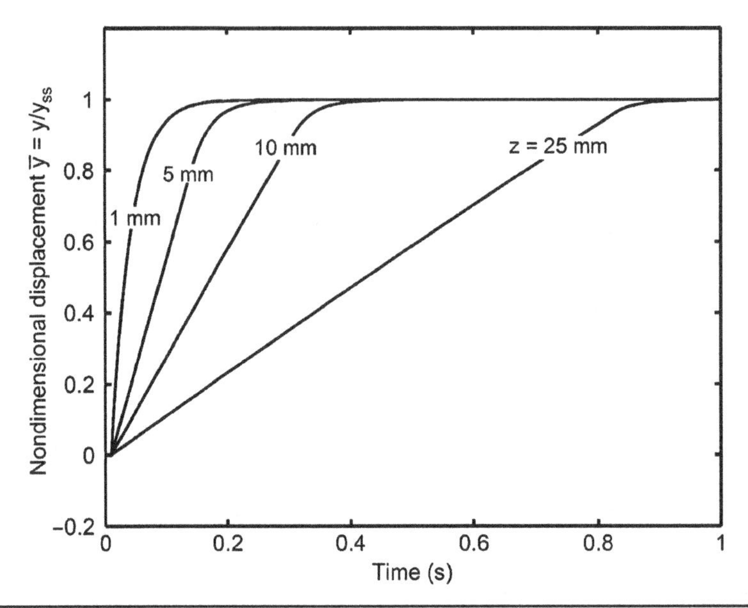

FIGURE 8A.5 A normalized step response of the HSA for step inputs of different magnitudes, applied at time $t = 10$ ms.

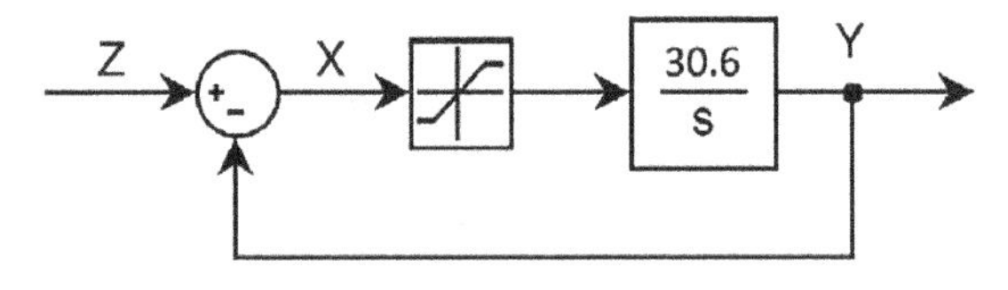

FIGURE 8A.6 A block diagram of the HSA representative model.

Nomenclature

A = Restriction areas, m^2
A_p = Piston area, m^2
A_r = Radial clearance area, m^2
B = Bulk's modulus of oil, Pa
c = Spool radial clearance, m
C_d = Discharge coefficient
F_L = External loading force, N
f_v = Friction coefficient, Ns/m
m = Reduced (equivalent) mass of the moving parts, kg
P_s = Supply pressure, Pa
P_t = Return pressure, Pa
Q = Flow rate, m^3/s
Q_e = External leakage flow rate, m^3/s
Q_i = Internal leakage flow rate, m^3/s
R_e = Resistance to external leakage, Pa s/m^3

R_i = Resistance to internal leakage, Pa s/m^3
V_o = Half of the volume of oil filling the cylinder, m^3
x = Valve opening distance, m
y = Piston displacement, m
z = Input displacement, m
ρ = Oil density, kg/m^3
ω = Width of the valve port, m

CHAPTER 9

Electrohydraulic Servovalve Technology

9.1 Introduction

The marriage between electronics and hydraulic power systems has led to many powerful and precise control systems, saving much energy and money. This concept is applied in the electrohydraulic proportional and servo systems. These systems have the same advantages as hydraulic power systems, particularly the maximum powerto-weight ratio and the high stiffness of hydraulic actuators. They also have the same advantages as electronic controllers, particularly in regard to high controllability and precision.

Fluid power engineering has four classes of control valves that use electric controllers. They include the following.

- *Ordinary or switching valves:* Widely used to turn valves on and off.
- *Electrohydraulic proportional valves:* Used usually in openloop control systems. They are controlled electronically to produce an output pressure or flow rate proportional to the input signal. They offer advantages such as control reversal, stepless variation of the controlled parameters, and reduction of the number of hydraulic devices required for particular control jobs.
- *Electrohydraulic servovalves:* Usually used in closed-loop (feedback) control systems. The "servo concept" is a widely used expression. Taken alone, it indicates a system in which a lowpower input signal is amplified to generate a controlled highpower output or signal. An input signal of low power—for example, 0.08 watts—can provide analog control of power reaching more than 100 kW. Electrohydraulic servo systems provide one of the best controllers from the point of view of precision and speed of response. They are used to control almost all hydraulic and mechanical parameters, such as the pressure, pressure difference, angular speed, displacement, angular displacement, strain, force, and others. Electrohydraulic servovalves were used in military equipment as early as the 1940s.
- *Digital control* valves consist of miniature on/off valves controlled through software. The on/off type of valves connected in parallel could replace the conventional servo and proportional valves. A typical digital valve system has 4-6 parallel connected on/off valves per control edge totaling 16-24 valves in

the four-way valve configuration. The flow capacities of the parallel connected valves are set according to the powers of two such that it is possible to achieve 2^n different flow rates with n valves, Linjama (2015).

9.2 Applications of Electrohydraulic Servos

Two dominant performance parameters are used to classify most of the electrohydraulic servos. One of these is the size, which is the power or the flow rate, and the second is the dynamic behavior. For convenience, the power is defined as the primary power required to control a device, while the dynamic behavior can be described by the natural frequency. The natural frequency of a drive and the resulting total gain are decisive for the closed-loop control accuracy of the relevant drive. Table 9.1 shows the major applications of electrohydraulic servos in the industrial and aerospace fields and indicative upper limits of their performance.

9.3 Electromagnetic Motors

Electrohydraulic systems are controlled by means of electronic or digital controllers. In these systems, the electronic and hydraulic subsystems should be interconnected by means of an element transforming the low-power electric control signal into a proportional mechanical signal that actuates the hydraulic power elements. This element is usually an electromagnetic motor.

In practice, several types of motors are used to transform the electric control signals into proportional mechanical signals, such as the following:

- Electromagnetic torque motors
- Single-acting proportional solenoids
- Double-acting proportional solenoid
- Linear force motor
- Electrodynamic motor of moving bobbin

The electromagnetic torque motors are usually used in the electrohydraulic servovalves. They convert electric input signals of low-level current into proportional mechanical torque. The motor is usually designed to be separately mountable and testable. The motor is hermetically sealed against hydraulic fluid, and its typical construction is shown by Figs. 9.1 and 9.2.

An armature (6), produced from a soft ferromagnetic material, is elastically mounted on a thin-walled spring tube (2). (See Figs. 9.1 and 9.2.)

This tube operates as a carrier and centering spring for the armature and flapper (4), and a sealing element, separating the electric and hydraulic portions. The flapper is physically a part of the motor, but functionally it belongs to the hydraulic amplifier. The air gaps (8) are of the same length when the armature is in the neutral position. This length is nonadjustable in some cases (Fig. 9.1), and adjustable in others (Fig. 9.2). The permanent magnets (1) are located symmetrically with respect to the air gaps, and their permanent magnetic field is set in the air gaps. When the armature is in its neutral position, the four air gaps are of equal dimensions. Then, the magnetic flux in the four air gaps is equal. Therefore, the mechanical forces attracting the armature extremity to the

Field	Frequency,[a] Hz	Flow,[b] L/min	Power,[c] kW
Vibration exciters	600	4	1.5
Missile fin positioning	400	4	1.5
Seekers antenna	300	2	0.75
Oil exploration	200	450	190
Aircraft nose wheel steering	150	4	1.5
Fatigue testing	100	115	40
Machine tool	100	40	15
Turbine control	100	11	4.5
Missile launchers	70	20	7.5
Injection molding	60	300	120
Die casting	50	1140	450
Flight simulators	50	190	75
Space shuttle	50	265	105
Airplane primary flight controls	40	115	45
Robots	40	57	22
Aircraft engine fuel control	30	15	6
Aircraft refueling boom	30	20	7.5
Rolling mills	30	570	225
Tank turret positioning	20	190	75
Agriculture equipment	15	40	15
Conveyers	15	25	10
Cranes	7	75	30
Crawler vehicles	7	378	150
Process controls	5	7.5	3

[a]Frequency corresponding to a phase-lag of 90°.
[b]Flow rate corresponding to a 70-bar pressure drop across the servovalve.
[c]Hydraulic power at about 210 bar pressure.

Table 9.1 Major Fields of Application of Electrohydraulic Servos

upper and lower polar pieces (5 and 9) are equal and their resultant is zero. At these conditions, the torque of the motor is null. When the coils (7) are excited by the control current, the armature becomes magnetized. The magnetic field of this electric magnet, according to its polarity, reinforces the resultant magnetic field in two diagonally opposite air gaps and weakens the field in the other two gaps. The resulting antisymmetry leads to a resultant torque acting on the armature. The change of polarity of input current changes the direction of torque. The torque is practically linearly proportional to the applied current as long as the armature displacement is too small with respect to the air gap length.

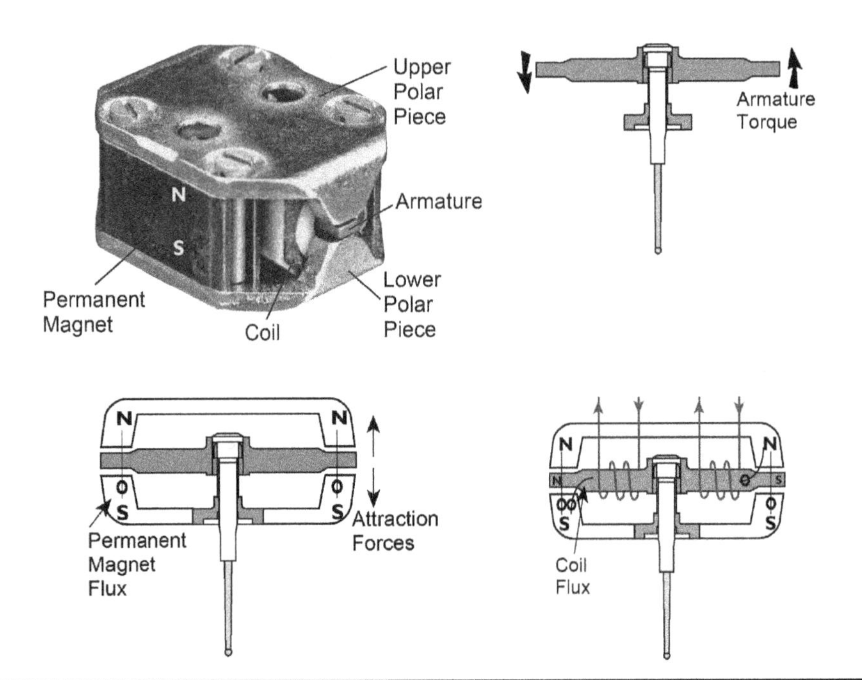

Figure 9.1 An electromagnetic torque motor with nonadjustable air gaps. (*Courtesy of Moog Inc.*)

The armature is manufactured from soft ferromagnetic material that reduces the effect of magnetic hysteresis. However, when the intensity of the applied coil current is reduced to zero, the armature does not become fully demagnetized due to its magnetic hysteresis. Therefore, a low-value torque exists. Figure 9.3a shows the torque-current relation of a typical torque motor. This figure shows a practically linear relation, for the low level input current, in addition to the effect of magnetic hysteresis.

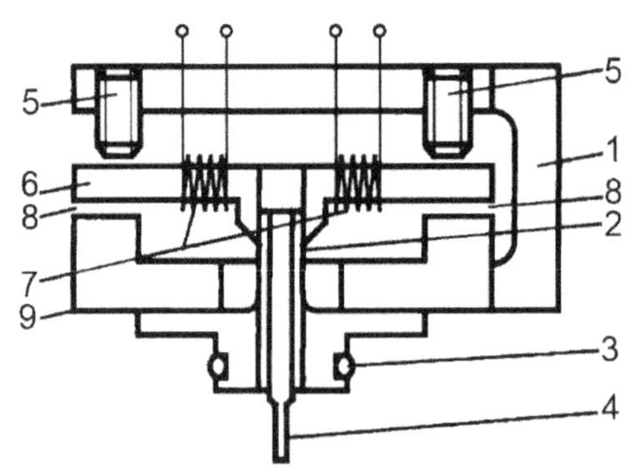

1. Permanent magnet, 2. Flexible tube, 3. Sealing ring,
4. Flapper plate, 5. Adjustable pole screws, 6. Armatures,
7. Control coils, 8. Air gap, 9. Polar piece

Figure 9.2 Electromagnetic torque motor with adjustable air gaps. (*Courtesy of Bosch Rexroth AG.*)

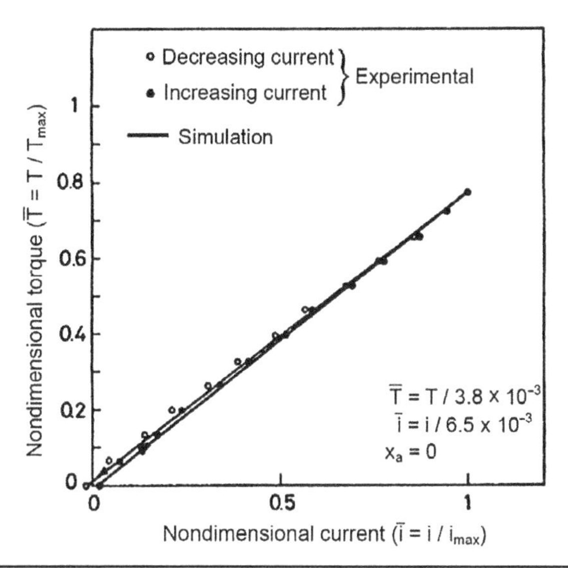

FIGURE 9.3(a) Torque-current relation of a typical electromagnetic torque motor, calculated.

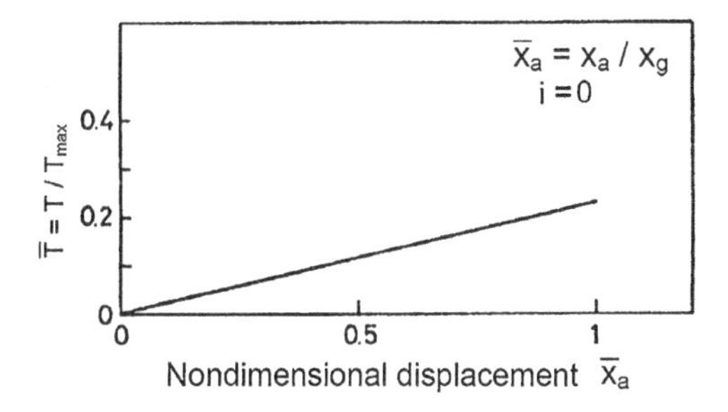

FIGURE 9.3(b) The torque-displacement relationship of a typical electromagnetic torque motor, calculated.

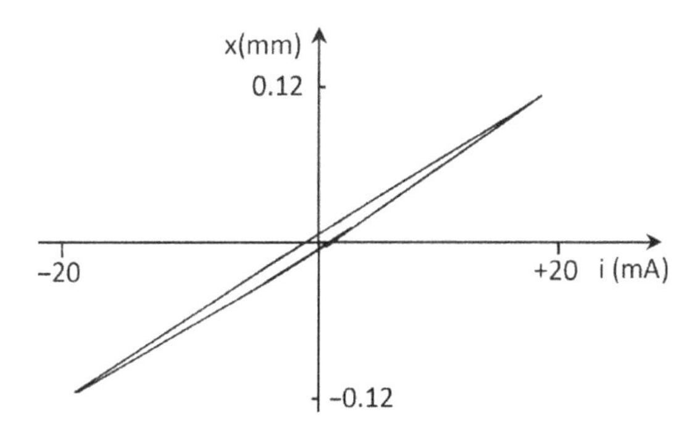

Max. electric power = 170 mW, Spring stiffness = 27 N/mm
Maximum displacement = 0.12 mm, Volume = 24 cm^3
Hysteresis = 5%, Air gap length = 0.4 mm, Natural frequency = 770 Hz

FIGURE 9.3(c) A characteristic curve and basic parameters of a typical electromagnetic torque motor (x is measured at armature extremity).

The power consumption of torque motors is within 20 to 200 mW. Exceptionally, the torque motors used for direct driving of spools are of much higher power; up to 5 W. Actually, the torque of the motor is slightly affected by the armature extremity displacement, x_a, even in the absence of any exciting current (see Fig. 9.3b). The torque is given by ($T = k_i i + k_x x$, where k_i and k_x are constants). The armature extremity displacement x_a is actually too small compared with the air gap thickness x_g. Therefore, the part ($k_x x$) is of negligible value relative to (k_i i). The resulting torque, together with the stiffness of the flexible tube, results in a displacement linearly proportional to the applied current, as shown in Fig. 9.3c.

The permanent magnets are placed outside of the electromagnetic circuit; therefore, they are not affected by the magnetic field of the electromagnet.

It is important to remember that the armature displacement should be limited to small values, compared with the length of air gaps; otherwise, the torque-current and displacement-current relations become nonlinear.

The electromagnetic torque motors have the following advantages:

- Friction-free construction
- Low effect of magnetic hysteresis
- A dry motor due to the perfect sealing of the flexible tube
- No magnetic field in the hydraulic medium
- A relatively high natural frequency and speed of response

9.4 Servovalves Incorporating Flapper Valve Amplifiers

9.4.1 Single-Stage Servovalves

The single-stage servovalve (see Fig. 9.4) controls the pressure difference between its two exit ports (A and B). It may be used as a singlestage pressure servovalve or as a pilot controller of a multistage servovalve. It consists of an electromagnetic torque motor, a hydraulic amplifier designed as a double-jet flapper valve, and an interchangeable filter element.

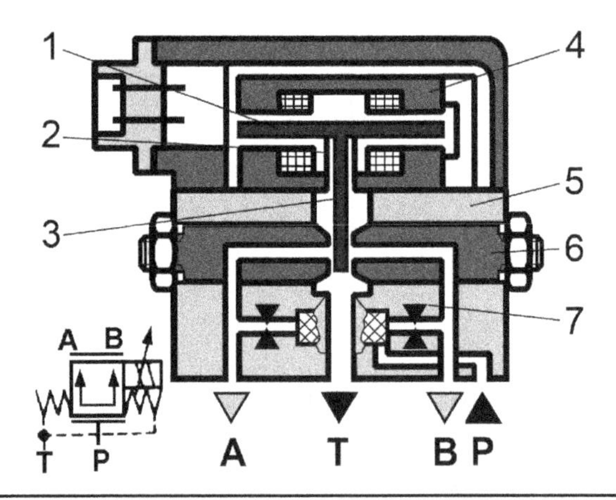

Figure 9.4 The construction of a single-stage servovalve. (*Courtesy of Bosch Rexroth AG.*)

The electromagnetic torque motor produces a torque proportional to the applied current. The armature (1) and the flapper (3) are held in the neutral position by the flexible tube (2). The neutral position of the armature may deviate slightly due to the effect of magnetic hysteresis of the armature material. By communicating the control current to the coils, the torque motor produces a torque proportional to the input current. This proportionality is practically linear, except for the observable hysteresis loops resulting from the effect of magnetic hysteresis on the armature material. The resulting torque rotates the armature and flapper by relatively small rotational angles, within 0.5 degree. This rotational angle is the actuating signal for the servovalve. It is a very low-power mechanical signal. Therefore, it is amplified by the hydraulic amplifier. Three basic types of hydraulic amplifiers are used in electrohydraulic servovalves: *flapper valve*, *jet pipe*, and *jet deflector* amplifiers.

Figure 9.4 shows a single-stage electrohydraulic valve incorporating a flapper valve hydraulic amplifier (also called a nozzle flapper amplifier). The high-pressure oil is supplied to the valve via the port (P) and the fine filter.

A double nozzle flapper valve is shown in Fig. 9.5a. This valve consists of two fixed orifices, N_1 and N_2, and two regulating flapper nozzles. The input control pressure P_s is

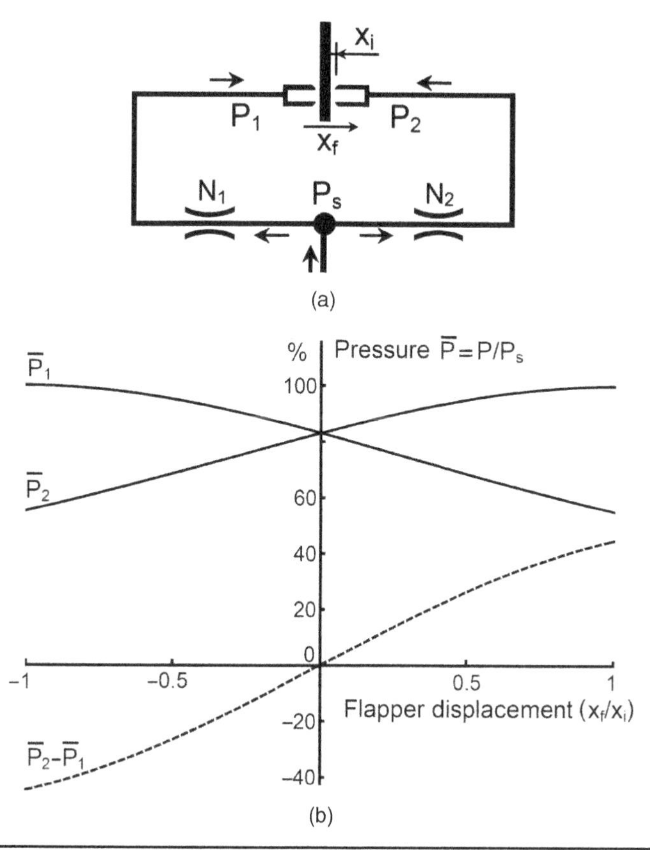

FIGURE 9.5 (a) The layout of the double jet flapper valve. (b) Pressure characteristics of a double jet flapper valve, calculated. (c) Pressure characteristics of two typical single-stage servovalves.

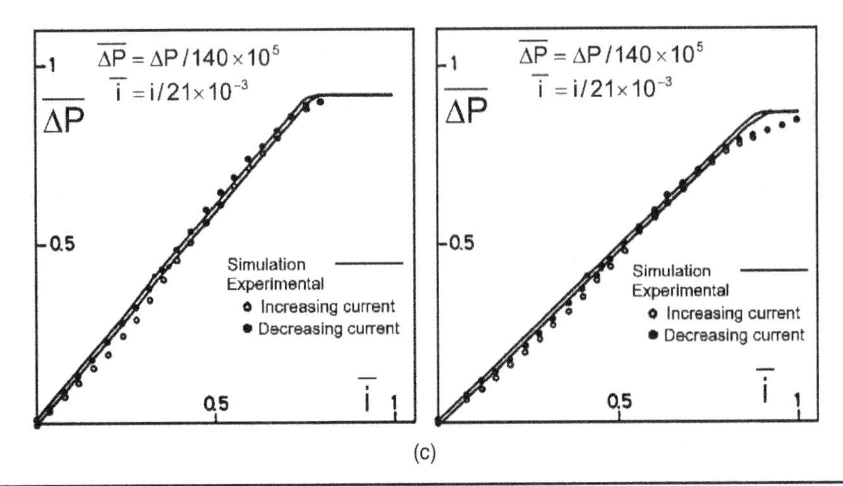

FIGURE 9.5 *(Continued)*

decreased via orifice N_1 and N_2 and the jet nozzles. If the cross-sectional areas of the nozzles of both sides are the same, then the same pressure drop occurs for both.

The displacement of the flapper plate changes the throttle area of the two regulating jet nozzles. The flapper motion to the right increases the area of the left nozzle and decreases the area of the right nozzle (see Fig. 9.5a). The pressure, P_1, decreases and P_2 increases. The pressure difference ($\Delta P = P_2 - P_1$) is proportional to the flapper displacement (see Fig. 9.5b). Figure 9.5c shows the variation of the valve pressures with the input current for two different single-stage valves. This figure shows practically linear behavior in most of the operating range. The effects of magnetic hysteresis and saturation are clearly indicated.

9.4.2 Two-Stage Electrohydraulic Servovalves

Valves with Mechanical Feedback

Figure 9.6 gives the construction of an electrohydraulic servovalve (EHSV) of two stages. The first stage of the servovalve includes a torque motor of an electromagnetic type and a double-nozzle flapper valve. The second stage consists of a spool valve driven hydraulically by the pressure difference developed by the flapper valve.

The operation of the two-stage EHSV with mechanical feedback (see Fig. 9.6) is explained by the block diagram of Fig. 9.7 and the functional schemes shown in Fig. 9.8. The feedback between the second and first stages of the valve is achieved by the feedback wire (8) attached to the flapper at one end and engaged in a groove in the spool (9) at its opposite end.

The displacement of the spool from the null position causes a torque on the flapper (feedback torque), which opposes the armature torque. When the spool displacement is such that the feedback torque equals the armature torque, the flapper returns, almost, to its neutral position and the spool movement ceases. Actually, the flapper is slightly displaced from its neutral position. The flapper valve produces a very small pressure difference, just sufficient to equilibrate the feedback spring force. This arrangement ensures the proportionality between the spool displacement and control current.

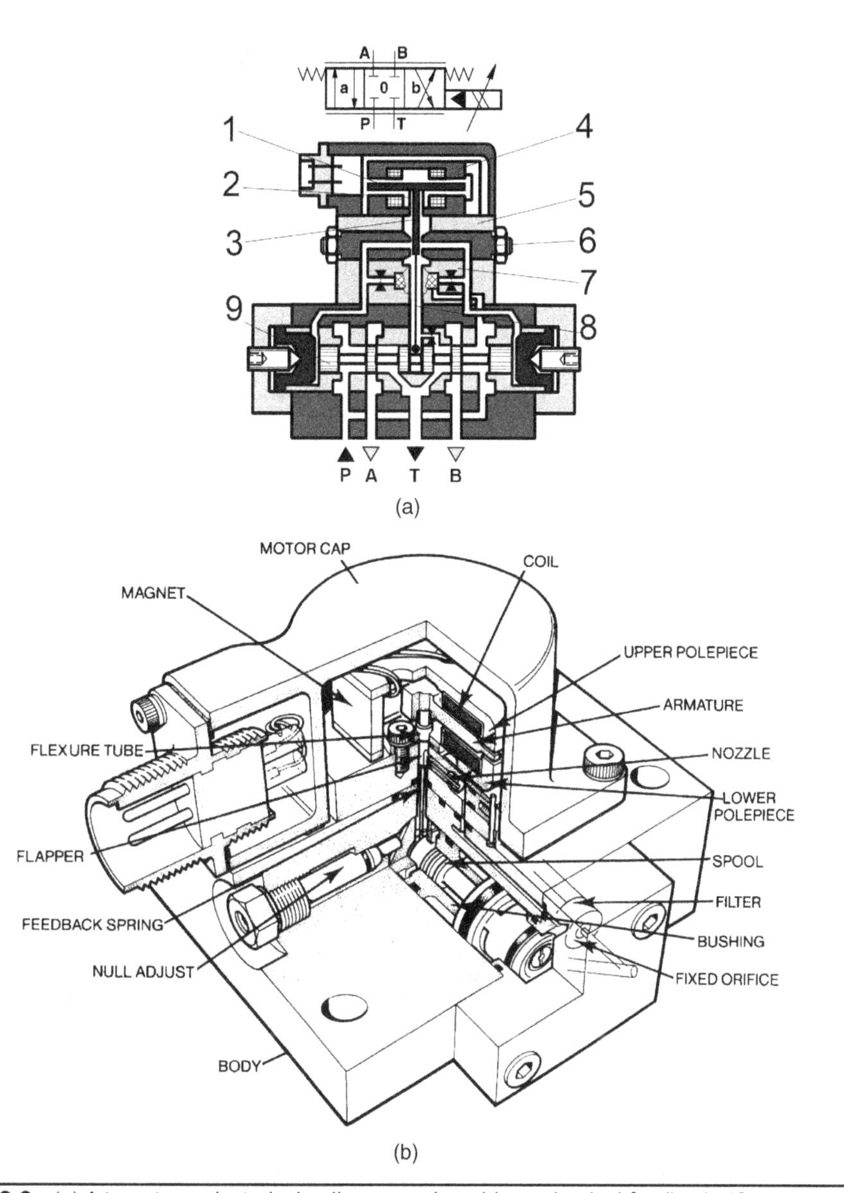

Figure 9.6 (a) A two-stage electrohydraulic servovalve with mechanical feedback. (*Courtesy of Bosch Rexroth AG.*) (b) Axonometric view of a two-stage electrohydraulic servovalve with mechanical feedback. (*Courtesy of Moog Inc.*)

Valves with Electrical Feedback

Figure 9.9 shows an electrohydraulic servovalve of two stages with electrical feedback on the second stage. The spool (9) is coupled to the core (11) of an inductive positional transducer (12). The core displaces within the coil system of the transducer, producing a voltage at the output of the measuring amplifier. Its value is proportional to the spool displacement. By comparing the feedback signal with the command signal value, any

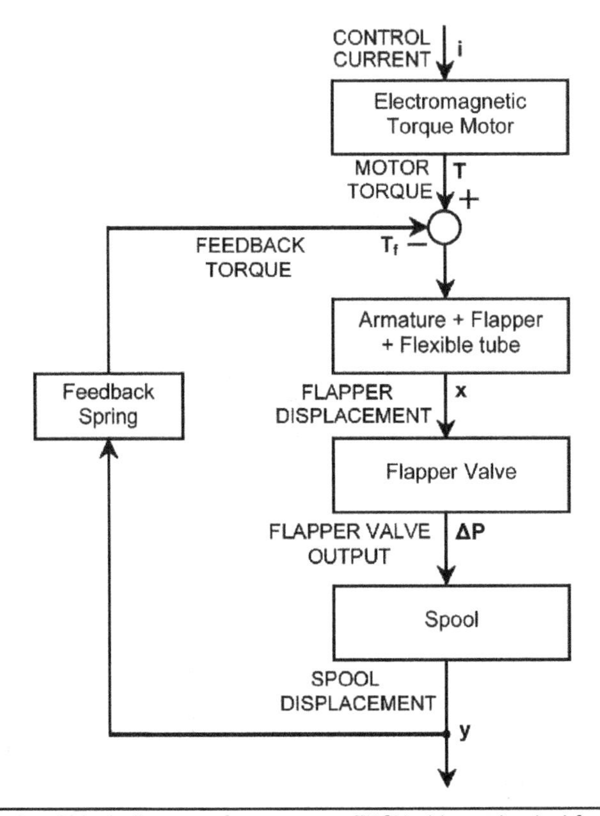

FIGURE 9.7 A functional block diagram of a two-stage EHSV with mechanical feedback.

deviation can be determined. The resulting deviation (error signal) is fed to the first stage via the electronic control system. This signal causes the flapper plate (3) to move between the jet nozzles (6). This in turn produces a proportional pressure difference between the spool side chambers (10). The spool (9) and its attached core (11) are then moved until the actual value agrees with the command value, and the control signal becomes once again almost zero.

For the control of oil flow rate, an opening is created by the displacement of the spool (9) relative to the sleeve (13). The opening area and the flow rate are proportional to the spool displacement and the command current.

Valves with Barometric Feedback

In valves with barometric feedback, the spool is spring centered (see Fig. 9.10). In the de-energized state, the spool (9) is pressure-balanced and is held in the neutral position by the springs (14). When the flapper plate (3) is offset by an electrical signal, a pressure difference is generated between the spool side-chambers (10), proportional to the input signal. The spool is then moved until the forces on it (due to the pressure difference in the side chambers and the control springs) become in equilibrium. The springs are of linear characteristics, for the actual small range of spool displacement. Therefore, the stroke of the control spool, and the flow rate through the servovalve, are proportional to the input current.

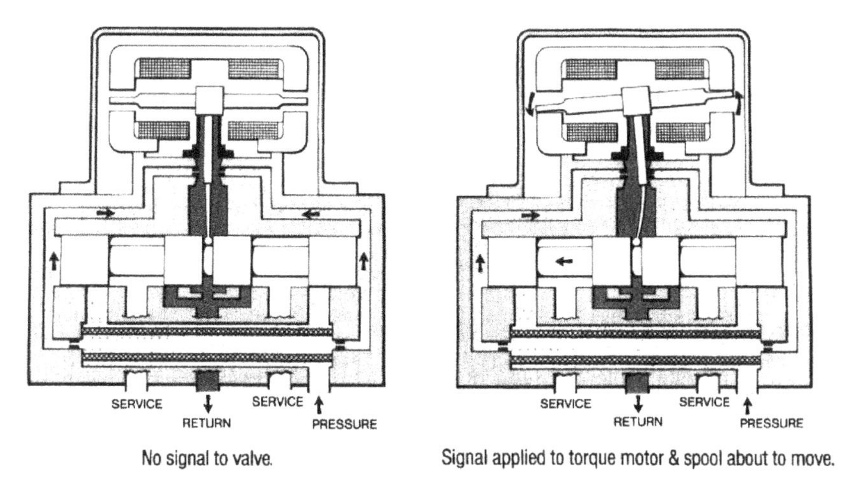

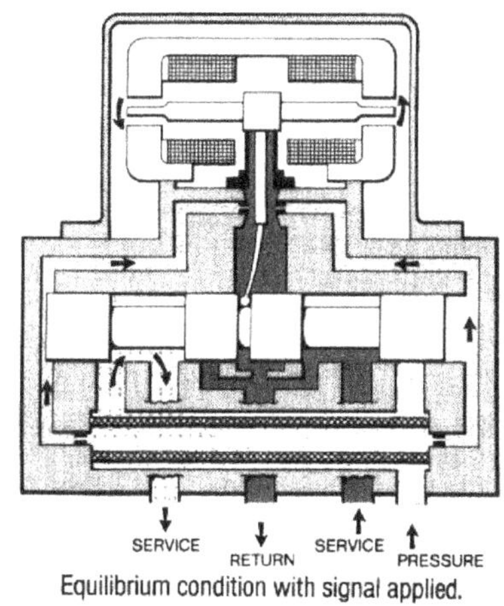

FIGURE 9.8 The operation of a two-stage servovalve with mechanical feedback. (*Courtesy of Moog Inc.*)

The Performance of Two-Stage Electrohydraulic Servovalves Valve Pressure Characteristics

Considering the servovalve in Fig. 9.9 and defining the load pressure to be $P_L = P_B - P_A$, then when blocking the two lines (A and B), the maximum load pressure is obtained. The variation of this pressure with the valve input current is given in Fig. 9.11. This figure shows clearly the effect of magnetic hysteresis on the armature and pressure saturation. The saturation is reached when any of the output pressures equals the supply pressure P_s.

PRESSURE GAIN (pressure amplification): The pressure gain, g_P, is the relation between the output load pressure ($P_L = P_B - P_A$) and input signal, $g_P = dP_L/di$, where i is the input current. The valve output pressure is influenced by the spool valve opening. The opening of the valve is, in turn, controlled by a closed-loop control circuit so that the pressure gain affects the closed-loop control accuracy and stability.

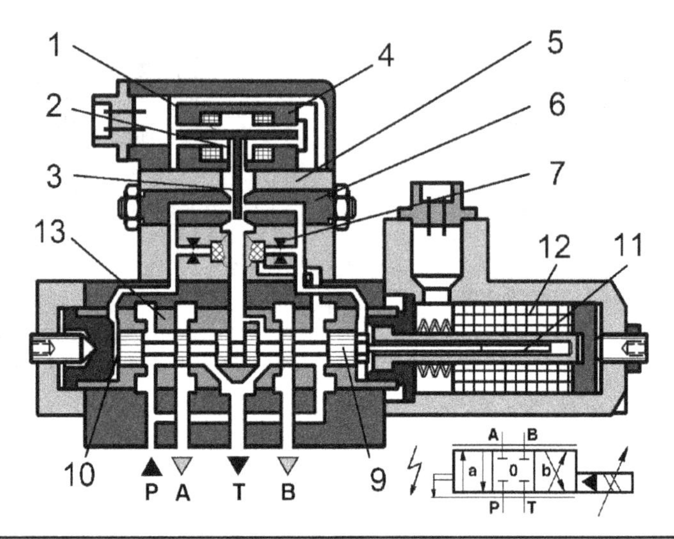

Figure 9.9 An electrohydraulic servovalve of two stages with electric feedback. (*Courtesy of Bosch Rexroth AG.*)

Valve Flow Characteristics

The electrohydraulic servovalves are usually used to control the motion of symmetrical cylinders and motors (see Fig. 9.12). The exit flow rate is affected by the load pressure, supply pressure, spool displacement, and the geometry of the spool valve opening. For a sharp-edged spool with rectangular, or annulus, throttling areas, the valve flow rate is given by: ($Q = Ky\sqrt{P_S - P_L}$) and $Q_{\max} = Ky\sqrt{P_s}$, where P_S and P_L are the supply and load

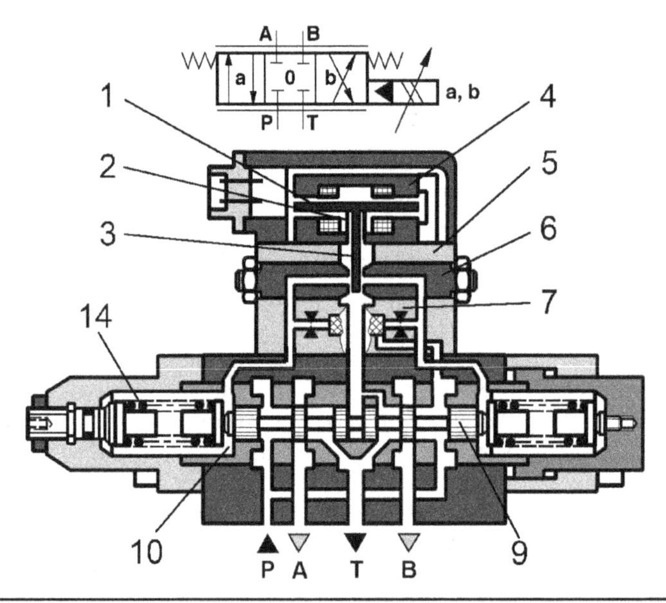

Figure 9.10 An electrohydraulic servovalve of two stages with barometric feedback. (*Courtesy of Bosch Rexroth AG.*)

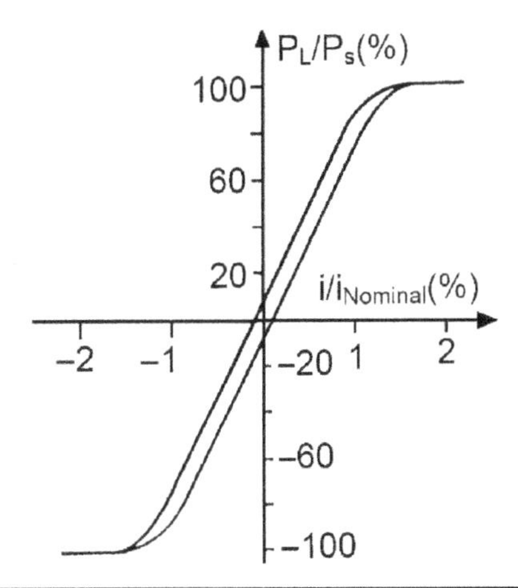

FIGURE 9.11 Typical pressure gain characteristics of the servovalve.

pressures respectively. K is a constant introducing the effect of valve geometry, discharge coefficient, and oil density. The spool displacement is proportional to the input current. The variation of the flow rate with the load pressure and input current, for a typical servovalve, is given in Fig. 9.13. This figure shows that the maximum valve flow rate corresponds to zero load pressure and maximum input current.

FLOW GAIN (flow rate amplification): The flow gain of the servovalve is defined as the variation of the valve flow rate with the input current, keeping the input and load pressures constant. This relation is usually evaluated by setting an input pressure of 70 bar and zero load pressure. The spool displaces inside a sleeve of rectangular windows, which are opened by the spool displacement, y. The width of these slots determines the valve flow amplification, flow gain g_F.

The flow rate is given by

$$Q = C_d \omega y(i)\sqrt{2\Delta P/\rho} \quad \text{and} \quad g_F = \frac{dQ}{di}$$

where y is the spool displacement and ω is the width of the throttling area.

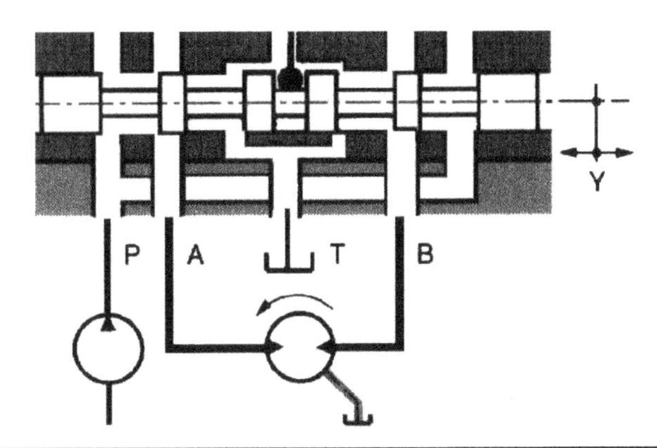

FIGURE 9.12 The typical connection of a hydraulic motor with the second stage of a servovalve.

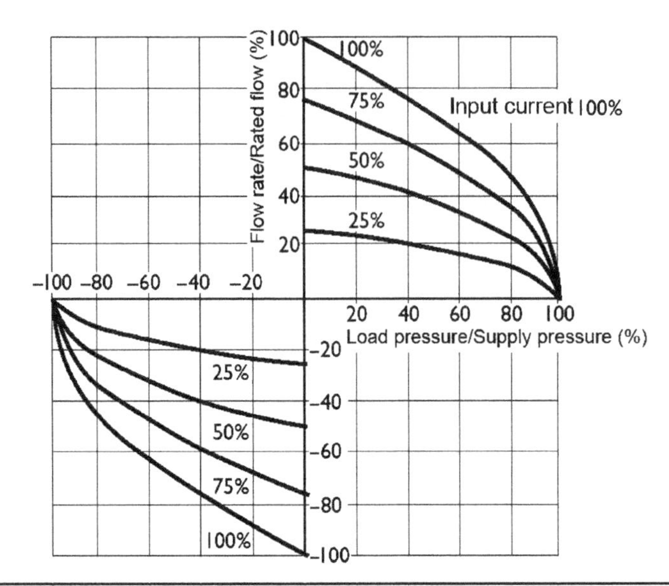

FIGURE 9.13 Servovalve flow characteristics, theoretical.

The valve flow gain is influenced by the spool displacement-current relation and spool valve geometry. Figure 9.14 shows typical servovalve flow characteristics. These characteristics are highly affected by the spool overlap. The general shape of the servovalve flow-current relation is shown by Fig. 9.15. This figure shows the effect of magnetic hysteresis and the slight under-lapping of the valve.

Valve Leakage Characteristics

The servovalves are characterized by the fine radial clearance of their spool valves, within 4 μm. In the case of valves with zero-lapping spools, the internal leakage is minimal, except when the spool is near its neutral position where the leakage increases due to the edge rounding (see Fig. 9.16).

Transient and Frequency Responses of Valves

The transient response is not sufficient to describe the dynamic characteristics. The most commonly used methods for examining the dynamic characteristics are based on the frequency response.

The frequency response is found by exciting the servovalve with sinusoidal signals and recording the valve output. For a well-designed servovalve, which presents linear characteristics, the output signal is also sinusoidal with modified amplitude and phase. Generally, the increase of frequency of the input signal decreases the magnitude and increases the phase shift. The magnitude ratio, also called the gain, is the ratio of the magnitude of the output sinusoidal signal to that of the input. The frequency response characteristics are described by the variation of the magnitude ratio and phase shift with the frequency of the input signal. A commonly used representation is the Bode diagram, where the magnitude ratio is expressed in decibels: Gain (dB) = $20 \log_{10}$ (magnitude ratio).

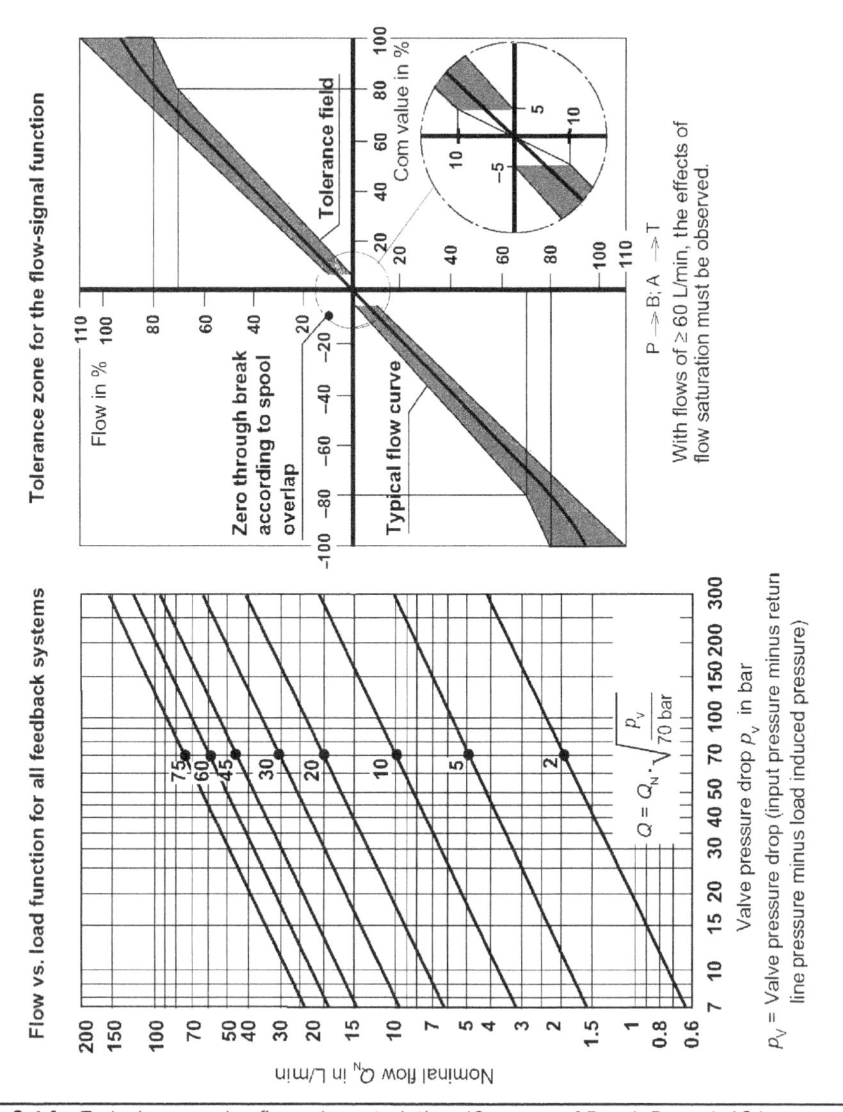

FIGURE 9.14 Typical servovalve fl ow characteristics. (*Courtesy of Bosch Rexroth AG.*)

To facilitate the quantitative description of the frequency response, two characteristic values for frequency have been defined as f_{-3dB} and $f_{-90°}$.

f_{-3dB} is the frequency at which the gain is equal to −3 dB. This point corresponds to a magnitude ratio of 0.707.

$f_{-90°}$ is a point on the phase-frequency characteristic curve at which the output signal lags behind the input signal by 90°.

The electrohydraulic servovalves have, in general, a high natural frequency and quick response. Figure 9.17 shows the step response and frequency response of an electrohydraulic servovalve of two stages with mechanical feedback. The valve presents a transient response of settling time less than 12 ms.

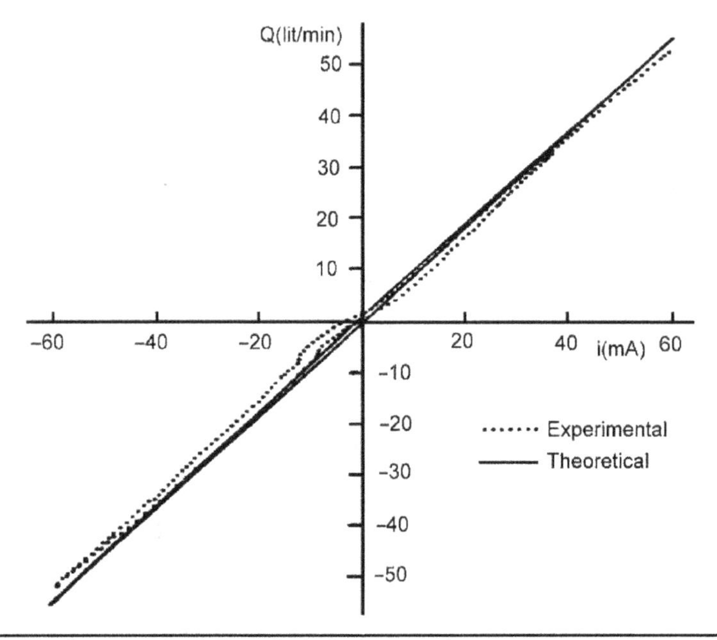

Figure 9.15 The flow-current relation of a typical servovalve.

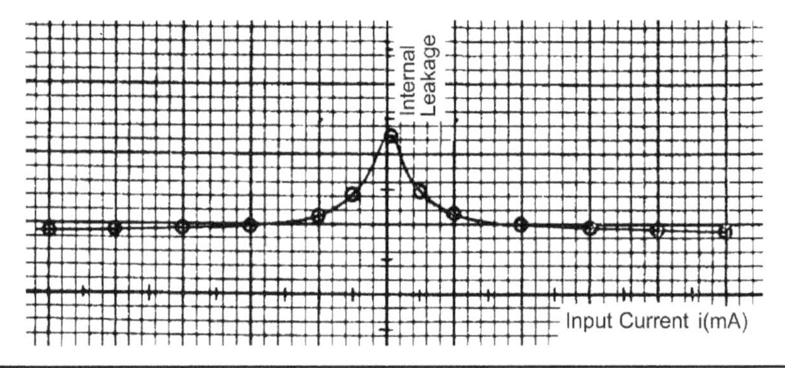

Figure 9.16 Typical leakage characteristics of a servovalve. (*Courtesy of Moog Inc.*)

Figure 9.18 shows the frequency response plots of commercial servovalves of two stages with different feedback types. The study of this figure shows that:

- The frequency response of the servovalves is considerably influenced by the system pressure and the magnitude of input signal.
- The valve with electric feedback has the best dynamic behavior, followed by that with mechanical feedback.
- Compared with the valves of mechanical and electric feedback, the valve with barometric feedback is of poor dynamic behavior. This is mainly due to the considerable mass of the spool and control springs.

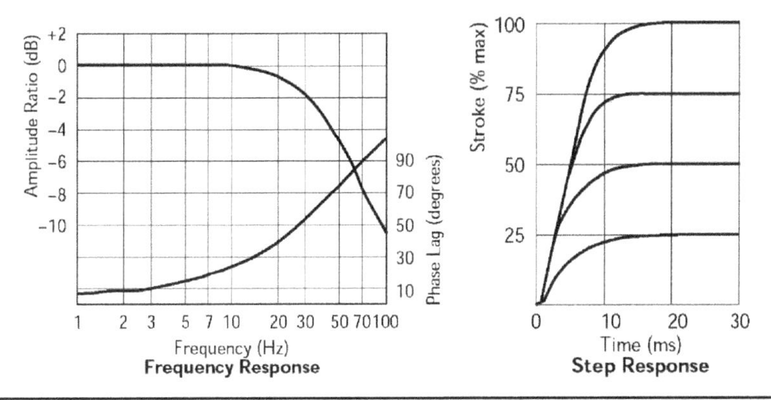

Figure 9.17 Frequency and step responses of a typical two-stage electrohydraulic servovalve.

9.5 Servovalves Incorporating Jet Pipe Amplifiers

Another well-known hydraulic amplifier is the jet valve shown in Figs. 9.19 and 9.21. The jet valve has relatively large clearances, compared with the flapper valve. Therefore, it is less sensitive to the effects of contamination.

The Jet Pipe is rotated by the torque motor. The high-pressure fluid flows out of the Jet Pipe and impinges on a receiver. Two small diameter holes (see Fig. 9.20) located side by side on the receiver are connected to either end of the spool. With the Jet Pipe centered over the two holes, equal pressures are developed on each side of the spool. When the torque motor causes the Jet Pipe to rotate over the center, the jet impinges more on one hole and less on the other. This creates a pressure difference across the spool. The exact shape of parts cannot be computed in the way that flapper valves are analyzed; however, the following guides hold practical:

- The clearance between the jet and receiver should be at least twice its diameter.
- The two receiver holes should be as close to each other as possible.
- If the previous constraints are satisfied and if the internal diameter of the nozzle is constant over a length equal the value of diameters up the nozzle, then, when the nozzle is centered, the load pressures are of the same value.

The two-stage electrohydraulic servovalve converts the input electrical signal into a precisely proportional spool displacement. It consists of two stages (see Fig. 9.21):

- The first stage is the pilot valve, which includes the torque motor, the Jet Pipe, and the two receivers.
- The second stage includes the spool and sleeve assembly.

The high-pressure hydraulic fluid is fed through a filter to the Jet Pipe that directs a fine stream of fluid to the two receivers. Each receiver is connected to one side of the spool of the second stage. At the null position, where no signal is connected to the torque motor, the jet is directed exactly between the two receivers, creating equal pressures on both sides of the spool. The force balance created by the equal pressures in both end chambers holds the spool in a midposition (see Fig. 9.19).

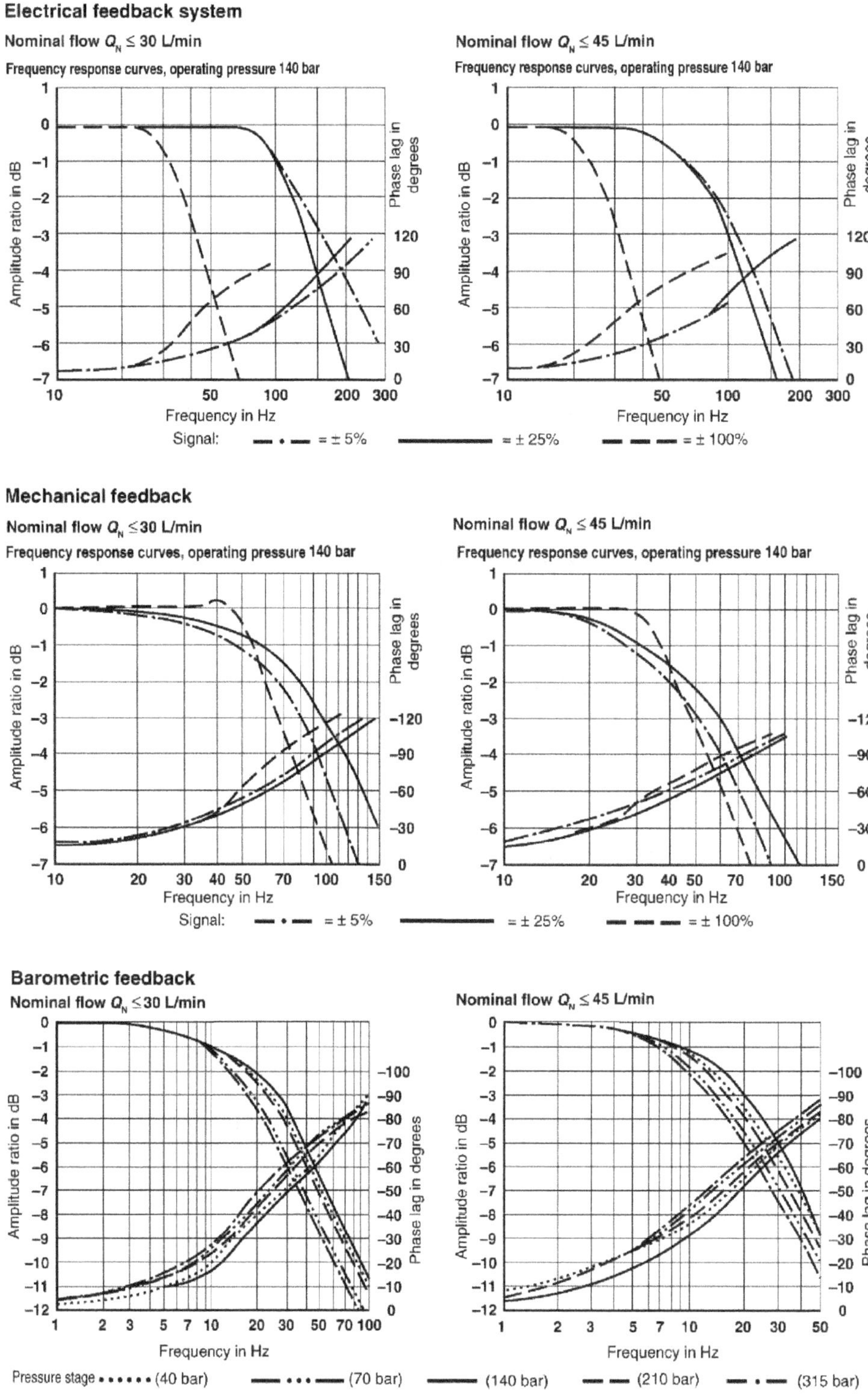

Electrical feedback system

Nominal flow $Q_N \leq 30$ L/min

Frequency response curves, operating pressure 140 bar

Nominal flow $Q_N \leq 45$ L/min

Frequency response curves, operating pressure 140 bar

Amplitude ratio in dB

Phase lag in degrees

Frequency in Hz

Signal: = ± 5% = ± 25% = ± 100%

Mechanical feedback

Nominal flow $Q_N \leq 30$ L/min

Frequency response curves, operating pressure 140 bar

Nominal flow $Q_N \leq 45$ L/min

Frequency response curves, operating pressure 140 bar

Amplitude ratio in dB

Phase lag in degrees

Frequency in Hz

Signal: = ± 5% = ± 25% = ± 100%

Barometric feedback

Nominal flow $Q_N \leq 30$ L/min

Nominal flow $Q_N \leq 45$ L/min

Amplitude ratio in dB

Phase lag in degrees

Frequency in Hz

Pressure stage (40 bar) (70 bar) (140 bar) (210 bar) (315 bar)

Figure 9.18 Frequency response of two-stage electrohydraulic servovalves with different types of feedback. (*Courtesy of Bosch Rexroth AG.*)

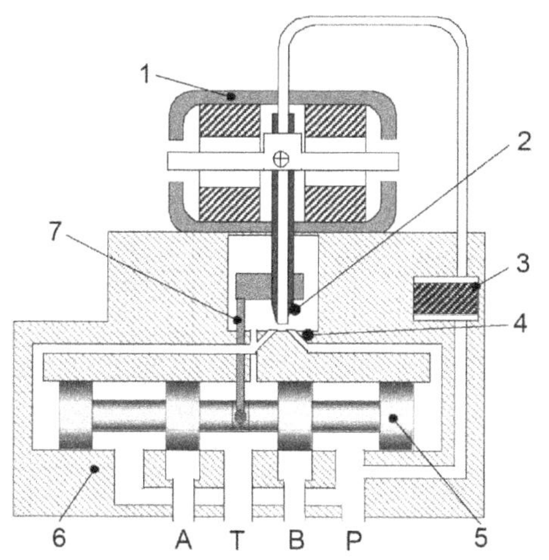

FIGURE 9.19 An electrohydraulic servovalve incorporating a Jet Pipe amplifier. (*Courtesy of Moog Inc.*)

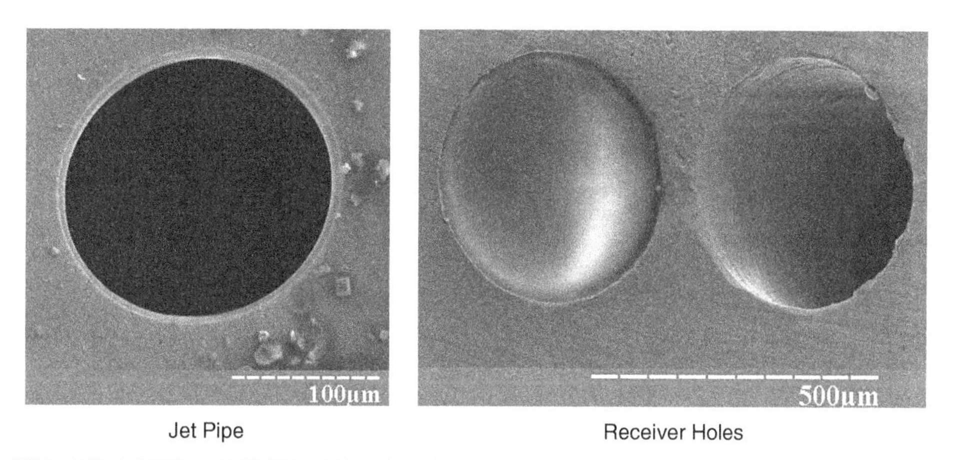

FIGURE 9.20 Photos of the jet nozzle and receivers of a typical servovalve (*photographed by a scanning electron microscope*).

As the Jet Pipe and armature of the torque motor rotate around the pivot point, the fluid jet is directed toward one of the two receivers, creating a higher pressure in the spool end chamber connected to that receiver. The created differential pressure moves the spool in the direction opposite to the jet displacement (see Fig. 9.21a).

In the case of a servovalve with mechanical feedback, a feedback spring is connected to the spool and Jet Pipe. The feedback spring translates the spool position into a force that is applied on the Jet Pipe in a proportional manner. The increased spool displacement, away from the null position, increases the force exerted on the Jet Pipe. The forces transmitted from the spool to the Jet Pipe create a feedback torque. When the feedback spring torque equals that of the torque motor, the jet is returned, almost, to its

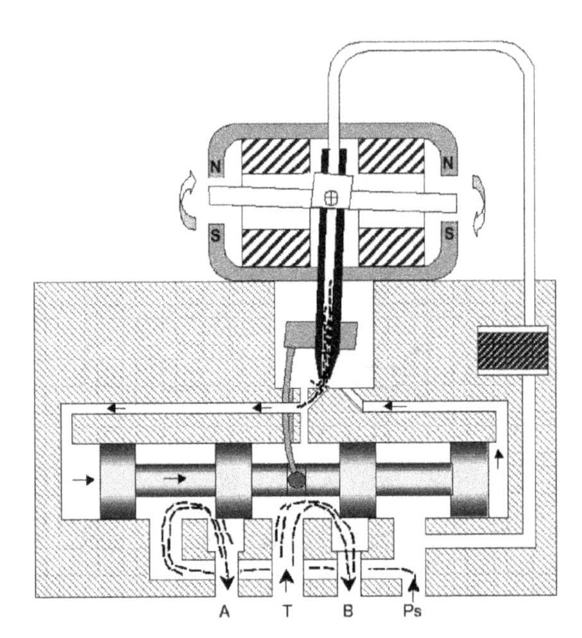

(a) Valve with input current

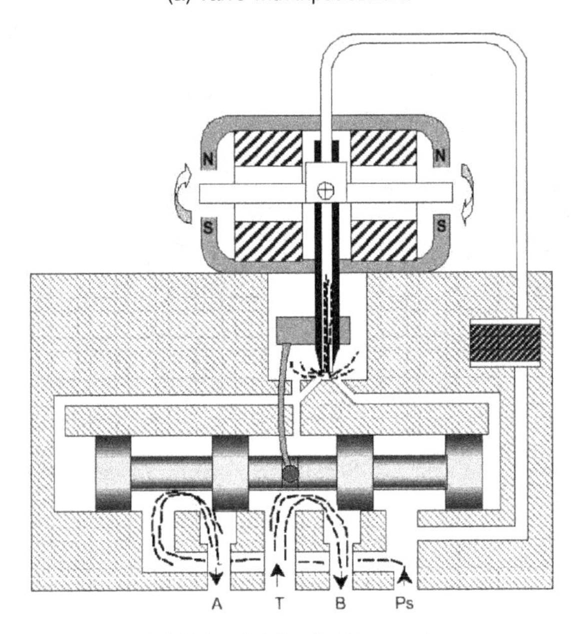

(b) Valve stabilized with current

FIGURE 9.21 Operation of a two-stage servovalve incorporating a Jet Pipe amplifier. (*Courtesy of Moog Inc.*)

null position between the two receivers. This position creates a pressure balance between the end chambers (see Fig. 9.21b).

The torque of the electromagnetic torque motor is proportional to the input current, and the feedback torque is proportional to the spool displacement. Then, the resulting spool displacement is proportional to the input current.

By reversing the polarity of the applied current, the armature torque is reversed. The hydraulic jet flow impinges on the other receiver, creating an imbalance in spool end chambers' pressures. The spool moves in the opposite direction until a first-stage force balance is achieved by the feedback spring. The jet flow is then directed between the receivers, and equal pressures hold the spool in the new position.

9.6 Servovalves Incorporating Jet Deflector Amplifiers

Principally, the jet deflector amplifier operates in a way similar to the Jet Pipe amplifier. The valve is equipped with a fixed jet nozzle and receiver holes. The jet of fluid is deflected toward one of the two receivers by means of a jet deflector. The jet deflector is displaced by the torque motor (see Figs. 9.22 and 9.23).

The function of the jet deflector is summarized in the following:

- The armature and deflector are rigidly joined and supported by a thin-wall flexure tube.
- The fluid flows continuously from the fixed inlet, through the moveable deflector into the two receivers (A and B). When the deflector is centered, it produces equal pressures in each receiver.
- The rocking motion of the armature-deflector assembly directs the flow to one of the two receiver holes. A pressure difference builds across the spool, which causes the spool motion.
- The electric current fed to the torque motor creates proportional magnetic forces on the ends of the armature.
- The armature and deflector assembly rotate about flexure tube support.

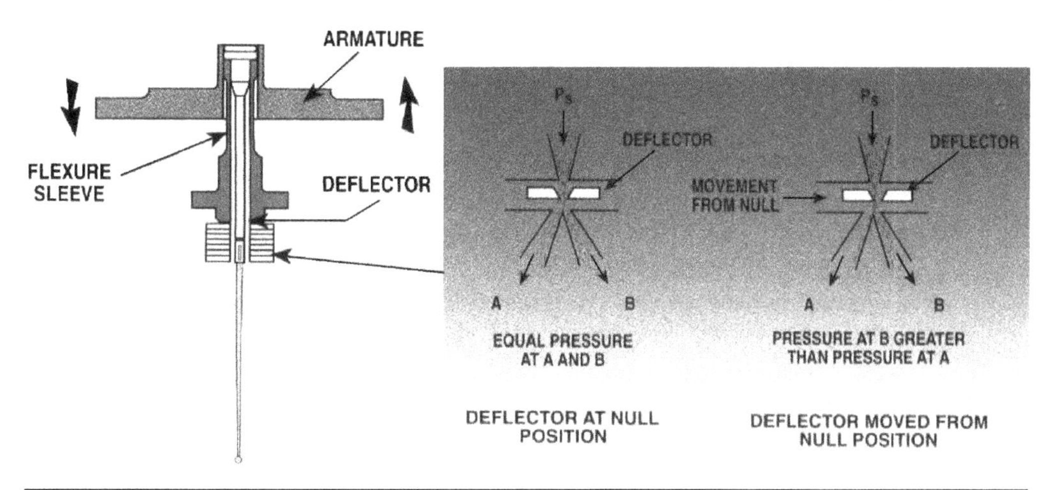

FIGURE 9.22 Operation of the jet deflector. (*Courtesy of Moog Inc.*)

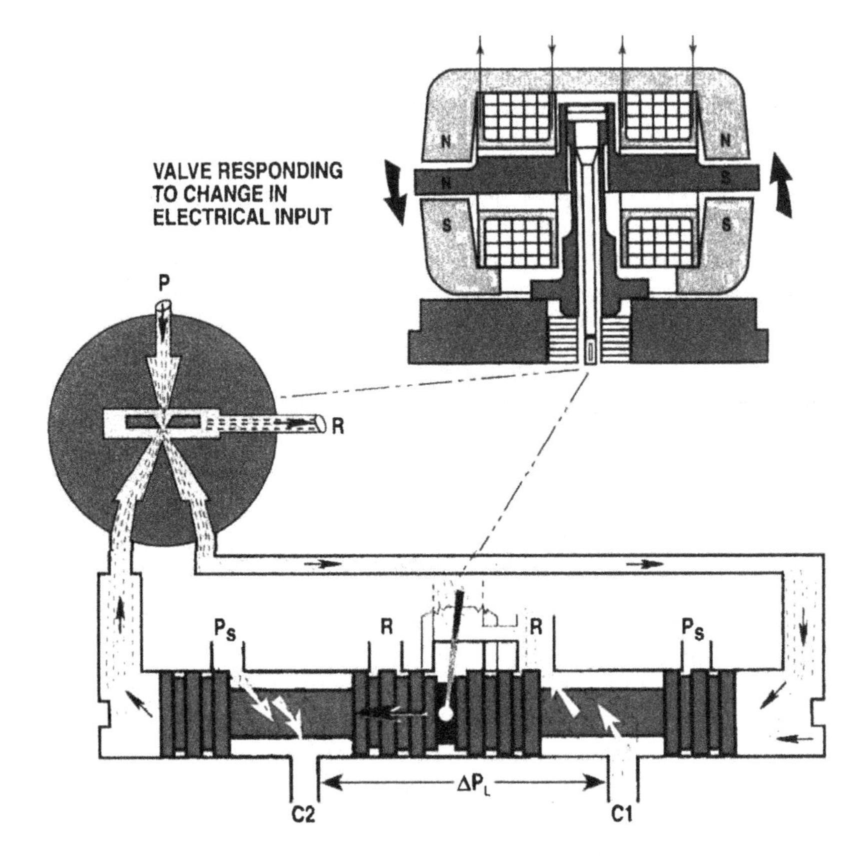

(a) Valve responding to change in control current.

FIGURE 9.23 The operation of a servovalve with a jet deflector. (*Courtesy of Moog Inc.*)

- The jet deflector diverts the flow through the receiver to the spool side-chambers.
- The spool moves and connects the pressure and return lines with the ports (C2) and (C1).
- The spool pushes the ball end of the feedback spring creating a restoring feedback torque on the armature/deflector.
- As the feedback torque becomes equal to the torque-motor torque, the armature–deflector assembly moves back to its centered position.
- The spool stops at a position where the feedback spring torque equals the torque-motor torque. The spool displacement is proportional to the input current.
- Keeping the pressures constant, the flow rate directed to the load is proportional to the spool displacement.

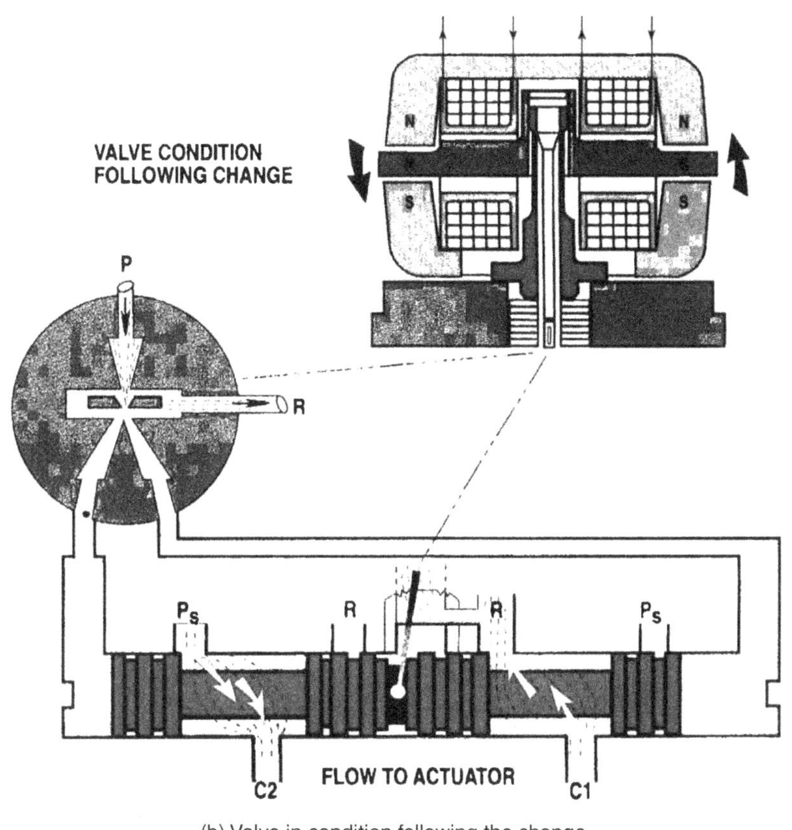

(b) Valve in condition following the change.

FIGURE 9.23 *(Continued)*

9.7 Jet Pipe Amplifiers Versus Nozzle Flapper Amplifiers

In many servovalve applications, the use of a Jet Pipe valve (JP) is preferred over the common nozzle flapper valves (NF). Unfortunately, the fluid flow characteristics in a JP servovalve are extremely complex. Therefore, these valves are usually configured experimentally and a suitable configuration is reached only after many trials. In some specific applications, a large motion of the jet in the JP amplifiers is needed to establish maximum power, which is considered undesirable. The selection of the required amplifier type for an application will depend mainly upon the application requirements.

JP amplifiers offer several advantages over NF amplifiers. The main differences are summarized as follows:

- The minimum flow area in a standard JP amplifier nozzle orifice area is of an order larger than that of the NF (the gap between the nozzle and flap) of the same power. Consequently, the JP amplifier is more reliable in operation since it has fewer tendencies to blocking and requires a lower fluid filtration level. This is important to redu ce the maintenance cost. Also, the design of its parts doesn't require strict tolerances, which reduces manufacturing costs.

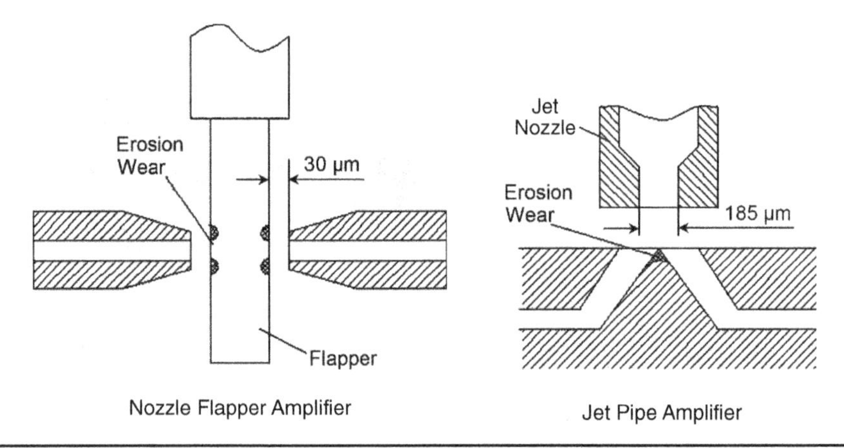

FIGURE 9.24 Long-term erosion patterns of typical flapper and jet deflector valves.

- If any of the elements of the hydraulic bridge are blocked in NF amplifiers, an active failure takes place, as a result of which a maximum signal occurs at the output in the absence of an input signal. In JP amplifiers, if there is blocking of the nozzle, a passive failure takes place, with the output signal being close to zero. Consequently, the operation of the JP amplifier is safer than the NF amplifier in the case of blocking.
- Due to long-term operation of the NF amplifier, the erosion of the flapper blade occurs, as shown in Fig. 9.24. The erosion of the flapper blade by 30 microns (the approximate width of the variable orifice), reduces the pressure gain by about 50 percent. Moreover, the leakage will be doubled and these will double the power losses in the amplifier. On the other hand, the JP erosion occurs symmetrically between the two receivers, as shown in Fig. 9.24. The JP amplifier erosion actually improves the amplifier linearity without noticeable effect on the pressure gain. Also the projector jet does not increase in size. Therefore, the amplifier leakage remains unchangeable.

9.8 Exercises

1. Discuss briefly the construction and performance of the electromagnetic torque motors. (See Figs. 9.1, 9.2, and 9.3.)

2. Explain the function of the single-stage electrohydraulic servovalve. (See Fig. 9.4.)

3. Discuss briefly the construction and performance of the flapper valve hydraulic amplifier. (See Fig. 9.5.)

4. Explain the function of a two-stage electrohydraulic servovalve with a flapper valve amplifier and mechanical feedback. (See Figs. 9.6, 9.7, and 9.8.)

5. Explain the function of the two-stage electrohydraulic servovalve with a flapper valve amplifier and electrical feedback. (See Fig. 9.9.)

6. Explain the function of the two-stage electrohydraulic servovalve with a flapper valve amplifier and barometric feedback. (See Fig. 9.10.)

7. Discuss the static and dynamic behavior of the electrohydraulic servovalves, incorporating different types of feedback systems. (See Figs. 9.11 to 9.18.)

8. Explain the construction and operation of the electrohydraulic servovalve, incorporating a Jet Pipe amplifier. (See Figs. 9.19, 9.20, and 9.21.)

9. Explain the construction and operation of the electrohydraulic servovalve, incorporating a jet defl ector amplifi er. (See Figs. 9.22 and 9.23.)

CHAPTER 10

Modeling and Simulation of Electrohydraulic Servosystems

10.1 Introduction

This chapter deals with the analysis of the static and dynamic performance of electrohydraulic servo actuators (EHSAs). The equations describing the behavior of the basic elements of EHSAs are deduced and the steady-state performance of these elements is discussed—mainly, the electromagnetic torque motor and the flapper valve. A mathematical model describing the dynamic behavior of the whole electrohydraulic servo actuator is deduced.

10.2 Electromagnetic Torque Motors

10.2.1 Introducing Magnetic Circuits

Figure 10.1 shows a coil with a soft ferromagnetic core of length L and cross-section area A, a toroidal coil. Whenever an electric current flows through the coil, it induces a magneto-motive force, λ, inside the iron core. This force sets up a magnetic flux φ. The magnetic flux density B is defined as the flux per unit area.

$$\lambda = Ni \tag{10.1}$$

$$B = \varphi / A \tag{10.2}$$

where λ = Magneto-motive force, A
φ = Magnetic flux, Vs (1 Vs = 1 weber)
A = Cross-sectional area of the core, m^2
N = Number of turns of the coil
B = Magnetic flux intensity, Vs/m^2
i = Electric current, A

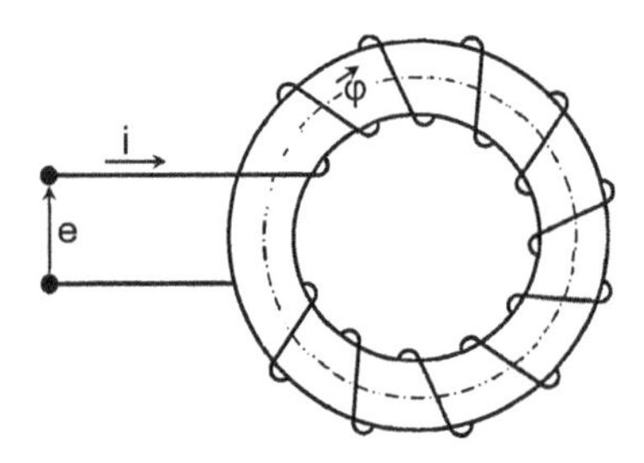

Figure 10.1 A toroidal coil.

On the other hand, a variable magnetic flux, $\overset{o}{\varphi}$, creates an induced electromotive force in a coil, according to Faraday's law.

$$e = -N\overset{o}{\varphi} \tag{10.3}$$

where e = Electromotive force, V.

The magnetic material is characterized by the relation between the magnetic flux intensity B and the magnetizing force H, which is the magneto-motive force per unit length ($H = \lambda/L$). A soft ferromagnetic material is one that is easily magnetized and demagnetized. This is a material for which H and B are related by a single valued curve so ($B = 0$) when ($H = 0$). Such materials exhibit a saturation effect as shown in Fig. 10.2a. As H increases, the increase of B slows down. There exist, practically, a limiting value of flux intensity that the core could attain when H is very large. The hard ferromagnetic materials show a hysteresis loop when H is cycled (see Fig. 10.2b). The initial slope of $B(H)$ relation is a characterizing parameter of the material. It is the permeability, μ. The free space has a permeability μ_0. Sometimes, the magnetic permeability is indicated by the relative permeability μ_r and the slope of the slope B-H is given by ($\mu = \mu_o\mu_r$).

$$\mu = B/H \tag{10.4}$$

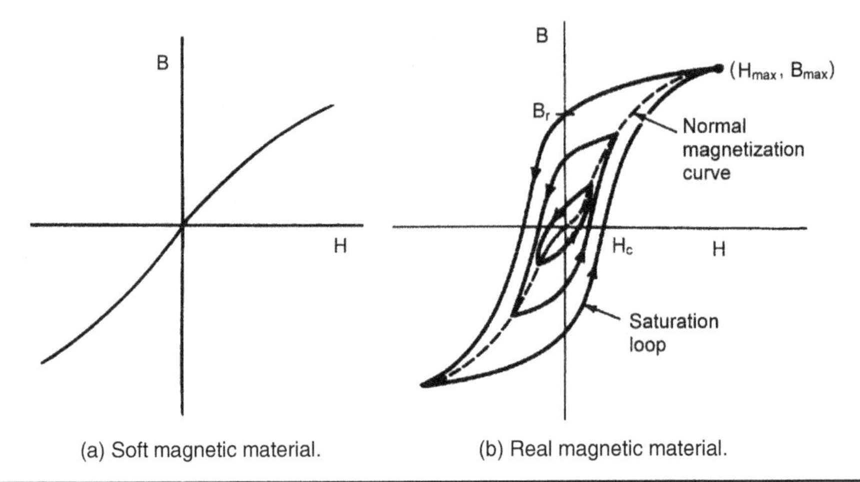

Figure 10.2 Relation between the magnetic flux intensity and the magneto-motive force per unit length.

The coil reluctance is defined as the resistance to magnetic flow. It is the magneto-motive force needed to set up a unit magnetic flux in the medium. For the entire coil, the reluctance R is found from the $\varphi(\lambda)$ relation, where

$$R = \frac{\lambda}{\varphi} = \frac{HL}{BA} \quad \text{or} \quad R = \frac{L}{\mu A} \tag{10.5}$$

The reluctance, R, is sometimes considered as analogous to the electric resistance, which implies that the flux, φ, is analogous to the electric current and the magneto-motive force, λ, is analogous to the electromotive force. This is just mathematical analogy. It can be used for the calculation of the magnetic circuits in steady-state conditions. However, it is not correct from the energetic point of view. An electric resistor dissipates energy, while the magetic reluctance stores energy. The reluctance is described by the following relation:

$$\lambda = R\varphi \tag{10.6}$$

or

$$\lambda = \frac{1}{1/R}\int^{o} \varphi\, dt \tag{10.7}$$

This expression is analogous to the following relation, which describes electric capacitance:

$$e = \frac{1}{C}\int i\, dt \tag{10.8}$$

Actually, the magnetic reluctance is analogous to the electric capacitance. Both of them store energy and are described by the same mathematical relations. The reciprocal of reluctance of a magnetic material is the permeance, P, where

$$P = \frac{\varphi}{\lambda} = \frac{1}{R} = \frac{\mu A}{L} \tag{10.9}$$

The attraction force between two magnetic poles (see Fig. 10.3) separated by an air gap of cross-sectional area, A, is given by the following expression:

$$F = \frac{\varphi^2}{2\mu_o A} \tag{10.10}$$

where F = Force, N
H = Magneto-motive force per unit length, A/m
L = Length, m
P = Magnetic permeance, Vs/A
R = Magnetic reluctance, A/Vs

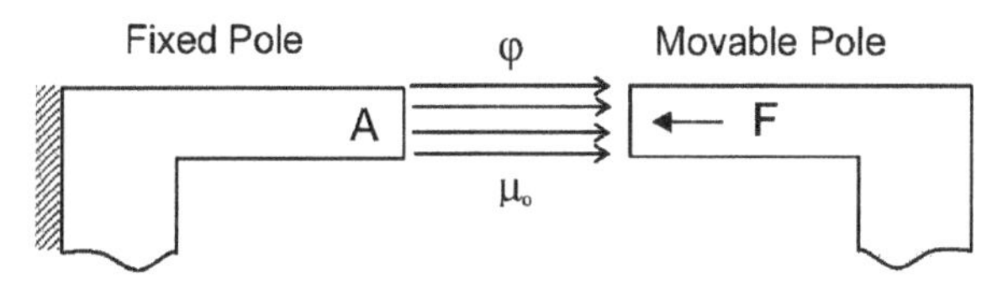

FIGURE 10.3 Illustration of a typical magneto-mechanical transducer.

λ = Magneto-motive force, A
μ = Permeability, Vs/Am
μ_o = Permeability of free space, Vs/Am
μ_r = Relative permeability

10.2.2 Magnetic Circuit of an Electromagnetic Torque Motor

Electromagnetic torque motors with permanent magnets are widely used in electrohydraulic devices due to their excellent dynamic characteristics. The torque motor (see Fig. 10.4) consists of an armature mounted on a flexible tube and suspended in the air gaps of a magnetic field. The two polar pieces form the framework around the armature and provide paths for the magnetic flux. The permanent magnet produces a permanent magnetic field in the four air gaps separating the armature extremities and the polar pieces. When the current is made to flow through the armature coils, the magnetic field is reinforced in two diagonally opposite air gaps and weakened in the other two. A torque is thus produced on the armature causing its angular displacement. The flexible tube, acting as a torsion spring, causes the rotational angle to be proportional to the torque. The permanent magnets are placed outside of the electromagnet's field and are not affected by it.

10.2.3 Analysis of Torque Motors

The four air gaps constitute the dominant reluctance in the magnetic circuit of the torque motor (see Fig. 10.5). The reluctance of the magnetic materials of polar pieces is negligible relative to that of the air gaps. The magnetic flux in the air gap may be found by evaluating separately the effects of the permanent magnet and the electric current.

The diagonally opposite air gaps are of equal length due to the symmetry of the torque motor; their reluctances are given by

$$R_1 = \frac{x_o - x_a}{\mu_o A} \tag{10.11}$$

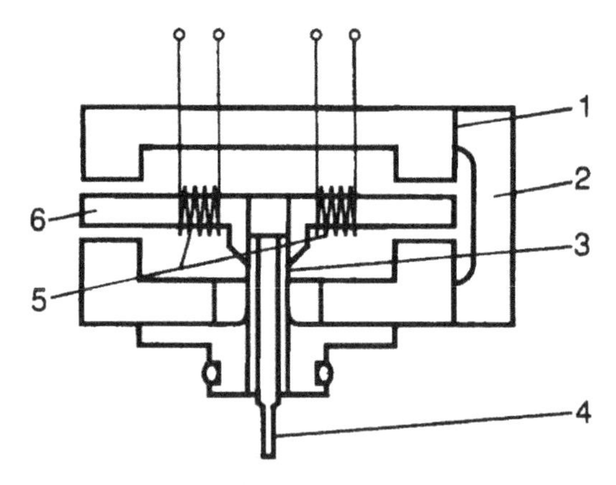

1. Polar pieces, 2. Permanent magnet, 3. Flexible tube, 4. Flapper, 5. Control coils, 6. Armature

FIGURE 10.4 Makeup of an electro-magnetic torque motor.

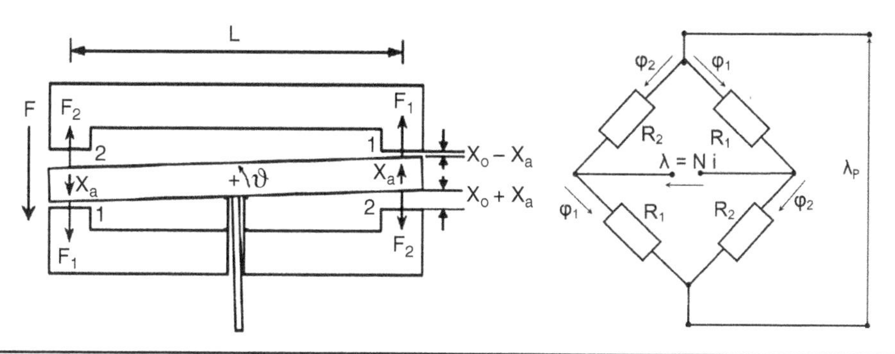

FIGURE 10.5 Magnetic circuit of an electromagnetic torque motor.

$$R_2 = \frac{x_o + x_a}{\mu_o A} \tag{10.12}$$

$$x_a = \vartheta L/2 \tag{10.13}$$

where x_o = Length of the air gap in the neutral position of the armature, m
x_a = Displacement of the armature end, m
A = Area of air gap, m²
L = Armature length, m
λ_p = Magneto-motive force of the permanent magnet, A
ϑ = Armature rotation angle, rad

The magnetic circuit of the electromagnetic torque motor is shown in Fig. 10.5. This circuit is symmetrical; the magnetic fluxes in the diagonally opposite arms are equal. There is a mathematical analogy between the electric circuit and the magnetic circuit. The magnetomotive force around each loop must be zero. This analogy compares the electric resistance and magnetic reluctance. It allows treating the magnetic circuit in the same way as the electric circuit. By equalizing the magneto-motive force across each loop to zero (neglecting the magnetic hysteresis of the armature), the expressions for the magnetic flux, φ_1 and φ_2, are deduced.

$$-iN + R_1\varphi_1 - R_2\varphi_2 = 0 \tag{10.14}$$

$$-\lambda_p + R_1\varphi_1 + R_2\varphi_2 = 0 \tag{10.15}$$

or

$$\varphi_1 = \frac{\lambda_p + iN}{2R_1} \quad \text{and} \quad \varphi_2 = \frac{\lambda_p - iN}{2R_2} \tag{10.16}$$

The following expressions for the magnetic flux in the air gaps are obtained by the treatment of Eqs. (10.11) through (10.16):

$$\varphi_1 = \frac{(\lambda_p + iN)\mu_o A}{2(x_o - x_a)} \tag{10.17}$$

$$\varphi_2 = \frac{(\lambda_p - iN)\mu_o A}{2(x_o + x_a)} \tag{10.18}$$

The mathematical expressions for the mechanical forces acting on the armature extremities, and the resultant torque acting on the armature are deduced as follows:

$$F_1 = \varphi_1^2/2\mu_o A \tag{10.19}$$

$$F_2 = \varphi_2^2/2\mu_o A \tag{10.20}$$

$$F = F_1 - F_2 = (\varphi_1^2 - \varphi_2^2)/2\mu_o A \tag{10.21}$$

$$T = FL \tag{10.22}$$

$$T = \frac{L}{2\mu_o A}\left[\frac{(\lambda_p + iN)^2\mu_o^2 A^2}{4(x_o - x_a)^2} - \frac{(\lambda_p - iN)^2\mu_o^2 A^2}{4(x_o + x_a)^2}\right] \tag{10.23}$$

$$T = \frac{\mu_o AL}{8(x_o^2 - x_a^2)^2}\left[(\lambda_p + iN)^2(x_o + x_a)^2 - (\lambda_p - iN)^2(x_o - x_a)^2\right] \tag{10.24}$$

Actually, the magneto-motive force produced by the coil current is too small compared with that of the permanent magnet ($iN << \lambda_p$). Moreover, the displacement of the armature extremity is too small compared with the air gap thickness, ($x_a << x_o$). Therefore, the values of x_a^2 and i^2N^2 are negligible with respect to x_o^2 and λ_p^2, respectively. The torque is then given by the following expressions:

$$T = \frac{\mu_o AL}{8x_o^4}\left[4x_o x_a \lambda_p^2 + 4iN\lambda_p x_o^2\right] \tag{10.25}$$

$$T = \frac{\lambda_p^2 \mu_o AL}{2x_o^3} x_a + \frac{N\lambda_p \mu_o AL}{2x_o^2} i \tag{10.26}$$

Hence,

$$T = K_x x_a + K_i i \quad \text{or} \quad T = K_\vartheta \vartheta + K_i i \tag{10.27}$$

where

$$\vartheta = \frac{2x_a}{L}, \quad K_x = \frac{\lambda_p^2 \mu_o AL}{2x_o^3}, \quad K_i = \frac{N\lambda_p \mu_o AL}{2x_o^2},$$

and

$$K_\vartheta = \frac{\lambda_p^2 \mu_o AL^2}{4x_o^3} \tag{10.28}$$

Considering the real servovalve parameters, the term $K_\vartheta \vartheta$ is too small compared to $K_i i$; therefore, the torque may be given by

$$T = K_i i \tag{10.29}$$

These expressions do not consider the magnetic hysteresis. However, it is preferred to take into account the effect of the magnetic saturation and magnetic hysterisis whenever possible. Figure 10.6 shows the simulation result of the torque of an electromagnetic torque motor. Note that the inner hysterisis loops can be generated and the percentage hysterisis can be preset.

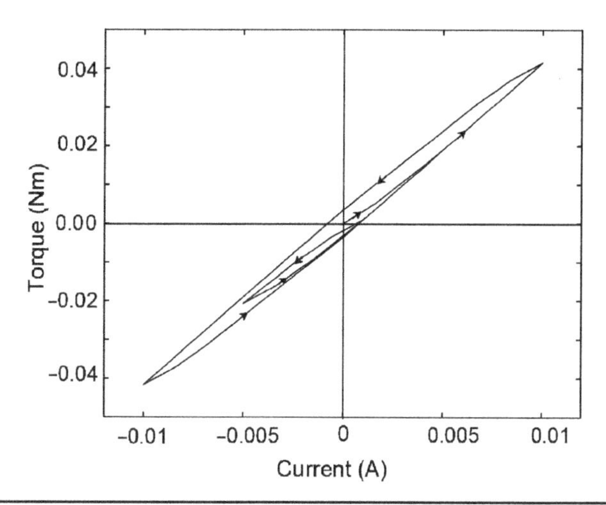

FIGURE 10.6 The typical torque-current steady-state relation of an electromagnetic torque motor, considering the effect of magnetic hysterisis, calculated.

10.3 Flapper Valves

The flapper valve acts as a hydraulic amplifier in electrohydraulic servovalves (EHSV). It is excited by the very small displacement of the flapper (usually within ±30 μm). It produces a considerably high-pressure difference in proportion with the flapper displacement. Figure 10.7 shows a double jet flapper valve. It consists of two fixed area orifices and two variable area jet nozzles. The areas of jet nozzles are controlled by the flapper displacement. The flapper valve is fed with high-pressure oil, at pressure P_s.

In the steady state, the ports C1 and C2 are closed by the stationary spool (as shown by Fig. 10.9), then $Q_1 = Q_3$ and $Q_2 = Q_4$. The valve flow rates are given by the following equations:

$$Q_1 = C_D A_o \sqrt{\frac{2}{\rho}(P_s - P_1)} \tag{10.30}$$

$$Q_2 = C_D A_o \sqrt{\frac{2}{\rho}(P_s - P_2)} \tag{10.31}$$

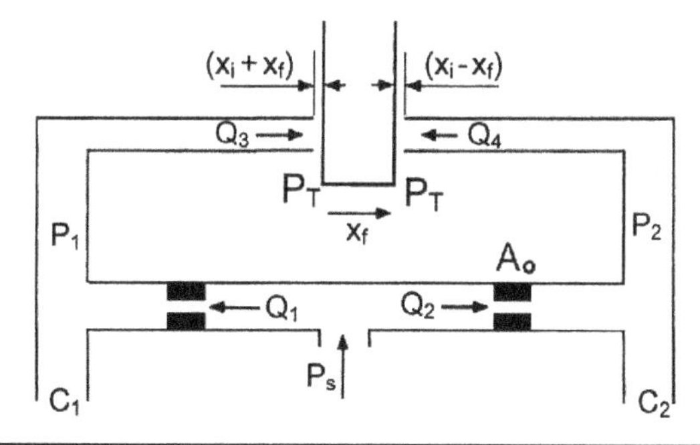

FIGURE 10.7 Schematic of a double jet flapper valve.

The throttling takes place in the jet nozzles of the flapper valve at the cylindrical opening separating the flapper face and the jet nozzles.

$$Q_3 = C_d \pi d_f (x_i + x_f)\sqrt{\frac{2}{\rho}(P_1 - P_T)} \tag{10.32}$$

$$Q_4 = C_d \pi d_f (x_i - x_f)\sqrt{\frac{2}{\rho}(P_2 - P_T)} \tag{10.33}$$

where A_o = Orifice area, m^2
C_d and C_D = Discharge coefficients
d_f = Flapper nozzle diameter, m
P_s = Supply pressure, Pa
P_T = Return line pressure, Pa
Q = Flow rate, m^3/s
x_f = Flapper displacement on the level of jet nozzles, m
x_i = Initial flapper nozzle opening; limiting flapper displacement, m
ρ = Oil density, kg/m^3

These equations are used to derive the following expressions for the nondimensional pressures, assuming zero return pressure: $P_T = 0$.

$$\overline{P}_2 = \frac{P_2}{P_s} = \frac{C_D^2 A_o^2}{C_D^2 A_o^2 + C_d^2 \pi^2 d_f^2 (x_i - x_f)^2} = \frac{1}{1 + a(x_i - x_f)^2} \tag{10.34}$$

$$\overline{P}_1 = \frac{P_1}{P_s} = \frac{C_D^2 A_o^2}{C_D^2 A_o^2 + C_d^2 \pi^2 d_f^2 (x_i + x_f)^2} = \frac{1}{1 + a(x_i + x_f)^2} \tag{10.35}$$

$$\overline{P}_L = \frac{P_2 - P_1}{P_s} = \frac{4a x_f x_i}{1 + 2a(x_i^2 + x_f^2) + a^2(x_i^2 - x_f^2)^2} \tag{10.36}$$

$$a = \left(\frac{C_d \pi d_f}{C_D A_o}\right)^2 \tag{10.37}$$

For small flapper displacement, $x_f << x_i$, then x_f^2 is negligible with respect to x_i^2 and the expression for $\overline{P}$ becomes

$$\overline{P}_L = \frac{4a x_i^2}{(1 + a x_i^2)^2}\,\overline{x}_f \qquad \text{where } \overline{x}_f = x_f / x_i \tag{10.38}$$

Figure 10.8 shows the steady-state pressure characteristics of a typical double jet flapper valve. The plot of $\overline{P}(\overline{x})$ is linear in the vicinity of the origin, which agrees with Eq. (10.38).

10.4 Modeling of an Electrohydraulic Servo Actuator

This section deals with the deduction of a mathematical model describing the dynamic performance of a symmetrical hydraulic cylinder controlled by means of a two-stage EHSV (see Fig. 10.9).

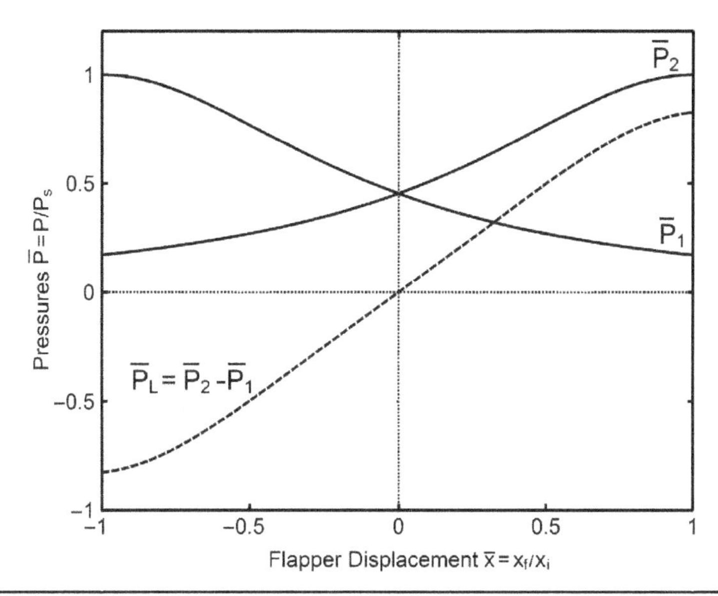

FIGURE 10.8 Steady-state characteristics of a double jet flapper valve, calculated.

Electromagnetic Torque Motors

The electromagnetic torque motor converts an electric input signal of low-level current (usually within 10 mA) into a proportional mechanical torque. The motor is usually designed to be separately mountable, testable, interchangeable, and hermetically sealed against the hydraulic fluid. The net torque depends on the effective input current and the flapper rotational angle. Neglecting the effect of the magnetic hysteresis, the following expression for the torque can be deduced:

$$T = K_i i_e + K_\theta \vartheta \tag{10.39}$$

where i_e = Torque motor input current, A
T = Torque of electromagnetic torque motor, Nm

The Equation of Motion of the Armature

The motion of the rotating armature and attached elements is governed by the following equation:

$$T = J\frac{d^2\vartheta}{dt^2} + f_\vartheta \frac{d\vartheta}{dt} + K_T\vartheta + T_L + T_P + T_F \tag{10.40}$$

$$T_p = \frac{\pi}{4} d_f^2 (P_2 - P_1) L_f \tag{10.41}$$

where J = Moment of inertia of the rotating parts, Nms^2
T_L = Torque due to flapper displacement limiter, Nm
f_ϑ = Damping coefficient, Nms/rad
K_T = Stiffness of flexure tube, Nm/rad
T_F = Feedback torque, Nm
T_p = Torque due to the pressure forces, Nm

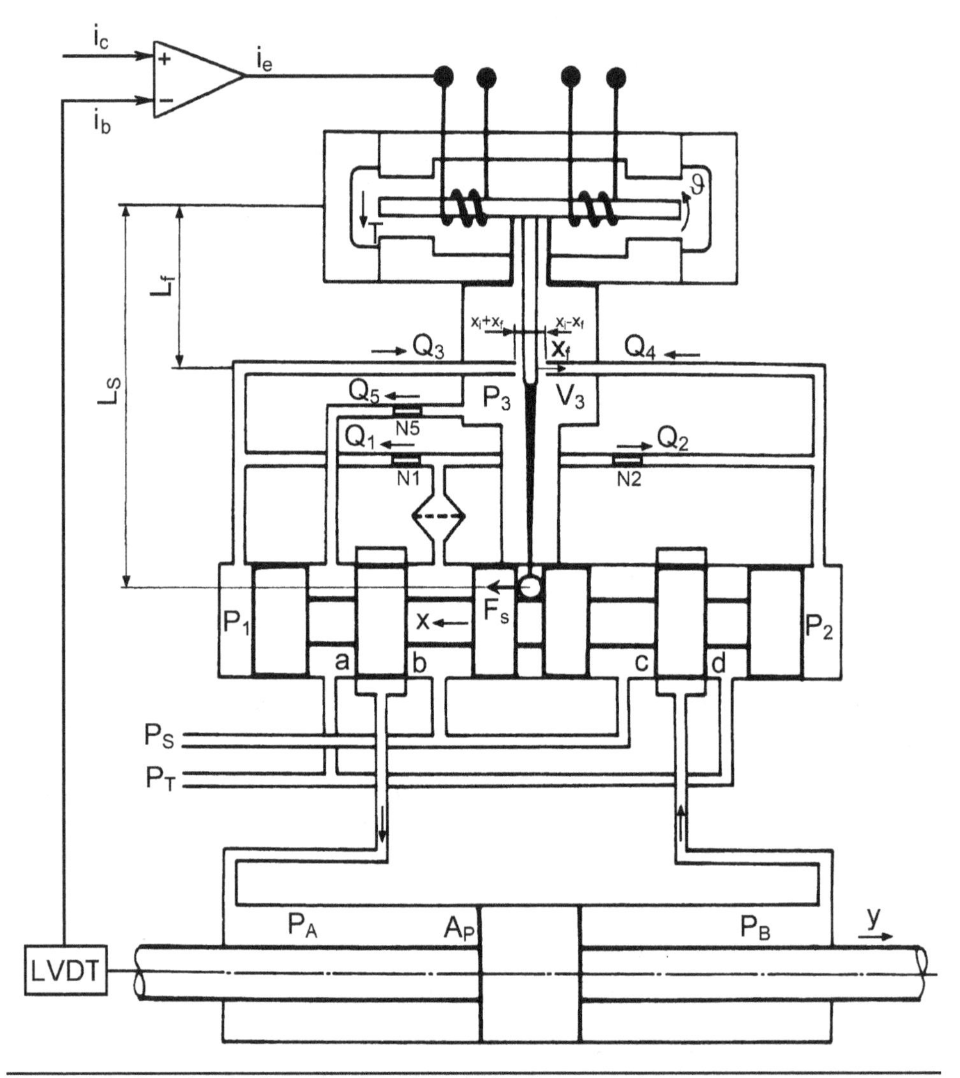

FIGURE 10.9 Functional schematic of an electrohydraulic servo actuator.

Feedback Torque

The feedback torque depends on the spool displacement and the flapper rotational angle as given by the following equations:

$$T_F = F_S L_S \tag{10.42}$$

$$F_S = K_S(L_S\vartheta + x) \tag{10.43}$$

$$T_F = K_S(L_S\vartheta + x)L_S \tag{10.44}$$

where K_s = Stiffness of the feedback spring, N/m
L_s = Length of the feedback spring and flapper, m

x = Spool displacement, m
F_s = Force acting at the extremity of the feedback spring, N

Flapper Position Limiter

The flapper displacement is limited mechanically by the jet nozzles. When reaching any of the side nozzles, the seat reaction develops a counter torque, T_L, given by the following equation:

$$T_L = \begin{cases} 0 & |x_f| < x_i \\ R_s \dfrac{d\vartheta}{dt} - (|x_f| - x_i) K_{Lf} L_f \operatorname{sign}(x_f) & |x_f| > x_i \end{cases} \tag{10.45}$$

where K_{Lf} = Equivalent flapper seat stiffness, N/m
R_s = Equivalent flapper seat damping coefficient, Nms/rad

Flow Rates Through the Flapper Valve Restrictions

The flow rates through the flapper valve restrictions are given by the following equations:

$$Q_1 = C_D A_o \sqrt{\frac{2}{\rho}(P_s - P_1)} = C_{12}\sqrt{(P_s - P_1)} \tag{10.46}$$

$$Q_2 = C_D A_o \sqrt{\frac{2}{\rho}(P_s - P_2)} = C_{12}\sqrt{P_s - P_2} \tag{10.47}$$

$$Q_3 = C_d \pi d_f (x_i + x_f) \sqrt{\frac{2}{\rho}(P_1 - P_3)} = C_{34}(x_i + x_f)\sqrt{(P_1 - P_3)} \tag{10.48}$$

$$Q_4 = C_d \pi d_f (x_i - x_f) \sqrt{\frac{2}{\rho}(P_2 - P_3)} = C_{34}(x_i - x_f)\sqrt{(P_2 - P_3)} \tag{10.49}$$

$$x_f = L_f \vartheta \tag{10.50}$$

$$Q_5 = C_d A_5 \sqrt{\frac{2}{\rho}(P_3 - P_T)} = C_5\sqrt{(P_3 - P_T)} \tag{10.51}$$

where A_5 = Drain orifice area, m^2
A_o = Orifice area, m^2
L_f = Flapper length, m
P_1 = Pressure in the left side of the flapper valve, Pa
P_2 = Pressure in the right side of the flapper valve, Pa
P_3 = Pressure in the flapper valve return chamber, Pa
P_T = Valve return pressure, Pa
Q_1 = Flow rate in the left orifice, m^3/s
Q_2 = Flow rate in the right orifice, m^3/s
Q_3 = Left flapper nozzle flow rate, m^3/s
Q_4 = Right flapper nozzle flow rate, m^3/s
Q_5 = Flapper valve drain flow rate, m^3/s

Continuity Equations Applied to the Flapper Chambers

The application of the continuity equation to the flapper valve chambers results in the following relations:

$$Q_1 - Q_3 + A_s \frac{dx}{dt} = \frac{V_o - A_s x}{B} \frac{dP_1}{dt} \tag{10.52}$$

$$Q_2 - Q_4 - A_s \frac{dx}{dt} = \frac{V_o + A_s x}{B} \frac{dP_2}{dt} \tag{10.53}$$

$$Q_3 + Q_4 - Q_5 = \frac{V_3}{B} \frac{dP_3}{dt} \tag{10.54}$$

where A_s = Spool cross-sectional area, m^2
B = Bulk modulus of oil, Pa
V_o = Initial volume of oil in the spool side chamber, m^3
V_3 = Volume of the flapper valve return chamber, m^3

Equation of Motion of the Spool

The motion of the spool is described by the following equations:

$$A_s(P_2 - P_1) = m_s \frac{d^2x}{dt^2} + f_s \frac{dx}{dt} + F_j + F_s \tag{10.55}$$

$$F_j = \begin{cases} \left(\dfrac{\rho Q_b^2}{C_c A_b} + \dfrac{\rho Q_d^2}{C_c A_d} \right) \text{sign}(x) & \text{For } x > 0 \\ \left(\dfrac{\rho Q_a^2}{C_c A_a} + \dfrac{\rho Q_c^2}{C_c A_c} \right) \text{sign}(x) & \text{For } x < 0 \end{cases} \tag{10.56}$$

where C_c = Contraction coefficient
F_j = Hydraulic momentum force, N
f_s = Spool friction coefficient, Ns/m
F_s = Force acting at the extremity of the feedback spring, N
m_s = Spool mass, kg

Flow Rates Through the Spool Valve

Neglecting the effect of transmission lines, connecting the valve to the symmetrical cylinder, the flow rates through the valve restriction areas are given by

$$Q_a = C_d A_a(x) \sqrt{\frac{2}{\rho}(P_A - P_T)} \tag{10.57}$$

$$Q_b = C_d A_b(x) \sqrt{\frac{2}{\rho}(P_s - P_A)} \tag{10.58}$$

$$Q_c = C_d A_c(x) \sqrt{\frac{2}{\rho}(P_s - P_B)} \tag{10.59}$$

$$Q_d = C_d A_d(x)\sqrt{\frac{2}{\rho}(P_B - P_T)} \tag{10.60}$$

The valve restrictions areas are given by

$$\left.\begin{aligned} A_a &= A_c = \omega c \\ A_b &= A_d = \omega\sqrt{(x^2 + c^2)} \end{aligned}\right\} \text{For } x \geq 0 \tag{10.61}$$

$$\left.\begin{aligned} A_a &= A_c = \omega\sqrt{(x^2 + c^2)} \\ A_b &= A_d = \omega c \end{aligned}\right\} \text{For } x \leq 0 \tag{10.62}$$

where c = Spool radial clearance, m
ω = Width of ports on the valve sleeve, m
P_A and P_B = Hydraulic cylinder pressures, Pa

Continuity Equations Applied to the Cylinder Chambers

Applying the continuity equation to the cylinder chambers, considering the internal leakage and neglecting the external leakage, the following equations were obtained:

$$Q_b - Q_a - A_P\frac{dy}{dt} - \frac{(P_A - P_B)}{R_i} = \frac{V_c + A_p y}{B}\frac{dP_A}{dt} \tag{10.63}$$

$$Q_c - Q_d + A_P\frac{dy}{dt} + \frac{(P_A - P_B)}{R_i} = \frac{V_c - A_p y}{B}\frac{dP_B}{dt} \tag{10.64}$$

where A_p = Piston area, m^2
R_i = Resistance to internal leakage, Ns/m^5
V_c = Half of the volume of oil filling the cylinder, m^3

Equation of Motion of the Piston

The motion of the piston under the action of pressure, viscous friction, inertia, and external forces is described by the following equation:

$$A_P(P_A - P_B) = m_P\frac{d^2y}{dt^2} + f_P\frac{dy}{dt} + K_b y \tag{10.65}$$

where K_b = Load coefficient N/m
m_p = Piston mass, kg
f_p = Piston friction coefficient, Ns/m

Feedback Equation

The piston displacement is picked up by a displacement transducer and fed back to the electronic controller, which generates the corresponding error signal. The feedback loop can be described by the following equations:

$$i_e = i_c - i_b \tag{10.66}$$

$$i_b = K_{FB}\, y \tag{10.67}$$

where i_c = Control current, A
i_b = Feedback current, A
K_{FB} = Feedback gain, A/m

10.5 Exercises

1. Draw a schematic of the electromagnetic torque motor, explain its function, and derive an expression for its torque.

2. Draw a schematic of a double jet flapper valve, explain its function, and derive the mathematical relations describing its static characteristics. Discuss the variation of the valve pressures with the flapper displacement.

3. Draw a schematic of a single-stage electrohydraulic servovalve, explain its function, and derive the mathematical model describing its static and dynamic behavior.

4. Draw a schematic of a two-stage electrohydraulic servovalve, explain its function, and derive the mathematical model describing its static and dynamic behavior.

5. Draw a schematic of an electrohydraulic servo actuator, explain its function, and derive the mathematical model describing its static and dynamic behavior.

10.6 Nomenclature

A = Area of air gap, m^2
A_5 = Drain orifice area, m^2
A_o = Orifice area, m^2
A_a, A_b, A_c, and A_d = Spool valve restrictions areas, m^2
A_p = Piston area, m^2
A_s = Spool cross-sectional area, m^2
B = Bulk modulus of oil, Pa
c = Spool radial clearance, m
C_c = Contraction coefficient
C_d and C_D = Discharge coefficients
d_f = Flapper nozzle diameter, m
F = Force, N
f_ϑ = Damping coefficient, Nms/rad
F_j = Hydraulic momentum force, N
f_p = Piston friction coefficient, Ns/m
f_s = Spool friction coefficient, Ns/m
F_s = Force acting at the extremity of the feedback spring, N
H = Magneto-motive force per unit length, A/m
i_b = Feedback current, A
i_c = Control current, A
i_e = Torque motor input current, A
J = Moment of inertia of rotating part, Nms^2
K_b = Load coefficient N/m
K_{FB} = Feedback gain, A/m
K_{Lf} = Equivalent flapper seat stiffness, N/m

K_i = Current-torque gain, Nm/A
K_s = Stiffness of the feedback spring, N/m
K_T = Stiffness of flexure tube, Nm/rad
K_x = Displacement-torque gain, Nm/m
K_ϑ = Rotational angle-torque gain, Nm/rad
L = Length; armature length, m
L_f = Flapper length, m
L_s = Length of the feedback spring and flapper, m
m_p = Piston mass, kg
m_s = Spool mass, kg
P = Magnetic permeance, Vs/A
P_1 = Pressure in the left side of the flapper valve, Pa
P_2 = Pressure in the right side of the flapper valve, Pa
P_3 = Pressure in the flapper valve return chamber, Pa
P_A and P_B = Hydraulic cylinder pressures, Pa
P_s = Supply pressure, Pa
P_T = Return line pressure, Pa
Q = Flow rate, m^3/s
Q_1 = Flow rate in the left orifice, m^3/s
Q_2 = Flow rate in the right orifice, m^3/s
Q_3 = Left flapper nozzle flow rate, m^3/s
Q_4 = Right flapper nozzle flow rate, m^3/s
Q_5 = Flapper valve drain flow rate, m^3/s
Q_a, Q_b, Q_c, and Q_d = Flow rates through the spool valve restrictions, m^3/s
R = Magnetic reluctance, A/Vs
R_f = Equivalent damping coefficient, Nms/rad
R_i = Resistance to internal leakage, Ns/m^5
R_s = Flapper seat damping coefficient, Nms/rad
T = Torque of electromagnetic torque motor, Nm
T_F = Feedback torque, Nm
T_L = Torque due to flapper displacement limiter, Nm
T_p = Torque due to the pressure forces, Nm
V_3 = Volume of the flapper valve return chamber, m^3
V_c = Half of the volume of oil filling the cylinder, m^3
V_o = Initial volume of oil in the spool side chamber, m^3
x = Spool displacement, m
x_a = Displacement of the armature end, m
x_f = Flapper displacement on the level of the jet nozzles, m
x_i = Initial flapper nozzle opening, flapper displacement limit, m
x_o = Length of the air gap in the neutral position of armature, m
λ = Magneto-motive force, A
λ_p = Magneto-motive force of the permanent magnet, A
μ = Permeability, Vs/Am
μ_o = Permeability of the air, Vs/Am
μ_r = Relative permeability
ρ = Oil density, kg/m^3
ω = Width of ports on the valve sleeve, m
ϑ = Armature rotation angle, rad

Appendix 10A Modeling and Simulation of an EHSA

This appendix presents mathematical models of an electrohydraulic servo actuator and its subassemblies—mainly the electromagnetic torque motor, the single-stage servovalve, and the two-stage servovalve. The effect of armature magnetic hysterisis and flow forces are neglected in this case study. However, considering the magnetic hysterisis (see Fig. 10.6) gives more accurate results, especially in the open-loop applications.

The deduced mathematical models were applied to develop computer simulation programs using SIMULINK. The step response of the systems is calculated and presented. The simulation programs are built considering the following numerical values of a typical electrohydraulic servoactuator incorporating a two-stage servovalve (see Fig. 10.9).

Numerical Values of the Studied System

Torque Motors

Current-torque gain, $K_i = 0.556$ Nm/A
Armature damping coefficient, $f_\vartheta = 0.002$ Nms/rad
Moment of inertia of the armature-flapper assembly, $J = 5 \times 10^{-7}$ kg m^2
Armature rotational angle-torque gain, $K_\vartheta = 9.45 \times 10^{-4}$ Nm/rad
Servo actuator feedback gain, $K_{FB} = 3$ A/m

Flapper Valves

Flapper length, $L_f = 9$ mm
Feedback spring and flapper length, $L_s = 30$ mm
Flapper limiting displacement, $x_i = 30$ μm
Flapper nozzle diameter, $d_f = 0.5$ mm
Equivalent flapper seat material damping coefficient, $R_s = 5000$ Nsm/rad
Flapper seat equivalent stiffness, $K_{Lf} = 5 \times 10^6$ N/m
Hydraulic amplifier nozzles (N1) and (N2) diameter, $d_f = 0.5$ mm
Diameter of the return orifice (N5), $d_5 = 0.6$ mm

Hydraulic Oil

Oil density, $\rho = 867$ kg/m^3
Bulk modulus of oil, $B = 1.5$ GPa

Spool Valves

Spool diameter, $d_s = 4.6$ mm
Initial volume of oil in the spool side chamber, $V_o = 2$ cm^3
Volume of oil in the return chamber, $V_3 = 5$ cm^3
Spool mass, $m_s = 0.02$ kg
Spool port width, $\omega = 2$ mm
Spool radial clearance, $c = 2$ μm
Spool friction coefficient, $f_s = 2$ Ns/m
Feedback spring stiffness, $K_s = 900$ N/m

System Pressures

Supply pressure, $P_s = 250$ bar
Return pressure, $P_T = 0$ bar

Hydraulic Cylinder

Piston area, $A_p = 12.5\ \text{cm}^2$
Initial volume of oil in the cylinder chamber, $V_c = 100\ \text{cm}^3$
Resistance to internal leakage, $R_i = 10^{14}$ Pa s/m^3
Piston mass, $m_p = 10$ kg
Friction coefficient on piston, $f_p = 1000$ Ns/m
Piston loading coefficient, $K_b = 0$ N/m; the piston is unloaded

Torque Motors

Figure 10A.1 shows a schematic of the electromagnetic torque motor. When separated from the valve, the armature displacement is limited by the air gap thickness δ. When operating within this displacement limit, the dynamic behavior of this system is described by Eqs. (10A.1) and (10A.2), neglecting the effect of magnetic hysterisis.

$$T = K_i i_e + K_\theta \vartheta \tag{10A.1}$$

$$T = J\frac{d^2\vartheta}{dt^2} + f_\vartheta \frac{d\vartheta}{dt} + K_T\vartheta \tag{10A.2}$$

Equations (10A.1) and (10A.2) were used to develop a computer simulation program. The step response of the torque motor was calculated using the simulation program and used to find the torque motor transfer function (see Fig. 10A.2). The torque motor showed an under-damped response with a settling time within 1.2 ms (see Fig. 10A.3). This behavior can be identified by a second order transfer function of $\omega_n = 711$ Hz, $\xi = 0.071$, and gain = 0.00556, Eq. (10A.3). The response calculated using the transfer function coincides with that of the detailed model (Fig. 10A.3). Actually, the same transfer function is obtained by the treatment of Eqs. (10A.1) and (10A.2) analytically.

$$G(s) = \frac{\theta(s)}{I_e(s)} = \frac{0.0556}{5\times10^{-8}s^2 + 2\times10^{-4}s + 1} \tag{10A.3}$$

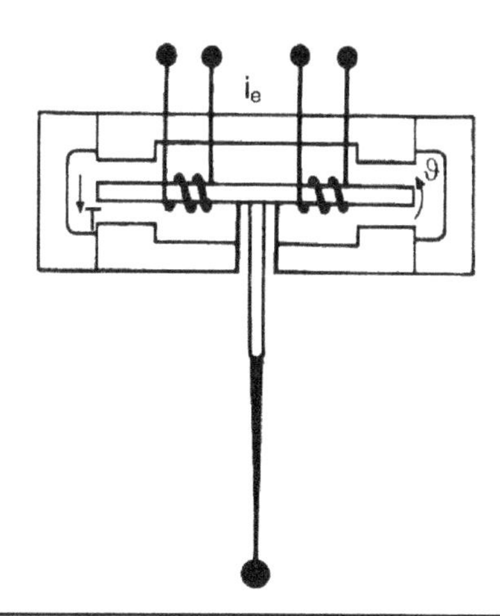

Figure 10A.1 Schematic of the electromagnetic torque motor.

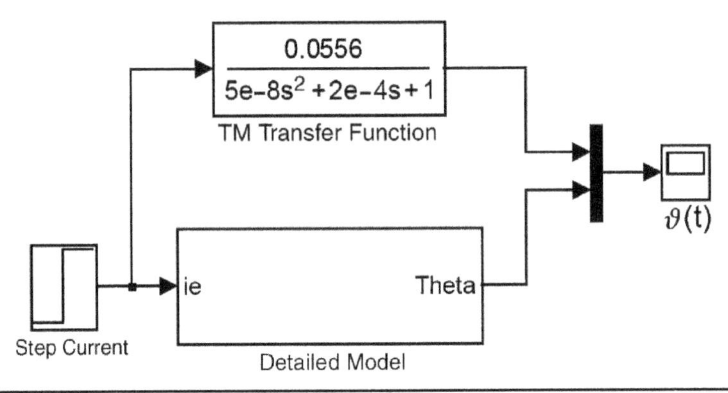

Figure 10A.2 A simulation block diagram of the torque motor, developed by the SIMULINK program.

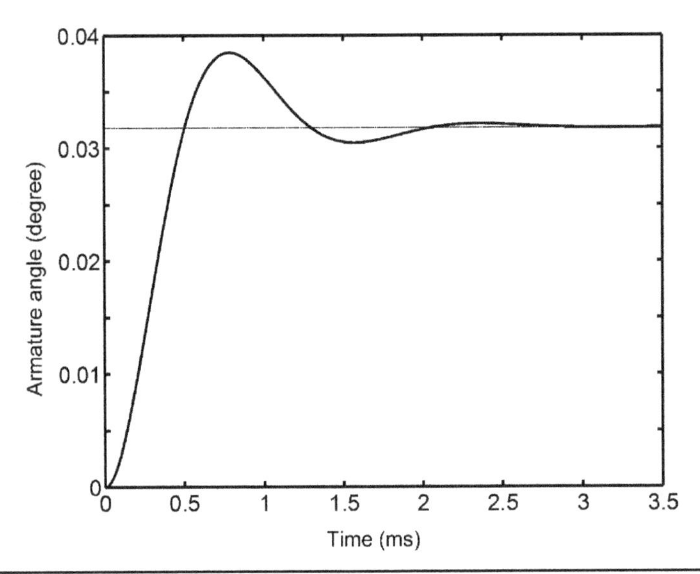

Figure 10A.3 Transient response of the armature rotational angle to a 10 mA step current.

Single-Stage Electrohydraulic Servovalves

The following equations describe the single-stage servovalve shown in Fig. 10A.4. It consists of an electromagnetic torque motor and a flapper valve hydraulic amplifier.

$$T = K_i i_e + K_\theta \vartheta \tag{10A.4}$$

$$T = J\frac{d^2\vartheta}{dt^2} + f_\vartheta \frac{d\vartheta}{dt} + K_T\vartheta + T_L + T_P \tag{10A.5}$$

$$T_p = \frac{\pi}{4} d_f^2 (P_2 - P_1) L_f \tag{10A.6}$$

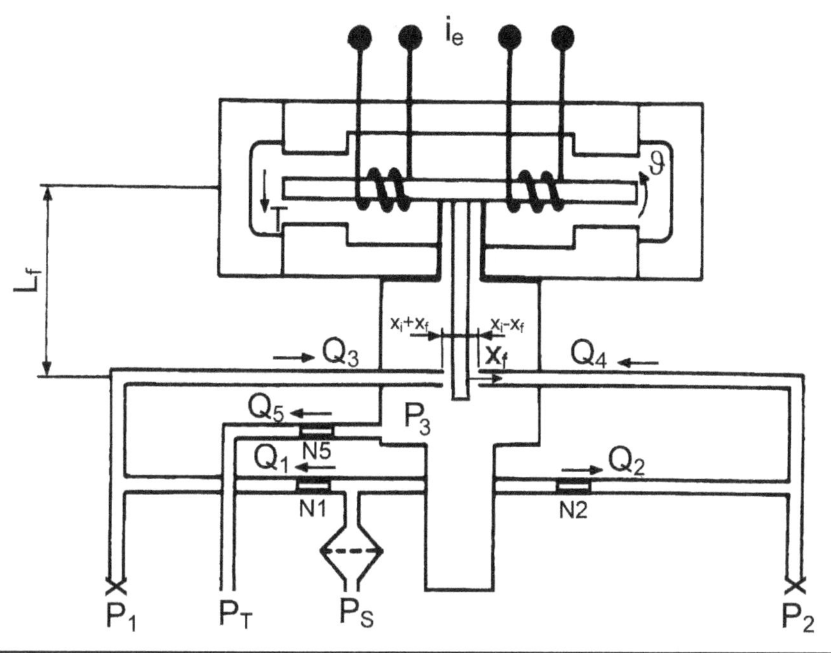

Figure 10A.4 Schematic of a single-stage electrohydraulic servovalve.

$$T_L = \begin{cases} 0 & |x_f| < x_i \\ R_s \dfrac{d\vartheta}{dt} - (|x_f| - x_i) K_{Lf} L_f \operatorname{sign}(\vartheta) & |x_f| > x_i \end{cases} \tag{10A.7}$$

$$Q_1 = C_d A_o \sqrt{\frac{2}{\rho}(P_s - P_1)} = C_{12}\sqrt{(P_s - P_1)} \tag{10A.8}$$

$$Q_2 = C_d A_o \sqrt{\frac{2}{\rho}(P_s - P_2)} = C_{12}\sqrt{P_s - P_2} \tag{10A.9}$$

$$Q_3 = C_d \pi d_f (x_i + x_f)\sqrt{\frac{2}{\rho}(P_1 - P_3)} = C_{34}(x_i + x_f)\sqrt{(P_1 - P_3)} \tag{10A.10}$$

$$Q_4 = C_d \pi d_f (x_i - x_f)\sqrt{\frac{2}{\rho}(P_2 - P_3)} = C_{34}(x_i - x_f)\sqrt{(P_2 - P_3)} \tag{10A.11}$$

$$x_f = L_f \vartheta \tag{10A.12}$$

$$Q_5 = C_d A_5 \sqrt{\frac{2}{\rho}(P_3 - P_T)} = C_5\sqrt{(P_3 - P_T)} \tag{10A.13}$$

$$Q_1 - Q_3 = \frac{V_o}{B}\frac{dP_1}{dt} \tag{10A.14}$$

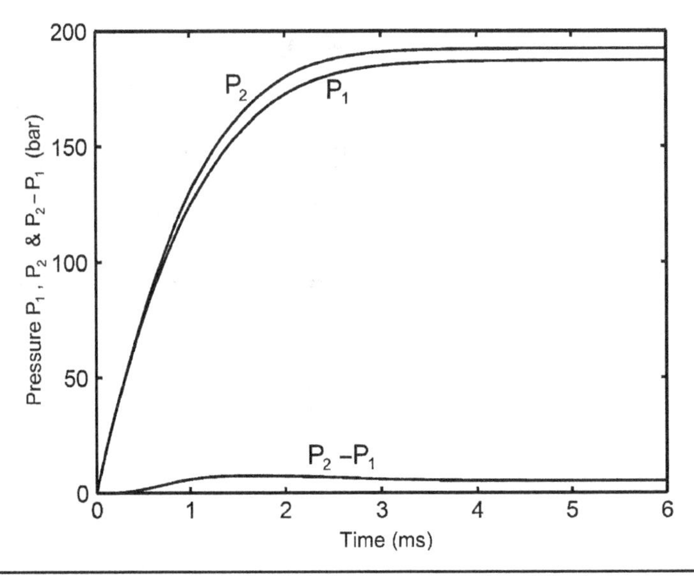

FIGURE 10A.5 Transient response of the first-stage pressures to a 10 mA step current, calculated using the SIMULINK program.

$$Q_2 - Q_4 = \frac{V_o}{B}\frac{dP_2}{dt} \tag{10A.15}$$

$$Q_3 + Q_4 - Q_5 = \frac{V_3}{B}\frac{dP_3}{dt} \tag{10A.16}$$

The single-stage servovalve controls the pressure difference: $\Delta P = P_2 - P_1$. The valve is described mathematically by Eqs. (10A.4) through (10A.16). These equations were used to develop a computer simulation program. The transient response of valve pressures to step input current, calculated using the simulation program, is shown in Fig. 10A.5. The valve presents a transient response of settling time within 2 ms.

Two-Stage Electrohydraulic Servovalves

Figure 10A.6 illustrates an electrohydraulic servovalve of two stages. Neglecting the jet reaction forces on the level of the spool valve, this valve is described mathematically as follows:

$$T = K_i i_e + K_\theta \vartheta \tag{10A.17}$$

$$T = J\frac{d^2\vartheta}{dt^2} + f_\vartheta \frac{d\vartheta}{dt} + K_T\vartheta + T_L + T_P + T_F \tag{10A.18}$$

$$T_p = \frac{\pi}{4} d_f^2 (P_2 - P_1) L_f \tag{10A.19}$$

$$F_S = K_S(L_S\vartheta + x) \tag{10A.20}$$

$$T_F = F_S L_S = K_S(L_S\vartheta + x)L_S \tag{10A.21}$$

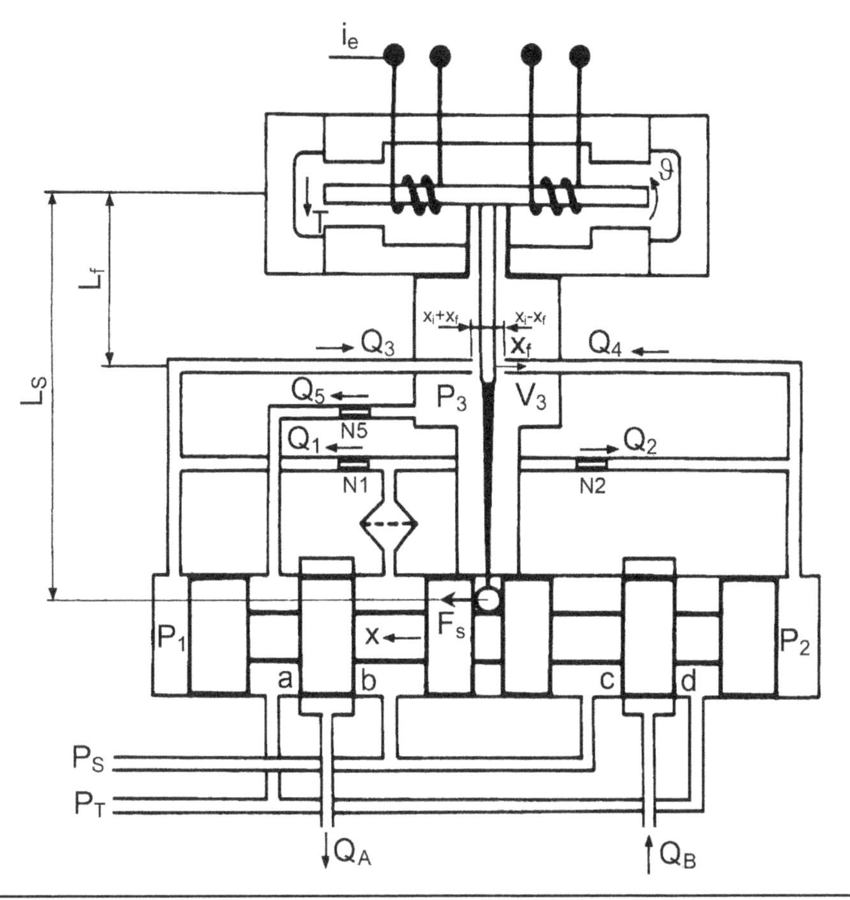

Figure 10A.6 Schematic of a two-stage electrohydraulic servovalve of two stages.

$$T_L = \begin{cases} 0 & |x_f| < x_i \\ R_s \dfrac{d\vartheta}{dt} - (|x_f| - x_i)K_L L_f \text{sign}(\vartheta) & |x_f| > x_i \end{cases} \tag{10A.22}$$

$$Q_1 = C_d A_o \sqrt{\frac{2}{\rho}(P_s - P_1)} = C_{12}\sqrt{(P_s - P_1)} \tag{10A.23}$$

$$Q_2 = C_d A_o \sqrt{\frac{2}{\rho}(P_s - P_2)} = C_{12}\sqrt{P_s - P_2} \tag{10A.24}$$

$$Q_3 = C_d \pi d_f (x_i + x_f)\sqrt{\frac{2}{\rho}(P_1 - P_3)} = C_{34}(x_i + x_f)\sqrt{(P_1 - P_3)} \tag{10A.25}$$

$$Q_4 = C_d \pi d_f (x_i - x_f)\sqrt{\frac{2}{\rho}(P_2 - P_3)} = C_{34}(x_i - x_f)\sqrt{(P_2 - P_3)} \tag{10A.26}$$

$$x_f = L_f \vartheta \tag{10A.27}$$

$$Q_5 = C_d A_5 \sqrt{\frac{2}{\rho}(P_3 - P_T)} = C_5 \sqrt{(P_3 - P_T)} \tag{10A.28}$$

$$Q_1 - Q_3 + A_s \frac{dx}{dt} = \frac{V_o - A_s x}{B} \frac{dP_1}{dt} \tag{10A.29}$$

$$Q_2 - Q_4 - A_s \frac{dx}{dt} = \frac{V_o + A_s x}{B} \frac{dP_2}{dt} \tag{10A.30}$$

$$Q_3 + Q_4 - Q_5 = \frac{V_3}{B} \frac{dP_3}{dt} \tag{10A.31}$$

$$A_s(P_2 - P_1) = m_s \frac{d^2x}{dt^2} + f_s \frac{dx}{dt} + F_s \tag{10A.32}$$

$$Q_a = C_d A_a(x) \sqrt{\frac{2}{\rho}(P_A - P_T)} \tag{10A.33}$$

$$Q_b = C_d A_b(x) \sqrt{\frac{2}{\rho}(P_s - P_A)} \tag{10A.34}$$

$$Q_c = C_d A_c(x) \sqrt{\frac{2}{\rho}(P_s - P_B)} \tag{10A.35}$$

$$Q_d = C_d A_d(x) \sqrt{\frac{2}{\rho}(P_B - P_T)} \tag{10A.36}$$

$$\left.\begin{aligned} A_a &= A_c = \omega c \\ A_b &= A_d = \omega\sqrt{(x^2 + c^2)} \end{aligned}\right\} \quad \text{For } x \geq 0 \tag{10A.37}$$

$$\left.\begin{aligned} A_a &= A_c = \omega\sqrt{(x^2 + c^2)} \\ A_b &= A_d = \omega c \end{aligned}\right\} \quad \text{For } x \leq 0 \tag{10A.38}$$

Figure 10A.7 shows the simulation block diagram of the two-stage servovalve, based upon Eqs. (10A.17) through (10A.38). The step response of the servovalve is shown in Fig. 10A.8. The transient response, calculated using the simulation program, was used to identify the servovalve by a second-order transfer function. The parameters of the identifying transfer function are $\omega_n = 112$ Hz, $\xi = 1.34$, and gain = 0.0202.

The spool responses, calculated using the detailed model and the transfer function, plotted in Fig. 10A.8, are almost coinciding. The valve response is overdamped with a 10 ms settling time.

Electrohydraulic Servo Actuators (EHSAs)

Figure 10.9 shows a functional schematic of an electrohydraulic servo actuator. Its dynamic behavior is described by the Eqs. (10A.39) through (10A.65).

$$T = K_i i_e + K_\theta \vartheta \tag{10A.39}$$

$$T = J \frac{d^2\vartheta}{dt^2} + f_\vartheta \frac{d\vartheta}{dt} + K_T \vartheta + T_L + T_P + T_F \tag{10A.40}$$

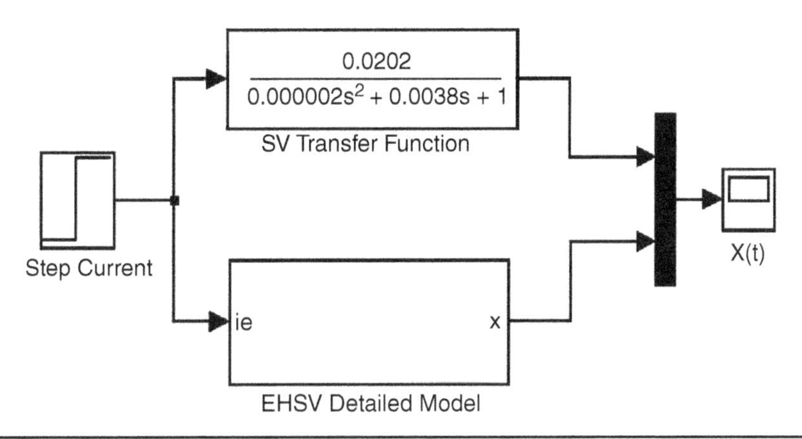

Figure 10A.7 A SIMULINK block diagram of the servovalve.

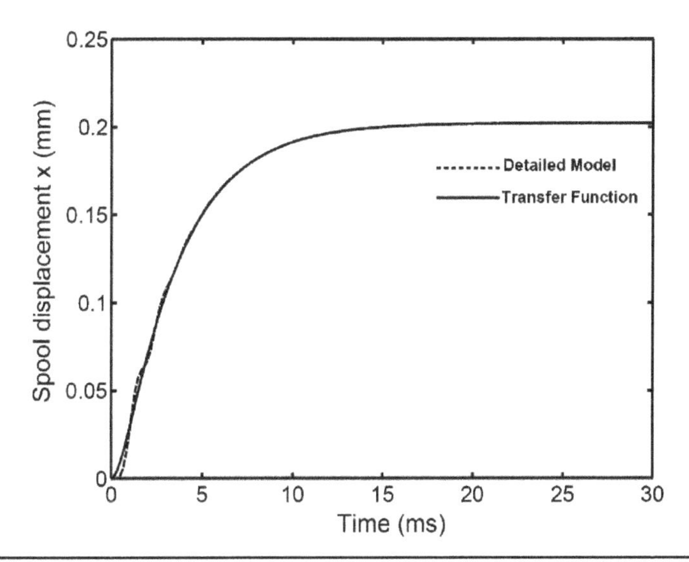

Figure 10A.8 The transient response of the servovalve spool displacement to a 10 mA step current, calculated by the detailed model and the identifying transfer function.

$$T_p = \frac{\pi}{4} d_f^2 (P_2 - P_1) L_f \tag{10A.41}$$

$$F_S = K_S (L_S \vartheta + x) \tag{10A.42}$$

$$T_F = F_S L_S = K_S (L_S \vartheta + x) L_S \tag{10A.43}$$

$$T_L = \begin{cases} 0 & |x_f| < x_i \\ R_s \dfrac{d\vartheta}{dt} - (|x_f| - x_i) K_L L_f \operatorname{sign}(\vartheta) & |x_f| > x_i \end{cases} \tag{10A.44}$$

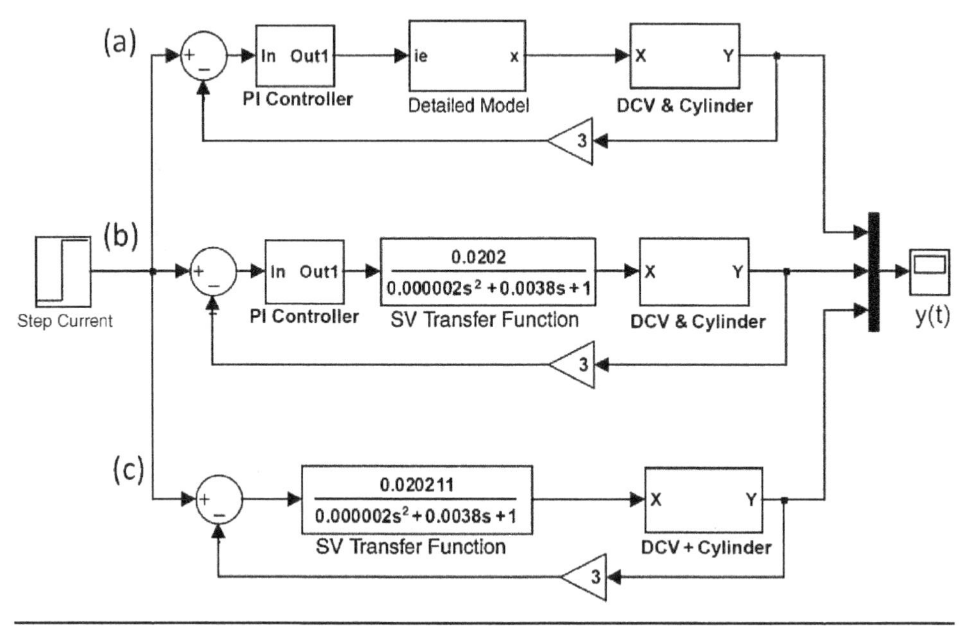

FIGURE 10A.9 A SIMULINK block diagram of the electrohydraulic servo actuator.

$$Q_1 = C_d A_o \sqrt{\frac{2}{\rho}(P_s - P_1)} = C_{12}\sqrt{(P_s - P_1)} \tag{10A.45}$$

$$Q_2 = C_d A_o \sqrt{\frac{2}{\rho}(P_s - P_2)} = C_{12}\sqrt{P_s - P_2} \tag{10A.46}$$

$$Q_3 = C_d \pi d_f (x_i + x_f)\sqrt{\frac{2}{\rho}(P_1 - P_3)} = C_{34}(x_i + x_f)\sqrt{(P_1 - P_3)} \tag{10A.47}$$

$$Q_4 = C_d \pi d_f (x_i - x_f)\sqrt{\frac{2}{\rho}(P_2 - P_3)} = C_{34}(x_i - x_f)\sqrt{(P_2 - P_3)} \tag{10A.48}$$

$$x_f = L_f \vartheta \tag{10A.49}$$

$$Q_5 = C_d A_5 \sqrt{\frac{2}{\rho}(P_3 - P_T)} = C_5\sqrt{(P_3 - P_T)} \tag{10A.50}$$

$$Q_1 - Q_3 + A_s\frac{dx}{dt} = \frac{V_o - A_s x}{B}\frac{dP_1}{dt} \tag{10A.51}$$

$$Q_2 - Q_4 - A_s\frac{dx}{dt} = \frac{V_o + A_s x}{B}\frac{dP_2}{dt} \tag{10A.52}$$

$$Q_3 + Q_4 - Q_5 = \frac{V_3}{B}\frac{dP_3}{dt} \tag{10A.53}$$

$$A_s(P_2 - P_1) = m_s\frac{d^2x}{dt^2} + f_s\frac{dx}{dt} + F_s \tag{10A.54}$$

$$Q_a = C_d A_a(x)\sqrt{\frac{2}{\rho}(P_A - P_T)} \tag{10A.55}$$

$$Q_b = C_d A_b(x)\sqrt{\frac{2}{\rho}(P_s - P_A)} \tag{10A.56}$$

$$Q_c = C_d A_c(x)\sqrt{\frac{2}{\rho}(P_s - P_B)} \tag{10A.57}$$

$$Q_d = C_d A_d(x)\sqrt{\frac{2}{\rho}(P_B - P_T)} \tag{10A.58}$$

$$\left.\begin{aligned} A_a &= A_c = \omega c \\ A_b &= A_d = \omega\sqrt{(x^2 + c^2)} \end{aligned}\right\} \text{For } x \ge 0 \tag{10A.59}$$

$$\left.\begin{aligned} A_a &= A_c = \omega\sqrt{(x^2 + c^2)} \\ A_b &= A_d = \omega c \end{aligned}\right\} \text{For } x \le 0 \tag{10A.60}$$

$$Q_b - Q_a - A_P\frac{dy}{dt} - \frac{(P_A - P_B)}{R_i} = \frac{V_c + A_p y}{B}\frac{dP_A}{dt} \tag{10A.61}$$

$$Q_c - Q_d + A_P\frac{dy}{dt} + \frac{(P_A - P_B)}{R_i} = \frac{V_c - A_p y}{B}\frac{dP_B}{dt} \tag{10A.62}$$

$$A_P(P_A - P_B) = m_P\frac{d^2y}{dt^2} + f_P\frac{dy}{dt} + K_b y \tag{10A.63}$$

$$i_e = i_c - i_b \tag{10A.64}$$

$$i_b = K_{FB}y \tag{10A.65}$$

Figure 10A.9 shows the simulation block diagram of the electrohydraulic servo actuator incorporating

(a) The PI controller and a detailed model of the servovalve.
(b) The PI controller and the transfer function of the servovalve.
(c) A simple proportional feedback controller and transfer function of the servovalve.

The transient responses of the EHSA, calculated using the simulation program for the three previously mentioned cases, are displayed in Fig. 10A.10. This figure shows that

- Cases (a) and (b) give practically identical responses due to the high accuracy of the representative model (the transfer function of the EHSV).

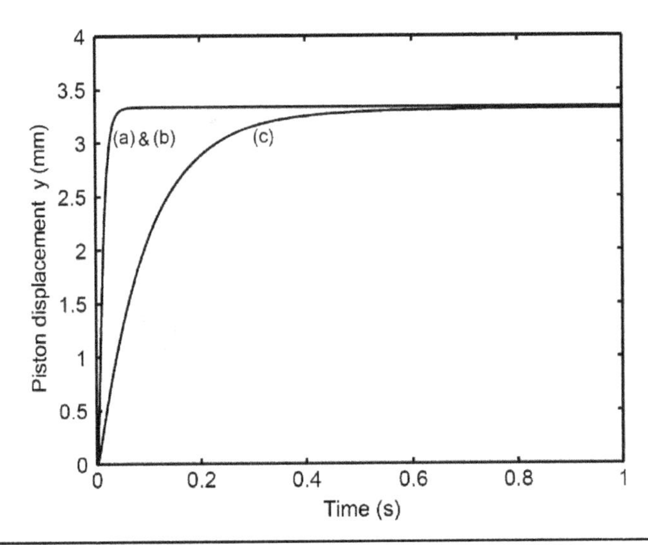

FIGURE 10A.10 Step response of the electrohydraulic servo actuator for the three configurations given by Fig. 10A.9.

- The transient response of the EHSA with simple proportional feedback (case c) shows a relatively longer settling time (t_s = 309 ms) compared with the cases using the proper PI controller (t_s = 33 ms).

Appendix 10B Design of P, PI, and PID Controllers

The proportional integral derivative (PID) controller is the most common form of feedback. It was an essential element of early governors. In process control today, more than 95 percent of the control loops are of PID or PI type. The PID controllers are today found in all areas where control is used. They have survived many changes in technology, from mechanics and pneumatics to microprocessors via electronic tubes, transistors, and integrated circuits. The microprocessor has had a dramatic influence on the PID controller. Practically all PID controllers made today are based on microprocessors. This has created opportunities to provide additional features like automatic tuning, gain scheduling, and continuous adaptation.

The PID algorithm is described by

$$u(t) = K\left\{e(t) + \frac{1}{T_i}\int_0^t e(\tau)d\tau + T_d\frac{de(t)}{dt}\right\} \tag{10B.1}$$

The error signal $e(t)$ is the difference between the instantaneous values of the input signal, $x(t)$, and the feedback signal $f(t)$, as illustrated by Fig. 10B.1.

$$e(t) = x(t) - f(t) \tag{10B.2}$$

The control signal is the sum of three terms:

- P-term; proportional to the error

$$P(t) = Ke(t) \tag{10B.3}$$

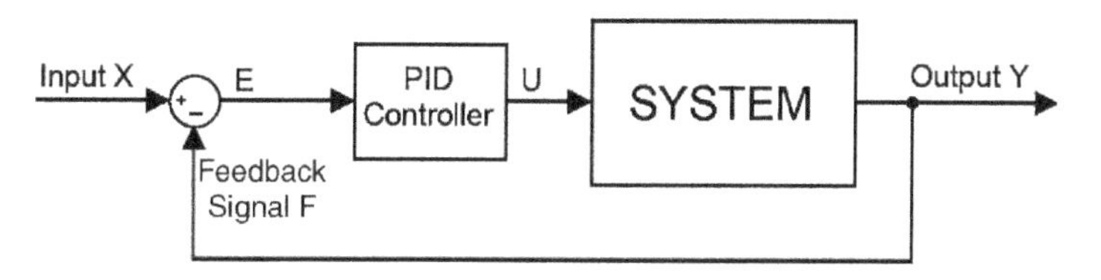

FIGURE 10B.1 The connection of the PID controller in the feedback system.

- *I*-term; proportional to the integral of the error

$$I(t) = K\frac{1}{T_i}\int_0^t e(\tau)d\tau \tag{10B.4}$$

- *D*-term; proportional to the derivative of the error

$$D(t) = KT_d\frac{de(t)}{dt} \tag{10B.5}$$

The controller parameters are the proportional gain, *K*, the integral time, T_i, and the derivative time, T_d. The most well-known methods for estimating and tuning the PID parameters are those developed by Ziegler and Nichols. They have had a major influence upon PID control for more than half a century. The methods are based on characterizations of process dynamics by a few parameters, and simple equations for the controller parameters. They can be designed, according to the Ziegler–Nichols rule. The process of designing a PID controller and its implementation is carried out according to the following four steps:

1. Find the Limiting Open Loop Gain for the Stability

- Connect the system as shown in Fig. 10B.2, and then apply a step input (with gain $K = 1$, for example).
- Calculate the step response, and then change the proportional gain, *K*, until continuous oscillations are observed. The resulting step responses are shown in Fig. 10B.3. The limiting value of gain, *K*, which makes the system marginally stable (the response is oscillatory), is called the ultimate gain, K_L. For the studied system, $K_L = 8$. The duration of one complete cycle is the ultimate period τ. For the studied system, $\tau = 3.63$ s. For proportional gain $K = 1$, the calculated step response shows that the system response has a considerable steady-state error ($\bar{e}_{SS} = 50\%$). For $K = 8$, the transient response converged to sustained oscillations.

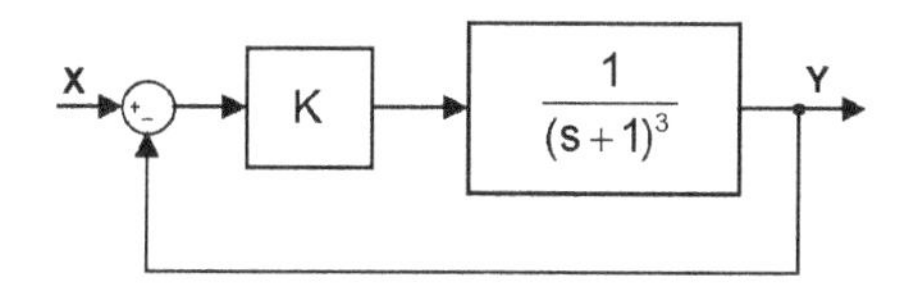

FIGURE 10B.2 A feedback system with a proportional controller of gain *K*.

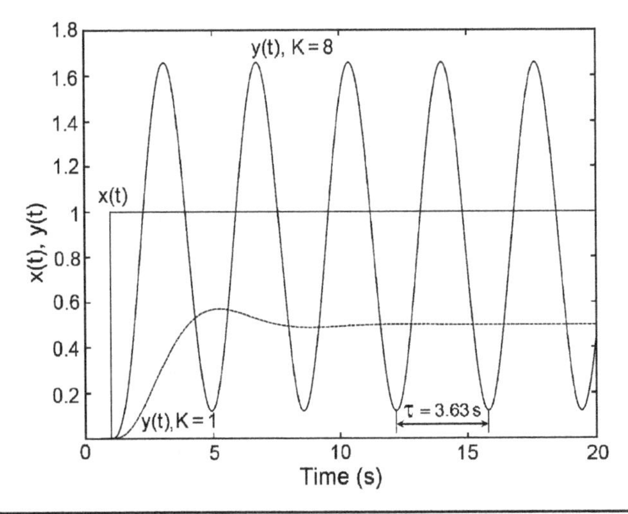

FIGURE 10B.3 Step response of the closed-loop system for open-loop gain $K = 1$ and limiting gain $K = 8$.

2. Design a P, PI, and PID Controller

The transfer functions of the proportional (P), proportional integral (PI), and proportional integral derivative (PID) controllers are the following:

P controller
$$D(s) = K = 0.5K_L \tag{10B.6}$$

PI controller
$$D(s) = K\left(1 + \frac{1}{T_i s}\right) \tag{10B.7}$$

PID controller
$$D(s) = K\left(1 + \frac{1}{T_i s} + T_d s\right) \tag{10B.8}$$

The first estimate of the PID controller parameters is calculated by applying the Ziegler and Nichols rule (see Table 10B.1). The calculation results are given in Table 10B.2.

Controller	Symbol	Gain	T_I	T_d
Proportional	P	$K = 0.5\ K_L$	—	—
Proportional integral	PI	$K = 0.45\ K_L$	$0.8\ \tau$	—
Proportional integral derivative	PID	$K = 0.6\ K_L$	$0.5\ \tau$	$0.125\ \tau$

TABLE 10B.1 Summary of Formulas Used to Calculate a First Estimate of the Parameters of the P, PI, and PID Controllers, According to Ziegler–Nichols Rules

Controller	First Estimate			Tuned Parameters		
	K	T_I	T_d	K	T_I	T_d
P	4	—	—	4	—	—
PI	3.6	2.9 s	—	1.5	3.03	—
PID	4.8	1.815 s	0.454 s	1.3	2.5 s	0.08 s

TABLE 10B.2 First Estimate and Tune Parameters of the P, PI, and PID Controllers, According to Ziegler–Nichols Rules

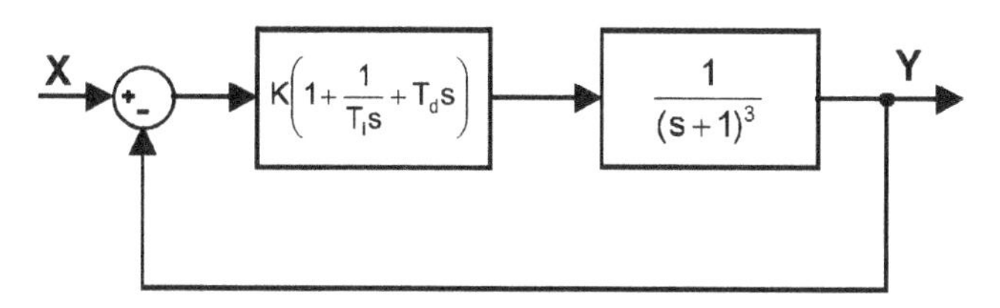

FIGURE 10B.4 Closed-loop system equipped with PID controller.

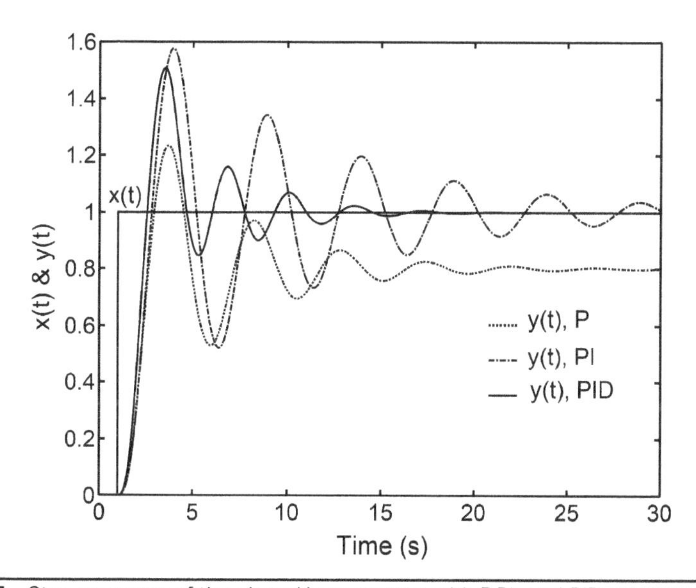

FIGURE 10B.5 Step response of the closed-loop system, with P, PI, and PID controllers.

3. Implementation of the P, PI, and PID Controllers

Connect the controllers as shown in Fig. 10B.4, and then calculate the step response. The resulting response is shown in Fig. 10B.5. This figure shows that the P controller response has a great steady-state error. The PI and PID controllers stabilized the system, with no steady-state error. The controller setting according to the Ziegler–Nichols rule improved the closed-loop system stability and precision. But a final tuning of the controller parameters must be done iteratively until a satisfactory response is obtained.

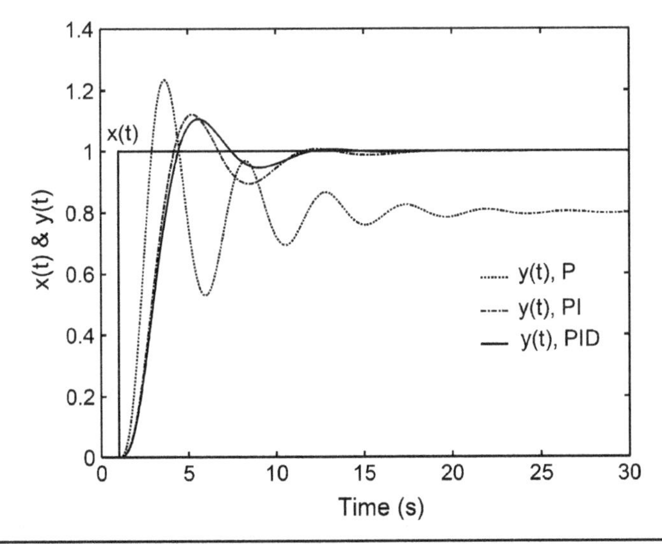

FIGURE 10B.6 The step response of a closed-loop system, equipped with P, PI, and PID controllers, with tuned parameters.

4. Tuning the Controllers' Parameters

The tuning of the gain of the proportional controller does not improve the system behavior, due to the contradiction between the stability and precision requirements. The tuning of the PI and PID controllers improved both the system's stability and precision radically. Figure 10B.6 shows that the response of the system with PI and PID controllers converges rapidly to the required steady-state value, without a steady-state error.

CHAPTER 11

Electrohydraulic Proportional Valves Technology

11.1 Introduction

By the 1960s, the need for higher response speed, control accuracy, and reasonable economy accelerated the development of electrohydraulic proportional valves. These valves consist of hydraulic control valves controlled electronically. They are usually connected in open-loop control systems. They have the advantages of hydraulic power systems, particularly the minimum power-to-weight ratio and high stiffness of hydraulic actuators. Moreover, they have the advantages of electronic controllers, especially high precision and controllability. They are controlled electronically to produce an output pressure or flow rate proportional to the input signal. Therefore, these valves are usually controlled using electronic cards and/or built-in electronics. Compared with ordinary switching valves, proportional valves have the advantages of stepless variation of the controlled parameters and the reduction of the number of hydraulic devices required for a particular control function. They can be placed anywhere on the machine, which improves the dynamic behavior of the system.

This chapter introduces the electrohydraulic proportional valves technology: construction, operation, characteristics, and possible applications. Figures 11.1 and 11.2 give a simple classification of the proportional solenoids and proportional valves covered by this text.

In electrohydraulic systems, the power elements are usually hydraulic while the controller is electronic, analog, or digital. In these systems, the electronic and hydraulic subsystems should be interconnected using an element transforming the low-power electric control signal into a proportional mechanical signal that actuates the hydraulic power elements. In practice, there are several types of motors used to transform the electric control signals into a proportional mechanical signal; mainly:

1. Electric proportional solenoids.
2. Electromagnetic torque motors, covered by Chaps. 9 and 10 of this text.

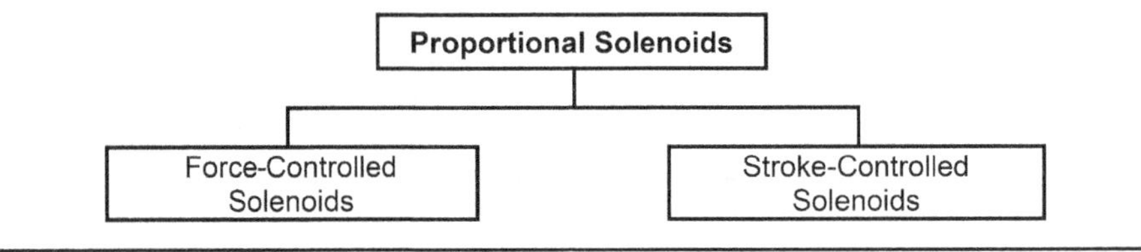

FIGURE 11.1 Proportional solenoids.

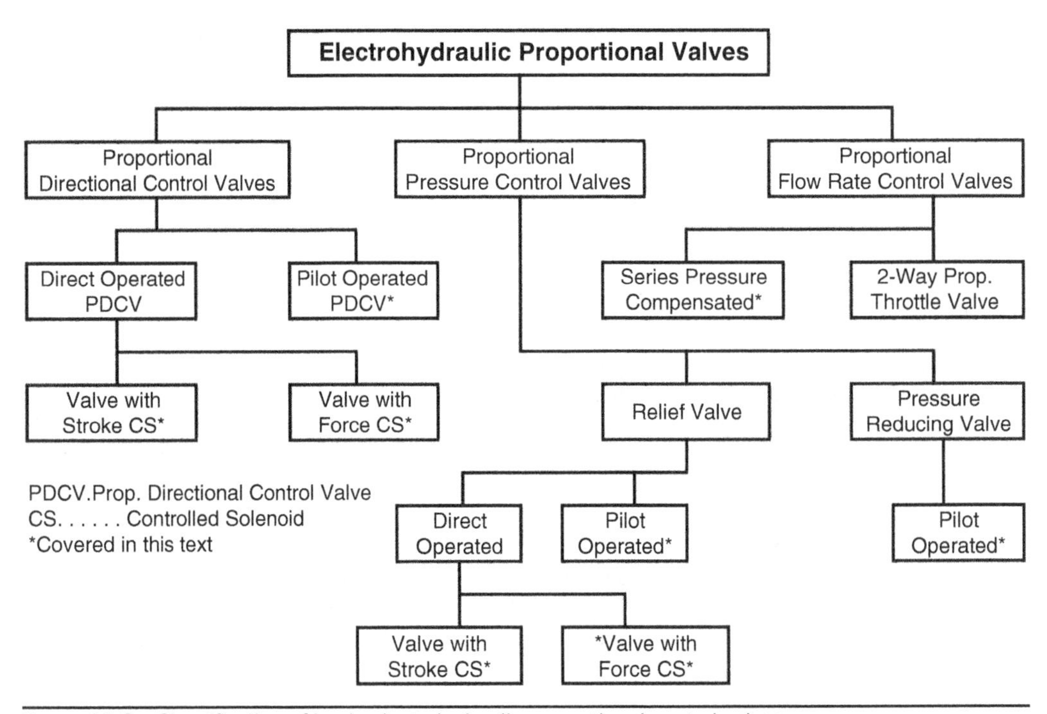

FIGURE 11.2 Classification of basic electrohydraulic proportional control valves.

11.2 Proportional Solenoids

Proportional electric solenoids have the same principle of operation as ordinary electric solenoids. They are a form of DC linear solenoids. These solenoids produce an output force or displacement proportional to the input electrical current. They are distinguished by shorter strokes, higher precision, and special force/stroke characteristics. They can produce sufficiently high forces to drive up to 4 to 6 mm diameter spools. The proportional solenoids are mostly single-acting. Figures 11.3 and 11.4 show typical force/stroke and force/current characteristics of commercial proportional solenoids (see Secs. 9.3 and 10.2).

11.2.1 Force-Controlled Proportional Solenoids

When a conventional solenoid is energized, the armature travels its full stroke. In the case of the force-controlled solenoid, the solenoid produces a force proportional to the

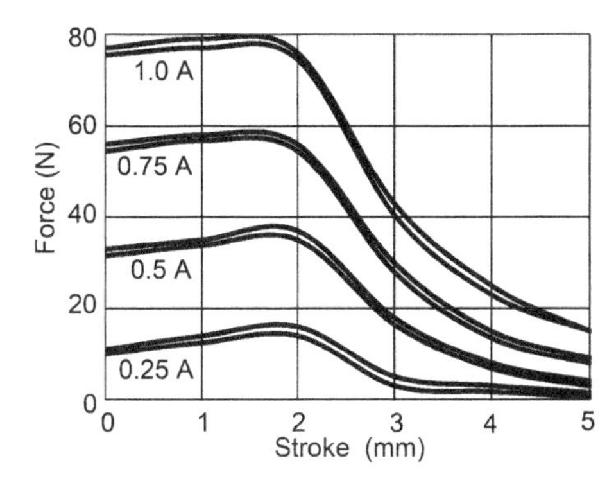

FIGURE 11.3 Typical force/stroke relation of a proportional solenoid at constant current. (*Courtesy Geeplus Inc., USA.*)

input current while the armature does not perform a measurable stroke. Due to current feedback in the control amplifier, the solenoid current and the solenoid force are kept constant even if the coil resistance changes due to temperature variations for example. Without the feedback control of the current, an increase of temperature by 80°C may increase the coil resistance by about 30%, which reduces the coil current.

Proportional force-controlled solenoids (Fig. 11.5) are wet pin DC units that tend to be similar to the conventional DC solenoids but have a modified internal construction that optimize the linearity of the solenoid performance.

Since an input current creates a certain force, the force-stroke curve exhibits this linear relationship at various current levels. When the current to the solenoid is held constant, the solenoid force will also remain constant over a limiting stroke. Therefore, the input current to the force-controlled solenoid is controlled using a closed-loop

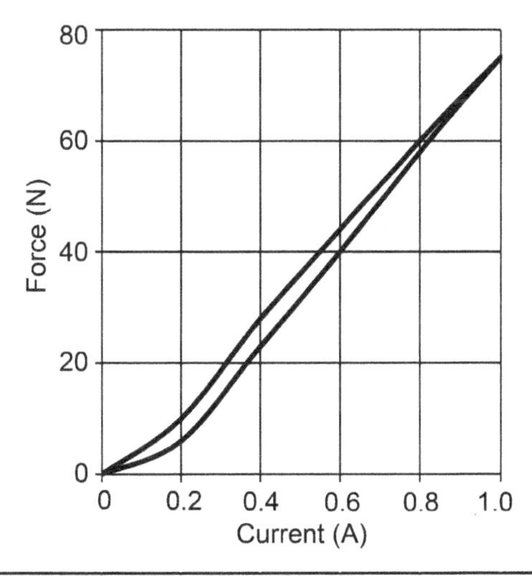

FIGURE 11.4 Typical force/current relation of a proportional solenoid at a constant stroke. (*Courtesy Geeplus Inc., USA.*)

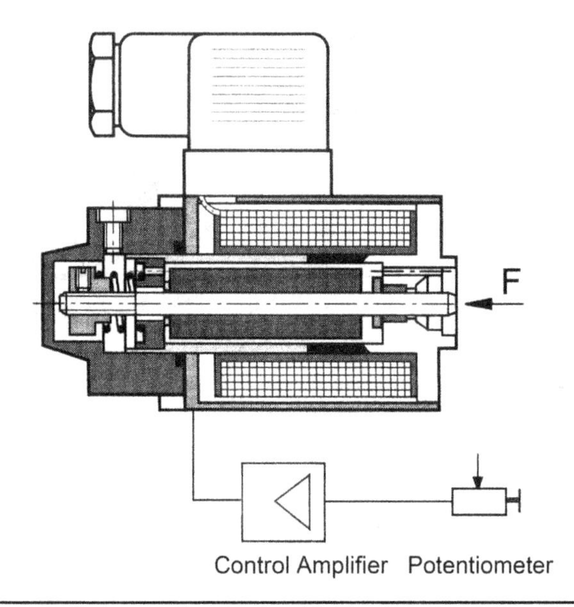

Figure 11.5 Force-controlled proportional solenoid. (*Courtesy Bosch Rexroth AG.*)

controller (Fig. 11.5). In this way, the input current to the solenoid is kept constant. This class of proportional solenoids is usually of limited stroke and compact size.

The main feature of the force-controlled proportional solenoids is the characteristic force/stroke curve. The solenoid force remains nearly constant over a defined stroke range for a constant current. The stroke for the solenoid shown in this example is within 1.5 mm (Fig. 11.6). Given this short stroke, the force-controlled solenoid is used for pilot-operated proportional directional and pressure control valves with the solenoid force being converted into a proportional pressure.

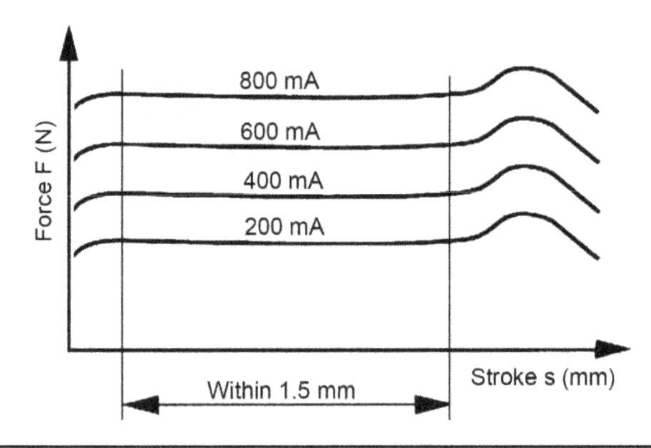

Figure 11.6 Typical steady force/stroke relation for a force-controlled solenoid. (*Courtesy Bosch Rexroth AG.*)

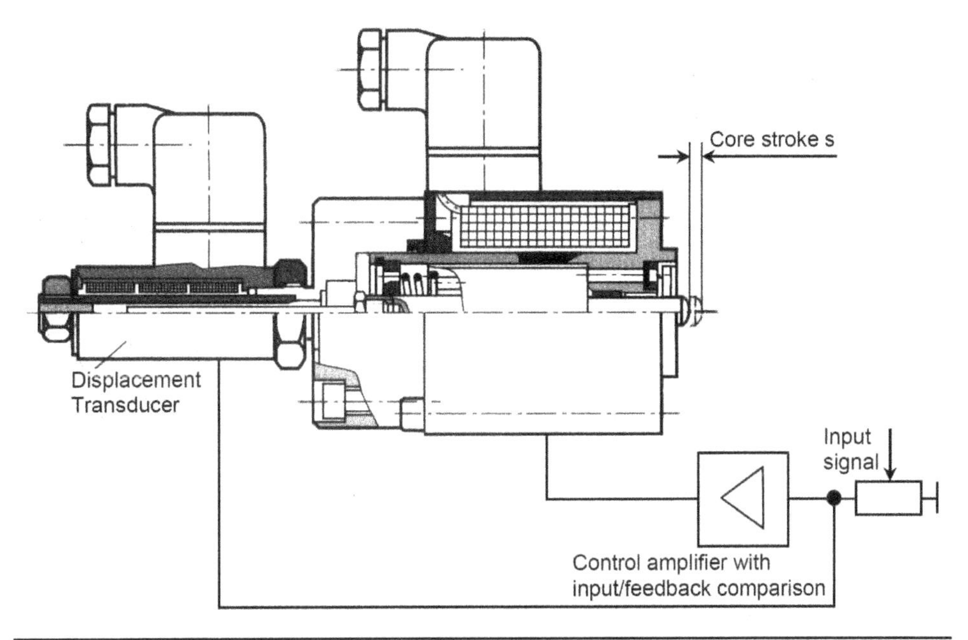

FIGURE 11.7 Stroke-controlled proportional solenoid. (*Courtesy Bosch Rexroth AG.*)

11.2.2 Stroke-Controlled Proportional Solenoids

A proportional solenoid with a displacement transducer (linear variable differential transformer, LVDT) is called a stroke-controlled solenoid (Fig. 11.7). The position of the armature is controlled by a closed-loop control circuit and maintained irrespective of the counter pressure, provided that it is within the rated working range of the solenoid. With the stroke-controlled solenoid, the spools of proportional directional control valves, flow control valves, as well as pressure control valves, can be controlled. The stroke of the solenoid is usually within 3 to 5 mm depending on the valve size. In conjunction with the electrical feedback, the hysteresis and the repetition error of the solenoid are maintained within tight tolerances. In addition, any flow forces (see Sec. 5.3.7), which act on the spool, are compensated.

11.3 Electrohydraulic Proportional Directional Control Valves

Electrohydraulic proportional control valves are controlled electronically. They are connected usually in an open-loop control arrangement. There are electrohydraulic proportional directional, pressure, and flow control valves.

11.3.1 Direct-Operated Electrohydraulic Proportional Directional Control Valves

Valve with Force-Controlled Solenoid

This class of valves is used to control both the direction and flow rates of hydraulic fluid. The valve is controlled using proportional force-controlled solenoids (Fig. 11.8). It consists of a body (3), one or two force-controlled proportional solenoids, depending

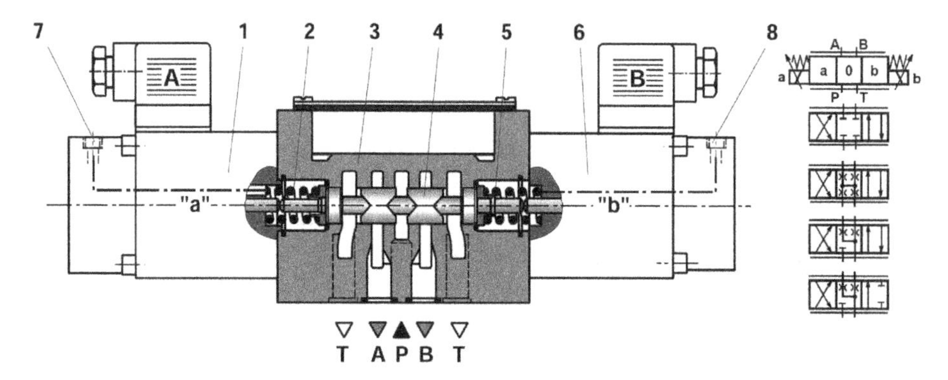

Figure 11.8 Direct-operated electrohydraulic proportional directional control valve with force-controlled solenoids. (*Courtesy Bosch Rexroth AG.*)

on the number of positions (1 and 6), spool (4), and two centering springs (2 and 5). In the case of the 3-position valve, shown in Fig. 11.8, if the coils are de-energized, the spool (4) is held in the center position by the centering springs (2 and 5). When the proportional solenoid (1) is supplied by an input current, the solenoid pushes the spool to the right. The spool displacement depends on the solenoid force and the stiffness of the centering springs. This displacement allows the fluid to flow from P to B and A to T. Likewise, when solenoid (6) is energized, the spool shifts to the left allowing the fluid to flow from P to A and B to T. Optional manual overrides are available for fault diagnostics and emergency operation of the valve. To achieve the proper functioning of the valve, it must be bled when commissioning. Plugs (7 and 8) are used for this purpose.

The spool displacement, which is in proportion to the solenoid force, causes the V-shaped grooves on the spool to open, clearing a corresponding opening area. The opening area and the corresponding flow rate are proportional to the square of the spool displacement. When the proportional solenoids are de-energized, the spool is returned to the center position under the action of the centering springs and the flow forces. This class of valves operates, generally, for flow rates within 100 L/min.

The typical steady-state and dynamic valve performance are illustrated by Figs. 11.9 through 11.11. Figure 11.9 shows the flow characteristics for different values of pressure difference applied to the valve, P_v. The flow rate increases nonlinearly with the input current due to the dead zone dictated by the proportional solenoid characteristics and the nonlinear variation of the valve restriction area.

The transient response is widely used to evaluate the system's transient behavior, precision, and stability. Figure 11.10 shows the transient response of the valve spool stroke to step input signals of different magnitudes. The plots show that the settling time of this valve is within 120 ms for the spool displacement under the action of the solenoid's forces, as opposed to the springs, friction, and flow forces. The settling time is the time required for the response to reach an end state within ±5% of the final steady-state value. This time decreases to values less than 80 ms for the spool return under the action of the springs and flow forces.

The frequency response is another efficient tool for the evaluation of the behavior of dynamic systems. It enables one to judge the system's stability and the possible range

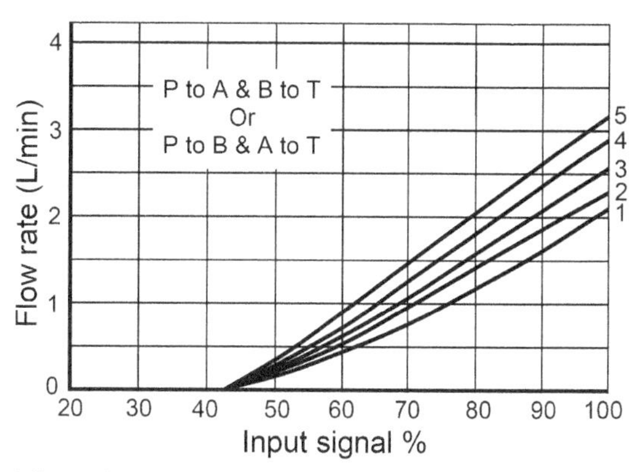

FIGURE 11.9 Typical steady-state flow-current characteristics of a direct-operated electrohydraulic proportional directional control valve, with force-controlled solenoids. (*Courtesy Bosch Rexroth AG.*)

of operation under certain constraints. It could be evaluated by subjecting the studied system to a sinusoidal input signal and measuring/calculating its output. If the system is linear, the steady-state output is also sinusoidal and has the same frequency as the input but with a shift in phase and change in magnitude. In the case of nonlinear systems, the output signal will be distorted from the sinusoidal. In this case, a signal treatment is necessary to separate the basic harmonic, having a frequency equal to that of the input signal. The separated basic harmonic is treated as the system's output and is used to calculate the frequency response.

The frequency response is based on the measurement/calculation of the gain and phase. The gain equals the amplitude of the output signal divided by that of the input one, and the phase is the phase shift between the input and output signals. The Bode

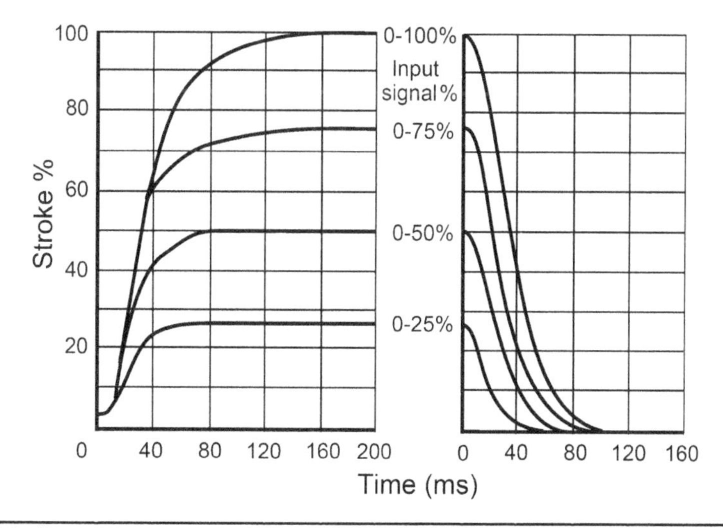

FIGURE 11.10 Transient response of a direct operated electrohydraulic proportional directional control valve, with force-controlled solenoids. (*Courtesy Bosch Rexroth AG.*)

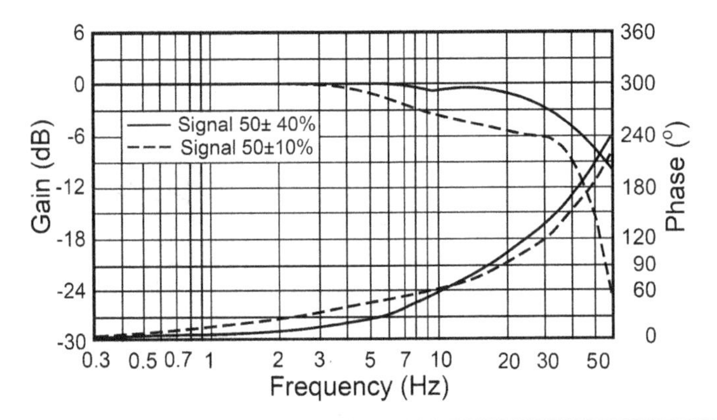

FIGURE 11.11 Typical frequency response; Bode diagram, of a direct operated electrohydraulic proportional directional control valve, with force-controlled solenoids. (*Courtesy Bosch Rexroth AG.*)

plot is one of the efficient frequency response plots. It consists of two simultaneous plots: the gain and phase versus the frequency of the input signal. In the gain plot, the magnitude ratio is plotted against the frequency of the input signal and the phase plot consists of a plot of the phase shift against the frequency of the input signal. The magnitude ratio is converted into decibels to plot the Bode diagram where the gain in decibels is:

$$\text{Gain} = 20\log_{10}\left(|\text{output}(t)|/|\text{input}(t)|\right)\text{ dB}$$

Figure 11.11 shows that the spool displacement can follow, precisely, the oscillating input signals of frequencies up to 3 Hz for signals of small amplitudes (50 ± 10%). Meanwhile, for inputs of larger amplitudes (50 ± 40%), the spool displacement can follow, precisely, the oscillating input signals of frequencies up to 7 Hz.

Valve with Stroke-Controlled Solenoid

For more precise control of the spool displacement, and consequently the restriction area and flow rate, the stroke-controlled proportional solenoids could be used. Figure 11.12 shows a typical design of a direct-operated electrohydraulic

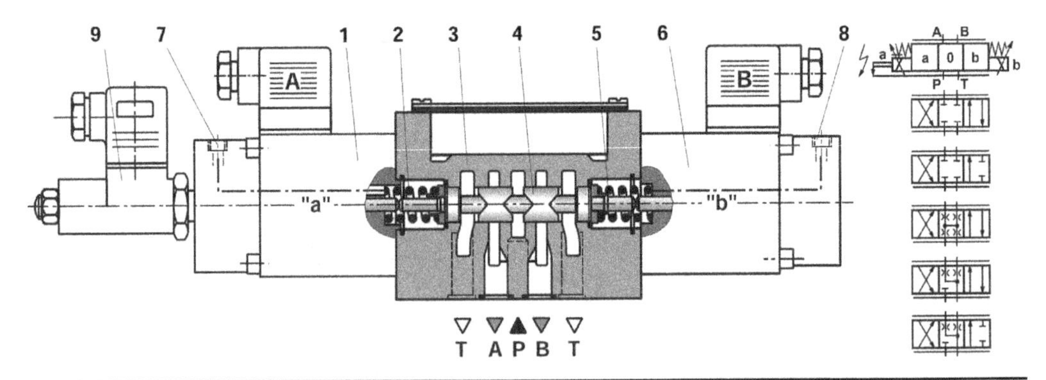

FIGURE 11.12 Direct-operated electrohydraulic proportional directional control valve with stroke-controlled solenoids. (*Courtesy Bosch Rexroth AG.*)

proportional directional control valve with a stroke-controlled solenoid. The solenoids control the start, stop, direction, and fluid flow rate for smooth acceleration and deceleration of an actuator. The direction control is carried out by displacing the spool to the desired position by applying input control signals to the relevant solenoid. The valve consists of a spool (4), housing (3), two centering springs (2 and 5), two stroke-controlled proportional solenoids (1 and 6), and a displacement transducer (9). The displacement transducer feeds back the picked-up spool displacement to the closed loop controllers of both of the solenoids. The valve is equipped with two plugs (7 and 8) for valve bleeding to achieve the proper functioning of the valve when commissioning.

The position of the spool is monitored by the displacement transducer (9), which is connected in a closed-loop circuit on the electronic control card. In the case of the absence of a control current, the cores of solenoids (1 and 6) are not displaced and the spool (4) is kept in the neutral position under the action of the centering springs (2 and 5). When any of the solenoids is energized, the spool displaces by a distance corresponding to the exciting electric signal. Any deviation from the desired position is then corrected by the closed-loop control. The valve ports are then interconnected through the resulting valve opening areas. The flow rate is in proportion to the spool displacement. When the solenoid is de-energized, the spool is returned to its neutral position by the centering springs and the flow forces.

The valve is usually produced in different sizes, allowing for different levels of flow rates. Size 6 operates for nominal flow rates within 65 L/min. Figure 11.13 shows the steady-state flow characteristics of this valve. There is no dead zone, which could be attributed to the effect of the spool valve geometry and the closed-loop spool displacement control. It shows also that the valve flow rate increases with the increase in input current. The proportionality in the flow-current relation is nonlinear, depending mainly on the stroke-throttle area relation.

The step response (Fig. 11.14) shows that the settling time is within 90 ms for a step increase in current and within 70 ms for a step decrease in current.

The Bode plot (Fig. 11.15) shows that spool displacement can follow, precisely, oscillating input signals of frequencies up to 2 Hz.

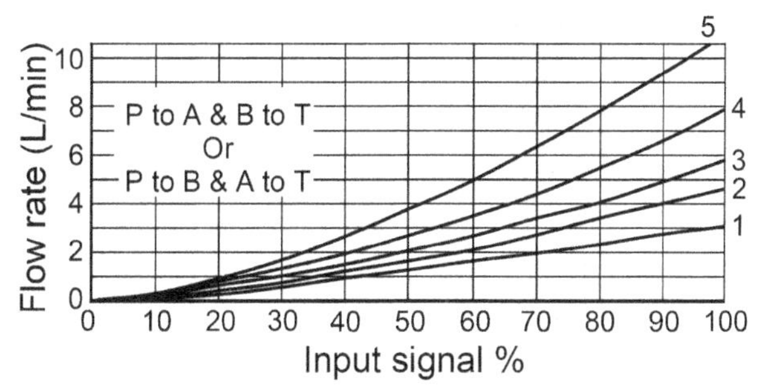

1-Pressure drop P_V=10 bar, 2-P_V=20 bar, 3-P_V=30 bar, 4-P_V=50 bar, 5-P_V=100 bar.

FIGURE 11.13 Typical steady-state flow-current characteristics of a direct-operated electrohydraulic proportional directional control valve, with stroke-controlled solenoids. (*Courtesy Bosch Rexroth AG.*)

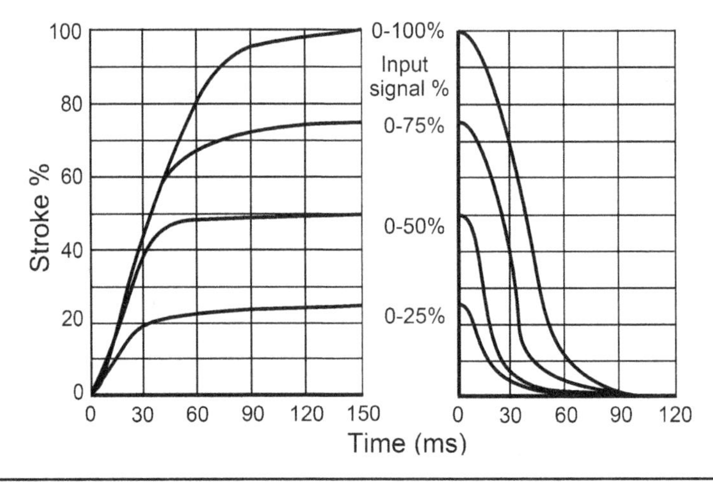

FIGURE 11.14 Typical transient response of a spool displacement of a direct-operated electrohydraulic proportional directional control valve size 6, with stroke-controlled solenoids.

11.3.2 Pilot-Operated Electrohydraulic Proportional Directional Control Valves

In the case of the ordinary pilot-operated directional control valves, Sec. 5.3.9, Fig. 5.45, the pilot stage consists of a simple direct-operated directional control valve. In this class of valves, the pilot valve spool is displaced to one of its extreme positions by energizing one of the electric solenoids. The pilot valve ports are directly communicated to the control chambers of the main spool, which displaces, in turn, to its end position under the action of pressures at the pilot valve output ports.

Figure 11.19 shows a pilot-operated electrohydraulic proportional directional control valve. In the case of the proportional directional valve, it is recommended to control

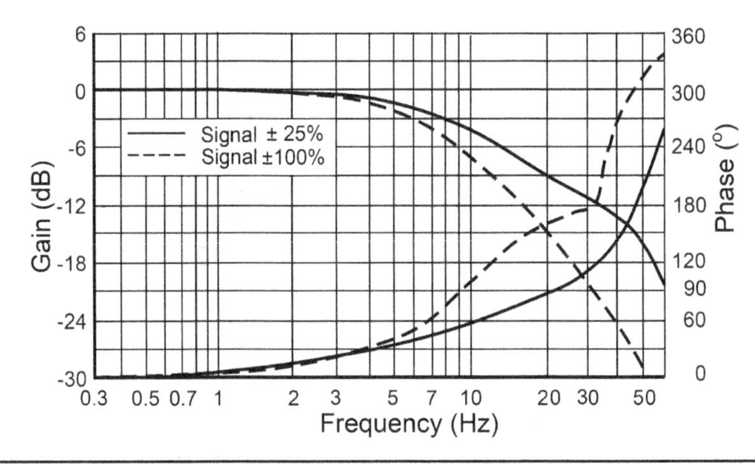

FIGURE 11.15 Typical frequency response of a direct-operated electrohydraulic proportional directional control valve, with stroke-controlled solenoids. (*Courtesy Bosch Rexroth AG.*)

both the magnitude and direction of the main spool displacement. Therefore, a dual proportional pressure-reducing valve is usually used as a pilot stage. In this way, the pressures in the side chamber of the main spool (13 and 16, Fig. 11.19), are controlled. They act on the main spool by a proportional force, which, together with the centering springs, result in a proportional displacement of the main spool.

Dual Three-Way Electrohydraulic Proportional Pressure Reducer

The principle of operation of hydraulic pressure reduction is explained in Sec. 5.2.3. Herein, a typical example of the pilot stage of a proportional pilot-operated direction control valve is explained. This pilot stage consists of a dual three-way proportional pressure-reducing valve, Fig. 11.16. Each valve produces an exit-reduced pressure of value proportional to the applied input current. The valve is equipped with two force-controlled proportional solenoids. The control current produces a proportional solenoid force. For a constant value of the input signal, a feedback loop serves to keep the solenoid current and force constant over the whole stroke of the solenoid core. The direct-operated proportional pressure-reducing valves are used to control the pressures and direction of fluid flows in the main stage. The required reduced pressure is set by applying a corresponding input control signal.

This valve consists of two force-controlled proportional solenoids (1) and (6), a spool (2), a housing (3), two centering springs (9) and (10), and two pressure-sensing spools (4 and 5). The proportional solenoids (1) control the reduced pressure in port (A) while the proportional solenoids (6) control the reduced pressure in port (B).

When both of the solenoids are de-energized, the solenoid forces are null and the spool (2) is kept at the center position under the action of the centering springs (9) and (10). At this position, the two ports (A) and (B) are connected to the tank, while

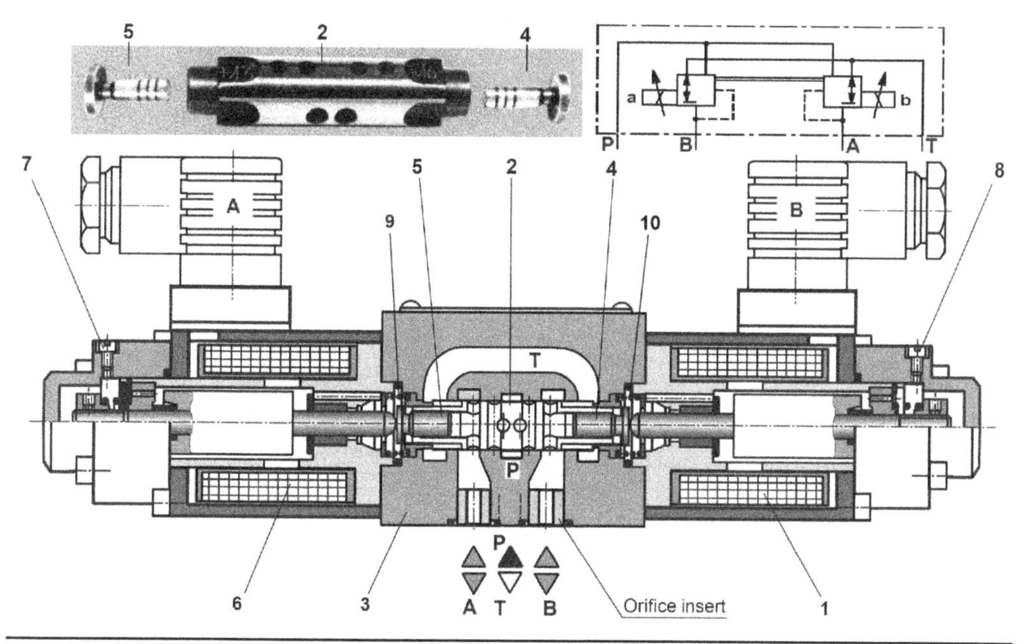

FIGURE 11.16 Dual three-way electrohydraulic proportional pressure reducer. (*Courtesy Bosch Rexroth AG.*)

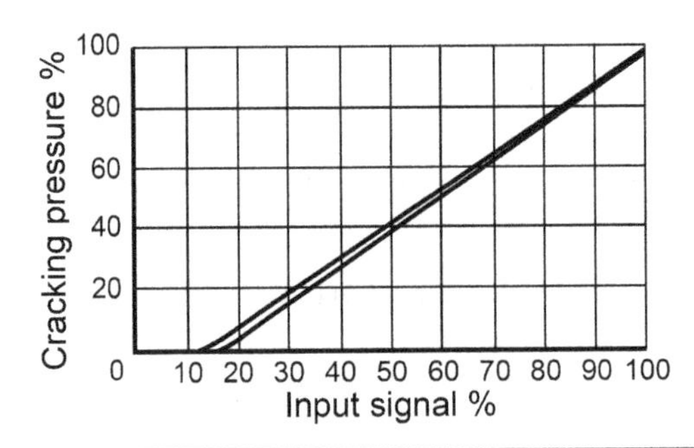

FIGURE 11.17 Typical steady-state relation between the cracking pressure and input signal for a three-way electrohydraulic proportional pressure reducer. (*Courtesy Bosch Rexroth AG.*)

the pressure port P is disconnected from both ports (A) and (B). The proportional solenoid forces are directly applied to the spool (2). When the solenoid (6) is energized, it produces a force proportional to the control current. This force acts directly on the spool (2), displacing it to the right. The oil may now flow from (P) to (B) while port (A) connects widely to the tank line. As the pressure in port (B) increases, the pressurized fluid is communicated through the radial drillings in the spool (2) to the inner left side of the pressure-sensing spool (4). The spool (4) moves to the right until it is supported by the pin of the core of the solenoid (1). The force generated by the pressure in port (B) acts on the spool against the force of the solenoid (6). It pushes the spool (2) to the left until a balance is achieved between the two forces. The balance is reached when the pressure in port B reaches a value corresponding to the input signal to solenoid (6). As a pressure reducer, the reduced pressure in port B corresponds to the force produced by solenoid 6 at zero flow conditions. This pressure value is called nominal pressure or cracking pressure. The cracking pressure varies with the control signal as shown in Fig. 11.17.

Figure 11.18 shows a typical steady-state performance of the studied twin-three-way proportional pressure reducer. The pressure in the reduced pressure port (A or B) varies not only with the control signal setting but also with the flow rate at these ports. If the pressure in port (B) is increased above the preset value, the pressure force overcomes the solenoid force. The spool moves to the left connecting port (B) to tank

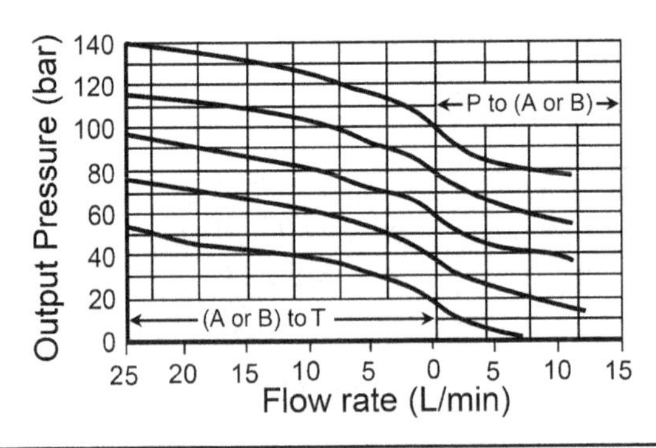

FIGURE 11.18 Typical steady-state outlet reduced pressure/flow rate relation for a three-way electrohydraulic proportional pressure reducer at different input signals. (*Courtesy Bosch Rexroth AG.*)

line (T). Then, the pressure in port (B) decreases until the forces balance is reached again. This case is illustrated by the left zone of Fig. 11.18. This Figure shows the pressure variations in the two reduced pressure ports, (A) and (B), with the variation of the flow rate for five different values of the control signal. Only one solenoid is energized at a time.

If the pressure in the output port (B) is decreased, the force of solenoid (6) overcomes the pressure force, and the spool moves to the right. The port (P) connects to (B). The high-pressure fluid flows from (P) to (B). As a result, the pressure in port (B) increases until it equals the preset cracking pressure, where the forces balance is reached again. This case is illustrated by the right zone of Fig. 11.18. Then, the increase or decrease in the pressure in port B due to variations of load pressure, for example, moves the spool in the direction to recover the cracking pressure.

The valve is equipped with two plugs (7 and 8) for valve bleeding to achieve the proper functioning of the valve when commissioning.

Pilot-Operated Electrohydraulic Proportional Directional Control Valve

Figure 11.19 shows a typical example of the pilot-operated electrohydraulic proportional directional control valve. This valve consists of a pilot stage (9) and the main valve (12). The pilot stage consists of a dual three-way electrohydraulic proportional pressure reducer, controlled by two force-controlled proportional solenoids (1 and 6). It is equipped with two plugs (7 and 8) for valve bleeding to achieve the proper functioning of the valve when commissioning.

The main stage consists of the main spool (14), a double-acting centering spring assembly (15), and two control chambers (13 and 16). The main spool is kept in its neutral position under the action of the centering spring (15). The spring is installed in

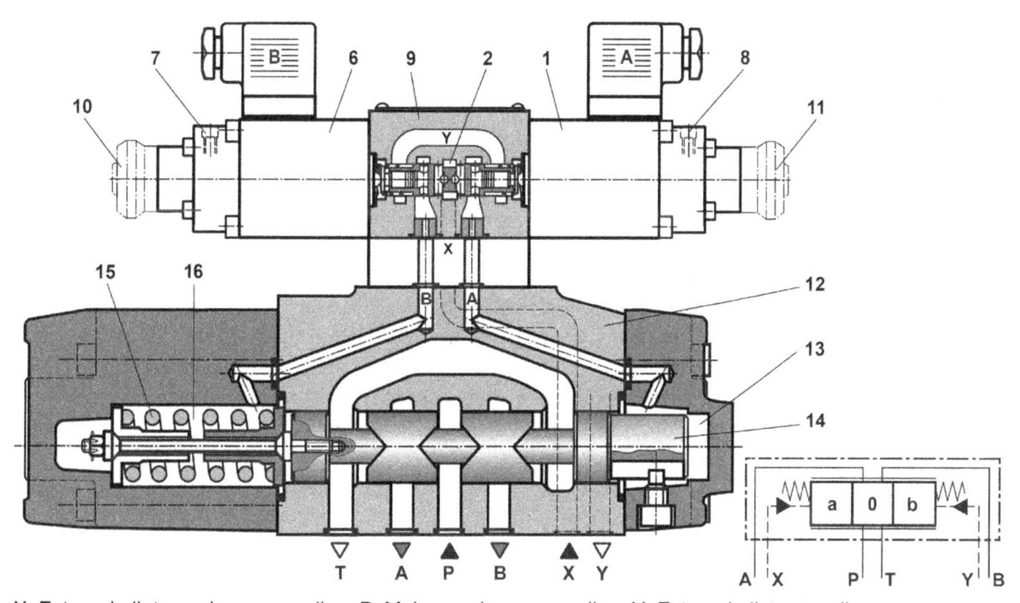

FIGURE 11.19 Pilot-operated electrohydraulic proportional directional control valve. (*Courtesy Bosch Rexroth AG.*)

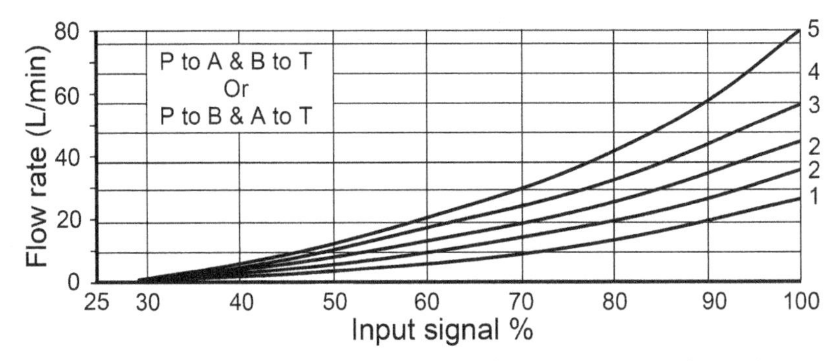

FIGURE 11.20 Typical flow rate/input signal relation of an electrohydraulic proportional pilot operated directional control valve. (*Courtesy Bosch Rexroth AG.*)

a way such that it is compressed by the spool displacement in any of its two directions of motion. At rest, the centering spring (15) holds the main spool (14) at the center position. If solenoid (6) is energized, the pilot spool (2) is moved to the right. The pilot oil is fed externally from port X via the pilot valve (9) into the control chamber (13), while the left chamber (16) is connected to the external drain line (Y). The reduced pressure in the chamber (13) increases to the pressure level preset at the solenoid (6). The main spool (14) moves to the left, against the centering spring, for a distance proportional to the reduced pressure in the chamber (13). The throttling grooves of the main spool open progressively as the control signal increases. The valve throttling areas, communicating the valve ports, (P-A) and (B-T), are changed in proportion to the valve displacement. Then, if the pressure drops across the main valve restrictions are controlled, using a pressure compensator, for example, the valve flow rate becomes proportional to the control signal. When the electrical signal is reduced to zero, both the pilot and main spools return to their neutral positions.

The valve is equipped with emergency hand operators (10 and 11), used for emergency operations and fault diagnostics.

Figure 11.20 shows the steady-state flow characteristics of the presented pilot-operated valve. It shows the variation of the valve flow rate with the input current for different values of the pressure drop across the main valve restriction. The nonlinearity of the flow-current relation is dictated mainly by the geometry of the spool, the precompression of the spring (15), and the pressure reducer characteristics.

The step response of the pilot-operated directional control valve is given in Fig. 11.21, which shows that the pilot-operated valve presents a settling time within 55 ms, for a step increase in control current. This value decreases to less than 60 ms for a step decrease in the input signal.

11.4 Electrohydraulic Proportional Pressure Control Valves

11.4.1 Direct-Operated Electrohydraulic Proportional Pressure Relief Valve

In the case of the direct-operated electrohydraulic proportional relief valves, the conventional prestressed loading spring of the ordinary valves is replaced by a proportional solenoid with or without a spring. There exist two types of direct-operated

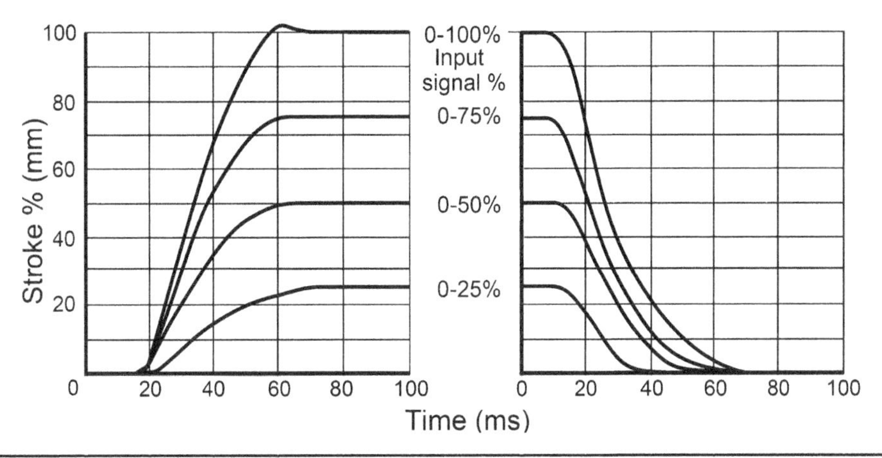

Figure 11.21 Typical transient response of an electrohydraulic proportional pilot-operated directional control valve size 10. (*Courtesy Bosch Rexroth AG.*)

proportional relief valves; a valve with a force-controlled solenoid and a valve with a stroke-controlled solenoid.

Valve with Stroke-Controlled Solenoid

The direct-operated proportional pressure relief valve (Fig. 11.22) consists basically of a body (1), stroke-controlled proportional solenoid (2), with a displacement transducer (3), poppet (5), poppet seat (4), spring (6) and spring plate (7). The input signal for the pressure setting is set by an input potentiometer on the electronic card. This input signal operates via an amplifier and proportional solenoid (2) to produce a proportional force. This force acts on the core to compress the spring (6). The compression distance of the spring is picked up by the displacement transducer (3) and feedback to the controller. In this way, the cracking pressure of the valve is determined. Greater control

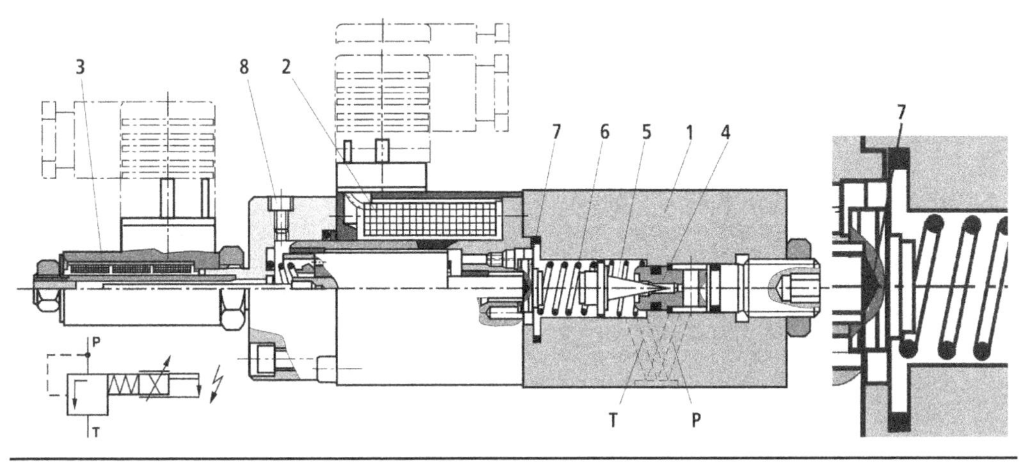

Figure 11.22 Direct-operated electrohydraulic proportional relief valve with stroke-controlled solenoid. (*Courtesy Bosch Rexroth AG.*)

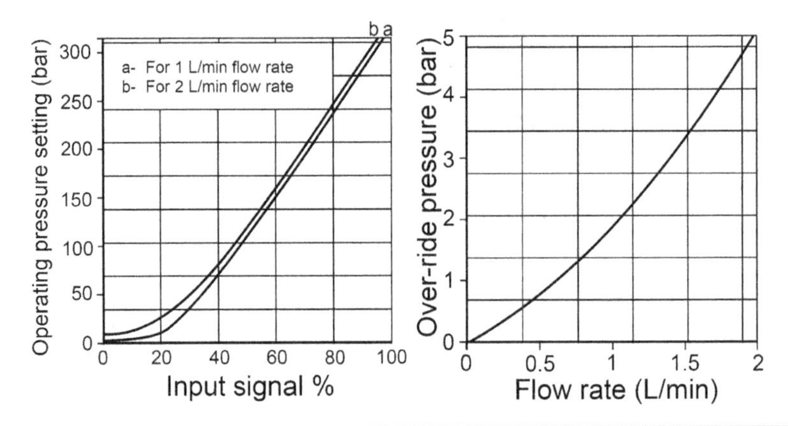

FIGURE 11.23 Typical steady-state characteristics of the direct-operated electrohydraulic proportional relief valve of stroke-controlled solenoid. (*Courtesy Bosch Rexroth AG.*)

current produces greater solenoid displacement and greater valve cracking pressure. Any difference between the input value and the actual displacement is then corrected by the feedback system. In this way, the spring compression distance, and consequently the relief valve cracking pressure, could be preset. The cracking pressure is the pressure level at which the poppet valve becomes about to open, while the valve flow rate is still zero. This valve is of low hysteresis and good repeatability. With zero input current, the spring displaces the core to the left position and the valve automatically sets to the lowest pressure. The valve is equipped with an air bleeding plug (8) for valve bleeding to achieve the proper functioning of the valve when commissioning.

Figure 11.23 shows a typical steady-state relation between the operating pressure and the command signal of a direct-operated relief valve with the stroke-controlled solenoid. The valve output-operating pressure depends on the input signal and relieved flow rate. It reaches up to 315 bars for maximum control signal in this case. Moreover, for certain input signals, this pressure increases for increased flow rate and the maximum override pressure reaches up to 5 bar for a maximum flow rate of up to 2 L/min.

At zero control signal, the poppet valve is fully open and the minimum pressure increases to a limited value with the increase of the valve flow rate from the port (P) to (T) through the poppet seat open area. Generally, the override pressure decreases with the decrease of the spring stiffness and increase of the poppet valve seat diameter, Sec. 5.2.1.

Valve with Force-Controlled Solenoid

Figure 11.24 shows a direct-operated proportional pressure relief valve with the force-controlled solenoid. This valve consists of a body (1), force-controlled proportional solenoid (2), poppet seat (3), and valve poppet (4). Initially, the spring (6) puts the valve in initial opening position. The input signal to the solenoid (2) produces a proportional force acting on the valve poppet. In this way, the cracking pressure of the valve is adjusted. Greater control current produces greater solenoid force and greater valve cracking pressure.

Figure 11.25 shows typical steady-state characteristics of the direct-operated relief valves with the force-controlled solenoid. It shows the relation between the limiting

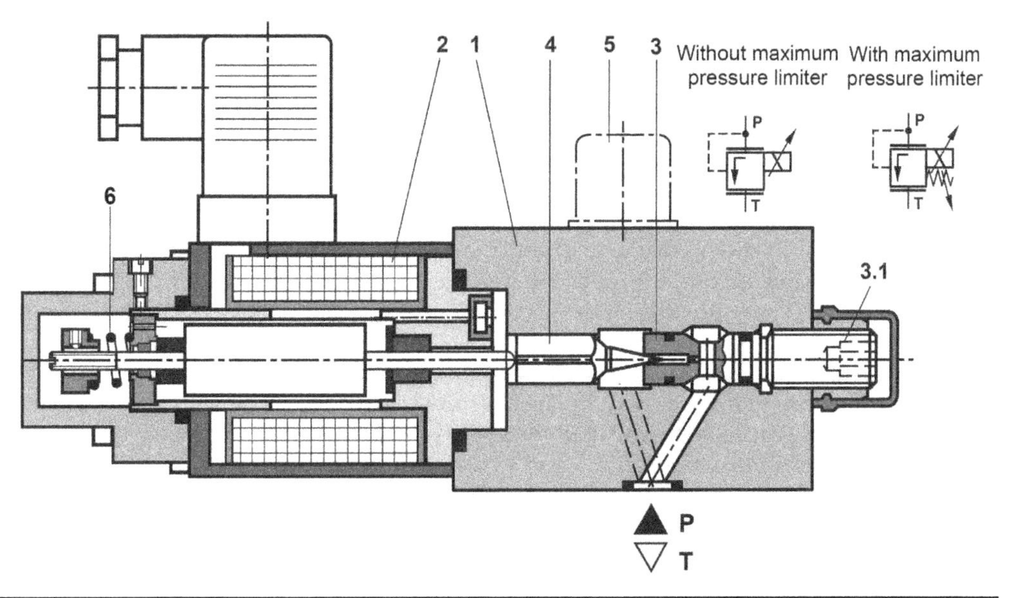

Figure 11.24 Direct-operated electrohydraulic proportional relief valve with force-controlled solenoid. (*Courtesy Bosch Rexroth AG.*)

pressure and the command signal. Moreover, the effect of magnetic hysteresis is within ±1.5% of the maximum pressure limit. At zero control signal, the poppet valve is fully open under the action of a low-stiffness spring (6) and the pressure level is minimal.

The direct-operated relief valves could be optionally equipped with an additional maximum pressure valve (5). This valve is simply another ordinary direct-operated relief valve that operates in the case of any unexpected solenoid current increase, due to possible defects in control electronics. The maximum pressure limiter (5) should be set at 15% higher than the maximum pressure set at the proportional solenoid (2). The maximum pressure limiter does not affect the valve characteristics during its normal operation.

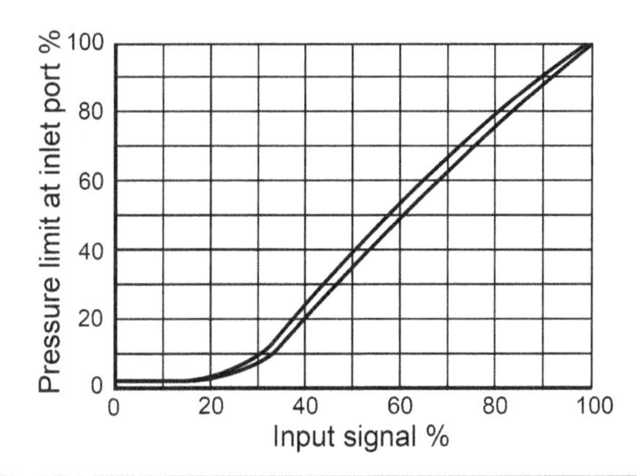

Figure 11.25 Typical steady-state characteristics of an electrohydraulic proportional direct-operated relief valve with force-controlled solenoid. (*Courtesy Bosch Rexroth AG.*)

11.4.2 Pilot-Operated Electrohydraulic Proportional Pressure Relief Valve

Generally, the direct-operated relief valves are applicable for relatively small flow rates. The increased flow rates introduce great override pressure. The override pressure could be reduced by increasing the poppet area and using a spring of low stiffness (see Secs. 5.2.1 and 5.2.2). These contradicting demands could be fulfilled by using the pilot-operated valves (Fig. 11.26). This valve could be used for relatively high flow rates, up to 600 L/min for example. The valve consists of the pilot stage (1) with its force-controlled proportional solenoid (2), and the main valve (3) with the main poppet insert (4). The pressure setting is proportional to the electric current applied to the solenoid (2). The circuit pressure acts on the lower side of the main poppet (4). At the same time, this pressure acts on the upper face of the main poppet via controlled drilling (8) and orifices (5, 6, and 7). This pressure is also communicated to the poppet valve (9) of the pilot stage, where it is opposed by the force of the proportional solenoid (2). When the pressure force exceeds the solenoid force, the pilot poppet (10) opens and allows oil to flow to the tank via external drain line Y (12). The fluid flows to the pilot valve through restrictions 5 and 6, which are of small diameters and operate at a very small pressure difference. Therefore, the flow rate through the pilot stage is very small; consequently, it operates with negligible override pressure. Thus, the pilot stage acts to limit the maximum pressure at its inlet (A) to values proportional to the applied current. For further increase of inlet pressure, the pressure difference across the main poppet increases due

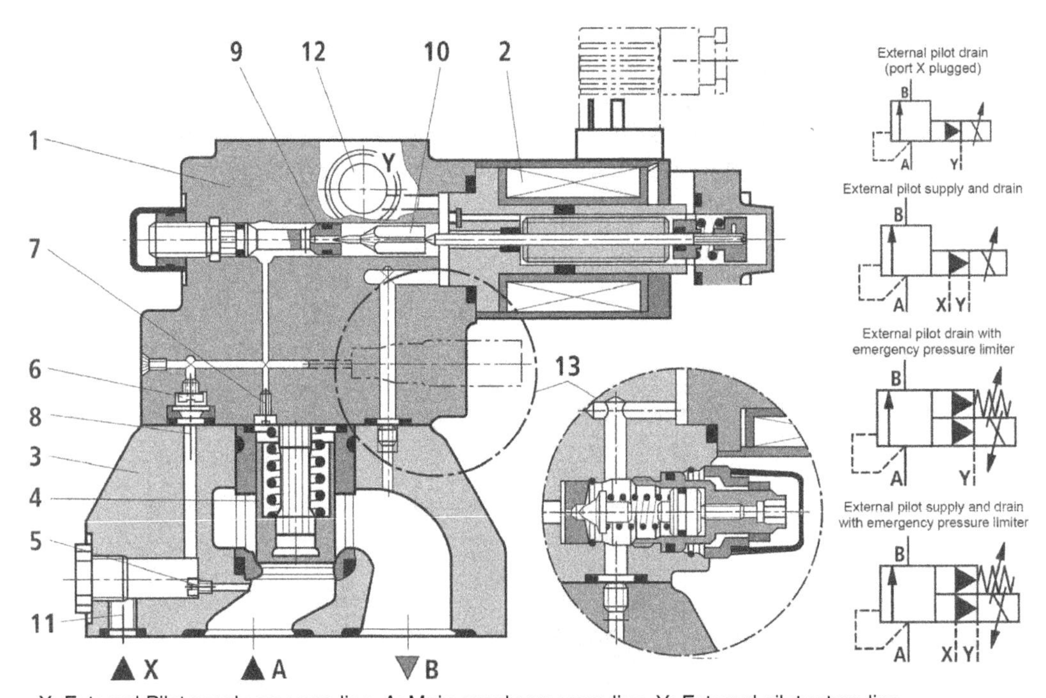

FIGURE 11.26 Pilot-operated electrohydraulic proportional relief valve with maximum pressure safety. (*Courtesy Bosch Rexroth AG.*)

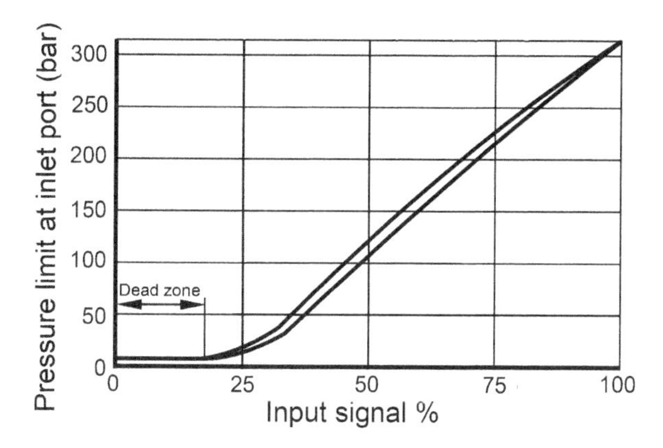

Figure 11.27 Steady-state relation between the limiting pressure and input signal for a typical pilot-operated electrohydraulic proportional pressure relief valve. (*Courtesy Bosch Rexroth AG.*)

to the orifices (5 and 6). The main valve lifts from its seat, opening a connection from A to B (pump to tank).

The pilot stage may be supplied internally by screwing in the plug (11) or externally through port (X) by removing this plug and plugging the orifice (5). Moreover, it can be drained internally to port (B) or externally to port Y (12) by removing or installing the relevant plug.

The pilot-operated relief valve could be optionally equipped with a maximum pressure limiting insert (13), Fig. 11.26. This insert is simply another ordinary direct-operated relief valve that operates in the case of any unexpected failure of operation of the pilot valve. This additional insert is installed in parallel with the pilot stage. It is preset to a pressure limit at least 15% higher than the maximum pressure set at the proportional solenoid (2).

Figure 11.27 shows the steady-state characteristics of a typical pilot-operated proportional pressure relief valve. Moreover, Fig. 11.28 shows the steady-state pressure-flow relation of a typical pilot-operated proportional pressure relief valve. Figure 11.28

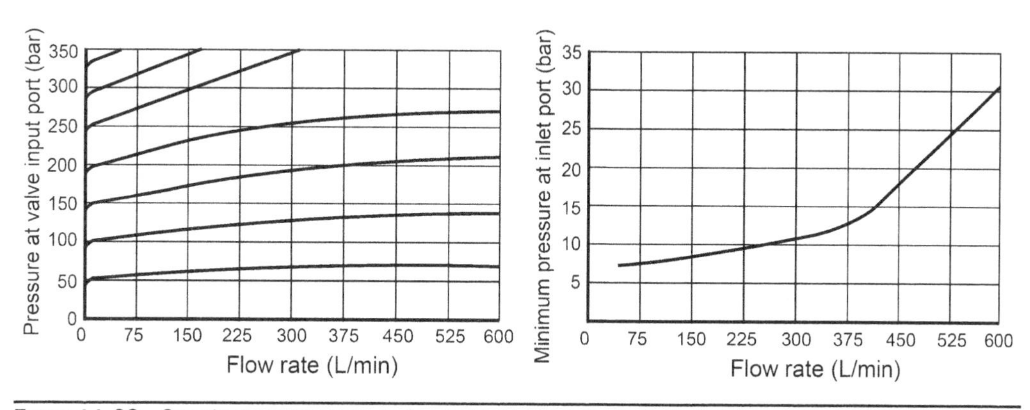

Figure 11.28 Steady-state pressure to flow rate relation of a typical pilot-operated electrohydraulic proportional pressure relief valve. (*Courtesy Bosch Rexroth AG.*)

shows this relation for different preset cracking pressures. The override pressure increases with the increase of the flow rate. For a cracking pressure of 150 bar at zero flow rate, the maximum pressure reached 215 bar at 600 L/min, with about 65 bar override pressure. Figure 11.28 shows also the minimum pressure characteristics at zero input signal. The main stage is loaded practically with its light spring precompression. Its cracking pressure ranges mostly from 4 to 10 bar depending on the valve design. The inlet pressure increases with the valve flow rate as illustrated in Fig. 11.28.

11.4.3 Pilot-Operated Electrohydraulic Proportional Pressure Reducer

The pilot-operated proportional pressure reducer operates for relatively high flow rates, like 300 L/min. It comprises the pilot valve (1) with force-controlled proportional solenoid (2), main valve (3) with main spool assembly (4), and an optional check valve (5), Fig. 11.29. The setting of the pressure in port (A) depends upon the proportional solenoid current. At rest, with no pressure in port B, the main spool (4) displaces downwards under the action of the law stiffness spring (11) and opens widely the

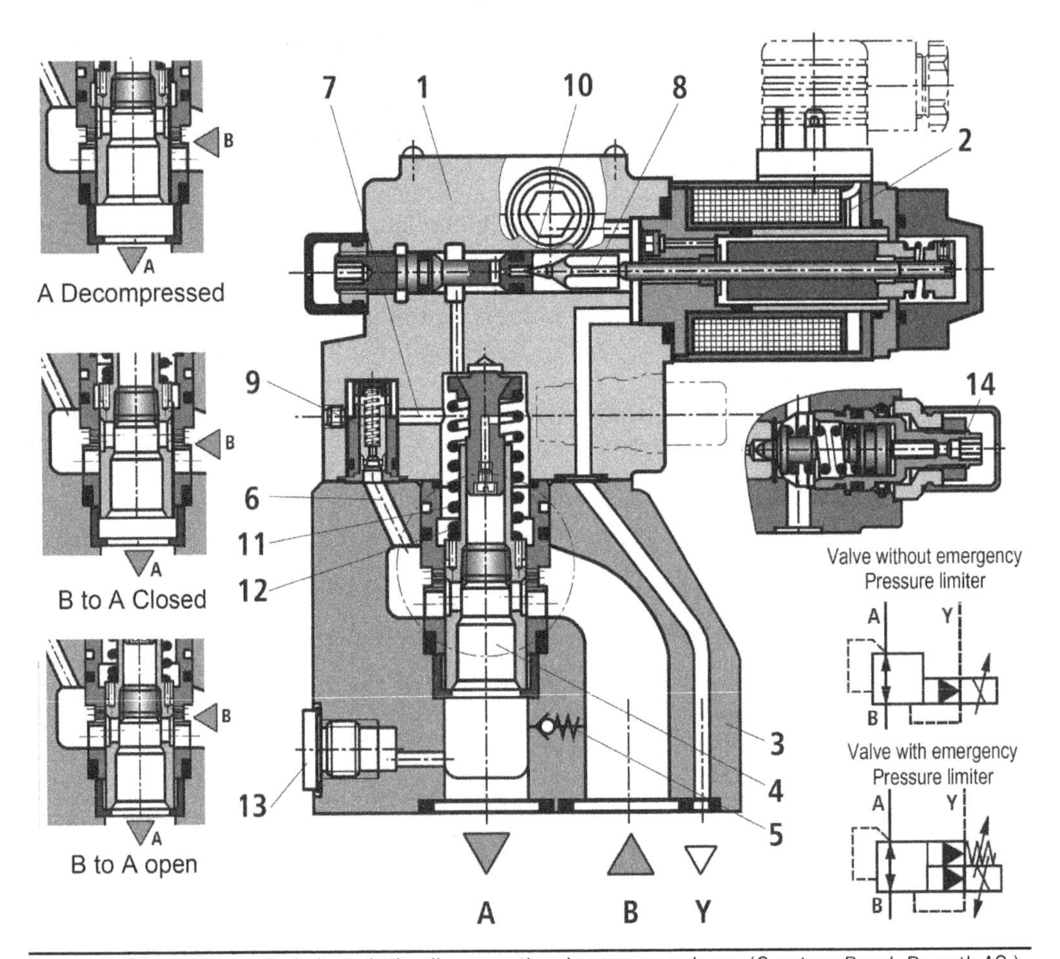

FIGURE 11.29 Pilot-operated electrohydraulic proportional pressure reducer. (*Courtesy Bosch Rexroth AG.*)

passage from the input port (B) to exit line (A). The pressure in line (A) acts on the lower area of the spool, (tending to displace it upwards in the direction to close the (B-A) connection. By Increasing the pressure in the inlet port (B), the high-pressure oil flows to the upper chamber of the main spool through the flow control valve (9). The pressure increases in the upper chamber and acts on the main spool downwards, in the direction to open widely the passage from (B) to (A).

If the proportional solenoid is not energized, the poppet (8) of the pilot valve is put in an initial opening position under the action of the core spring. The pilot oil flows from port (B), through the drilling (6), flow control valve (9), drilling (7), the spring chamber (12) and pilot poppet valve (10), then via port Y to the tank. The pressure in the upper spool chamber equals nearly the return line pressure. The fluid flows from port B to A and the pressure in port (A) increases, acting on the spool upwards. A limited pressure increase in port (A) will overcome the force of the spring (11), displacing the spool upwards to interrupt the fluid flow from port (B) to (A). This state is illustrated by the dead zone in Fig. 11.30.

When the proportional solenoid (2) is energized, it acts on the pilot poppet (8) by a force proportional to the solenoid current. The pressure at the inlet port of the pilot stage increases in proportion to this force. The oil flows from (B) to (A) through the main spool (4) until the pressure in (A) rises to a value equal to the pressure set at the pilot valve plus the pressure needed to overcome the main valve spring (11). Consequently, the main spool displaces upwards until the pressure in (A) reaches a final steady-state value corresponding to the solenoid current. If no more flow is required in port A, or the proportional solenoid is set at lower pressure, the main spool (4) shifts upwards closing the connection from B to A. Further upper displacement of the main spool connects port (A) to the spring chamber of the main spool. In this position, the compression volume in port A can relieve via the pilot valve (1) and port Y to the tank.

An optional check valve (5) is installed to allow free flow from port A to port B. The pressure gauge connection can monitor the reduced pressure in port A (13). Moreover,

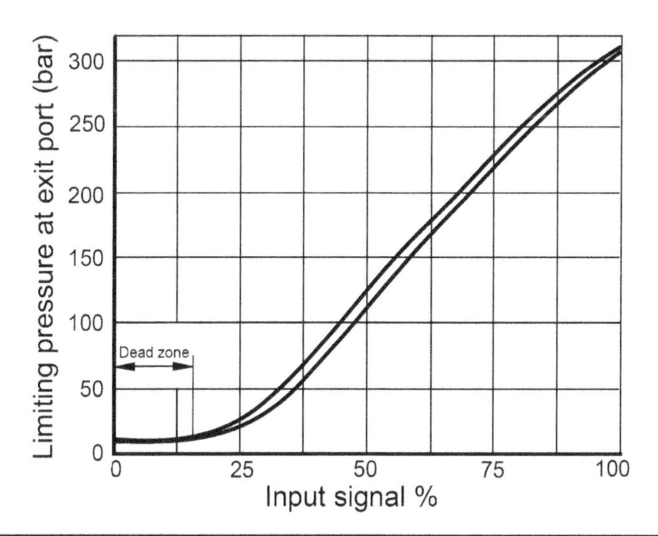

FIGURE 11.30 Typical steady-state reduced pressure/current relation of a pilot-operated electrohydraulic proportional pressure reducer. (*Courtesy Bosch Rexroth AG.*)

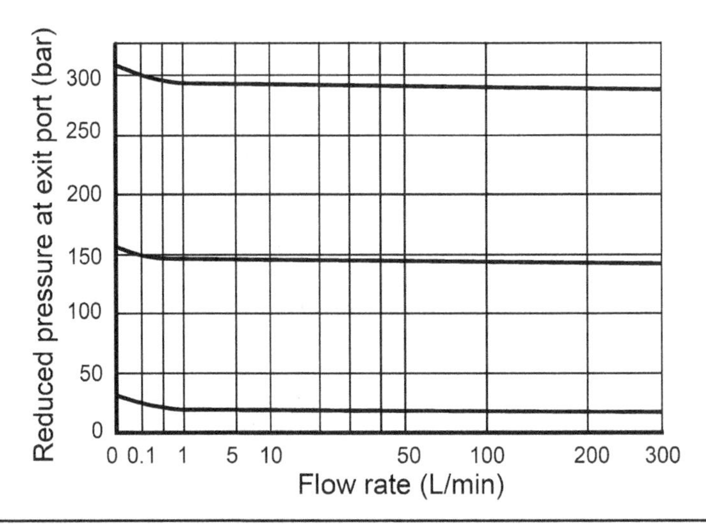

FIGURE 11.31 Typical steady-state reduced pressure/exit flow rate, for a pilot-operated electrohydraulic proportional pressure reducer. (*Courtesy Bosch Rexroth AG.*)

optionally, a direct-operated pressure relief valve (14) can be fitted as maximum pressure protection to protect the system against abnormal voltage surges at the proportional solenoid, which would cause excessive pressures.

The steady-state performance of this valve is shown in Figs. 11.30 and 11.31. Figure 11.30 shows the reduced pressure variation with the input signal's magnitude for zero flow rate in port A. This figure shows the effect of magnetic hysteresis, the dead zone of the proportional solenoid, and the minimum exit pressure, imposed by the main valve spring. For a typical design of this class of valves, the relation between the reduced pressure and the exit flow rate, for different input signals, is given in Fig. 11.31. This figure shows that the reduced pressure decreases slightly for flow rates greater than 1 L/min. The pressure reduction is less than 2.5% for a flow rate increase from 1 to 300 L/min, at 304 bar preset reduced pressure. While for 30 bar preset reduced pressure, the pressure reduction is less than 1%.

11.5 Electrohydraulic Proportional Flow Control Valves

The construction, operation, and performance of the ordinary series pressure compensated flow control valves are discussed in Sec. 5.5.3. Herein, this section deals with a series-pressure-compensated electrohydraulic proportional flow-control valve. This valve consists of a main sharp-edge throttle with a pressure compensator connected in series. The main throttle area is controlled using a stroke-controlled proportional solenoid. This valve can keep the flow rate nearly constant for pressure difference across the valve greater than a prescribed value, within 5 to 10 bar. The valve flow rate is determined by controlling the input electrical signal.

Figure 11.32 shows a typical design of this valve. It consists mainly of a housing (8), a stroke-controlled proportional solenoid (2) with its displacement transducer (1), a spool controlling the main valve throttling area (9), a pressure compensator spool (7), and a check valve (4). The spool (7) controls the area of the pressure compensator

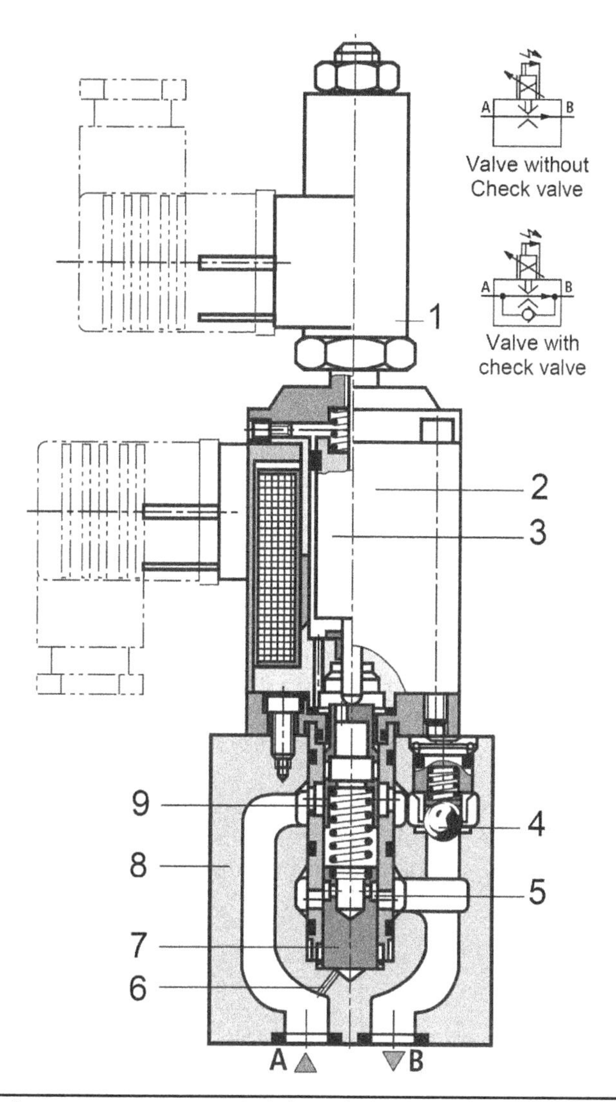

Figure 11.32 Series pressure compensated electrohydraulic proportional flow control valve. (*Courtesy Bosch Rexroth AG.*)

restriction (5). The pressure compensator is fitted downstream of the main orifice. The input signal actuates the stroke-controlled proportional solenoid to produce a proportional core (3) displacement. Consequently, the core displaces the main spool (9), against the spring force, to open the main throttle by a proportional area. The opened main orifice area is precisely controlled by the stroke-controlled proportional solenoid. The pressure compensator maintains a constant pressure difference across the main orifice. The pressure compensator spool is subjected to the input pressure force through the hole (6). This force acts upward against the spring force and the spring chamber pressure force which act downwards.

The pressure compensator spool displaces in the direction to compensate for the effect of variation of pressure difference across the main orifice. Starting from a steady-state operating mode, if the pressure difference across the main orifice increases, the valve flow rate will increase. Simultaneously, the same pressure difference acts upwards on the pressure compensator spool. This spool displaces upwards, decreasing the area of the pressure compensator restriction (5), which is connected in series with the main orifice. This action increases the pressure downstream of the main orifice, decreasing its flow rate. Therefore, the oil flow rate is practically independent of the load due to the action of the pressure compensator. This compensator acts to keep an almost constant pressure drop across the main orifice.

The sharp edges of the main orifice reduce the effect of viscosity variation resulting from the possible changes in oil temperature. The starting without jump is ensured electronically by controlling the rate of variation of solenoid current by the electronic amplifier.

The check valve (5) ensures free return flow from B to A. In this way, the actuator speed is controlled in one direction only. Actually, there exist several smart solutions for the bidirectional actuators' speed control.

Figure 11.33 shows the steady-state characteristics of the presented valve. It shows the steady state dependence of the valve flow rate on the control signal. The observed nonlinearity depends on the main orifice shape.

Figure 11.34 shows a typical transient response of the electrohydraulic proportional flow control valve for different initial conditions and different magnitudes of the applied step inputs. For the maximum increase of input step current, this valve shows a transient response of 58 ms settling time and less than 66 ms for step current reduction.

The frequency response of the presented two-way electrohydraulic proportional flow control valve for two different amplitudes of input sinusoidal signals is shown in Fig. 11.35. The valve shows zero dB magnitude for input sinusoidal current of frequency up to 3 Hz and amplitude of (50% + 5%) of the maximum current. The valve

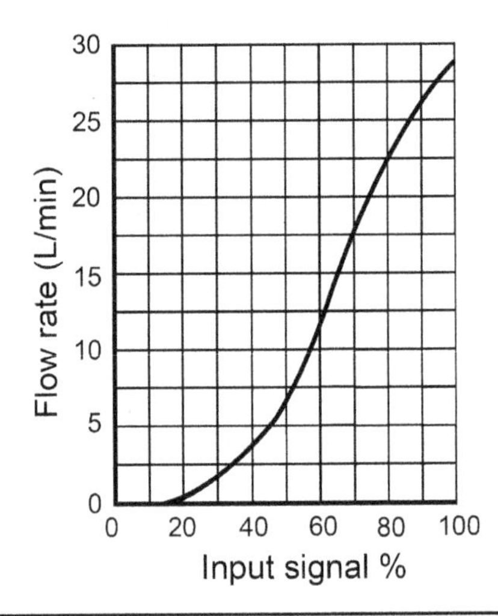

FIGURE 11.33 Typical steady-state flow characteristics of a series pressure compensated electrohydraulic proportional flow control valve. (*Courtesy Bosch Rexroth AG.*)

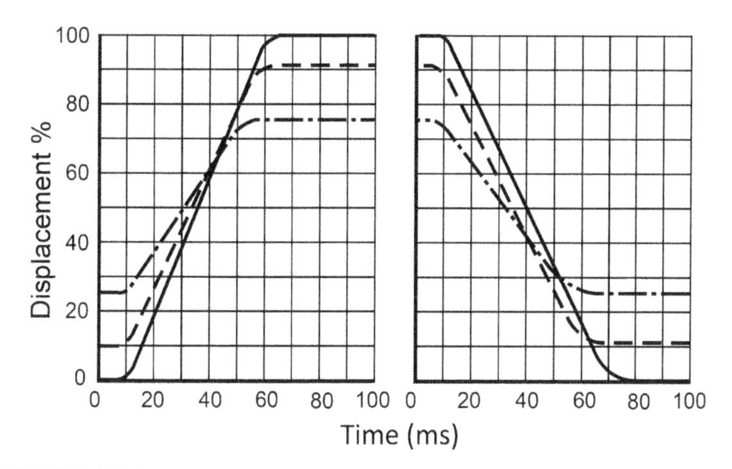

FIGURE 11.34 Typical transient response of a series pressure compensated electrohydraulic proportional flow control valve. (*Courtesy Bosch Rexroth AG.*)

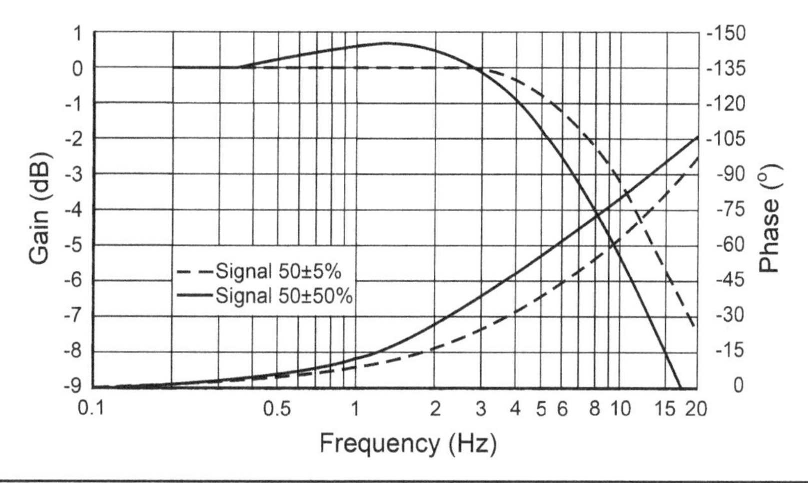

FIGURE 11.35 Typical frequency response of a series pressure compensated electrohydraulic proportional flow control valve. (*Courtesy Bosch Rexroth AG.*)

response follows precisely the input harmonic signal of up to 3 hz frequency. While, for greater signal amplitude; (50% + 50%), signals of up to 0.4 Hz can be precisely followed by this valve.

11.6 Exercises

1. Explain briefly the construction and operation of proportional solenoids illustrated by Figs. 11.5 through 11.7.

2. Explain the construction and operation of the direct operated proportional directional control valve with force-controlled solenoids (Fig. 11.8) and discuss its steady-state and dynamic performance illustrated by Figs. 11.9 through 11.11.

3. Explain the construction and operation of the direct operated proportional directional control valve with stroke-controlled solenoids (Fig. 11.12) and discuss its steady-state and dynamic performance illustrated by Figs. 11.13 through 11.15.

4. Explain the construction and operation of the dual three-way proportional pressure reducer (Fig. 11.16) and discuss its steady-state performance illustrated in Figs. 11.17 and 11.18. Compare the construction and operation of this valve with the pilot stage of the ordinary pilot-operated directional control valve (Fig. 5.45).

5. Explain the construction and operation of the pilot-operated proportional directional control valve (Fig. 11.19) and discuss its steady-state performance illustrated by Figs. 11.20 and 11.21.

6. Explain the construction and operation of the direct-operated proportional relief valve with stroke-controlled solenoids (Fig. 11.22).

7. Explain the construction and operation of the direct-operated proportional relief valve with force-controlled solenoids (Fig. 11.24) and discuss its steady-state performance illustrated by (Fig. 11.25).

8. Explain the construction and operation of the pilot-operated proportional relief (Fig. 11.26) and discuss its steady-state performance illustrated by Figs. 11.27 and 11.28.

9. Explain the construction and operation of the pilot-operated proportional pressure-reducing valve (Fig. 11.29) and discuss its steady-state performance illustrated by Figs. 11.30 and 11.31.

10. Explain the construction and operation of the series pressure compensated proportional flow control valve (Fig. 11.32) and discuss its steady-state and dynamic performance illustrated by Figs. 11.33 through 11.35.

CHAPTER 12

Introduction to Pneumatic Systems

12.1 Introduction

Pneumatic systems are power systems using compressed air as a working medium for the power transmission. Their principle of operation is similar to that of the hydraulic power systems. An air compressor converts the mechanical energy of the prime mover into, mainly, pressure energy of the compressed air. This transformation facilitates the transmission, storage, and control of energy. After compression, the compressed air should be prepared for use. The air preparation includes filtration, cooling, water separation, drying, and adding lubricating oil mist. The compressed air is stored in compressed air reservoirs and transmitted through transmission lines: pipes and hoses. The pneumatic power is controlled by means of a set of valves such as the pressure, flow, and directional control valves. Then, the pressure energy is converted to the required mechanical energy by means of the pneumatic cylinders and motors. (See Fig. 1.5, Chapter 1.)

12.2 Peculiarities of Pneumatic Systems

The static and dynamic characteristics of the pneumatic systems differ from those of the hydraulic systems due to the difference in the physical properties of the energy transmitting fluid, mainly the high compressibility, low density, and low viscosity of air.

12.2.1 Effects of Air Compressibility

The fluid compressibility is the ability of fluid to change its volume due to pressure variation. It is evaluated by the bulk modulus, B, or the compressibility coefficient, β: ($\beta = 1/B$). The bulk modulus is defined by the following relation:

$$B = -\frac{dp}{dV/V} \tag{12.1}$$

where p = Applied pressure, Pa (abs)
V = Fluid volume, m^3

The negative sign is introduced since the volume decreases as the pressure increases.

For real gas, the following law is valid for the polytropic compression process:

$$pV^n = \text{constant} \tag{12.2}$$

$$V^n dp + npV^{n-1}dV = 0 \tag{12.3}$$

or

$$V\frac{dp}{dV} = -np \tag{12.4}$$

Hence, the bulk modulus of compressed air is given by

$$B = np \tag{12.5}$$

At 10 MPa pressure, the air has a bulk modulus $B_a = 1.4 \times 10^7$ Pa, for $n = 1.4$. This value is too small, compared with that of the hydraulic liquid ($B_{\text{oil}} = 1$ to 2 GPa). Therefore, the air, even when compressed to high pressures, is much more compressible than the hydraulic liquids.

This compressibility allows for energy to be stored. Considering that a volume V_o of air at pressure p_o, is allowed to expand to a low pressure p, an expression for the energy released during the expansion process is deduced as follows:

$$E_a = \int_{V_o}^{V} p\,dV \tag{12.6}$$

$$p_o V_o^n = pV^n = \text{constant} \tag{12.7}$$

$$V = \left(\frac{p_o}{p}\right)^{1/n} V_o \tag{12.8}$$

$$dV = -\frac{V}{np}dp \tag{12.9}$$

Then

$$E_a = -\int_{p_o}^{p} \frac{V}{np} p\,dp = -\frac{V_o p_o^{1/n}}{n} \int_{p_o}^{p} p^{-1/n}\,dp \tag{12.10}$$

or

$$E_a = \frac{p_o^{1/n} V_o}{n-1}\left[p_o^{\frac{n-1}{n}} - p^{\frac{n-1}{n}}\right] \tag{12.11}$$

A similar expression for energy stored in a volume of liquid is deduced as follows:

$$E_L = \int_{V_o}^{V} p\,dV \tag{12.12}$$

For liquids, if the initial volume is V_o, then

$$B = -\frac{dp}{dV/V_o} \text{ or } B = -\frac{dp}{dV/V_o} \tag{12.13}$$

or

$$dV = -\frac{V_o}{B} dp \tag{12.14}$$

$$V = V_o + \Delta V = V_o - V_o \frac{\Delta p}{B} = V_o\left(1 - \frac{p_o - p}{B}\right) \tag{12.15}$$

$$E_L = \int_{V_o}^{V} p\,dV = \int_{p_o}^{p} p\left(-\frac{V}{B}dp\right) = -\int_{p_o}^{p} \frac{V_o p}{B}\left(1 - \frac{p_o - p}{B}\right)dp \tag{12.16}$$

or

$$E_L = -\frac{V_o}{B^2}\left\{\frac{1}{2}(B - p_o)(p^2 - p_o^2) + \frac{1}{3}(p^3 - p_o^3)\right\} \tag{12.17}$$

Calculation of Energy Stored in One Liter of Compressed Air and Liquid

For $p_o = 15$ MPa, $V_o = 10^{-3}$ m^3, $p = 0$, and $n = \gamma = 1.4$, the energy stored in one liter of air is $E_a = 28.7$ kJ, which is calculated using Eq. (12.11). For the same conditions, one liter of liquid of $B = 1.4 \times 10^9$ Pa stores an energy $E_L = 0.08$ kJ, which is calculated using Eq. (12.17). These numerical results show that the energy stored in a certain volume of liquid is less than 0.3 percent of the energy stored in the same volume of compressed air at the same conditions.

Calculation of Energy Stored per Kilogram of Compressed Air and Liquid

The air density is given by the following relation:

$$\rho = \frac{p}{RT} \tag{12.18}$$

where R = Universal gas constant, (287.1 J/kgK for air)
T = Absolute temperature, K
p = Absolute pressure, Pa
ρ = Air density, kg/m^3

For $T = 288$ K, $R = 287.1$ J/kgK, and $p = 15$ MPa, the air density is $\rho = 181$ kg/m^3. The density of typical hydraulic oil is 850 kg/m^3. Relating the accumulated energy to the unit of mass of fluid, at 15 MPa absolute pressure, and 288 K temperature, the following results could be concluded:

- The mass of one liter of air at these conditions = 0.1814 kg
- The mass of one liter of hydraulic liquid = 0.85 kg
- Energy accumulated per kilogram of air = 157000 J
- Energy accumulated per kilogram of liquid = 94 J

This comparison shows that the compressed air reservoirs are able to store a considerable amount of energy. Therefore, they can be used as a source of energy in pneumatic systems.

The following peculiarities of operations of pneumatic systems are due to the high compressibility of air:

Time Delay of Response

The time delay is the time interval between the moment of opening the control valve and the beginning of motion of the working organ. This delay is caused by the gradual increase of pressure in the transmission lines and actuator chamber. The piston starts to move only when the pressure reaches the value needed to drive the load. The time delay depends upon the volume of the line and actuator chamber, the flow rate of the air, and the loading conditions, including the friction.

The Nonuniform Motion of a Pneumatic Cylinder Piston

The nonuniform motion of a pneumatic cylinder piston is caused by the variable friction in the cylinder and volumetric variation of inlet chamber due to piston displacement.

Pneumatic Systems Are Not Subject to Hydraulic Shocks

Hydraulic shocks result from the rapid change of liquid velocity in transmission lines. This change occurs due to the sudden closure or opening of the line by control valves as well as the sudden stopping of a piston at its end position. In the case of compressible fluids, the sudden closure of valves results in a gradual increase of fluid pressure. Consequently, the fluid speed decreases gradually. Taking into consideration the low air density and high compressibility, the pneumatic transmission lines are not subjected to these shocks.

Pneumatic Systems Can Supply Great Energy During a Short Period

This is insured by storing the required volume of compressed air in the air reservoirs.

Pneumatic Cylinders Need a Braking System for Position Locking

The variation of load affects the air pressure and volume in the actuating elements. Therefore, it is difficult to fix any intermediate position of the piston without using a mechanical locking element or by using an efficient electro-pneumatic servo system.

Limited Effect of Fluid Thermal Expansion on the Air Pressure

The pressure variation of a trapped volume of air due to temperature variation is too small compared with that of the hydraulic systems. In the case of hydraulic systems, the variation of pressure ($\Delta p = p_2 - p_1$) of a volume of trapped liquid of bulk modulus, B, and thermal expansion coefficient, α, is given by the following expression (see Sec. 2.2.4):

$$p_2 = p_1 + \alpha B \Delta T \tag{12.19}$$

where ΔT = Temperature variation, K
p_1 = Pressure at temperature T_1, Pa
p_2 = Pressure at temperature T_2, Pa

For a constant volume of gas, the pressure varies with temperature according to the following relation:

$$p_2 = p_1 \frac{T_2}{T_1} \tag{12.20}$$

Temperature T (°C)	Pressure (bar) Gauge	
	Air	Mineral Oil
0	0	0
50	0.18	490

TABLE 12.1 Calculated Pressure Increment in Oil and Air Due to Thermal Expansion

Table 12.1 shows the pressure variation of a certain volume of air and of liquid of bulk modulus $B = 1.4 \times 10^9$ Pa and volumetric thermal expansion coefficient $\alpha = 7 \times 10^{-4}\ K^{-1}$. The air pressure increase due to thermal expansion is within 0.037 percent of the liquid pressure increment at the same conditions. This small variation of air pressure with temperature variation gives a good advantage to the pneumatic systems, especially in the case of systems subjected to a considerable variation in temperature.

12.2.2 The Effect of Air Density

The air density changes with the pressure and temperature.

$$\rho = \frac{p}{RT} \qquad (12.21)$$

For $T = 288$ K and $p = 15$ MPa, the compressed air density is $\rho = 181\ kg/m^3$, and at the atmospheric pressure, the air density is $1.21\ kg/m^3$. Generally, the density of compressed air is much smaller than that of hydraulic liquids (for mineral hydraulic oils $\rho = 800–900\ kg/m^3$). This small density gives several advantages to the pneumatic systems such as

- Protection against hydraulic shocks, due to small inertia forces and the high compressibility of air.
- Reduction of the total weight of the system.
- The air speed in transmission lines is greater than that of liquids for the same pressure difference. Therefore, small line diameters can be used, which lead to an additional reduction of the system weight.

12.2.3 The Effect of Air Viscosity

The dynamic viscosity, μ, of compressed air is very small, compared with that of hydraulic liquids. At atmospheric temperature and pressure, typical mineral hydraulic fluids have a dynamic viscosity $\mu_{oil} = 2 \times 10^{-2}$ Pa s. Under the same conditions, the air viscosity is $\mu_{air} = 2 \times 10^{-5}$ Pa s. The friction losses in pneumatic transmission lines are very small, which allows reduction of the line diameter. On the other hand, the air is able to leak through the smallest clearance, mainly due to its small viscosity and density. Therefore, it is difficult to achieve full tightness of pneumatic systems.

12.2.4 Other Peculiarities of Pneumatic Systems

(a) After its expansion in pneumatic cylinders or motors, the air is expelled into the atmosphere. Therefore, only supply lines are used. There are no return lines.

(b) Compressed air reservoirs are of considerable volume and weight.

(c) The air is of poor lubricity. Therefore, friction surfaces need special lubrication. Lubricators are installed in the air preparation process to introduce a lubricating oil mist in the compressed air.

(d) The air contains a certain amount of water vapor. After compression and cooling, the vapor condenses. The condensed water should be removed to avoid filling the compressed air reservoir with condensed water and rust formation. To do this, different types of air dryers are used.

(e) Pneumatic systems are not fire hazards. However, their air reservoirs have the potential to explode.

12.3 Advantages and Disadvantages of Pneumatic Systems

12.3.1 Basic Advantages of Pneumatic Systems

(a) Small weight of transmission lines due to

- The small diameter of lines. (Hydraulic losses due to air flow are small, which allows the reduction of the line diameter.)
- The low density of energy transmitting fluid; the air.
- There are no return lines; used air is expelled into the atmosphere.

(b) Availability of the energy transmission fluid, the air.

(c) The system is fireproof.

(d) Pneumatic systems are able to supply a great amount of energy during a short time period, from the compressed air reservoir.

12.3.2 Basic Disadvantages of Pneumatic Systems

(a) Difficult system tightness.

(b) Low working pressure compared with the hydraulic systems due to the tightness problems and compressor design (within 10 bar for industrial systems and more than 200 bar for aerospace systems).

(c) Difficulty of holding pneumatic actuators at intermediate positions.

(d) Delay of actuators' response due to the time needed for filling the long lines with compressed air.

(e) The variation of pressure in air reservoirs with temperature.

(f) The possibility of the condensation of humidity and the freezing of condensed water at low temperatures.

(g) Special lubricators are needed due to the poor lubricity of air.

(h) Danger of explosion.

12.4 Basic Elements of Pneumatic Systems

12.4.1 Basic Pneumatic Circuits

Figure 12.1 shows the basic circuit of a pneumatic system. The compressed air is prepared by means of the air preparation unit, including the compressor, filters, air drier, compressed air reservoir, cooler, and pressure control elements. The mechanical energy provided by the prime mover is converted by the compressor to, mainly, pressure energy. The compressed air is stored in an air reservoir of sufficient capacity. The maximum pressure at the compressor exit line is limited by a relief valve. The pressure in the air reservoir should be greater than that needed for system operation. Therefore, a pressure reducer is used to control the driving forces and save the compressed air. Compact air preparation units are commercially available, and are comprised of a pressure reducer, an air filter, a lubricator, and a pressure gauge indicating its exit pressure. Finally, the cylinder is fed by the compressed air by means of a directional control valve.

12.4.2 Air Compressors

The function of a compressor is to compress a gas and deliver it to the user through piping. The main parameters characterizing the performance of a compressor are the volumetric flow rate, Q, intake pressure, p_1, and dis charge pressure, p_2 (or the compression ratio $\pi = p_2/p_1$), rotating speed, and the compressor shaft power, N.

The two basic classes of air compressors are displacement and dynamic. In displacement-type compressors, the air pressure increases because of the change in the volume of air trapped in a confined space. However, in dynamic compressors, the pressure rise is due to the acceleration of the moving gas and converting its energy into, mainly, pressure energy in an exit diffuser or stator. The typical parameters of commonly used compressors are given in Table 12.2. Figure 12.2 shows the classification of

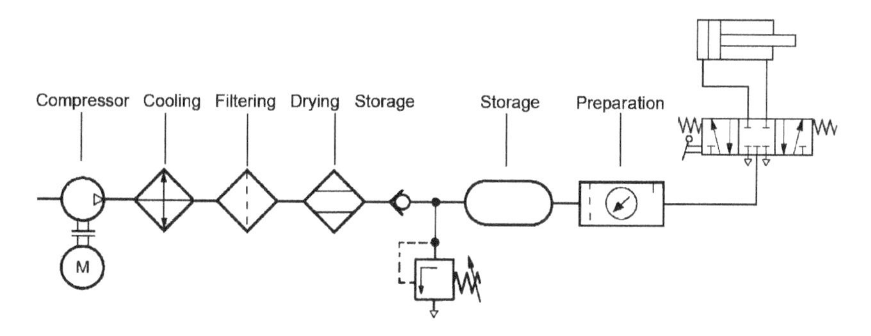

FIGURE 12.1 Circuit diagram of a simple pneumatic system.

Type		Flow Q, m^3/min	Compression Ratio π	Speed n, rpm
Displacement	Reciprocating	0–500	2.5–1000	100–3000
	Rotary	0–500	3–12	300–15,000
Dynamic	Radial (centrifugal)	60–3000	3–20	1500–60,000
	Axial	100–9,000	2–25	500–20,000

TABLE 12.2 Typical Parameters of Commonly Used Compressors

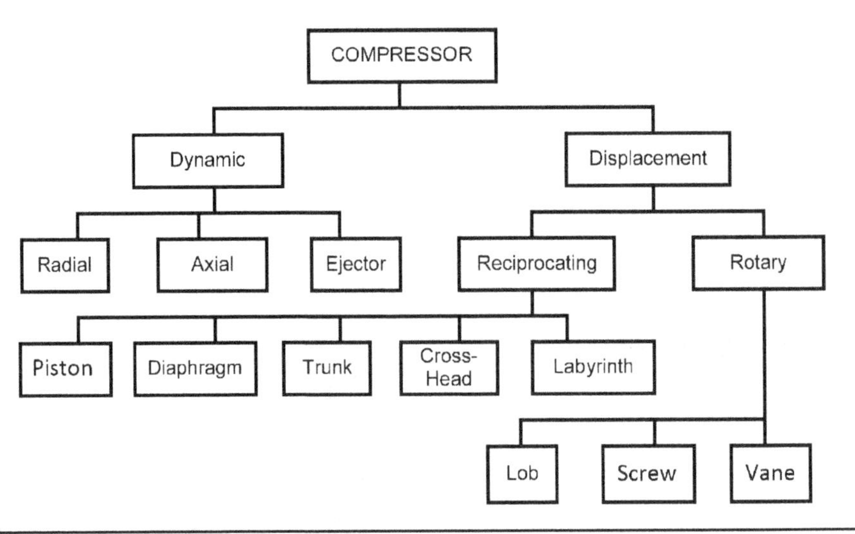

FIGURE 12.2 Classification of air compressors.

commonly used air compressors. The construction and operation of piston-type compressors are presented in the following section. More details about the compressors' construction and selection are given in the work of R.N. Brown (1997).

Piston Compressors

In the piston class of compressors, the process phases of expansion, suction, compression, and discharge are accomplished by the reciprocating motion of a piston. The compression process is based on displacing the gas by the piston. A functional schematic of a piston-type compressor and the associated theoretical *p-V* diagram are presented in Fig. 12.3, where *V* is the volume trapped by the piston in the cylinder.

As the piston moves from the bottom dead point to its top dead point, it compresses the gas contained in the cylinder. The inlet valve is closed during the entire compression stroke. The discharge valve remains closed until the pressure in the cylinder overcomes both the load pressure and the exit valve cracking pressure. Then, the discharge valve opens and the piston displaces the gas into the discharge line. In the *p-V* diagram, the building pressure is represented by the line 1-2, and the gas discharge stroke by line 2-3. If p_2 is the pressure in the cylinder during discharge, the volume of gas delivered by

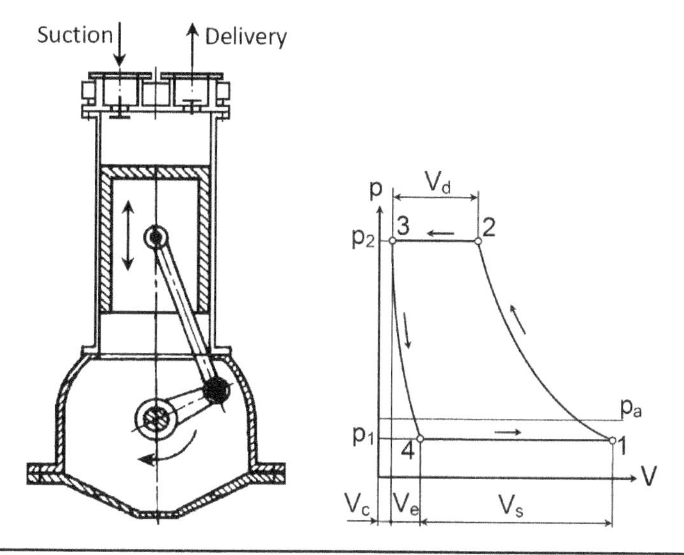

FIGURE 12.3 A single-cylinder piston compressor and its *p-V* diagram.

the compressor at this pressure will be V_d. The compression line is a polytrope, given in the *p-V* diagram by the equation: pV^n = Constant. Theoretically, the discharge line 2-3 is an isobar, p_2 = const. Actually, the gas is not discharged at a strictly constant pressure due to the effect of the inertia of gaseous masses and the effect of valves and their springs.

The clearance, V_c, of the cylinder is the volume of gas present in the cylinder minus the swept volume of the piston at its top dead center. At the beginning of the expansion stroke, the discharge valve will close and the clearance gas will expand along the line 3-4. By the end of the expansion process, this volume occupies the volume $(V_c + V_e)$. The expansion is polytropic. The gas expands until the pressure in the cylinder lowers to $p_1 < p_a$, where p_a is the pressure in the compressor intake line. The intake valve will open, against the spring force, under the action of the force due to the pressure difference $(p_a - p_1)$. The piston will draw gas into the cylinder during the expansion stroke.

The pressure, p_1, is always below p_a due to the fluid resistance in the intake system. The suction is represented by the isobar 4-1. The resulting closed line 1-2-3-4-1 is the theoretical indicator diagram of a compressor. The actual indicator diagram differs somewhat from the theoretical (mainly in the suction and discharge lines). In the case of a compressor without clearance, $V_c = 0$, the points 3 and 4 lie on the vertical axis.

The temperature of the gas increases due to the compression process. The increase in the compressed air temperature can be calculated as follows:

$$p_1 V_1^n = p_2 V_2^n \tag{12.22}$$

$$\pi = p_2 / p_1 \tag{12.23}$$

$$\frac{p_1V_1}{T_1} = \frac{p_2V_2}{T_2} \tag{12.24}$$

$$\Delta T = T_2 - T_1 = T_1\left(\frac{T_2}{T_1} - 1\right) = T_1\left(\frac{p_2V_2}{p_1V_1} - 1\right) \tag{12.25}$$

Since
$$p_1V_1^n = p_2V_2^n, \qquad \text{then} \left(\frac{V_2}{V_1}\right)^n = \pi^{-1} \tag{12.26}$$

or
$$\Delta T = T_1\left(\pi^{\frac{n-1}{n}} - 1\right) \tag{12.27}$$

For an inlet air temperature of 15°C, the calculated temperature increment due to the compression is plotted in Fig. 12.4 for different values of the compression ratio π (assuming $n = 1.3$).

Considering the pressure losses in the inlet and outlet valves and the heat dissipation in the cylinder head, two correction factors k and η are added to the deduced relation. Thus, the deduced relation becomes

$$\Delta T = T_1\left(k\pi^{\frac{n-1}{\eta n}} - 1\right) \tag{12.28}$$

where k is a valve loss correction factor ($k = 1$ to 1.13, for up to 6 percent loss in pressure through each set of valves), and η is the heat leak factor. For a water jacket cylinder, $\eta = 1$ to 1.11, depending on the cylinder cooling.

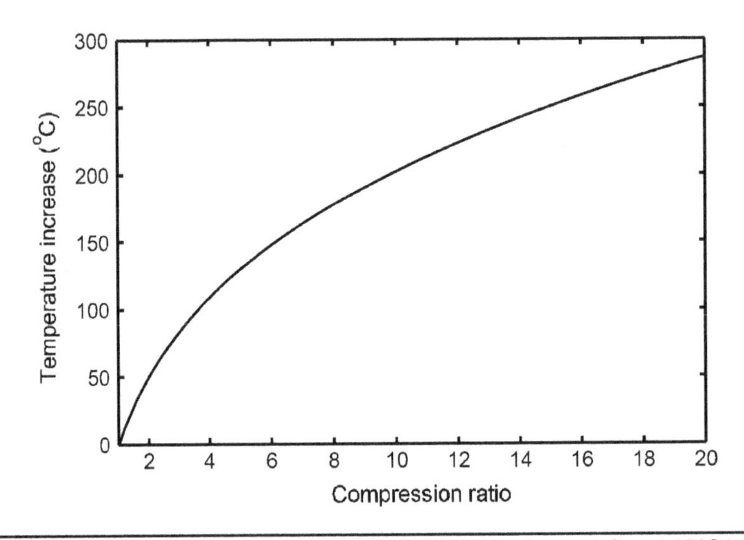

FIGURE 12.4 The air temperature increase due to compression for $n = 1.3$ and 15°C inlet temperature, calculated.

12.4.3 Pneumatic Reservoirs

Generally, the air compressor serves to charge the compressed air reservoir. The compressor operation can be controlled by a governor to keep the air reservoir pressure within certain limits. The pneumatic system is directly fed from the reservoir. However, in some special cases, the pneumatic system does not include an air compressor, such as in some aircrafts and missiles. They use pre-charged high-pressure compressed air bottles as a source of pneumatic energy.

12.4.4 Air Filters

The solid particles and liquid droplets are removed from the compressed air by using air filters with either surface or depth filtering elements. The filtration process is carried out in two stages. The inlet guides force the inlet stream to rotate. The centrifugal force acting on the solids and water separates them from the air stream. They are then collected in the lower part of the filter. In the second stage, a fine filter is added to separate additional impurities. Condensed water is collected by means of drain valves. Certain air filters come with or without a water collector. Moreover, filters with water traps are drained automatically and/or manually (see Fig. 12.5).

12.4.5 Air Lubricators

Oil fog lubricators are used to lubricate the compressed air by adding a fine fog of oil. Oil fog lubricators operate, mostly, according to the Venturi principle (see Fig. 12.6). The reduction of area in the air path produces a vacuum. The oil is drawn up through a narrow pipe which reaches into the lubricating oil reservoir. The oil then drips into the flowing compressed air and forms a fine oil fog.

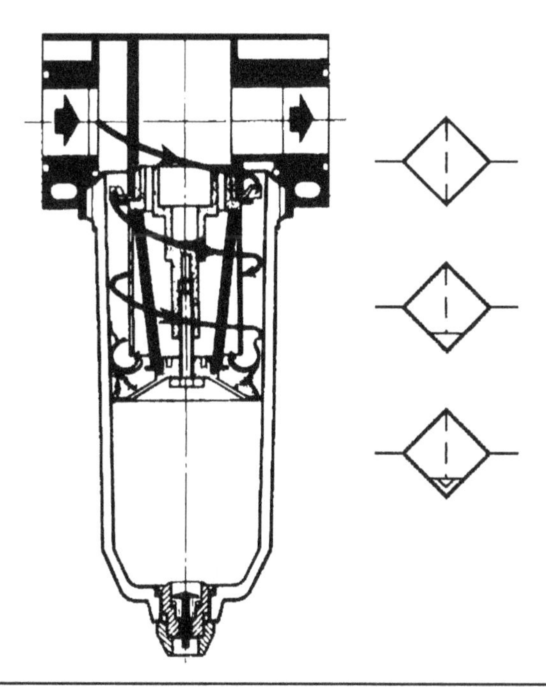

Figure 12.5 An air filter with a water trap. (*Courtesy of Bosch Rexroth AG.*)

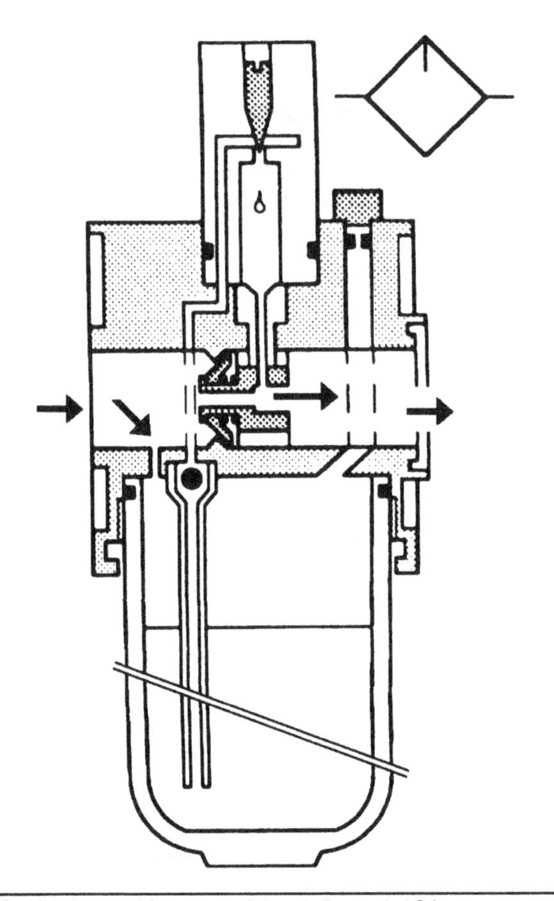

Figure 12.6 An oil fog lubricator. (*Courtesy of Bosch Rexroth AG.*)

12.4.6 Pneumatic Control Valves

Relief Valves

Usually, simple poppet-type relief valves are used to limit the maximum system pressure (see Fig. 12.7). The dimensions of pneumatic valves are much smaller compared with corresponding hydraulic valves because of the low-pressure losses.

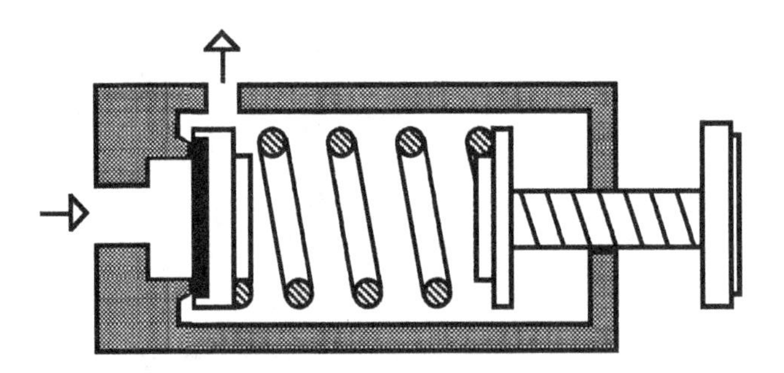

Figure 12.7 A pneumatic relief valve.

Therefore, direct-operated relief valves are usually used in pneumatic systems. The poppet, or its seat, is rubberized to reach the required tightness.

Pressure Reducers

The pressure reducer is positioned downstream of the high-pressure compressed air reservoir. It is used to control the pressure of compressed air supplied to a subsystem.

Ordinary Pressure Reducers The ordinary pressure reducer (see Fig. 12.8) has a rubberized poppet with a corresponding seat. The poppet controls the throttle area connecting the inlet (high pressure) with the outlet (reduced pressure) lines. The high-pressure air supplied to the valve is allowed to flow (expand) through the poppet valve. When the exit pressure rises, it acts on the piston through an internal pilot orifice. The piston and poppet move upward against the spring force, reducing the throttling area. When the exit pressure is increased to the required value, the poppet rests against its seat and the flow of air is stopped. If for any reason the pressure in the exit line is increased, this class of valve does not interfere, and an additional relief valve should be installed on the reduced pressure line.

Venting-Type Pressure Reducers A venting-type pressure reducer (see Fig. 12.9) acts as an ordinary pressure reducer as well as a relief valve. It produces the required reduced pressure and limits the downstream pressure to a pre-selected maximum value. The reduced exit pressure is proportional to the force applied by the spring. When the exit pressure reaches the required value, the poppet seats and the supply of compressed air to the exit port is stopped. When the downstream pressure is increased, the increased pressure force displaces the diaphragm upward and the venting hole is cleared. The exit chamber is then connected to the outer air through the venting hole and the compressed air in the exit line is discharged. The exit pressure reaches a steady-state value proportional to the applied force.

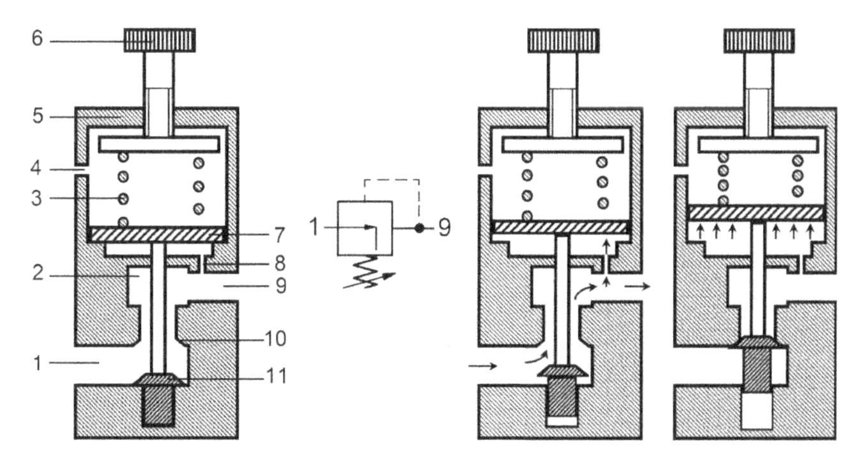

1. High pressure air inlet port, 2. Exit chamber, 3. Spring, 4. Venting port, 5. Housing, 6. Adjusting knob, 7. Piston or diaphragm, 8. Pilot orifice, 9. Exit, reduced pressure port, 10. Poppet seat, 11. Poppet.

FIGURE 12.8 Construction and operation of an ordinary pneumatic pressure reducer.

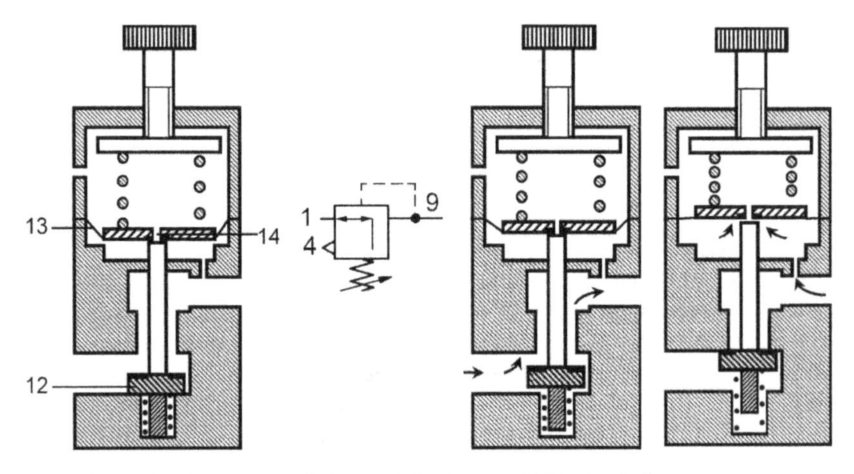

12. Rubberized poppet, 13. Reinforced diaphragm, 14. Venting hole

Figure 12.9 Construction and operation of venting-type pressure reducers.

Directional Control Valves

Poppet-Type Directional Control Valves Figure 12.10 shows a schematic of a 3/2 directional control valve (DCV) of the poppet type. The double face poppet is rubberized. The poppet is held to the left under the action of the spring and pressure forces. It closes the pressure line (P) and connects line (A) to exhaust line (R). When the pilot line (X) is pressurized, the poppet displaces to the right and rests against its right seat, connecting line (A) to the pressure line (P) and closing the venting line (R). The poppet-type DCV is of high resistance to leakage.

Spool-Type Directional Control Valves The spool-type DCV can be designed for a greater number of service ports, but the radial spool clearance is of very low resistance to air leakage. Therefore, this class of valves is equipped with sealing rings spaced by perforated metallic spacing rings (see Fig. 12.11). The sealing rings

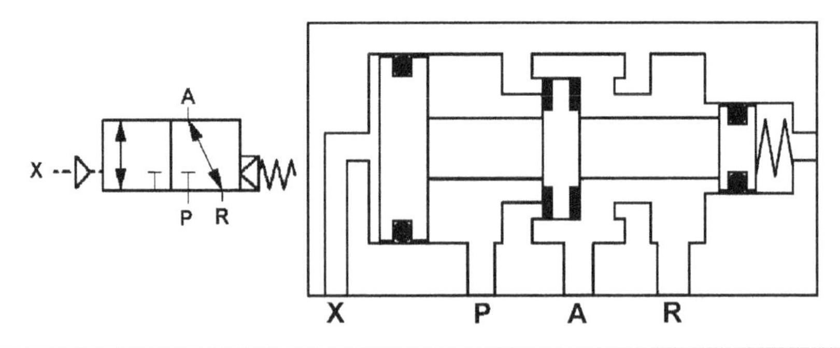

Figure 12.10 A poppet-type DCV.

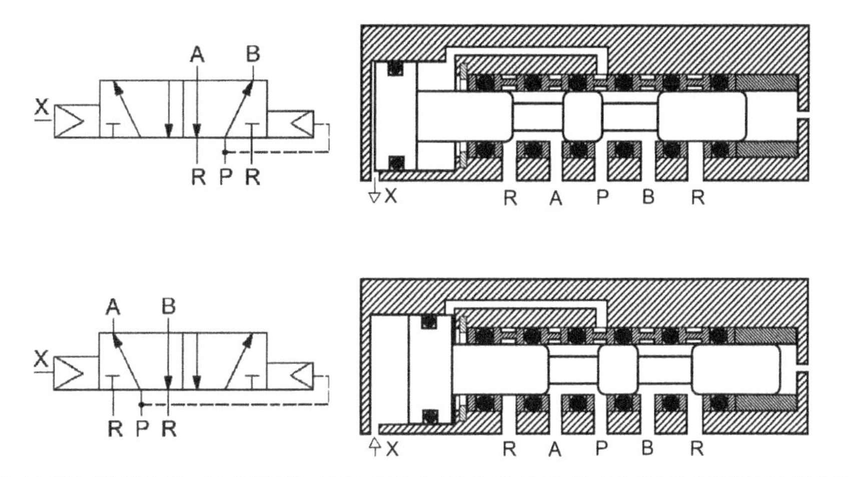

FIGURE 12.11 A spool-type pneumatic directional control valve.

introduce a considerable resistance force. Thus, this class of valves is usually pilot-operated when controlled electrically. This is due to the small force of the solenoid, which should be of limited volume.

Pilot Stage with an Electric Solenoid The pneumatic valves are, in general, of small dimensions. Then, when using electric solenoids, they have small dimensions and limited forces. Therefore, the electrically controlled pneumatic spool valves are usually pilot-operated. Figure 12.12 illustrates a pilot stage incorporating an electric solenoid. When the solenoid is de-energized, the poppet rests against its right seat, closing the pressure line and venting the pilot port (X). When the solenoid is energized, the core and attached poppet move to the left, closing the venting port and connecting the pressure and pilot ports. The pilot valve is equipped with a mechanical override assembly (6). By depressing the plunger to the left, it performs the same action of solenoid energizing.

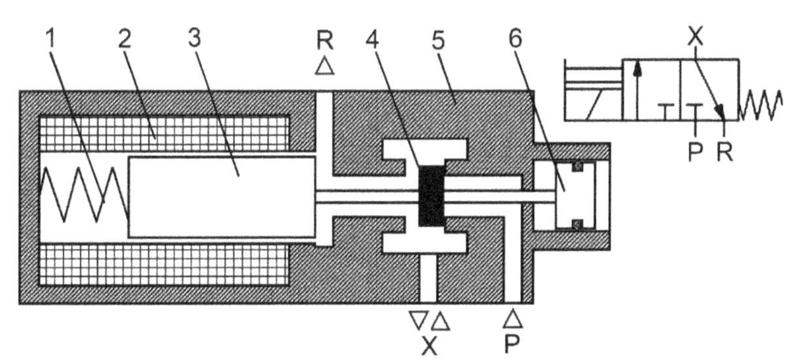

1. Spring, 2. Electric coil, 3. Core, 4. Poppet, rubberized at both sides, 5. Housing, 6. Mechanical override plunger, P. Inlet pressure port, X. Pilot port, R. Venting port

FIGURE 12.12 Pilot stage of a pneumatic directional control valve.

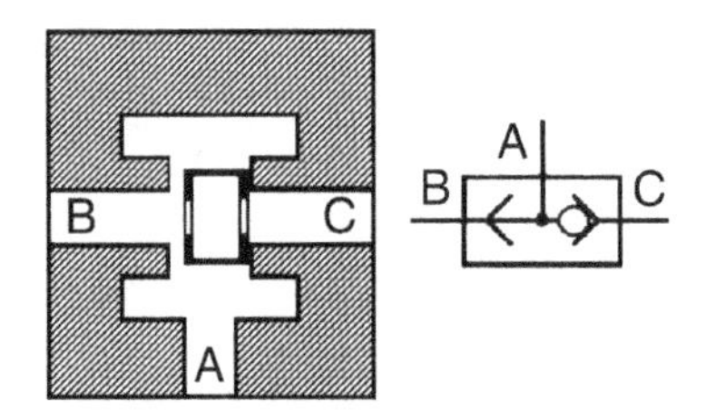

FIGURE 12.13 A shuttle valve.

Shuttle Valves; Logic OR

The construction of a shuttle valve is illustrated by Fig. 12.13. Port (A) is pressurized if port (B) OR port (C) is pressurized.

Pressure Shuttle Valves; Logic AND

Port (A) is pressurized if ports (B) AND (C) are both pressurized (see Fig. 12.14).

Flow Control Valves

In general, the flow rate is controlled by using a simple throttling element, of fixed or variable area, with or without a parallel connected check valve as shown in Fig. 12.15.

Quick Exhaust Valves

Quick exhaust valves are used wherever it is recommended to rapidly discharge the compressed air to maximize the cylinder speed, or for the quick release of brakes, for example. The air flows from port P to A with the exhaust line R closed (see Fig. 12.16). When flowing from A, the line P is closed, and the air flows directly through exhaust port R.

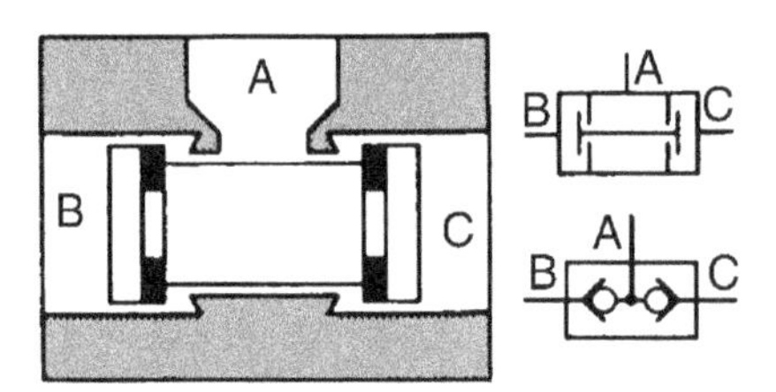

FIGURE 12.14 A pressure shuttle valve.

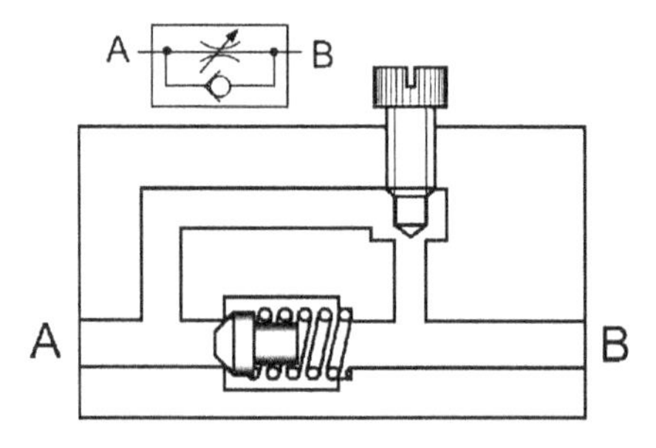

FIGURE 12.15 A variable-area throttle with check valve. (*Courtesy of Bosch Rexroth AG.*)

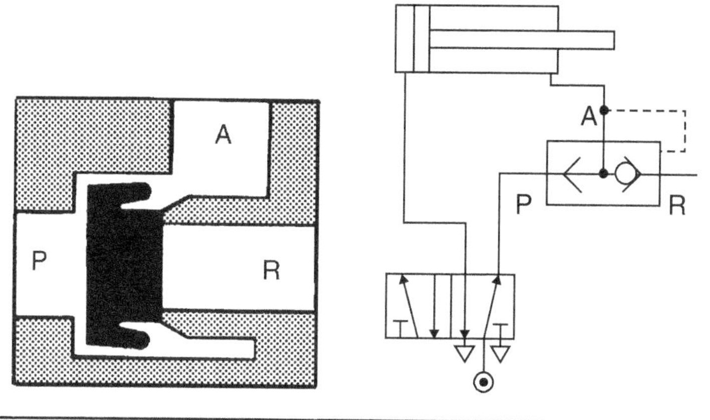

FIGURE 12.16 A quick exhaust valve.

12.5 Case Studies: Basic Pneumatic Circuits

This section presents basic pneumatic circuits performing different logical and sequential operations. The principal circuit drawings were generated by the Pneusoft software, developed by Norgren, UK.

12.5.1 Manual Control of a Single-Acting Cylinder

Initially, the directional control valve is under the action of the spring (10) (see Fig. 12.17a). The piston chamber is vented and the cylinder is fully retracted. By pushing and holding the button (12), the piston chamber is pressurized and the cylinder extends (see Fig. 12.17b). The cylinder retracts by releasing the button.

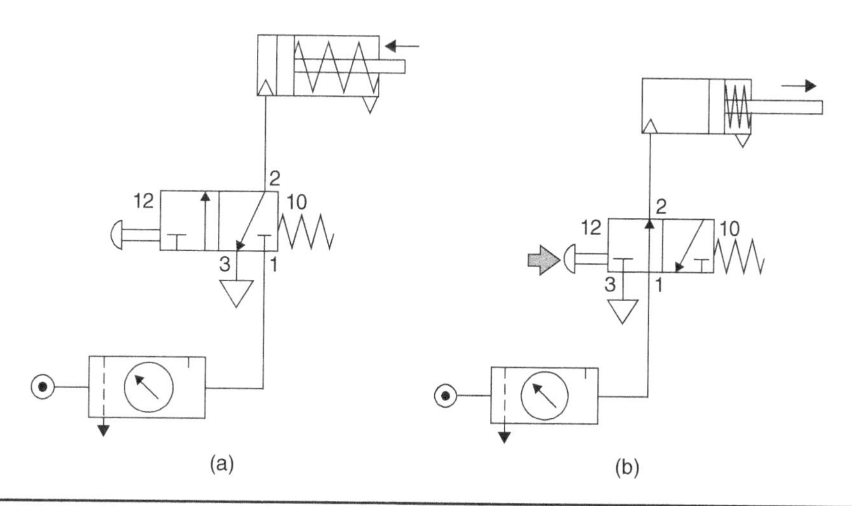

FIGURE 12.17 Manual control of a single-acting cylinder.

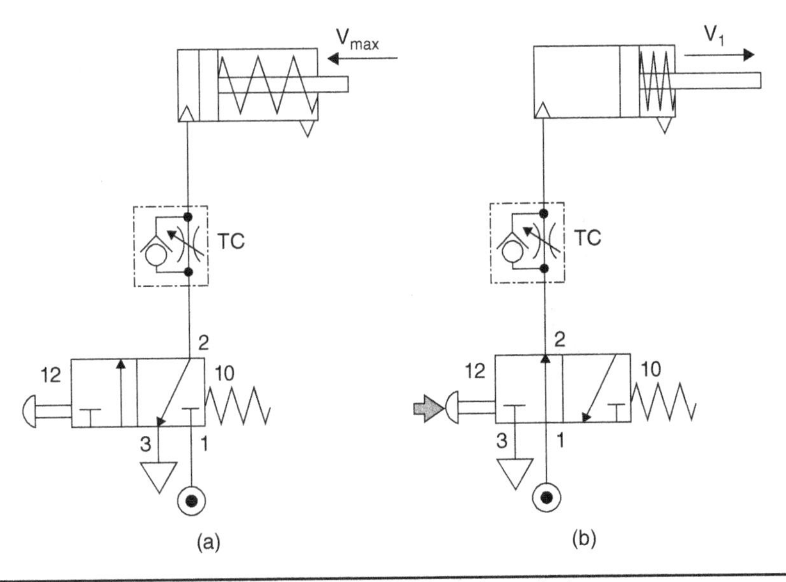

Figure 12.18 Unidirectional speed control of a single-acting cylinder.

12.5.2 Unidirectional Speed Control of a Single-Acting Cylinder

Initially, the cylinder is fully retracted (see Fig. 12.18a). By pushing and holding the button (12) (see Fig. 12.18b), the cylinder extends. Its speed is controlled by the throttle valve (TC). When the button is released, the cylinder retracts without any speed control as the air flows through the check valve out to the atmosphere.

12.5.3 Bidirectional Speed Control of a Single-Acting Cylinder

This system includes a 3/2 directional control valve, manually operated with mechanical position locking. Two throttle check valves are installed to control the extension and retraction speeds. Initially, the cylinder is retracted (see Fig. 12.19a). By operating the lever (12), the cylinder extends (see Fig. 12.19b). Its speed, v_2, is controlled by the throttle valve (TC1). When the lever is reset, the cylinder retracts and its speed, v_1, is controlled by the throttle valve (TC2).

12.5.4 OR Control of a Single-Acting Cylinder

Initially, the cylinder is fully retracted, with both push buttons PB1 and PB2 released (see Fig. 12.20a). The system is equipped with an OR element. The cylinder can be operated if any of the two push buttons, PB1 or PB2, are depressed—positions (b) and (c).

12.5.5 AND Control of a Single-Acting Cylinder

To out-stroke the cylinder, both directional control valves must be operated (see Fig. 12.21a). This provides an AND logic function. The cylinder retracts if any or both of the two directional control valves are released (see Fig. 12.21b, 12.21c, and 12.21d).

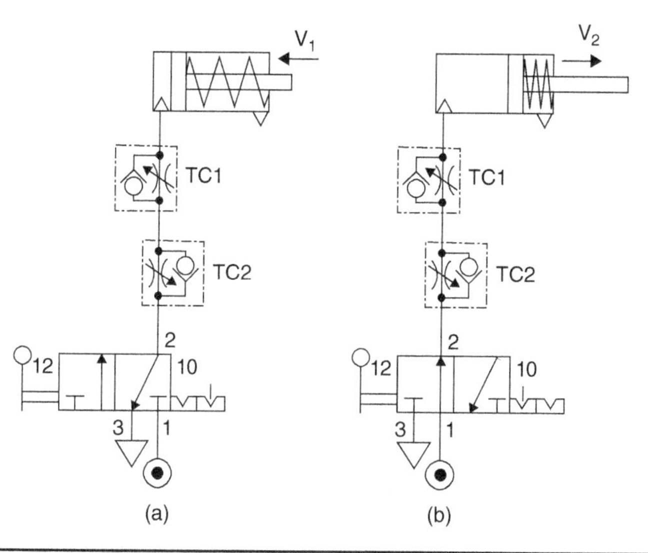

FIGURE 12.19 Bidirectional speed control of a single-acting cylinder.

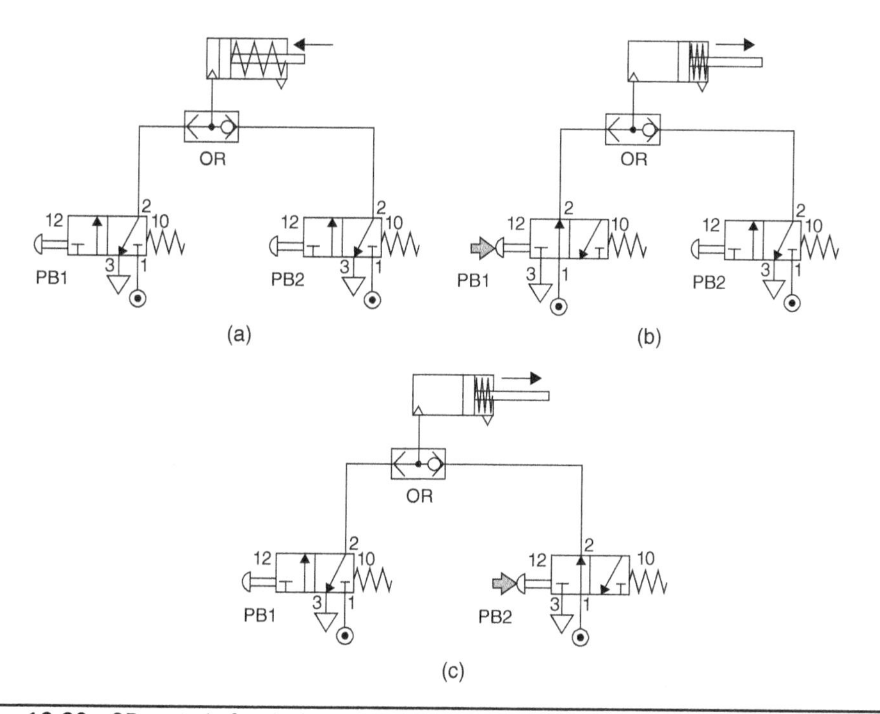

FIGURE 12.20 OR control of a single-acting cylinder.

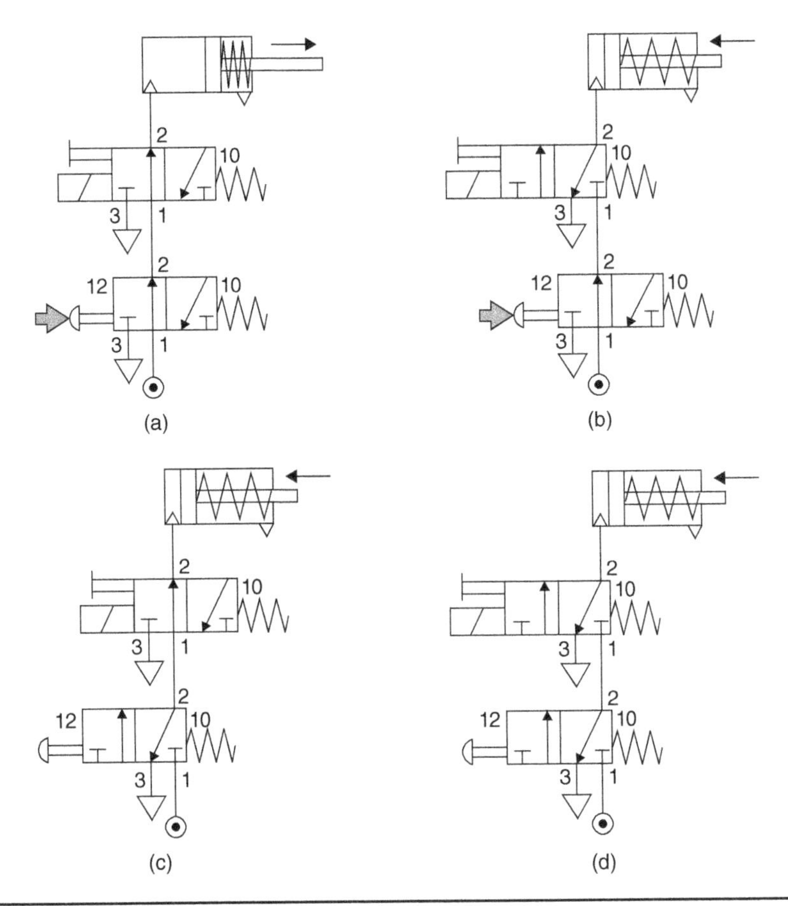

FIGURE 12.21 AND control of a single-acting cylinder.

12.5.6 AND Control of Single-Acting Cylinders; Logic AND Control

This system is equipped with pressure shuttle valves (Logic AND). (See Fig. 12.14.) To extend the cylinder, both directional control valves must be held operated which satisfies the "and" logic shuttle valve (see Fig. 12.22a). The cylinder retracts by releasing either or both valves (see Fig. 12.22b, 12.22c, and 12.22d).

12.5.7 Logic NOT Control

The logic NOT function is provided by a normally open directional control valve (see Fig. 12.23a). Thus, the cylinder extends if the DCV is NOT operated. The cylinder retracts by switching the DCV (see Fig. 12.23b).

12.5.8 Logic MEMORY Control

The cylinder extends by push button PB1, which in turn operates a bi-stable 3/2 DCV (see Fig. 12.24b). When the push button is released, this condition is remembered; the DCV keeps its position, and the cylinder stays in position (see Fig. 12.24c). The cylinder retracts by push button PB2, which resets the 3/2 DCV (see Fig. 12.24d).

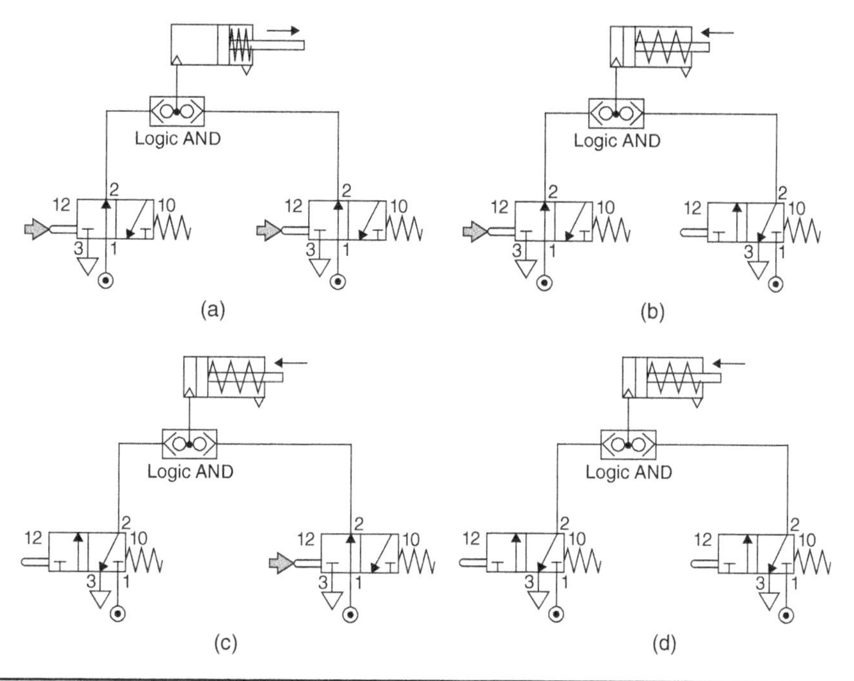

FIGURE 12.22 AND control of single-acting cylinders; logic and control.

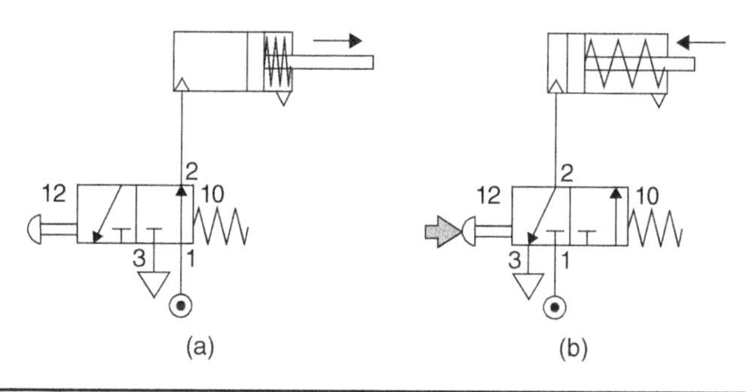

FIGURE 12.23 Logic NOT control.

12.5.9 Bidirectional Speed Control of a Double-Acting Cylinder

The double-acting actuator is controlled by a 5/2 directional control valve. When the push button (14) is pressed and held, the actuator extends (see Fig. 12.25a). Its extension speed, v_1, is controlled by the throttle check valve TC2, which restricts the exhausting air. When the push button is released, the actuator retracts. Its retraction speed, v_2, is controlled by the valve TC1 (see Fig. 12.25b).

12.5.10 Unidirectional and Quick Return Control of a Double-Acting Cylinder

The double-acting cylinder is controlled by a 5/2 DCV. When the push button (14) is pressed and held, the compressed air flows through the quick exhaust valve (QE) to

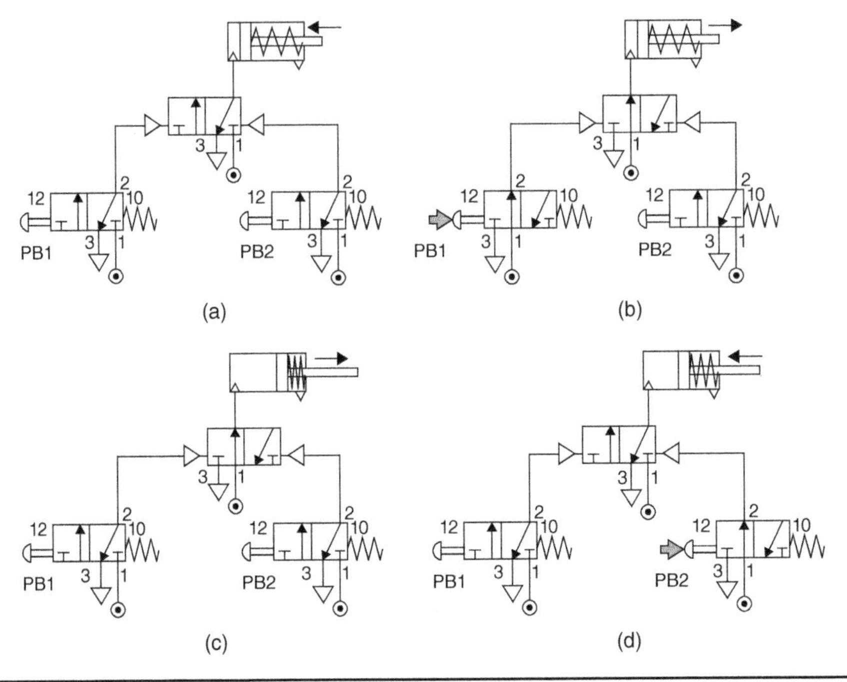

FIGURE 12.24 Logic MEMORY control.

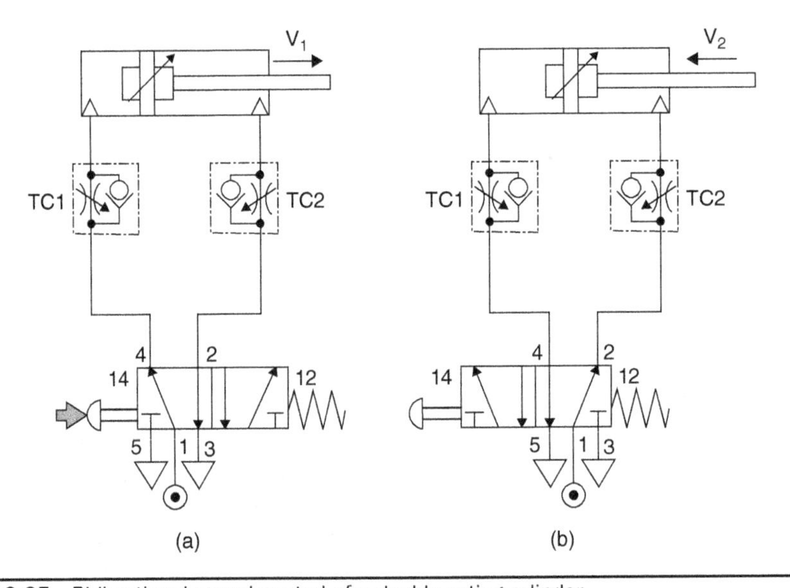

FIGURE 12.25 Bidirectional speed control of a double-acting cylinder.

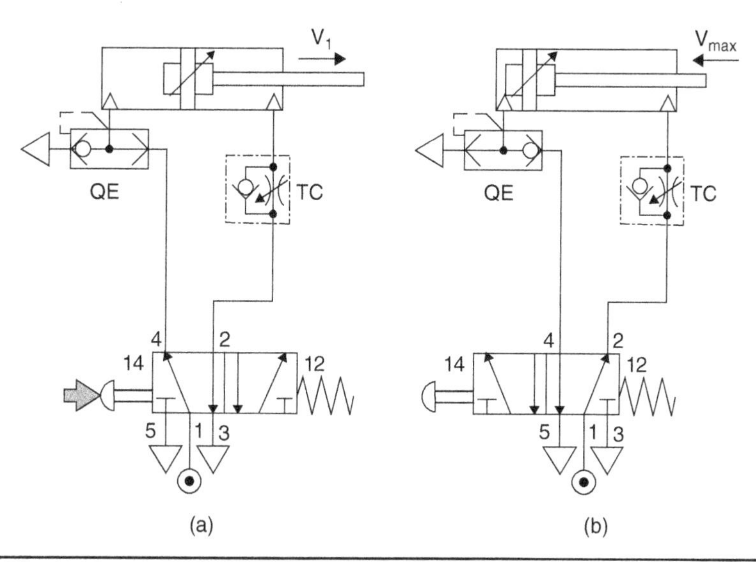

Figure 12.26 Unidirectional speed control and quick return of a double-acting cylinder.

extend the cylinder (see Fig. 12.26a). Its speed is controlled by the throttle check valve TC. When the push button is released, the cylinder retracts rapidly. The exhausting air is dumped directly to the atmosphere through the quick exhaust valve (see Fig. 12.26b).

12.5.11 Dual Pressure Control of a Double-Acting Cylinder

During the extension mode (see Fig. 12.27a), the cylinder is supplied by the compressed air with maximum supply pressure, acting on the whole piston area. The piston is driven by the maximum force. When the button (14) is released, the cylinder will retract under the action of the reduced pressure, acting on the smaller rod-side area (see Fig. 12.27b).

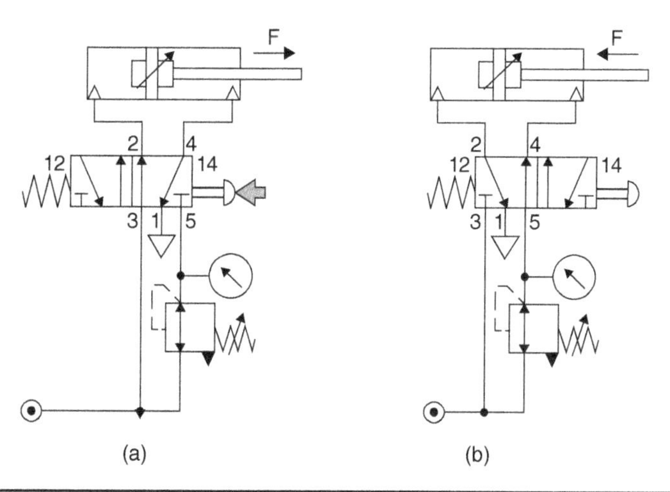

Figure 12.27 Dual pressure control of a double-acting cylinder.

12.5.12 Semi-Automatic Control

A manual action gives a single extension–retraction cycle (see Fig. 12.28a). Push and release the button (15) to send a signal to operate the 5/2 DCV. The cylinder will extend, with speed v_1 controlled by the throttle check valve TC2 (see Fig. 12.28b). At the end of its extension stroke, it operates the roller valve (RV) (see Fig. 12.28c). This sends a signal to reset the 5/2 DCV, causing the cylinder to retract. Its retraction speed, v_2, controlled by the throttle check valve TC1 (see Fig. 12.28d).

12.5.13 Fully Automatic Control of a Double-Acting Cylinder

Manual action of the lever (15) gives continuous reciprocation. With the cylinder fully retracted (see Fig. 12.29a), the roller valve (RV1) is held operated so that when the lever (15) is operated, air will pass through to the 5/2 DCV. This will cause the cylinder to extend (see Fig. 12.29b). When the piston moves out, RV1 is released. At the end of the extension stroke, RV2 is operated, causing the 5/2 DCV to reset (see Fig. 12.29c). The cylinder retracts, releasing RV2 (see Fig. 12.29d). When fully retracted, RV1 will again be operated to repeat the cycle. This will continue until the lever (15) is reset to prevent the next out-stroke. The speed is controlled in both directions by the throttle check valves TC1 and TC2.

12.5.14 Timed Control of a Double-Acting Cylinder

By pushing the button (15), the 5/2 DCV is operated, causing the piston to move to the right (see Fig. 12.30b). By the end of its stroke, the roller RV1 is depressed, which feeds a time delay group of components consisting of a flow regulator, air reservoir (40) and relay valve (see Fig. 12.30c). When the set time has elapsed, the 3/2 DCV resets the 5/2 DCV, which then directs the compressed air to move the piston to the left (see Fig. 12.30d). Then, RV1 is released and quickly exhausts the time delay components to be ready for the next operation (see Fig. 12.30a).

12.5.15 Basic Positional Control of a Double-Acting Cylinder

This system provides a visually set position control. By pushing and holding PB1, the 5/3 directional control valve will be set to the lefthand state and the cylinder will extend (see Fig. 12.31b). When the piston reaches the desired position, the push button is released. The 5/3 valve is put in its neutral position under the action of its centering springs, locking the cylinder ports and stopping the cylinder (see Fig. 12.31a). By operating PB1 again, the actuator will continue to extend. The actuator can be stopped in this way as many times as required. The activation of PB2 controls the position in the opposite direction (see Fig. 12.31c). The speed control is insured by the throttle check valves TC1 and TC2.

12.5.16 Electro-Pneumatic Logic AND

A single-acting cylinder is controlled by a single-solenoid-operated spring-returned 3/2 DCV (see Fig. 12.32). The relay circuit controlling the valve provides an AND logic function. To operate the solenoid Y, and extend the cylinder, it is necessary to push buttons S1 AND S2 simultaneously (see Fig. 12.32c). If any of these two buttons is released, the cylinder will retract (see Fig. 12.32a and 12.32b).

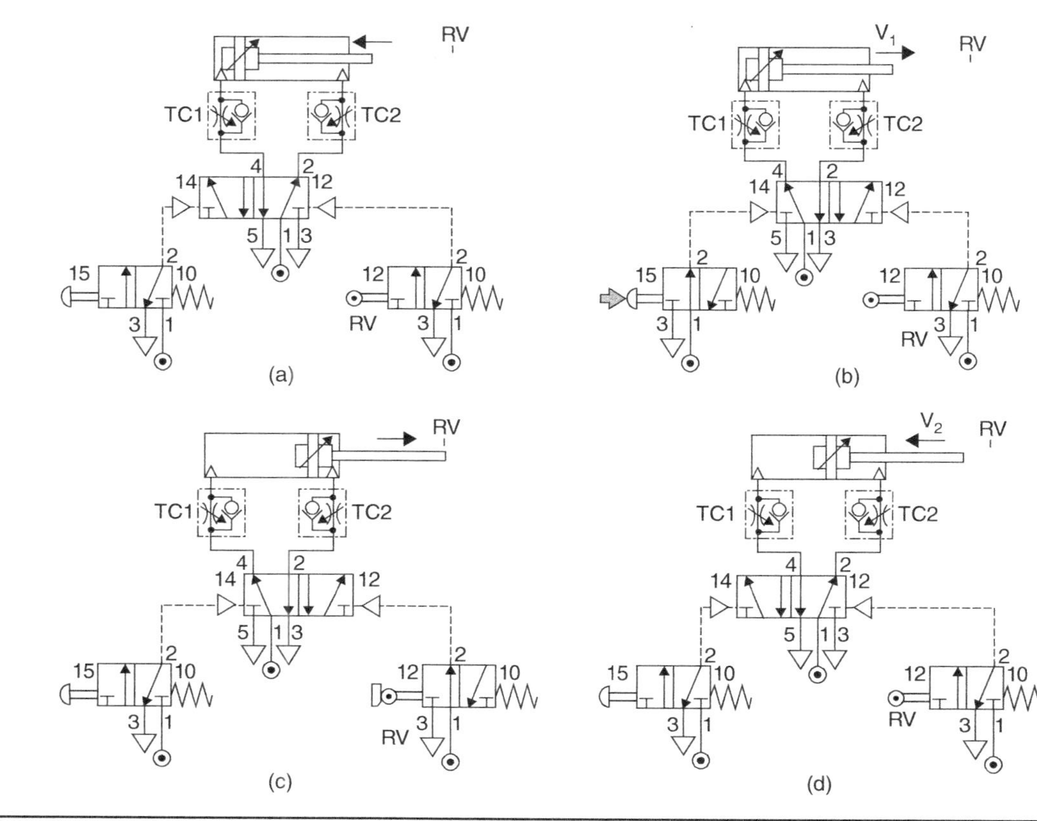

Figure 12.28 Semi-automatic control.

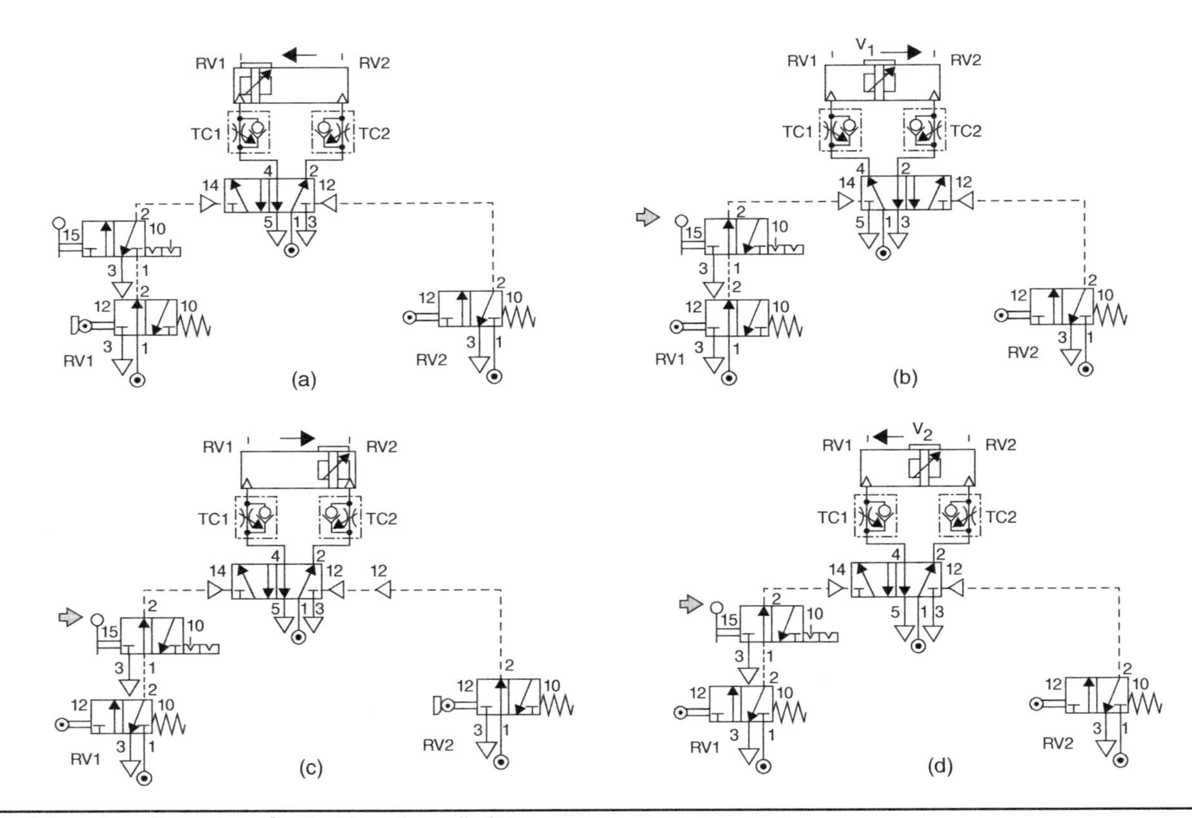

FIGURE 12.29 Fully automatic control of a double-acting cylinder.

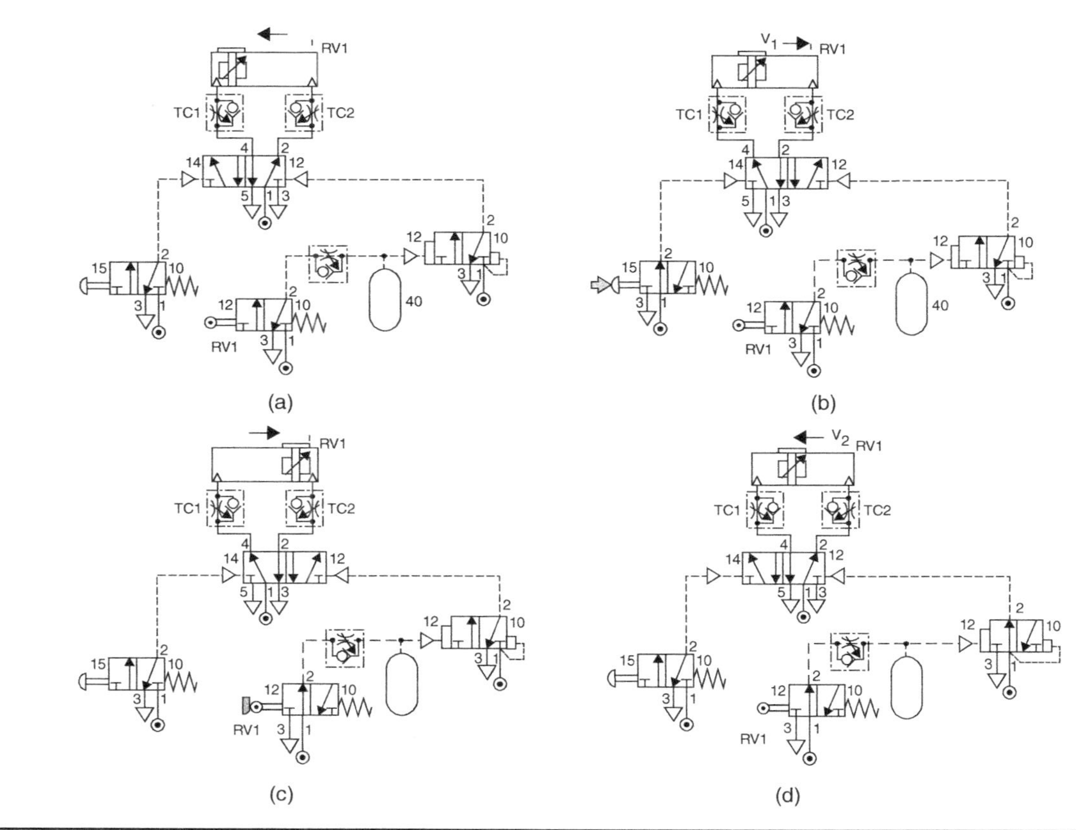

Figure 12.30 Timed control of a double-acting cylinder.

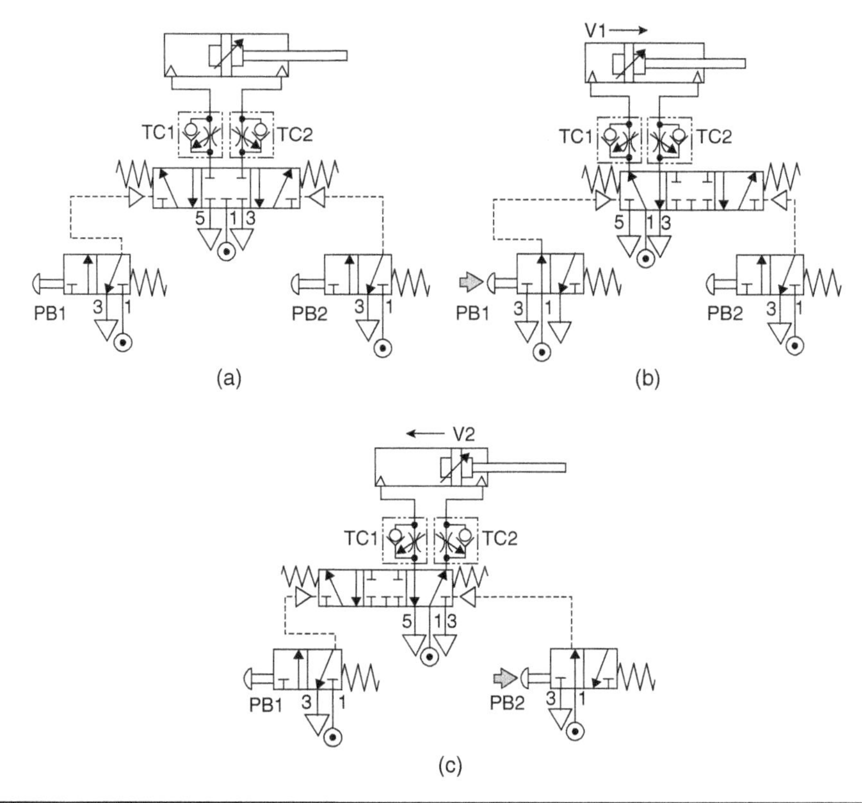

FIGURE 12.31 Basic positional control of a double-acting cylinder.

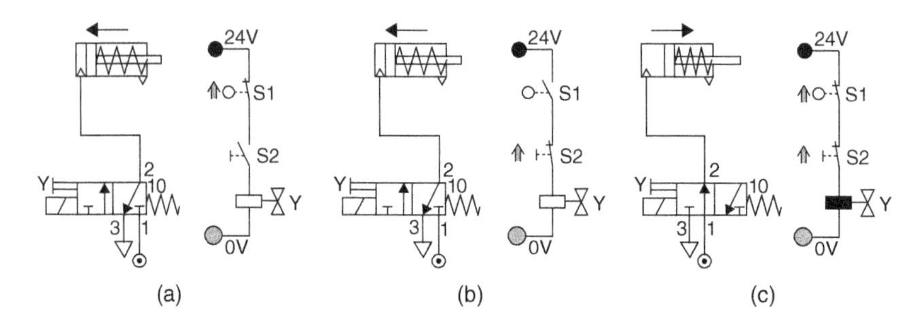

FIGURE 12.32 Electro-pneumatic logic AND.

12.5.17 Electro-Pneumatic Logic OR

A single-acting cylinder is controlled by a single-solenoid-operated spring-returned 3/2 DCV (see Fig. 12.33). The relay circuit controlling the valve provides an OR logic function. To operate solenoid Y and extend the cylinder, it is necessary to push button S1 OR S2 or both (see Fig. 12.33b and 12.33c). However, both must be released for the cylinder to retract (see Fig. 12.33a).

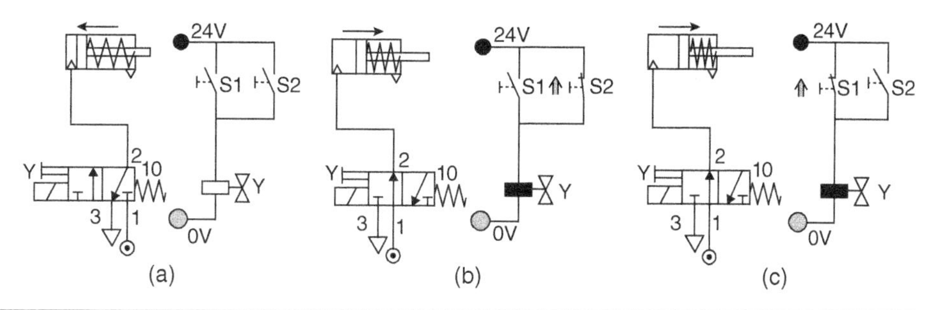

Figure 12.33 Electro-pneumatic logic OR.

12.5.18 Electro-Pneumatic Logic MEMORY

The single-acting cylinder is controlled by a single-solenoid-operated spring-returned 3/2 DCV (see Fig. 12.34). The relay circuit controlling the valve provides a MEMORY logic function. The push button A+ is operated to turn on the relay RR1, which closes both contacts R1 and R2 (see Fig. 12.34b). The push button A+ can be released as the first contact (R1) is providing continuity to hold the relay on. At the same time, the second contact (R2) operates the solenoid Y to extend the cylinder. This state will continue until the normally closed push button contact A− is opened, whereupon the relay will de-energize and both contacts R1 and R2 will open. The push button A− can now be released and the relay remains off. At the same time, the solenoid is de-energized, causing the cylinder to retract (see Fig. 12.34a).

12.5.19 Electro-Pneumatic Logic NOT

The single-acting cylinder is controlled by a single-solenoid-operated spring-returned 3/2 DCV. The relay circuit controlling the valve provides a NOT logic function. The

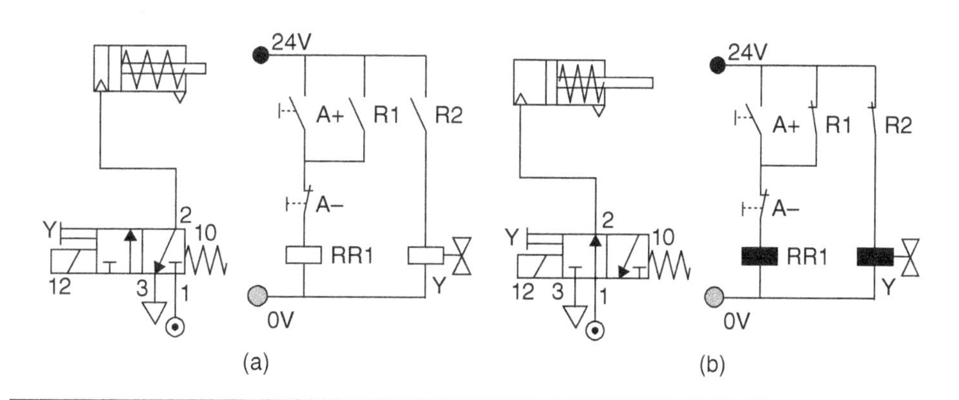

Figure 12.34 Electro-pneumatic logic MEMORY.

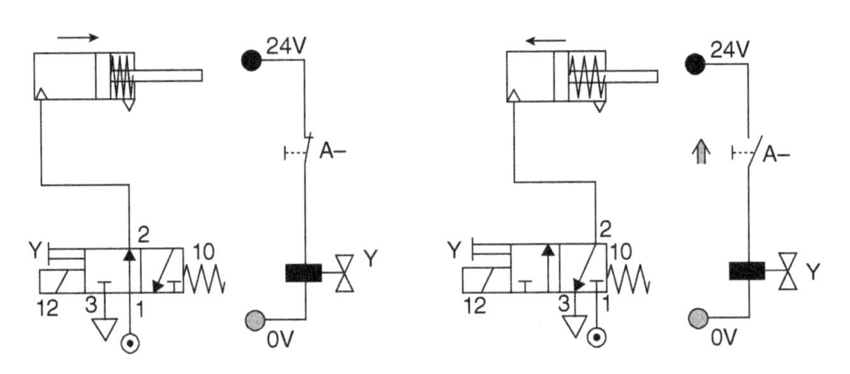

FIGURE 12.35 Electro-pneumatic logic NOT.

normally closed push button A– when operated will open, switching the solenoid Y off, and causing the cylinder to retract (see Fig. 12.35). The cylinder will extend if A– is NOT actuated.

12.6 Exercises

1. Explain the principal of operation of pneumatic power systems.

2. Derive an expression for the bulk modulus of air at different pressure levels.

3. Derive an expression for the energy stored in a volume of compressed air.

4. Discuss in detail the effect of air compressibility on the function of pneumatic systems.

5. Discuss the effect of ambient temperature on the operation of pneumatic power systems using compressed air bottles as an energy source.

6. Deal with the effect of air density on the operation of pneumatic systems.

7. Deal with the effect of air viscosity on the operation of pneumatic systems.

8. State and discuss briefly the advantages and disadvantages of pneumatic systems.

9. Draw the circuit of a simple pneumatic system and explain the function of its elements.

10. Explain the operation of ordinary pressure reducers and venting-type pressure reducers, giving the necessary drawings.

11. Explain the function of piston-type air compressors.

12. Explain the construction and operation of the different control valves using the illustrations given by Figs. 12.7 to 12.16.

13. Discuss the operation of the basic pneumatic circuits given by Figs. 12.17 through 12.35.

12.7 Nomenclature

B = Bulk modulus, Pa
E = Energy, J
n = Speed, rps
p = Absolute pressure, Pa
p_1 = Pressure at temperature T_1, Pa
p_2 = Pressure at temperature T_2, Pa
Q = Flow rate, m^3/s
R = Universal gas constant, (287.1 J/K kg for air)
T = Absolute temperature, K
V = Fluid volume, m^3
ΔT = Temperature variation, K
α = Volumetric thermal expansion coefficient
ρ = Air density, kg/m^3
η_v = Capacity coefficient (volumetric efficiency)

References

2021 Roman M. R., Shaaban S. M., Rabie M. G., and Aly M. H. (2021) "System Identification and Sliding Mode Tracking Control for Electro-Hydraulic Steer-By-Wire System," *Engineering Research Journal*, Helwan University, Cairo, Egypt, Faculty of Engineering, Matareya Branch, Vol. 172, No. 0, pp. 121–130.

2021 Yousef M. A., Rabie M. G., and Rateb R. A. (2021) "System Identification and Controller Design for a Typical Electro-Hydraulic Servo Motor," The 2021 International Telecommunications Conference, ITC-Egypt, 2021 July 13–15, 2021, ADC, Egypt, 978-1-6654-4574-0/21/$31.00 ©2021 IEEE.

2021 Yousef M. A., El-Sanabawy M., Rabie G., and Rateb R. A. (2021) "Modeling, Simulation and Controller Design for a Typical Bent Axis Electrohydraulic Servo Motor," *19th ASAT Conference* organized by the Military Technical College, Cairo, April 9–11, paper 021, IOP Conf. Series: *Materials Science and Engineering* 1172 (2021) 012038, IOP Publishing doi:10. 1088/1757-899X/1172/1/012038.

2019 Fadel M. Z., Rabie M. G., and Youssef A. M. (2019) "Motion Control of an Aircraft Electro-hydraulic Servo Actuator," *18th ASAT Conference* organized by the Military Technical College, Cairo, April 9–11, paper 021, IOP Conf. Series: *Materials Science and Engineering*, 610 (2019) 012073, IOP Publishing doi: 10.1088/1757-899X/610/1/0120732.

2019 Fadel M. Z., Rabie M. G., and Youssef A. M. (2019) "Modeling, Simulation and Control of a Fly-by-Wire Flight Control System Using Classical PID and Modified PI-D Controllers," *Journal Européen des Systèmes Automatisés*, Vol. 52, No. 3, pp. 267–276.

2018 Fadel M. Z., Rabie M. G., and Youssef A. M. (2018) "Investigation of the Operation of an Electro-Hydraulic Control System of a Flying Vehicle under Failure Conditions," *18th AMME Conference* organized by the Military Technical College, Cairo, April 3–5, paper MP-6.

2017 Ali M.H., Roman M. R., Rabie M. G., and Shaaban S. M. (2017) "Observer-Based Optimal Position Control for Electro-Hydraulic Steer-by-Wire System Using Grey-Box System Identified Model," *ASME, Journal of Dyn. Sys. Meas. and Control*, Vol. 139, No. 12, pp. 443–451.

2017 Abu Alnagah M. G., Rabie M. G., and Elfaisal Elrefaie M. E. (2017) "On the Dynamic Behavior of Series Hydraulic Hybrid Vehicle," *Journal of Al Azhar University Engineering Sector*, JAUES, ISSN: 1110-6409, Vol. 139, No. 12, pp. 443–451.

2016 Elattar Y. M., Rabie M. G., and Metwalli S. M. (2016) "On the Dynamics of Vehicle Passive Suspension Incorporating Twin-Tube Pressurized Shock Absorber," *Proceedings of the 17th AMME Conf.*, M.T.C., Cairo, April 19–21, paper No. 27.

2016 Elattar Y. M., Metwalli S. M., and Rabie M. G. (2016) "Pdf versus PID Controller for Active Vehicle Suspension," *Proceedings of the 17th AMME Conf.*, M.T.C., Cairo, April 19–21, paper No. 28.

2015 Linjama M., Paloniitty M., Tiainen L., and Huhtala, K. (2015) "Mechatronic design of digital hydraulic micro valve package," *ScienceDirect, Procedia Engineering, Dynamics and Vibroacoustics of Machines 107*, Elsevier.

2015 Peter Chapple P. (2015) *Principles of Hydraulic Systems Design*, 2nd edition, New York: Momentum Press.

2015 Semeda M., Rabie M. G., and Sabry A. (2015) "Performance Investigation of Axial Piston Pumps of Constant Power Regulation," *Proceedings of the 16th ASAT Conf.*, M.T.C., Cairo, May, paper 062-HF.

2015 Swidan M. A., Adam I. G., Shawky M., and Rabie M. G. (2015) "Implementing of a Hydraulic Cylinder as a New Isobaric Energy Recovery Device for Ro Desalination Units," *Proceedings of the IDA2015 World Congress*, San Diego, California, USA, August 30–September 4.

2014 Paloniitty M., Linjama M., Huhtala K. (2014) "Equal coded digital hydraulic valve system – improving tracking," *ScienceDirect, Procedia Engineering, Dynamics and Vibroacoustics of Machines 106*, Elsevier.

2013 Esposito A. (2013) *Fluid Power with Applications*, 7th edition, Ohio: Prentice Hall.

2012 Tayea M. S., Emam M. A., Shaaban S. M., El-Demerdash S. M., and Rabie M. G. (2012) "Dynamic Performance of Electro-hydraulic Steering System for off-road Vehicles," *Int. J. of Vehicle Structures & Systems*, Chennai, India, Vol. 4, No. 1, ISSN: 0975-3060 (Print), 0975-3540 (Online), pp. 1–9.

2012 Draz A. H., Rabie M. G., Abdel-Aziz A. I., and Dwidar M. S. (2012) "Dynamic Behavior of an Antiskid Braking System," *Proceedings of the 15th AMME Conf.*, M.T.C., Cairo, May 29–31, paper MP-6.

2012 Saleh I., and Rabie M. G. (2012) *Fluid Mechanics for Engineers*, Lecture Notes. Cairo: Published by the authors, ISBN 9789775092007.

2010 Rabie M. G. (2010) *Automatic Control for Mechanical Engineers*, Cairo: Published by the author, ISBN 9771798693.

2009 Sallam A. M. F., Abdel-Aziz A. I., Rabie M. G., and Dwidar M. S. (2009) "Dynamic Behavior of a Hydraulic Motor Controlled by a Poppet Type Directional Control Valve," *Proceedings of the 13th ASAT Conf.*, M.T.C., Cairo, May26–28, paper HC03.

2009 Draz A. H., Rabie M. G., Abdel-Aziz A. I., and Dwidar M. S. (2009) "'Dynamic Behavior of a Hydraulic Braking Valve Incorporating a Hydraulic Servo Actuator," *Proceedings of the 13th ASAT Conf.*, M.T.C., Cairo, May 26–28, paper HC05.

2008 Rateb R. A., Rabie M. G., and Elsenbawy M. A. (2008) "Implementation of Electrohydraulic Servo Valve for Modulation of the Rocket Thrust," *Proceedings of the 13th AMME Conf.*, M.T.C., Cairo, May 27–29, paper MP-8.

2007 RABIE M. G. (2007) "On the Application of Oleo-Pneumatic Accumulators for the Protection of Hydraulic Transmission Lines against Water Hammer—A Theoretical Study," *International Journal of Fluid Power*, Vol. 8, No. 1, pp. 39–49.

2007 Metwally M., Saleh I., Rabie M. G., and Girgis N. (2007) "Effect of the Air and Fuel Flow on the Performance of a Turboshaft Gas Turbine Engine, Part II: Analysis of the Engine Dynamic Performance," *Proceedings of the 12th ASAT Conf.*, M.T.C., Cairo, May 29–31, paper TMN-02.

2007 Metwally M., Saleh I., Rabie M. G., and Girgis N. (2007) "Effect of the Air and Fuel Flow on the Performance of a Turboshaft Gas Turbine Engine, Part I: Modeling and

Simulation," *Proceedings of the 12th ASAT Conf.*, M.T.C., Cairo, May 29–31, paper TMN-01.

2006 Metwally M., Rabie M. G., Girgis N., and Saleh I (2006) "Dynamic Performance of an Electrohydraulic Servo Actuator with Contactless controlled Spool," *Proceedings of the 12th AMME Conf.*, M.T.C., Cairo, May 16–18, pp. 467–489.

2006 Akers A., Gassman M., and Smith R. (2006) *Hydraulic Power Systems Analysis*, UK: Taylor & Francis Group, LLC.

2006 Mahmoud T. N., Rabie M. G., and M. ElSenbawi A. (2006) "Dynamic Behavior of Control Valves of a Liquid Rocket Engine Feeding System," *Proceedings of the 12th AMME Conf.*, M.T.C., Cairo, May 16–18, paper PW-02, pp. 467–489.

2006 Kassab S. Z., Adam I. G., Swidan M. A., and Rabie M. G. (2006) "Influence of Operating and Construction Parameters on the Behavior of Hydraulic Cylinder Subjected to Jerky Motion," *Proceedings of the 8th International Conference of Fluid Dynamics & Propulsion*, December 14-17, Sharm El-Sheikh, Sinai, Egypt, pp. 1–8, Copyright © by ASME.

2005 Rexroth, "Hydraulic and Electronic Components for Proportional and Servo Systems"; Rexroth, Germany, Data Sheets No. RE 29142/05.05, RA 29054/5.94, RA 29057/6.94, RA 29060/5.94, RA 29113/2.96, RA 29160/5.94, RA 29165/3.96, RA 29166/5.94, RA 29 204/5.94.

2005 Ibrahim Z. A., Rabie M. G., Hegazy S. A., and ElSherif I. A. (2005) "Experimental and Theoretical Investigation of Dynamic Behavior of Oleo-Pneumatic Car Suspension," *Proceedings of the 11th ASAT Conf.*, M.T.C., Cairo, May 17–19, paper (FH-03), pp. 131–146.

2005 Ibrahim Z. A., M. ElSherif I. A., Rabie M. G., and Hegazy S. A. (2005) "Design and Analysis of Dynamic Performance of A vehicle Active Suspension System," *Proceedings of the 11th ASAT Conf.*, M.T.C., Cairo, May 17–19, paper FH-04, pp. 147–163.

2004 Kassab S. Z., Swidan M. A., Adam I. G., and Rabie M. G. (2004) "Dynamic Behavior of Hydraulic Cylinder Subjected to Jerky Motion During Load Lowering, Part 2: Modeling and Simulation," *Proceedings of the 11th AMME Conf.*, M.T.C., Cairo, November 23–25, Vol. 1, paper DV-02, pp. 53–72.

2004 Kassab S. Z., Swidan M. A., Adam I. G., and Rabie M. G. (2004) "Dynamic Behavior of Hydraulic Cylinder Subjected to Jerky Motion during Load Lowering, Part 1 Experimental Work," *Proceedings of the 11th AMME Conf.*, M.T.C., Cairo, November 23–25, Vol. 1, paper DV-01, pp. 38–52.

2003 Ibrahim Z. A., Hegazy S. A., ElSherif I. A. and Rabie M. G. (2003) "Dynamic Behavior of Gas Charged Single Tube Shock Absorber," *Proceedings of the 10th ASAT Conf.*, M.T.C., Cairo, May 13–15, paper HF-08, pp. 185–201.

2003 Esposito A. (2003) *Fluid Power with Applications*, 6th edition, Upper Saddle River: Prentice Hall, N.J.

2003 El-Sayed A. A., Gobran M. H., Rabie M. G., and Shalaan M. R. (2003) "Investigation of Characteristics of an Electrohydraulic Servo-actuator Incorporating Jet Pipe Amplifier," *Proceedings of the 10th ASAT Conf.*, M.T.C., Cairo, May 13–15, paper HF-03, pp. 109–123.

2001 Osama G., El-Sayed, Rabie M. G., and El-Taher R. (2001) "Experimental and Theoretical Study of the Static and Dynamic Behavior of a Variable-Displacement Pump with Power Control," *AMSE Periodicals; Modeling, Measurement & Control*, B—(B: Mechanics and Thermic), Vol. 70, No. 5, pp. 11–30.

2001 Eaton Corp., Industrial, and Mobile Fluid Power, Product Literature 900, Release 1.1, CD from Eaton Corp., Eden Prairie, Minn., 2001.

1999 Zayed A. N. (1999) *Summary for Engineers*, Alexandria, Egypt: Ziad Press.

1999 Vickers Incorporated, Electronic Catalogue, Vol. X, CD from Vickers Incorporated, Hampshire, U.K., 1999.

1999 Mannesmann Rexroth A. G. (1999) *Product Catalogue of Axial Piston Units*, CD from Brueninghaus Hydromatik, Lohr am Main, Germany.

1999 Mannesmann Rexroth A. G. (1999) *Interactive Hydraulic Designer*, CD from Brueninghaus Hydromatic & Rexroth Hydraulics, Lohr am Main, Germany.

1999 Karkub M. A., Gad O. E., and Rabie M. G., (1999) "Predicting Axial Piston Pump Performance Using Neural Networks," *Int. J. of Mechanism and Machine Theory*, Vol. 34, pp. 1211–1226.

1999 Du Y., Mamishev A. V. Lesicutre B. C., Zahn M., and Kang S. H. (1999) "Measurement of Moisture Solubility for Differently Conditioned Transfer Oils," *Proceedings of the 13th International Conference on Dielectric Liquids* (ICDL99), Nara, Japan, July 20–25, 1999, pp. 357–360.

1998 Sauer Sundstrand Co. (1998) *All Product Technical, Application, and Service/Repair Information*, CD from Sauer Sundstrand Co., Ames, Iowa.

1998 Rexroth-Bosch GmbH. (1998) *Electronic Reference Library, Hydraulics*, CD from Bosch Automation Technology, Stuttgart, Germany.

1998 Parker Hannifin GmbH. (1998) *Hydraulic Control, Catalogue 2500/GB*, CD from Parker Hydraulics, Kaarst, Germany.

1998 Ibrahim S. Y., Rabie M. G., and Lotfy A.H. (1998) "Experimental and Theoretical Investigation of the Performance of the Servo actuator of a Hydraulic Power Steering System," *Proceedings of the 8th AMME Conf.*, M.T.C., Cairo, May, pp. 395–407.

1997 Rabie M. G. (1997) "On the Validity of the Lumped Parameter Models for Fluid Flow between Coaxial Pipes," *AMSE Periodicals; Modeling, Measurement & Control, B*, AMSE Press, Lyon, France, Vol. 63, Nos. 1, 2, pp. 31–47.

1997 Brown R. N. (1997) *Compressors: Selection and Sizing*, Houston, Texas: Gulf Professional Publishing.

1996 Mannesmann Rexroth A. G. (1996) *Industrial Hydraulic Valve Catalogue and Pump and Motor Catalogue*, CD from Rexroth USA, Bethlehem, Pa.

1996 Hodges P. K. B. (1996) *Hydraulic Fluids*, New York: John Wiley & Sons, Inc.

1996 ELSAID M., Rabie M. G., and Khattab A. A. (1996) "Investigation of the Transient Behavior of End Position Cushioning in Hydraulic Cylinders," *Proceedings of the 7th AMME Conf.*, M.T.C., Cairo, May 28–31, paper PE-1, pp. 351–364.

1994 Sayed A., Gadallah N. A., Saleh I., Rabie M. G., and Hanna H. A. (1994) "Experimental Investigation of the Effect of Wear on the Performance of Hydraulic Directional Control Valves," *Proceedings of 6th AMME Conf.*, M.T.C., Cairo, May 3–5, Vol. 1, paper PE-7, pp. 563–576.

1993 Rabie M. G., Metwally S. M., and Abdou S. E. (1993) "Design of a New Controller for a Hydraulic Servomechanism," *Proceedings of 4th ASAT Conf.*, M.T.C., Cairo, May 4–6, Vol. 1, paper FM-3, pp. 175–188.

1993 Rabie M. G., Awad M. A., Imam E. I., and Gadallah N. A. (1993) "Experimental and Theoretical Investigation of the Transient Response of Hydraulic Valve Controlled Actuators," *Alexandria Engineering Journal*, Alexandria University, Vol. 32, No. 1, pp. A7–A16.

1993 Rabie M. G., and Saleh I. (1993) "On the Effect of Location of Damping Orifice on the Performance of Hydraulic Systems with Counterbalance Valve," *Bulletin of Fac. of Eng.*, Ain Shams University., Vol. 28, No.1, *Mech. Eng.*, Mars, pp. 537–553.

1991 Rabie M. G., and Younis Y. (1991) "Investigation of performance of a Direct Operated Hydraulic Pressure Reducer, Approach by Block Bond Graph," *Alexandria Engineering Journal*, Vol. 30, No. 2, pp. 115–120.

1991 Rabie M. G., and Hafez H. E. (1991) "Modeling by Block Bond Graph and Investigation of Dynamic Performance of a Hydraulic Pressure Reducer," *Bulletin of Fac. of Eng.*, Ain Shams University, Vol. 26, No.1, *Mech. Eng.*, Mars, pp. 492–509.

1991 Rabie M. G. (1991) "On the Dynamics of Fluid flow Between Two Coaxial Pipes; a Lumped Parameter Model," *Eng. Res. Bulletin of Fac. of Eng. &Tech.*, Mataria, University of Helwan, Vol. 3, Mars, pp. 1–11.

1990 Yeaple F. (1990) *Fluid Power Design Handbook*, 2nd edition, New York: M. Dekker.

1990 Rabie M. G., and Ibrahim U. M. (1990) "Modeling and Simulation of a Compact Electrohydraulic Servoactuator," *J. of Faculty of Eng.*, Cairo University, Vol. 37, No. 4, pp. 1003–1018.

1990 Nasca (1990) *Testing Fluid Power Components*, New York: Industrial Press.

1990 Burrows C. R., and Edges K. A. (1990) *Fluid Power Components and Systems*, New York: RSP, U.K., and John Wiley & Sons, Inc.

1989 Watton J. (1989) *Fluid Power Systems: Modeling, Simulation, Analogue and Microprocessor Control*, Hertfordshire, U.K.: Prentice Hall.

1989 Rabie M. G., Kassem S. A., ElSayed S. A., Aziz M. A., and ElSayed. (1989) "Block Bond Graph and TUTSIM; a Powerful Tool for Nonlinear Dynamic System Modeling and Simulation," *Alexandria Engineering Journal*, Alexandria University, Vol. 28, No. 3, pp. 519–537.

1988 Kassem S. A., Rabie M. G., El Adawy E. S., Younis Y. I., and Ahmed H. (1988) "Dynamic Analysis of Hydraulic Transmission Lines by a Lumped Model," *Bulletin of Fac. of Eng.*, Ain Shams University, Vol. 22, No.2, *Mech. Eng.*, pp. 1–18.

1987 Mannesmann Rexroth A. G. (1987) *Hydraulic Components, General Catalogue*, Germany: Lohr am Main.

1985 Tonyan M. J. (1985) *Electronically Controlled Proportional Valves*, New York: M. Dekker.

1985 Reed E. W., and Larman I. S. (1985) *Fluid Power with Microprocessor Control*, Prentice Hall.

1985 Dorr H et al. (1986) "Proportional and Servo Valve Technology," Mannesmann Rexroth GmbH, Germany.

1984 Pippenger J. J. (1984) *Hydraulic Valves and Controls, Selection and Application*, New York: Marcel Dekker, Inc.

1984 Henn A. (1984) *Fluid Power Trouble Shooting*, New York: M. Dekker.

1984 Gotz W. (1984) *Hydraulics: Theory and Applications from Bosch*, Robert Bosch GmbH. Hydraulics Division, Stuttgart, Germany.

1983 Lambeck R. P. (1983) *Hydraulic Pumps and Motors*, New York: M. Dekker.

1982 Sullivan J. A. (1982) *Fluid Power: Theory and Applications*, 2nd edition, Reston, Va.

1981 Rabie M. G., and Lebrun M. (1981) "Modeling by Bond Graph and Simulation of a Bi-Axial Electrohydraulic Fatigue Machine," *IASTED Symp. Modeling, Ident.*, Cont. Davos, Switzerland, February 18–21 (in French).

1981 Rabie M. G., and Lebrun M. (1981) "Modeling by Bond Graph and Simulation of an Electrohydraulic Servovalve of Two Stages," *RAIRO Automatic Systems Analysis and Control*, Vol. 15, No. 2, pp. 97–129 (in French).

1980 McCloy D. and Martin H. R. (1980) *The Control of Fluid Power: Analysis and Design*, 2nd edition, West Sussex, U.K.: Ellis Haward.
1979 Baz A., Rabie M. G., Zaki H., and Barakat A. (1979) "Hydraulic Servo with Built-in Tuned Damper," *Fluidic Quarterly*, Vol. 11, No. 2, pp. 1–24.
1977 Keller G. R. (1977) "Hydraulic System Analysis," Published by the editors of Hydraulics & Pneumatics Magazine.
1977 Braillon S. A. (1977) *Electro-Magnetic Proportional Solenoids as Control Elements of Hydraulic Valves*, MSM Division, France, pp. 81–89 (in French).
1976 Lallement J. (1976) "Study of the Dynamic Behavior of Hydraulic Lines" (in French), Les Memoir Techniques du Centre Technique des Industries Mecaniques, Senlis, France, No. 27, Sept.
1976 Goodwin A. B. (1976) *Fluid Power Systems*, London: Macmillan Press Ltd.
1975 Karnopp D., and Rosenberg R. (1975) *System Dynamics: A Unified Approach*, New York: John Wiley & Sons, Inc.
1973 McCloy D., and Martin H. R. (1973) *The Control of Fluid Power*, London: Longman.
1969 Nekrasov B. (1969) *Hydraulics for Aeronautical Engineers*, Moscow: Mir Publisher.
1967 Merrit H. E. (1967) *Hydraulic Control Systems*, New York: John Wiley & Sons, Inc.
1957 Conway H. G. (1957) *Aircraft Hydraulics*, Vol. 2, London: Chapman and Hall Ltd.
1957 Conway H. G. (1957) *Aircraft Hydraulics*, Vol. 1, London, Chapman and Hall Ltd.

Index

Note: Page numbers followed by *f* denote figures; by *t*, tables.

F

M

N

O

P

T